PROGRESS IN

Molecular Biology and Translational Science

Volume 105

PROGRESS IN

Molecular Biology and Translational Science

Animal Models of Molecular Pathology

edited by

P. Michael Conn

Divisions of Reproductive Sciences and Neurosciences (ONPRC)
Departments of Pharmacology and Physiology
Cell and Developmental Biology, and Obstetrics and Gynecology (OHSU)
Beavorton, Oregon, USA

Volume 105

ELSEVIER

AMSTERDAM • BOSTON • HEIDELBERG • LONDON
NEW YORK • OXFORD • PARIS • SAN DIEGO
SAN FRANCISCO • SINGAPORE • SYDNEY • TOKYO
Academic Press is an imprint of Elsevier

Academic Press is an imprint of Elsevier
32 Jamestown Road, London, NW1 7BY, UK
Radarweg 29, PO Box 211, 1000 AE Amsterdam, The Netherlands
225 Wyman Street, Waltham, MA 02451, USA
525 B Street, Suite 1900, San Diego, CA 92101-4495, USA

This book is printed on acid-free paper. ♾

Library of Congress Cataloging-in-Publication Data
A catalog record for this book is available from the Library of Congress

British Library Cataloguing in Publication Data
A catalogue record for this book is available from the British Library

ISBN: 978-0-12-394596-9
ISSN: 1877-1173

For information on all Academic Press publications
visit our website at elsevierdirect.com

Printed and bound by CPI Group (UK) Ltd, Croydon, CR0 4YY

Transferred to digital print 2013

Contents

Contributors

Numbers in parentheses indicate the pages on which the authors' contributions begin.

Vicente Andrés, Department of Epidemiology, Atherothrombosis and Imaging, Centro Nacional de Investigaciones Cardiovasculares, Madrid, Spain (1)

Cenk Ayata, Department of Radiology, Neurovascular Research Laboratory; and Stroke Service and Neuroscience Intensive Care Unit, Department of Neurology, Massachusetts General Hospital, Harvard Medical School, Charlestown, Massachusetts, USA (25)

Brian J. Bacskai, Department of Neurology, MassGeneral Institute for Neurodegenerative Disease, Massachusetts General Hospital, Harvard Medical School, Charlestown, Massachusetts, USA (25)

Glen B. Banks, Division of Medical Genetics, Department of Neurology, University of Washington, Seattle, Washington, USA (83)

Dwight A. Bellinger, Francis Owen Blood Research Laboratory, University of North Carolina, Chapel Hill, North Carolina, USA (151)

René M. Botnar, Division of Imaging Sciences, King's College London, London, United Kingdom (227)

Amy R. Brooks-Kayal, Division of Neurology, Department of Pediatrics, School of Medicine, University of Colorado Denver; and Children's Hospital Colorado, Department of Neurology, Aurora, Colorado, USA (57)

Ana I. Castillo, Department of Epidemiology, Atherothrombosis and Imaging, Centro Nacional de Investigaciones Cardiovasculares, Madrid, Spain (1)

Jeffrey S. Chamberlain, Division of Medical Genetics, Department of Neurology, University of Washington, Seattle, Washington, USA (83)

Joel R. Chamberlain, Division of Medical Genetics, Department of Medicine, University of Washington, Seattle, Washington, USA (83)

Carolyn Cray, Division of Comparative Pathology, Department of Pathology, University of Miami Miller School of Medicine, Miami, Florida, USA (113)

José J. Fuster, Department of Epidemiology, Atherothrombosis and Imaging, Centro Nacional de Investigaciones Cardiovasculares, Madrid, Spain (1)

John K. Hall, Division of Medical Genetics, Department of Neurology, University of Washington, Seattle, Washington, USA (83)

Roland W. Herzog, Department of Pediatrics, University of Florida, Gainesville, Florida, USA (151)

Borja Ibáñez, Department of Epidemiology, Atherothrombosis and Imaging, Centro Nacional de Investigaciones Cardiovasculares, Madrid, Spain (1)

Patryk Konieczny, Division of Medical Genetics, Department of Neurology, University of Washington, Seattle, Washington, USA (83)

Jeong Hyun Lee, Department of Radiology, Neurovascular Research Laboratory, Massachusetts General Hospital, Harvard Medical School, Charlestown, Massachusetts, USA (25)

Jacob Lister, Department of Psychiatry, UT Southwestern Medical School, Dallas, Texas, USA (411)

Ronald Lubet, Chemoprevention Branch, National Cancer Institute, Bethesda, Maryland, USA (211)

Marcus R. Makowski, Division of Imaging Sciences, King's College London, London, United Kingdom; and Department of Radiology, Charite, Berlin, Germany (227)

Elizabeth Merricks, Francis Owen Blood Research Laboratory, University of North Carolina, Chapel Hill, North Carolina, USA (151)

Atsushi Mizoguchi, Molecular Pathology Unit, Department of Pathology, Center for the Study of Inflammatory Bowel Disease, Massachusetts General Hospital, Harvard Medical School, Boston, Massachusetts, USA (263)

Paul E. Monahan, Department of Pediatrics, University of North Carolina, Chapel Hill, North Carolina, USA (151)

Laurence Morel, Department of Pathology, Immunology, and Laboratory Medicine, University of Florida, Gainesville, Florida, USA (321)

Lindsey A. Muir, Program in Cellular and Molecular Biology, Department of Neurology, University of Washington, Seattle, Washington, USA (83)

Rainer Ng, Division of Medical Genetics, Department of Neurology, University of Washington, Seattle, Washington, USA (83)

Timothy C. Nichols, Francis Owen Blood Research Laboratory, University of North Carolina, Chapel Hill, North Carolina, USA (151)

Guy L. Odom, Division of Medical Genetics, Department of Neurology, University of Washington, Seattle, Washington, USA (83)

Julian N. Ramos, Program in Cellular and Molecular Biology, Department of Neurology, University of Washington, Seattle, Washington, USA (83)

Yogendra H. Raol, Division of Neurology, Department of Pediatrics, School of Medicine, University of Colorado Denver, Aurora, Colorado, USA (57)

Lucina Rouggly, Department of Surgery, Washington University School of Medicine, St. Louis, Missouri, USA (211)

Denise E. Sabatino, Division of Hematology, The Children's Hospital of Philadelphia, Philadelphia, Pennsylvania, USA (151)

Allison Sang, Department of Pathology, Immunology, and Laboratory Medicine, University of Florida, Gainesville, Florida, USA (321)

Jane Seto, Division of Medical Genetics, Department of Neurology, University of Washington, Seattle, Washington, USA (83)

Gideon P. Smith, Department of Dermatology, Massachusetts General Hospital, Harvard University, Boston, Massachusetts, USA (371)

Sarah Southcott, Department of Psychiatry, UT Southwestern Medical School, Dallas, Texas, USA (411)

Carol A. Tamminga, Department of Psychiatry, UT Southwestern Medical School, Dallas, Texas, USA (411)

Yian Wang, Department of Surgery, Washington University School of Medicine, St. Louis, Missouri, USA (211)

Jacqueline Wicki, Division of Medical Genetics, Department of Neurology, University of Washington, Seattle, Washington, USA (83)

Masaya Yanagi, Department of Psychiatry, UT Southwestern Medical School, Dallas, Texas, USA (411)

Yiming Yin, Department of Pathology, Immunology, and Laboratory Medicine, University of Florida, Gainesville, Florida, USA (321)

Ming You, Cancer Center, Medical College of Wisconsin, Milwaukee, Wisconsin, USA (211)

Carlos Zaragoza, Department of Epidemiology, Atherothrombosis and Imaging, Centro Nacional de Investigaciones Cardiovasculares, Madrid, Spain (1)

Ying-Yi Zheng, Department of Pathology, Immunology, and Laboratory Medicine, University of Florida, Gainesville, Florida, USA (321)

Preface

Picking the right model for studies is at the core of successful science. The idea, of course, is that the model selected reacts to a disease or its treatment as does the human. Models are often classified into three groups: homologous, isomorphic, and predictive. *Homologous* animals have identical causes, symptoms, and treatment as do humans with the same disease. *Isomorphic* animals share the same symptoms and treatments but need not have the identical cause and form the largest number of models. *Predictive* models show only the treatment characteristics of a disease. This method is commonly used when the causes of a disease is unknown.

In this volume, the authors have been tasked with the goal of describing models of human molecular pathology. The result has been a diverse set of animal models with a description of why each was chose and the advantages and disadvantages of each.

P. Michael Conn

Animal Models of Atherosclerosis

José J. Fuster, Ana I. Castillo, Carlos Zaragoza, Borja Ibáñez, and Vicente Andrés

Department of Epidemiology, Atherothrombosis and Imaging, Centro Nacional de Investigaciones Cardiovasculares, Madrid, Spain

Cardiovascular disease is currently the predominant cause of mortality worldwide and its incidence is expected to increase significantly during the next decades owing to the unhealthy effects of modern lifestyle habits (e.g., obesity and lack of physical exercise). Cardiovascular death is frequently associated with acute myocardial infarction or stroke, which are generally the ultimate consequence of an underlying atherosclerotic process. Small and big animal models are valuable tools to understand the molecular mechanisms underlying atherosclerotic plaque formation and progression, as well as the occurrence of associated ischemic events. Moreover, animal models of atherosclerosis are pivotal for testing mechanistic hypothesis and for translational research, including the assessment of dietary and/or pharmacological interventions and the development of imaging technologies and interventional devices. In this chapter, we will describe the most widely used animal models that have permitted major advances in atherosclerosis research and significant improvements in the treatment and diagnosis of atherosclerotic disease.

I. Atherosclerosis Development: Basic Concepts

Cardiovascular disease (CVD) is currently the predominant cause of mortality worldwide with more than 17 million annual deaths, and global cardiovascular deaths are predicted to increase to more than 23 million by 2030.[1]

Progress in Molecular Biology
and Translational Science, Vol. 105
DOI: 10.1016/B978-0-12-394596-9.00001-9

1877-1173/12 $35.00

In most cases, cardiovascular death is provoked by acute myocardial infarction or stroke, which are generally the ultimate consequence of an underlying atherothrombotic process. Atherosclerosis is a complex inflammatory disease characterized by the progressive hardening and thickening of the vessel wall owing to the development of complex lesions that narrow the arterial lumen (atherosclerotic plaque, atheroma, or neointimal lesion) (Fig. 1). Atherosclerotic plaque formation is triggered and sustained by the dysfunction of the endothelium associated with the chronic exposure to cardiovascular risk factors (e.g., hyperlipidemia, hypertension, diabetes, smoking, etc.). A major cardiovascular risk factor is the existence in the blood of high levels of low-density lipoproteins (LDL) that accumulate in the subendothelial space of the arterial wall and undergo oxidative modifications that trigger an inflammatory response.[2,3] Locally produced oxidized LDL (oxLDL) induce the expression in the endothelial cells of chemotactic proteins, such as monocyte chemoattractant protein-1 (MCP-1),[4,5] and adhesion molecules, such as vascular cell adhesion molecule-1 (VCAM-1), E-selectin, and P-selectin.[6,7] This leads to the recruitment of blood-borne monocytes in the injured arterial wall, which subsequently differentiate into macrophages of different subtypes (macrophage polarization).[8] Neointimal macrophages avidly internalize oxLDL and become foam cells that critically contribute to plaque development by secreting a plethora of mediators that perpetuate the inflammatory process in the vessel wall.[9] This maladaptive, nonresolving inflammatory response is the driving force of atherosclerotic plaque development. It further promotes the recruitment of circulating monocytes and T cells that boost inflammation in the arterial wall, and also stimulates the migration of vascular smooth muscle cells (VSMCs) from the tunica media into the subendothelial space, where they exhibit abnormally high proliferation and secrete extracellular matrix proteins that also contribute to atheroma growth.[10–13] Moreover, the excessive accumulation of oxidized lipids leads to endoplasmic-reticulum stress-associated apoptosis of macrophages, generating apoptotic bodies that cannot be properly disposed and therefore enhance plaque inflammation and instability.[14] At advanced stages of the disease, rupture of high-risk vulnerable plaques exposes their thrombogenic compounds to the circulating blood, leading to luminal thrombosis and eventually to life-threatening acute ischemic events (e.g., myocardial infarction, stroke).[15]

II. Animal Models of Atherosclerosis

The first evidence that atherosclerosis can be induced in laboratory animals was provided in 1908 by Ignatowski, who demonstrated the formation of lesions in the aortic wall of rabbits that were fed a diet enriched in animal proteins (mainly from meat, milk, and egg yolks).[16] Since then, various animal

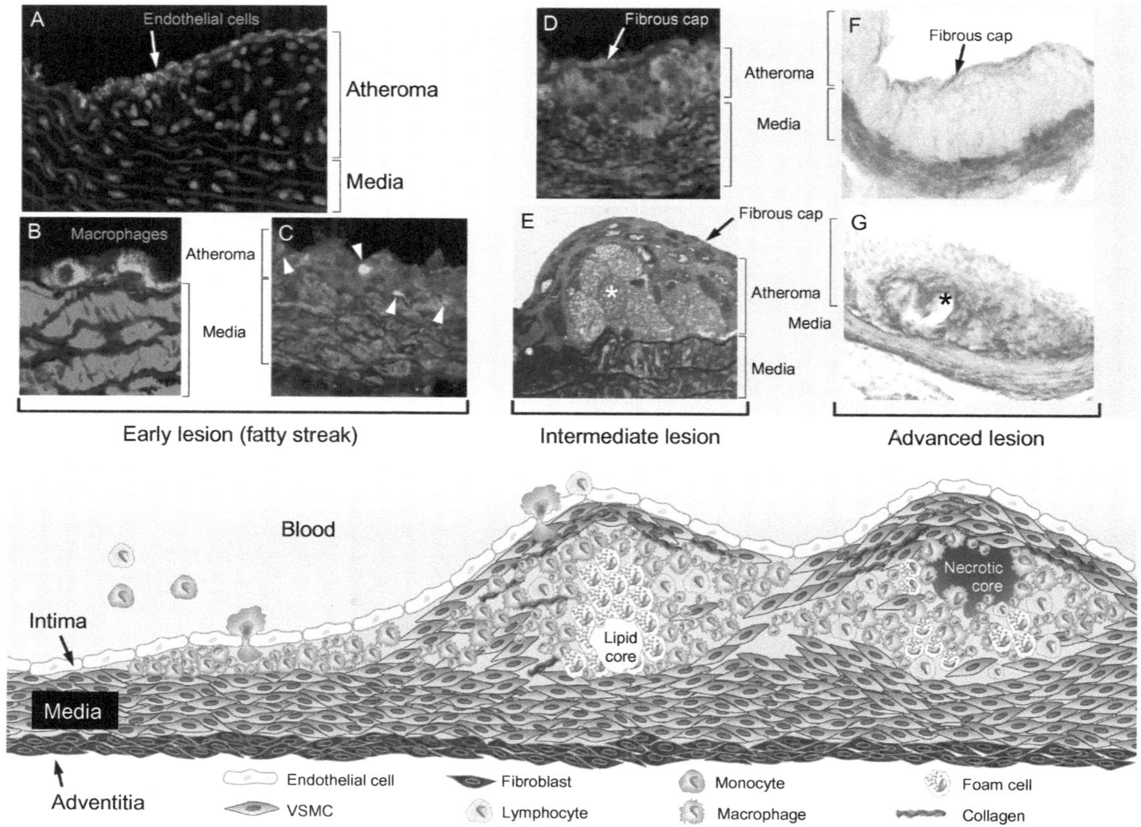
A
Endothelial cells
Atheroma
Media
B
Macrophages
Atheroma
Media
C
Early lesion (fatty streak)
D
Fibrous cap
Atheroma
Media
E
Fibrous cap
Atheroma
Media
Intermediate lesion
F
Fibrous cap
G
Atheroma
Media
Advanced lesion
Blood
Intima
Media
Adventitia
Lipid core
Necrotic core
Endothelial cell
VSMC
Fibroblast
Lymphocyte
Monocyte
Macrophage
Foam cell
Collagen

species have been used as experimental models of atherosclerosis, including rabbits, mice, rats, guinea pigs, hamsters, birds, swines, dogs, and nonhuman primates. A thorough examination of these animal models—especially rodents, swines, and rabbits—has greatly increased our understanding of the pathophysiological mechanisms that lead to the formation and development of the atherosclerotic plaque. In spite of the many differences existing between these models, all of them share the requirement of high levels of blood plasma cholesterol for atherosclerotic plaque to develop. The initial observations of this remarkable characteristic of every animal model of atherosclerosis were crucial for the discovery of the important role of cholesterol in atherosclerosis development. The next sections discuss the main advantages and disadvantages of the most commonly utilized animal models of atherosclerosis, which are summarized in Tables I–IV. Other available animal models of atherosclerosis that are used less frequently are discussed elsewhere.[17,18]

A. Rabbit Models of Atherosclerosis

The rabbit was the first animal model used in atherosclerosis research[16] and was pivotal in the initial experiments that led to the discovery of the role of elevated plasma cholesterol as a critical factor in the initiation of atherosclerosis.[19,20]

FIG. 1. Atherosclerotic plaque development. The scheme shows the main events that occur in the vessel wall during atherosclerosis development, starting with a "healthy" artery wall (left) and progression to an advanced atherosclerotic lesion (right). Not shown in the scheme is plaque rupture and thrombus formation at advanced disease stages, which can provoke life-threatening acute ischemic events (e.g., myocardial infarction and stroke). The photomicrographs in A–G show representative immunohistological staining in the aortic sinus and the ascending aorta of apoE-deficient mice, the most widely used mouse model of atherosclerosis. (A) Cell accumulation within the subendothelial space in an early atheroma. Red: nuclei; green: endothelial cells detected by von Willebrand immunofluorescent staining. Elastic laminae in the tunica media exhibit green autofluorescence. (B) Macrophage infiltration into the subendothelial space of the arterial wall. Blue: nuclei; green: neointimal macrophages detected by F4/80 immunofluorescent staining; red: medial VSMCs detected by smooth-muscle-α-actin immunofluorescent staining. (C) Cell proliferation in a fatty streak. Blue: nuclei; red: medial VSMCs detected by smooth-muscle-α-actin immunofluorescent staining; green: neointimal macrophages detected by F4/80 immunofluorescent staining; white: nuclei in proliferating cells detected by Ki67 immunofluorescent staining. Arrows mark proliferating macrophages in the atheroma. (D) Intermediate vascular lesion with a thin fibrous cap. Blue: nuclei; red: VSMCs detected by smooth-muscle-α-actin immunofluorescent staining; green: macrophages detected by F4/80 immunofluorescent staining. (E) Electron microscopy image of an intermediate atherosclerotic lesion with a fibrous cap and lipid core (asterisk). (F) Advanced atherosclerotic plaque with fibrous cap rich in VSMCs (detected by smooth-muscle-α-actin immunohistochemical red staining). (G) Advanced atherosclerotic plaque with necrotic core (asterisk). Macrophages detected by Mac3 immunohistochemical staining are shown in brown. Nuclei are counterstained with hematoxylin. (See Color Insert.)

TABLE I
ADVANTAGES AND DISADVANTAGES OF RABBIT MODELS OF ATHEROSCLEROSIS

Advantages	Easy to maintain and handle, no special requirements Low economical cost High availability Lipoprotein metabolism relatively similar to humans (except for hepatic lipase deficiency in rabbits) Good response to dietary cholesterol Availability of hyperlipidemic mutant strains
Disadvantages	Highly abnormal diet required for the development of hypercholesterolemia and atherosclerosis Long-term high-cholesterol feeding induces massive inflammation and hepatic toxicity

TABLE II
ADVANTAGES AND DISADVANTAGES OF SWINE MODELS OF ATHEROSCLEROSIS

Advantages	Cardiovascular anatomy very similar to humans Spontaneous formation of atherosclerosis lesions, even in low-fat standard chow Morphology of vascular lesions similar to humans Lesion distribution similar to humans, predominantly in aorta, coronary, and carotid arteries Lipoprotein metabolism similar to humans (except for apolipoprotein II deficiency in swines)
Disadvantages	High cost of purchase and maintenance Difficulty in handling (except for minipig strains) Atheroma formation requires longer time than in other species

TABLE III
ADVANTAGES AND DISADVANTAGES OF NONHUMAN PRIMATE MODELS OF ATHEROSCLEROSIS

Advantages	Phylogenetically close to humans Spontaneous formation of atherosclerosis lesions (in some strains, even in low-fat standard chow) Vascular lesions similar to humans, including vulnerability features and thrombosis
Disadvantages	High cost of purchase and maintenance Limited availability Requirement of special animal facilities Ethical concerns

Although currently less used than mice, the rabbit is still one of the most frequently employed animal models for atherosclerosis research due to its easy handling as well as its inexpensive maintenance and high availability.

TABLE IV
ADVANTAGES AND DISADVANTAGES OF MOUSE MODELS OF ATHEROSCLEROSIS

Advantages	Easy breeding and handling Short generation time Well-defined genetics and availability of inbred strains Well-established protocols for directed genetic manipulation
Disadvantages	High resistance to atherosclerosis development in wild-type mice. Requirement of genetically modified mice (e.g., apoE-deficient, LRLD-deficient) Plasma lipid profile markedly different to humans Differences in the morphology of the arterial wall due to the small size of murine vessels (e.g., reduced thickness of the medial layer, lack of vasa vasorum) Absence of plaque rupture and luminal thrombosis in most vessels

The most common rabbit strain that is used in atherosclerosis research is the New Zealand White (NZW), which is not inherently prone to atherosclerosis due to its low levels of plasma cholesterol when maintained under standard chow (around 50 mg/dl).[20] Induction of vascular lesions in NZW rabbits generally requires feeding a high-cholesterol diet (ranging from 0.2% to 2% cholesterol), which rapidly increases plasma cholesterol levels by up to eight times[21] and leads to the formation of foam cells-enriched fatty streaks in several vascular regions (especially the aortic arch and the thoracic aorta).[22] However, some studies have raised doubts on the mechanisms underlying the formation of atherosclerotic plaques in this herbivorous animal model, especially because of the highly abnormal diet that is required to induce vascular lesions. Indeed, other dietary manipulations, such as the substitution of casein for soy protein in the diet, are atherogenic in the rabbit in the absence of a high-cholesterol diet.[23] In addition, the development of advanced and complex atherosclerotic plaques containing a lipid core surrounded by VSMCs generally requires long periods of cholesterol feeding in NZW rabbits (from 6 months to several years).[24–27] However, long-term fat feeding of rabbits is discouraging, because it is frequently accompanied by noxious side effects and increased mortality owing to hepatic toxicity. Moreover, it induces a massive inflammatory response that does not resemble the chronic low-grade inflammatory response associated to human atherosclerosis. Some studies support the notion that the formation of advanced lesions in the rabbit depends on the age of the animals that are challenged with a cholesterol-rich diet. Old rabbits (3–4.5 years old) frequently develop fibrous plaques after being fed a high-cholesterol diet, while young rabbits (4-month-old) do not exhibit such advanced lesions after the same period of cholesterol feeding.[28] However, it must be noted that rabbits show high biological variability with respect to individual responsiveness to dietary cholesterol and that lesion morphology varies significantly depending on the cholesterol content of the diet.

Short-term feeding with diets containing high amounts of cholesterol (1–2%) mostly induces macrophage-rich fatty streaks, while long-term diets with lower amounts of cholesterol (0.2–0.75%) generate more complex lesions, with VSMC infiltration and cholesterol deposits.[20,24–27,29] In addition, it has been shown that complex atherosclerotic lesions can be induced in NZW rabbits by intermittent cycles of fat feeding with periods of normal diet.[27,30] A widely used alternative method to accelerate the development of advanced lesions in rabbits combines a high-fat/high-cholesterol diet with angioplasty-induced aortic denudation of the aorta, generally from the aortic arch to the iliac arteries.[31–38] This combined protocol greatly accelerates the formation of atherosclerotic lesions and, importantly, produces plaques that exhibit a lipid core surrounded by a fibrous cap due to increased proliferation of VSMCs, thus resembling more closely human advanced plaques than those produced by feeding rabbits with a high-cholesterol diet alone.

Some rabbit strains carry genetic mutations that lead to hyperlipidemia and atherosclerosis. This is the case of Watanabe, St Thomas, and Houston RT strains, which present genetic abnormalities in lipid metabolism.[39–42] The most widely used hyperlipidemic rabbit strain in atherosclerosis research is the Watanabe heritable hyperlipidemic (WHHL) rabbit,[39] which has been particularly important for the development of several cholesterol-lowering antiatherosclerotic drugs.[43] WHHL rabbits carry an inactivating mutation in the gene encoding the LDL receptor (LDLR), a major mediator of the hepatic clearance of circulating lipoproteins.[44,45] Consistently, WHHL rabbits spontaneously exhibit hypercholesterolemia, with increased plasma levels of atherogenic very low density lipoproteins (VLDL) and reduced levels of atheroprotective high-density lipoproteins (HDL), together with visceral fat accumulation and hyperinsulinemia. Furthermore, WHHL rabbits develop aortic atherosclerotic plaques, ranging from fatty streaks to more advanced, fibrous lesions,[46] and selective breeding has allowed the generation of WHHL rabbits that exhibit severe coronary atherosclerosis[47] and develop spontaneous myocardial infarctions.[48]

The generation of transgenic rabbits with altered expression of specific genes involved in cholesterol and lipoprotein metabolism has provided notable insight into the important role of cholesterol trafficking in atherosclerosis development. For instance, NZW rabbits with liver-specific expression of human apolipoprotein A-I (the major component of plasma HDL) exhibit significantly smaller atherosclerotic lesions than nontransgenic controls.[49,50] Similarly, transgenic rabbits that overexpress human lecithin–cholesterol acyltransferase, an enzyme that catalyzes the conversion of free cholesterol into cholesteryl esters, have markedly reduced atherosclerosis compared to control rabbits when fed a high-cholesterol diet.[51] WHHL rabbits have also been subjected to genetic manipulation. For example, WHHL rabbits that express human apolipoprotein(a), which assembles into atherogenic lipoprotein(a) particles in plasma, exhibit accelerated atherosclerosis development and more complex lesions than nontransgenic controls.[52] Likewise, human lipoprotein lipase overexpression in this rabbit strain leads to enhanced aortic atherosclerosis.[53]

B. Swine Models of Atherosclerosis

The development, morphology, and function of the cardiovascular system in swine as well as their lipoprotein profiles and metabolism closely resemble that of humans. Gottlieb and Lalich reported the first large quantitative data upon the spontaneous occurrence of atherosclerosis in the porcine aorta by analyzing more than 2000 specimens obtained at slaughterhouses.[54] Several strains of large farm pigs that bear mutant alleles for the LDLR or apolipoprotein B loci have been available for more than two decades.[55–58] These strains spontaneously develop humanoid atherosclerosis in most arterial beds, including the coronary arteries, and disease progresses in time periods inversely proportional to cholesterol levels. Like in humans, early lesions in swine models of familial hypercholesterolemia consist of fatty streaks rich in lipid-laden macrophages, which progress to complicated lesions with calcification and features of vulnerable plaques, including abundant neovascularization, necrotic cores, a thin fibrous cap, and intraplaque hemorrhage. Although these characteristics make familial hypercholesterolemic pigs a suitable preclinical model for testing new antiatherogenic and antithrombotic drugs and for developing imaging technologies and interventional devices,[59] their use is limited due to the long periods of time that are required to develop complex atherosclerotic lesions even when challenged with atherogenic diets (2–3 years) and the large size and weight they reach (> 200 kg). These difficulties in care and high maintenance cost are reduced with the use of smaller swine strains, such as the Yucatan miniature pig and several sublines that have been derived from the primary population, which also develop humanoid complicated lesions with abundant necrosis and cholesterol deposits and extensive calcification.[60–65] Very recently, Thim *et al.*[66] reported the generation of a downsized hypercholesterolemic pig strain named FBM that was produced by crossing the Rapacz familial hypercholesterolemic pig bearing the R84C LDLR mutation with a smaller pig (Chinese Meishan) and then crossing the offspring with an even smaller minipig from Brentocelles, France. FBM pigs breed like normal pigs, develop atherosclerotic lesions on standard diet, and disease progression is aggravated by atherogenic diet feeding, whereby plasma total cholesterol rose to > 20 mmol/l (> 800 mg/dl) and plaques mirrored human-like features, including a large necrotic core covered by a thin and inflamed fibrous cap, neovascularization, intraplaque hemorrhage, and expansive remodeling.[66]

Like in other animal models, atherosclerotic plaque development in swine can be accelerated by combining high-cholesterol diet with locally produced vascular injury inflicted by different means, including guide-wire-induced injury,[67] endovascular balloon inflation with or without stent deployment,[59,62,66] partial vessel ligation,[68] and balloon angioplasty followed 2 weeks later by percutaneous intramural injection of a mixture of cholesteryl esters and human oxLDL.[69,70] Atherogenic diets plus vascular injury protocols not only reduce the difficulties in care and high maintenance cost associated with the use of swine models by reducing the duration of the study but are also highly relevant models for translational research in the field of percutaneous interventions and cardiovascular imaging.

Diabetes and hypercholesterolemia frequently coexist in patients with metabolic syndrome, who are at significantly higher risk for atherosclerosis and its complications than the general population. The incidence of diabetes, particularly type 2 diabetes, is expected to increase significantly during the next decades owing to the unhealthy effects of modern lifestyle habits (e.g., obesity and lack of physical exercise). However, studies examining mechanisms underlying diabetes-accelerated atherosclerosis are scant in part due to the lack of suitable humanoid preclinical models. The combination of hypercholesterolemia and diabetes (induced by intravenous administration of streptozotocin, which destroys over 80% of pancreatic beta cells) accelerates atherosclerotic lesions in the aorta and coronary and femoral arteries of Yorkshire pigs.[71,72] In addition to reducing the duration of experiments, the combination of diabetes and hypercholesterolemia in the pig makes animal handling easier (diabetic pigs gain weight slower than nondiabetic controls) and leads to the formation of advanced human-like atherosclerotic lesions. However, treatment of complications associated with diabetes (e.g., hypoglycemia, hyperglycemia, gastroparesis, and infections) is required when working with swine models of diabetes-induced atherosclerosis. The combination of diabetes and hypercholesterolemia has proven relevant in several settings, including the demonstration that selective inhibition of lipoprotein-associated phospholipase A(2) (Lp-PLA(2)) with darapladib substantially reduces inflammation and the development of advanced coronary atherosclerosis.[73]

C. Nonhuman Primate Models of Atherosclerosis

Nonhuman primates are attractive for atherosclerosis research because their general anatomical resemblance and their phylogenetic proximity to humans, thus making likely that data derived from these animals are more directly applicable to the clinical scenario. However, New World monkeys have been of limited value in atherosclerosis research primarily because of the complications that are associated with their chronic renal disease, which affects both their lipoprotein metabolism and atherosclerotic plaque development.[74–76]

In contrast, extensive research has been conducted in Rhesus monkeys, which have been shown to develop spontaneous atherosclerosis in several vessels, including the coronary arteries. Moreover, feeding a high-cholesterol diet greatly accelerates the development of atherosclerosis and frequently induces fatal myocardial infarction in these primates.[77,78] Atherosclerosis regression upon low-fat feeding has also been demonstrated in this model.[79,80] Similarly, cynomolgus and Cebus monkeys develop atherosclerosis and have been used as experimental models of this disease.[81–83] Like in humans, age has a striking effect on the susceptibility to atherosclerosis in nonhuman primates; young animals are reasonably resistant to the development of atherosclerotic lesions, but become much more susceptible as they age.[83,84] In addition, the same gender differences that exist in human atherosclerosis are evident in nonhuman primates, with males being more affected than females.[84] Unfortunately, in spite of these striking similarities between human and primate atherosclerosis, the difficult handling of these animals, their high economical cost, and the important ethical concerns strongly limit their utility in research.

D. Rodent Models of Atherosclerosis

Pioneer studies of experimental atherosclerosis have been mainly performed in rabbits, pigs, and nonhuman primates. Although research with these species certainly provided valuable insight into the pathophysiology of atherosclerosis, it also encountered a number of important obstacles that limited their value as experimental models, as discussed above. In contrast, rodents, in particular, mice, have many advantages for experimental studies and are currently the most frequently used laboratory animals in atherosclerosis research.

1. Mice

Due to their small size, mice can be handled and maintained easily. Moreover, mice can live up to 3 years and exhibit a quick generation time (3 weeks for gestation and 6–8 weeks to reach sexual maturity), allowing the breeding of large cohorts in a short time frame. Moreover, the mouse is the most commonly used mammal for genetic manipulation and a vast diversity of transgenic and knockout mice is readily available to conduct atherosclerosis research. However, the use of mice as an experimental model of atherosclerosis faces one major obstacle: this species is extremely resistant to atherosclerosis. The reasons for this may be various. Cholesterol metabolism is very different in mice and humans, and plasma cholesterol and lipoprotein patterns are particularly dissimilar in both species. Plasma cholesterol levels are generally low in mice-fed standard chow, ranging from 60 to 100 mg/dl depending on the background strain. Moreover, while the major circulating lipoprotein in humans is pro-atherogenic LDL, plasma LDL levels in wild-type mice are extremely low and more than 85% of cholesterol is carried on atheroprotective HDL.[85] This remarkable difference is probably mostly due to the absence in mice of plasma cholesterol ester transfer protein (CETP), an enzyme that catalyzes the transfer of cholesteryl esters and triglycerides between different lipoproteins.[86] In addition, mice and men have several differences in cardiovascular physiology and anatomy. The human heart rate is normally 60–90 beats/min, while in the mouse is around 300 beats/min. Furthermore, the postural differences between mice and men affect hemodynamics, an important determinant of atherosclerosis susceptibility.[87–89] In spite of the natural resistance of mice to atherosclerosis, the combination of dietary challenges and genetic manipulation has allowed the generation of several murine models of atherosclerosis, which are discussed below.

The first mouse model of atherosclerosis was initially characterized during 1960s by Wissler and coworkers, who used an experimental diet that contained very high amounts of fat (30%) and cholesterol (5%), and 2% cholic acid, which promotes hypercholesterolemia by blocking cholesterol conversion to bile acids.[90] Although this extreme diet induced the formation of fatty streaks in different vascular regions of the mouse, it was also pro-inflammatory and highly toxic, leading frequently to weight loss and high susceptibility to infections.[90] Using a less toxic diet that contained 15% fat, 1.25% cholesterol, and 0.5% cholic acid, Paigen and collaborators found high variability in diet-induced atherosclerosis when comparing 10 different inbred strains of mice.[91] The most susceptible strain was C57BL/6, which developed mild hypercholesterolemia—around 200 mg/ml—and fatty streak lesions in the aortic root after 3–9 months of fat feeding, whereas other resistant inbred strains, such as the C3H, did not exhibit atherosclerotic lesions under the same dietary regimen.[91] Of note, lesions in the aorta of fat-fed C57BL/6 mice were small, consisted almost exclusively of macrophages and did not progress to fibrous plaques, unlike the situation in humans. This shortcoming, together with the toxicity and the pro-inflammatory actions of cholic acid, challenged the utility of wild-type mice for atherosclerosis research. Nevertheless, wild-type mice have proven useful to identify several genes involved in the etiopathogenesis of atherosclerosis.[92–94]

A major breakthrough in atherosclerosis research occurred in 1992 with the generation of mouse strains with genetic disruption of apolipoprotein E (apoE).[95,96] apoE is a structural component of all lipoproteins other than LDL and is a critical ligand for the hepatic clearance of plasma lipoproteins mediated by LDLR and LDLR-related proteins.[97–99] Consistently, even on standard low-fat chow, apoE-deficient mice exhibit marked hypercholesterolemia (around 400 mg/dl, which represents an approximately fivefold increase in plasma cholesterol levels compared to wild-type mice), and a dramatic shift in the plasma lipoprotein profile, with pro-atherogenic VLDL as the most abundant circulating particles, similar to the situation in human type III hyperlipidemia.[100] In addition, apoE-deficient mice spontaneously develop atherosclerotic plaques in several vascular beds, predominantly in the aortic root, the aortic arch, and the different branch points along the aorta. Atherosclerotic lesions in apoE-null mice are heterogeneous, ranging from fatty streaks rich in foam cells to complex lesions that exhibit macrophage-enriched shoulders and necrotic cores with a well-formed fibrous cap that includes VSMCs and extracellular matrix (Fig. 1). Moreover, atherosclerosis is greatly accelerated in apoE-deficient mice challenged with high-cholesterol high-fat diets that are not supplemented with cholic acid.[101,102] The most frequently used atherogenic diet in mice is the Western-type diet, which contains 0.15% cholesterol and 21% fat derived from milk fat. When fed this diet, apoE-deficient mice exhibit over a threefold elevation in plasma cholesterol levels compared to animals fed regular chow and develop complex fibrous plaques in the aortic sinus after 10–14 weeks of fat feeding. This rapidity of lesion progression is an important advantage in the experimental setting in comparison with other animal models of atherosclerosis. Although apoE-deficient mice have turned into the gold standard for experimental atherosclerosis studies, this model has also some important limitations. A major drawback is that most plasma cholesterol in apoE-deficient mice is mostly confined to VLDL particles and not to LDL particles, as it occurs in humans. Furthermore, there is increasing evidence that apoE has extra atheroprotective properties in addition to mediating lipoprotein clearance, including antioxidant, antiproliferative, and anti-inflammatory actions. Therefore, because atherosclerosis is an inflammatory disease, the potential immunomodulatory actions of apoE must be considered when analyzing data obtained in this animal model.

In addition to apoE-deficient mice, other genetically manipulated mouse strains have been generated that are prone to atherosclerosis and show particular advantages in certain experimental settings. A widely used model is the LDLR-deficient mouse, which has a milder lipoprotein alteration than apoE-deficient mice when fed standard low-fat chow, with plasma cholesterol levels around 250 mg/dl due mainly to the accumulation of LDL.[103,104] Although LDLR-deficient mice do not develop significant atherosclerosis on a normal chow diet, high-fat feeding leads to severe hypercholesterolemia (~ 900 mg/dl) with accumulation of VLDL and LDL and extensive atherosclerosis.[104] A major disadvantage of this murine model is that the development of human-like complex fibrous lesions typically requires longer periods of fat feeding than in apoE-deficient mice. However, LDLR-deficient mice are advantageous compared to apoE-null mice when performing bone marrow transplantation studies. apoE is locally produced by macrophages in the atheroma[105]; therefore, bone marrow transplantation studies with apoE-deficient mice require that the donor also carries apoE deficiency to avoid the atheroprotective action of apoE secreted by bone marrow-derived macrophages. In contrast, LDLR expression in hematopoietic cells is minimal and irrelevant in the process of atherogenesis, and this allows the utilization of wild-type mice as bone marrow donors in experiments using LDLR-null mice as recipients. Another important difference is that LDLR-deficient mice are more prone to the development of obesity and insulin resistance than apoE-deficient mice,[106] a particularly relevant issue when analyzing the role of genes involved in metabolic control and atherosclerosis development. In contrast, apoE-deficient mice are more susceptible to injury-induced neointimal formation,[107,108] and therefore may represent a better experimental model for the investigation of the molecular mechanisms underlying restenosis post angioplasty. Although the pros and cons of each model are frequently debated, results with apoE-deficient and LDLR-deficient mice are generally comparable. Another genetically modified mouse that has been used as an experimental model of atherosclerosis—although less frequently than apoE-null and LDLR-null—is the apoE*3Leiden (apoE3L) transgenic mouse, which carries a mutated form of the human *APOE3* gene. Although apoE3L mice express the endogenous murine wild-type apoE, forced expression of the human *APOE3* mutant gene highly impairs the hepatic clearance of lipoproteins and this leads to hypercholesterolemia and atherosclerosis upon cholesterol feeding.[109]

Even though genetically engineered mice are currently the most widely used animal models of atherosclerosis, they have been classically considered to lack the most pathologically relevant feature of human atherosclerosis, that is, plaque rupture, which is the leading cause of life-threatening acute ischemic events (e.g., myocardial infarction and stroke). However, while this appears to be generally true in the case of atherosclerotic plaques in the aortic sinus—the arterial region most typically analyzed in murine atherosclerosis studies—it might not be the case in other vascular regions in the mouse, because a number of studies have reported that plaque rupture spontaneously occurs in the brachiocephalic artery (also known as the innominate artery) of apoE-deficient mice.[110,111] However, this conclusion often relies exclusively on some particular histopathological markers of plaque rupture or on a definition of plaque rupture that differs from the concept in humans in the sense that it includes neither luminal thrombosis nor intraplaque hemorrhage.[112] Certainly, high controversy still exists about the definition and occurrence of spontaneous plaque rupture in murine models of atherosclerosis and further studies are required to clarify this important aspect. However, a number of models of artificially induced atherosclerotic plaque rupture and/or atherothrombosis in hypercholesterolemic mice have been proposed, and some of them have provided important insight into the molecular mechanisms that determine plaque vulnerability. One such approach is to mechanically provoke rupture of the atherosclerotic plaque. For example, platelet- and fibrin-rich thrombi can be induced within atheromata of apoE-deficient mice by compressing the aorta between forceps.[113] Similarly, carotid artery ligation followed by the application of a polyethylene cuff has been shown to induce intraplaque hemorrhage and luminal thrombus formation in apoE-deficient mice.[114] Photochemical injury has also been used to induce atherothrombosis in this murine model.[115] In addition, a number of genetic manipulations have been shown to promote plaque instability and rupture, including macrophage-specific expression of active metalloprotease-9 (MMP-9), which induces fibrin deposition and intraplaque hemorrhage in apoE-deficient mice,[116] as well as adenovirus-mediated overexpression of p53 in combination with treatment with the vasopressor compound phenylephrine.[117]

2. Rats

The rat has been pivotal for physiological and metabolic research since the development of the first defined rat strain at the Wistar Institute in the 1920s.[118] Like wild-type mice, rats lack plasma CETP and carry most plasma cholesterol in HDL particles. Consistently, the rat is highly resistant to atherosclerosis, with the exception of some particular strains that develop arterial fatty lesions that, however, do not resemble human atherosclerotic plaques.[119–121] In addition, rats are poorly responsive to fat feeding, and hyperlipidemia and atherosclerosis can only be induced using diets that contain an extremely high content of cholesterol and fat, together with cholic acid and thiouracil, which induces hypothyroidism and thereby lowers hepatic clearance of lipoproteins.[122,123] Therefore, wild-type rats are not generally considered a suitable model of experimental atherosclerosis. However, some rat strains that carry mutations which cause hyperlipidemia have been reported to develop atherosclerotic plaques in different vascular beds.[121,124–126] The JCR:LA-corpulent strain is of particular interest because it develops extensive atherosclerosis and ischemic lesions of the heart, as well as insulin resistance and other metabolic dysfunctions.[127–133] In addition, the recent development of molecular techniques for the manipulation of the rat genome[134] is expected to hasten the generation of transgenic and knockout rats of interest for atherosclerosis research. Of note in this regard, rats that are genetically engineered to overexpress human CETP exhibit dyslipidemia and coronary artery lesions with a strong male-to-female bias, like in humans.[135,136] In spite of this, the mouse remains the most widely used rodent model of atherosclerosis, mostly because of the easier breeding and genetic manipulation of this species.

3. Other Rodents

Hamsters and guinea pigs have been used as experimental models of atherosclerosis. Unlike mice and rats, hamsters express plasma CETP and carry a significant fraction of plasma cholesterol in LDL particles and thus are metabolically closer to humans than other rodents.[137,138] Several studies have demonstrated the development of hypercholesterolemia and atherosclerotic plaques in the aorta of Golden Syrian hamsters, ranging from fatty streaks to more complex lesions.[139–141] However, more recent work has not consistently replicated the extension and/or morphology of atherosclerotic lesions in hamsters,[142] and therefore this model is not frequently used in atherosclerosis research. Guinea pigs also express CETP and transport the majority of circulating cholesterol in LDL particles,[137] and high-cholesterol diets induce the development of early atherosclerotic plaques in the arterial wall of these animals.[143] However, advanced atherosclerotic lesions are not generally observed in guinea pigs, thus limiting the usefulness of this model in atherosclerosis research.

III. Concluding Remarks

Since the initial evidence that atherosclerosis development can be induced in laboratory animals was provided by Ignatowski in 1908, various animal species have been used as experimental models of this disease. While most animal models are still being used to a greater or lesser extent, none of them can be considered an ideal model of the human pathology. The choice of the most appropriate animal model depends on the nature of the research that is to be performed. Atherosclerosis in pigs and nonhuman primates closely recapitulates the main morphological and biochemical characteristics of human atherosclerosis, therefore the results obtained in large animal models are more easily extrapolated to humans in translational research studies, such as the development of imaging techniques and interventional devices and the assessment of novel therapeutic strategies. However, research with large animal species encounters a number of important obstacles, such as difficulties in handling, the high economical cost of maintenance, and important ethical considerations in the case of studies with nonhuman primates. In contrast, small animals, such as mice, rats, and rabbits, have many advantages for experimental studies (e.g., easy handling, low cost), but do not typically develop the advanced vulnerable atherosclerotic plaques that are characteristic of human individuals suffering severe CVD. Genetically engineered mouse models have been crucial to elucidate the molecular mechanisms underlying atherosclerotic plaque initiation and progression, and the recent advent of strategies to tissue specifically and/or temporally control the genetic manipulation offers added value to murine models. The uses of all animal models available nowadays will undoubtedly continue to permit major advances in atherosclerosis research that should translate into improved treatment, prevention, and diagnosis of atherosclerosis and associated ischemic diseases.

Acknowledgments

Work in the author's laboratories is supported by the Spanish Ministry of Science and Innovation (MICINN) (grants SAF2010-16044 and SAF2008-04629) and Fondo de Investigación Sanitaria—Instituto de Salud Carlos III (RECAVA: grant RD06/0014/0021 and PI10/02268). The CNIC is supported by the MICINN and the Pro-CNIC Foundation.

References

1. Mathers CD, Loncar D. Projections of global mortality and burden of disease from 2002 to 2030. *PLoS Med* 2006;**3**:e442.
2. Witztum JL. The oxidation hypothesis of atherosclerosis. *Lancet* 1994;**344**:793–5.
3. Steinberg D. The LDL modification hypothesis of atherogenesis: an update. *J Lipid Res* 2009;**50**:S376–81.
4. Cushing SD, Berliner JA, Valente AJ, Territo MC, Navab M, Parhami F, et al. Minimally modified low density lipoprotein induces monocyte chemotactic protein 1 in human endothelial cells and smooth muscle cells. *Proc Natl Acad Sci USA* 1990;**87**:5134–8.
5. Li D, Mehta JL. Antisense to LOX-1 inhibits oxidized LDL-mediated upregulation of monocyte chemoattractant protein-1 and monocyte adhesion to human coronary artery endothelial cells. *Circulation* 2000;**101**:2889–95.
6. Vora DK, Fang Z-T, Liva SM, Tyner TR, Parhami F, Watson AD, et al. Induction of P-selectin by oxidized lipoproteins: separate effects on synthesis and surface expression. *Circ Res* 1997;**80**:810–8.
7. Khan BV, Parthasarathy SS, Alexander RW, Medford RM. Modified low density lipoprotein and its constituents augment cytokine-activated vascular cell adhesion molecule-1 gene expression in human vascular endothelial cells. *J Clin Invest* 1995;**95**:1262–70.
8. Pello OM, Silvestre C, De Pizzol M, Andrés V. A glimpse on the phenomenon of macrophage polarization during atherosclerosis. *Immunobiology* 2011;**216**:1172–76.
9. Choudhury RP, Lee JM, Greaves DR. Mechanisms of disease: macrophage-derived foam cells emerging as therapeutic targets in atherosclerosis. *Nat Clin Pract Cardiovasc Med* 2005;**2**:309–15.
10. Fuster JJ, Fernández P, González-Navarro H, Silvestre C, Nabah YN, Andrés V. Control of cell proliferation in atherosclerosis: insights from animal models and human studies. *Cardiovasc Res* 2010;**86**:254–64.
11. Glass CK, Witztum JL. Atherosclerosis: the road ahead. *Cell* 2001;**104**:503–16.
12. Lusis AJ. Atherosclerosis. *Nature* 2000;**407**:233–41.
13. Libby P, Ridker PM, Maseri A. Inflammation and atherosclerosis. *Circulation* 2002;**105**:1135–43.
14. Tabas I. The role of endoplasmic reticulum stress in the progression of atherosclerosis. *Circ Res* 2010;**107**:839–50.
15. Fuster V, Moreno PR, Fayad ZA, Corti R, Badimón JJ. Atherothrombosis and high-risk plaque: Part I: evolving concepts. *J Am Coll Cardiol* 2005;**46**:937–54.
16. Ignatowski AC. Influence of animal food on the organism of rabbits. *ST Petersburg Izvest Imp Voyenno-Med Akad* 1908;**16**:154–76.
17. Mahley RW, Weisgraber KH, Innerarity T. Canine lipoproteins and atherosclerosis: II. Characterization of the plasma lipoproteins associated with atherogenic and nonatherogenic hyperlipidemia. *Circ Res* 1974;**35**:722–33.

18. Clair RWS. The contribution of avian models to our understanding of atherosclerosis and their promise for the future. *Lab Anim Sci* 1998;**48**:565–8.
19. Anichkov N, Chalatov S. Ueber experimentelle cholester-insteatose: ihre bedeutung fur die enstehung einiger pathologischer prozessen. *Centrbl Allg Pathol Pathol Anat* 1913;**24**:1–9.
20. Finking G, Hanke H. Nikolaj Nikolajewitsch Anitschkow (1885–1964) established the cholesterol-fed rabbit as a model for atherosclerosis research. *Atherosclerosis* 1997;**135**:1–7.
21. Bocan TM, Mueller SB, Mazur MJ, Uhlendorf PD, Brown EQ, Kieft KA. The relationship between the degree of dietary-induced hypercholesterolemia in the rabbit and atherosclerotic lesion formation. *Atherosclerosis* 1993;**102**:9–22.
22. Kamimura R, Suzuki S, Sakamoto H, Miura N, Misumi K, Miyahara K. Development of atherosclerotic lesions in cholesterol-loaded rabbits. *Exp Anim* 1999;**48**:1–7.
23. Richardson M, Kurowska EM, Carroll KK. Early lesion development in the aortas of rabbits fed low-fat, cholesterol-free, semipurified casein diet. *Atherosclerosis* 1994;**107**.165–78.
24. Adams CWM, Miller NE, Morgan RS, Rao SN. Lipoprotein levels and tissue lipids in fatty-fibrous atherosclerosis induced in rabbits by two years' cholesterol feeding at a low level. *Atherosclerosis* 1982;**44**:1–8.
25. Cornhill JF, Roach MR. A quantitative study of the localization of atherosclerotic lesions in the rabbit aorta. *Atherosclerosis* 1976;**23**:489–501.
26. Roach MR, Cornhill JF, Fletcher J. A quantitative study of the development of sudanophilic lesions in the aorta of rabbits fed a low-cholesterol diet for up to six months. *Atherosclerosis* 1978;**29**:259–64.
27. Kolodgie FD, Katocs Jr. AS, Largis EE, Wrenn SM, Cornhill JF, Herderick EE, et al. Hypercholesterolemia in the rabbit induced by feeding graded amounts of low-level cholesterol. Methodological considerations regarding individual variability in response to dietary cholesterol and development of lesion type. *Arterioscler Thromb Vasc Biol* 1996;**16**:1454–64.
28. Spagnoli LG, Orlandi A, Mauriello A, Santeusanio G, de Angelis C, Lucreziotti R, et al. Aging and atherosclerosis in the rabbit: 1. Distribution, prevalence and morphology of atherosclerotic lesions. *Atherosclerosis* 1991;**89**:11–24.
29. Böger RH, Bode-Böger SM, Kienke S, Stan AC, Nafe R, Frölich JC. Dietary L-arginine decreases myointimal cell proliferation and vascular monocyte accumulation in cholesterol-fed rabbits. *Atherosclerosis* 1998;**136**:67–77.
30. Constantinides P, Booth J, Carlson G, Cox B, Nakashima R, Williams K. Production of advanced cholesterol atherosclerosis in the rabbit. *Arch Pathol* 1960;**70**:712–24.
31. Abela GS, Picon PD, Friedl SE, Gebara OC, Miyamoto A, Federman M, et al. Triggering of plaque disruption and arterial thrombosis in an atherosclerotic rabbit model. *Circulation* 1995;**91**:776–84.
32. Corti R, Osende JI, Fallon JT, Fuster V, Mizsei G, Jneid H, et al. The selective peroxisomal proliferator-activated receptor-gamma agonist has an additive effect on plaque regression in combination with simvastatin in experimental atherosclerosis: *in vivo* study by high-resolution magnetic resonance imaging. *J Am Coll Cardiol* 2004;**43**:464–73.
33. Helft G, Worthley SG, Fuster V, Fayad ZA, Zaman AG, Corti R, et al. Progression and regression of atherosclerotic lesions: monitoring with serial noninvasive magnetic resonance imaging. *Circulation* 2002;**105**:993–8.
34. Skinner MP, Yuan C, Mitsumori L, Hayes CE, Nelson JA, Ross R. Serial magnetic resonance imaging of experimental atherosclerosis detects lesion fine structure, progression and complications *in vivo*. *Nat Med* 1995;**1**:69–73.
35. Aikawa M, Rabkin E, Okada Y, Voglic SJ, Clinton SK, Brinckerhoff CE, et al. Lipid lowering by diet reduces matrix metalloproteinase activity and increases collagen content of rabbit atheroma: a potential mechanism of lesion stabilization. *Circulation* 1998;**97**:2433–44.

36. Rosenfeld ME, Khoo JC, Miller E, Parthasarathy S, Palinski W, Witztum JL. Macrophage-derived foam cells freshly isolated from rabbit atherosclerotic lesions degrade modified lipoproteins, promote oxidation of low-density lipoproteins, and contain oxidation-specific lipid-protein adducts. *J Clin Invest* 1991;**87**:90–9.
37. Worthley SG, Helft G, Fuster V, Zaman AG, Fayad ZA, Fallon JT, et al. Serial *in vivo* MRI documents arterial remodeling in experimental atherosclerosis. *Circulation* 2000;**101**:586–9.
38. Ibáñez B, Vilahur G, Cimmino G, Speidl WS, Pinero A, Choi BG, et al. Rapid change in plaque size, composition, and molecular footprint after recombinant apolipoprotein A-I Milano (ETC-216) administration: magnetic resonance imaging study in an experimental model of atherosclerosis. *J Am Coll Cardiol* 2008;**51**:1104–9.
39. Watanabe Y. Serial inbreeding of rabbits with hereditary hyperlipidemia (WHHL-rabbit). *Atherosclerosis* 1980;**36**:261–8.
40. La Ville A, Turner PR, Pittilo RM, Martini S, Marenah CB, Rowles PM, et al. Hereditary hyperlipidemia in the rabbit due to overproduction of lipoproteins. I. Biochemical studies. *Arterioscler Thromb Vasc Biol* 1987;**7**:105–12.
41. Seddon AM, Woolf N, La Ville A, Pittilo RM, Rowles PM, Turner PR, et al. Hereditary hyperlipidemia and atherosclerosis in the rabbit due to overproduction of lipoproteins. II. Preliminary report of arterial pathology. *Arterioscler Thromb Vasc Biol* 1987;**7**:113–24.
42. Yanni AE. The laboratory rabbit: an animal model of atherosclerosis research. *Lab Anim* 2004;**38**:246–56.
43. Shiomi M, Ito T. The Watanabe heritable hyperlipidemic (WHHL) rabbit, its characteristics and history of development: a tribute to the late Dr. Yoshio Watanabe. *Atherosclerosis* 2009;**207**:1–7.
44. Kita T, Brown MS, Watanabe Y, Goldstein JL. Deficiency of low density lipoprotein receptors in liver and adrenal gland of the WHHL rabbit, an animal model of familial hypercholesterolemia. *Proc Natl Acad Sci USA* 1981;**78**:2268–72.
45. Attie AD, Pittman RC, Watanabe Y, Steinberg D. Low density lipoprotein receptor deficiency in cultured hepatocytes of the WHHL rabbit. Further evidence of two pathways for catabolism of exogenous proteins. *J Biol Chem* 1981;**256**:9789–92.
46. Buja LM, Kita T, Goldstein JL, Watanabe Y, Brown MS. Cellular pathology of progressive atherosclerosis in the WHHL rabbit. An animal model of familial hypercholesterolemia. *Arterioscler Thromb Vasc Biol* 1983;**3**:87–101.
47. Shiomi M, Ito T, Shiraishi M, Watanabe Y. Inheritability of atherosclerosis and the role of lipoproteins as risk factors in the development of atherosclerosis in WHHL rabbits: risk factors related to coronary atherosclerosis are different from those related to aortic atherosclerosis. *Atherosclerosis* 1992;**96**:43–52.
48. Shiomi M, Ito T, Yamada S, Kawashima S, Fan J. Development of an animal model for spontaneous myocardial infarction (WHHLMI rabbit). *Arterioscler Thromb Vasc Biol* 2003;**23**:1239–44.
49. Duverger N, Kruth H, Emmanuel F, Caillaud JM, Viglietta C, Castro G, et al. Inhibition of atherosclerosis development in cholesterol-fed human apolipoprotein A-I-transgenic rabbits. *Circulation* 1996;**94**:713–7.
50. Duverger N, Viglietta C, Berthou L, Emmanuel F, Tailleux A, Parmentier-Nihoul L, et al. Transgenic rabbits expressing human apolipoprotein A-I in the liver. *Arterioscler Thromb Vasc Biol* 1996;**16**:1424–9.
51. Hoeg JM, Santamarina-Fojo S, Bérard AM, Cornhill JF, Herderick EE, Feldman SH, et al. Overexpression of lecithin: cholesterol acyltransferase in transgenic rabbits prevents diet-induced atherosclerosis. *Proc Natl Acad Sci USA* 1996;**93**:11448–53.
52. Fan J, Shimoyamada H, Sun H, Marcovina S, Honda K, Watanabe T. Transgenic rabbits expressing human apolipoprotein (a) develop more extensive atherosclerotic lesions in response to a cholesterol-rich diet. *Arterioscler Thromb Vasc Biol* 2001;**21**:88–94.

53. Koike T, Liang J, Wang X, Ichikawa T, Shiomi M, Sun H, et al. Enhanced aortic atherosclerosis in transgenic Watanabe heritable hyperlipidemic rabbits expressing lipoprotein lipase. *Cardiovasc Res* 2005;**65**:524–34.
54. Gottlieb H, Lalich JJ. The occurrence of arteriosclerosis in the aorta of swine. *Am J Pathol* 1954;**30**:851–5.
55. Hasler-Rapacz J, Ellegren H, Fridolfsson AK, Kirkpatrick B, Kirk S, Andersson L, et al. Identification of a mutation in the low density lipoprotein receptor gene associated with recessive familial hypercholesterolemia in swine. *Am J Med Genet* 1998;**76**:379–86.
56. Rapacz J, Hasler-Rapacz J, Taylor KM, Checovich WJ, Attie AD. Lipoprotein mutations in pigs are associated with elevated plasma cholesterol and atherosclerosis. *Science* 1986;**234**:1573–7.
57. Prescott MF, Hasler-Rapacz J, von Linden-Reed J, Rapacz J. Familial hypercholesterolemia associated with coronary atherosclerosis in swine bearing different alleles for apolipoprotein B. *Ann N Y Acad Sci* 1995;**748**:283–92 discussion 292–283.
58. Prescott MF, McBride CH, Hasler-Rapacz J, Von Linden J, Rapacz J. Development of complex atherosclerotic lesions in pigs with inherited hyper-LDL cholesterolemia bearing mutant alleles for apolipoprotein B. *Am J Pathol* 1991;**139**:139–47.
59. Granada JF, Kaluza GL, Wilensky RL, Biedermann BC, Schwartz RS, Falk E. Porcine models of coronary atherosclerosis and vulnerable plaque for imaging and interventional research. *EuroIntervention* 2009;**5**:140–8.
60. Panepinto LM, Phillips RW. The Yucatan miniature pig: characterization and utilization in biomedical research. *Lab Anim Sci* 1986;**36**:344–7.
61. Barbeau ML, Klemp KF, Guyton JR, Rogers KA. Dietary fish oil: influence on lesion regression in the porcine model of atherosclerosis. *Arterioscler Thromb Vasc Biol* 1997;**17**:688–94.
62. de Smet BJ, van der Zande J, van der Helm YJ, Kuntz RE, Borst C, Post MJ. The atherosclerotic Yucatan animal model to study the arterial response after balloon angioplasty: the natural history of remodeling. *Cardiovasc Res* 1998;**39**:224–32.
63. Gal D, Rongione AJ, Slovenkai GA, DeJesus ST, Lucas A, Fields CD, et al. Atherosclerotic Yucatan microswine: an animal model with high-grade, fibrocalcific, nonfatty lesions suitable for testing catheter-based interventions. *Am Heart J* 1990;**119**:291–300.
64. Reitman JS, Mahley RW, Fry DL. Yucatan miniature swine as a model for diet-induced atherosclerosis. *Atherosclerosis* 1982;**43**:119–32.
65. Holvoet P, Theilmeier G, Shivalkar B, Flameng W, Collen D. LDL hypercholesterolemia is associated with accumulation of oxidized LDL, atherosclerotic plaque growth, and compensatory vessel enlargement in coronary arteries of miniature pigs. *Arterioscler Thromb Vasc Biol* 1998;**18**:415–22.
66. Thim T, Hagensen MK, Drouet L, Bal Dit Sollier C, Bonneau M, Granada JF, et al. Familial hypercholesterolaemic downsized pig with human-like coronary atherosclerosis: a model for preclinical studies. *EuroIntervention* 2010;**6**:261–8.
67. Mihaylov D, van Luyn MJ, Rakhorst G. Development of an animal model of selective coronary atherosclerosis. *Coron Artery Dis* 2000;**11**:145–9.
68. Ishii A, Vinuela F, Murayama Y, Yuki I, Nien YL, Yeh DT, et al. Swine model of carotid artery atherosclerosis: experimental induction by surgical partial ligation and dietary hypercholesterolemia. *AJNR Am J Neuroradiol* 2006;**27**:1893–9.
69. Granada JF, Moreno PR, Burke AP, Schulz DG, Raizner AE, Kaluza GL. Endovascular needle injection of cholesteryl linoleate into the arterial wall produces complex vascular lesions identifiable by intravascular ultrasound: early development in a porcine model of vulnerable plaque. *Coron Artery Dis* 2005;**16**:217–24.

70. Granada JF, Wallace-Bradley D, Win HK, Alviar CL, Builes A, Lev EI, et al. *In vivo* plaque characterization using intravascular ultrasound-virtual histology in a porcine model of complex coronary lesions. *Arterioscler Thromb Vasc Biol* 2007;**27**:387–93.
71. Gerrity RG, Natarajan R, Nadler JL, Kimsey T. Diabetes-induced accelerated atherosclerosis in swine. *Diabetes* 2001;**50**:1654–65.
72. Mohler 3rd ER, Sarov-Blat L, Shi Y, Hamamdzic D, Zalewski A, Macphee C, et al. Site-specific atherogenic gene expression correlates with subsequent variable lesion development in coronary and peripheral vasculature. *Arterioscler Thromb Vasc Biol* 2008;**28**:850–5.
73. Wilensky RL, Shi Y, Mohler 3rd ER, Hamamdzic D, Burgert ME, Li J, et al. Inhibition of lipoprotein-associated phospholipase A2 reduces complex coronary atherosclerotic plaque development. *Nat Med* 2008;**14**:1059–66.
74. Clarkson TB, Lehner ND, Bullock BC, Lofland HB, Wagner WD. Atherosclerosis in new world monkeys. *Primates Med* 1976;**9**:90–144.
75. Middleton CC, Clarkson TB, Lofland HB, Prichard RW. Atherosclerosis in the squirrel monkey. Naturally occurring lesions of the aorta and coronary arteries. *Arch Pathol* 1964;**78**:16–23.
76. Middleton CC, Rosal J, Clarkson TB, Newman 3rd WP, McGill Jr. HC. Arterial lesions in squirrel monkeys. *Arch Pathol* 1967;**83**:352–8.
77. Taylor CB, Cox GE, Hall-Taylor BJ, Nelson LG. Atherosclerosis in areas of vascular injury in monkeys with mild hypercholesterolemia. *Circulation* 1954;**10**:613.
78. Taylor CB, Cox GE, Manalo-Estrella P, Southworth J, Patton DE, Cathcart C. Atherosclerosis in rhesus monkeys. II. Arterial lesions associated with hypercholesteremia induced by dietary fat and cholesterol. *Arch Pathol* 1962;**74**:16–34.
79. Clarkson TB, Lehner NDM, Wagner WD, Clair RWS, Bond MG, Bullock BC. A study of atherosclerosis regression in Macaca mulatta. I. Design of experiment and lesion induction. *Exp Mol Pathol* 1979;**30**:360–85.
80. Clarkson TB, Bond MG, Bullock BC, McLaughlin KJ, Sawyer JK. A study of atherosclerosis regression in Macaca mulatta. V. Changes in abdominal aorta and carotid and coronary arteries from animals with atherosclerosis induced for 38 months and then regressed for 24 or 48 months at plasma cholesterol concentrations of 300 or 200 mg/dl. *Exp Mol Pathol* 1984;**41**:96–118.
81. Beere PA, Glagov S, Zarins CK. Experimental atherosclerosis at the carotid bifurcation of the cynomolgus monkey. Localization, compensatory enlargement, and the sparing effect of lowered heart rate. *Arterioscler Thromb Vasc Biol* 1992;**12**:1245–53.
82. Mann GV, Andrus SB, McNally A, Stare FJ. Experimental atherosclerosis in Cebus monkeys. *J Exp Med* 1953;**98**:195–218.
83. Bullock BC, Clarkson TB, Lehner NDM, Lofland Jr. HB, Clair RWS. Atherosclerosis in Cebus albifrons monkeys. III. Clinical and pathologic studies. *Exp Mol Pathol* 1969;**10**:39–62.
84. Clarkson TB. Nonhuman primate models of atherosclerosis. *Lab Anim Sci* 1998;**48**:569–72.
85. De Silva HV, Mas-Oliva J, Taylor JM, Mahley RW. Identification of apolipoprotein B-100 low density lipoproteins, apolipoprotein B-48 remnants, and apolipoprotein E-rich high density lipoproteins in the mouse. *J Lipid Res* 1994;**35**:1297–310.
86. Barter PJ, Brewer HB, Chapman MJ, Hennekens CH, Rader DJ, Tall AR. Cholesteryl ester transfer protein. *Arterioscler Thromb Vasc Biol* 2003;**23**:160–7.
87. Suo J, Ferrara DE, Sorescu D, Guldberg RE, Taylor WR, Giddens DP. Hemodynamic shear stresses in mouse aortas. *Arterioscler Thromb Vasc Biol* 2007;**27**:346–51.
88. Giannoglou GD, Chatzizisis YS, Zamboulis C, Parcharidis GE, Mikhailidis DP, Louridas GE. Elevated heart rate and atherosclerosis: an overview of the pathogenetic mechanisms. *Int J Cardiol* 2008;**126**:302–12.
89. Cunningham KS, Gotlieb AI. The role of shear stress in the pathogenesis of atherosclerosis. *Lab Invest* 2004;**85**:9–23.

90. Vesselinovitch D, Wissler RW. Experimental production of atherosclerosis in mice: Part 2. Effects of atherogenic and high-fat diets on vascular changes in chronically and acutely irradiated mice. *J Atheroscler Res* 1968;**8**:497–523.
91. Paigen B, Morrow A, Brandon C, Mitchell D, Holmes P. Variation in susceptibility to atherosclerosis among inbred strains of mice. *Atherosclerosis* 1985;**57**:65–73.
92. Babaev VR, Fazio S, Gleaves LA, Carter KJ, Semenkovich CF, Linton MF. Macrophage lipoprotein lipase promotes foam cell formation and atherosclerosis *in vivo*. *J Clin Invest* 1999;**103**:1697–705.
93. Rubin EM, Krauss RM, Spangler EA, Verstuyft JG, Clift SM. Inhibition of early atherogenesis in transgenic mice by human apolipoprotein AI. *Nature* 1991;**353**:265–7.
94. Schreyer SA, Peschon JJ, LeBoeuf RC. Accelerated atherosclerosis in mice lacking tumor necrosis factor receptor p55. *J Biol Chem* 1996;**271**:26174–8.
95. Zhang SH, Reddick RL, Piedrahita JA, Maeda N. Spontaneous hypercholesterolemia and arterial lesions in mice lacking apolipoprotein E. *Science* 1992;**258**:468–71.
96. Plump AS, Smith JD, Hayek T, Aalto-Setala K, Walsh A, Verstuyft JG, et al. Severe hypercholesterolemia and atherosclerosis in apolipoprotein E-deficient mice created by homologous recombination in ES cells. *Cell* 1992;**71**:343–53.
97. Mahley RW. Apolipoprotein E: cholesterol transport protein with expanding role in cell biology. *Science* 1988;**240**:622–30.
98. Krieger M. Structures and functions of multiligand lipoprotein receptors: macrophage scavenger receptors and LDL receptor-related protein (LRP). *Annu Rev Biochem* 1994;**63**:601–37.
99. Mahley RW, Ji ZS. Remnant lipoprotein metabolism: key pathways involving cell-surface heparan sulfate proteoglycans and apolipoprotein E. *J Lipid Res* 1999;**40**:1–16.
100. Ghiselli G, Schaefer EJ, Gascon P, Breser HB. Type III hyperlipoproteinemia associated with apolipoprotein E deficiency. *Science* 1981;**214**:1239–41.
101. Nakashima Y, Plump AS, Raines EW, Breslow JL, Ross R. ApoE-deficient mice develop lesions of all phases of atherosclerosis throughout the arterial tree. *Arterioscler Thromb Vasc Biol* 1994;**14**:133–40.
102. Reddick RL, Zhang SH, Maeda N. Atherosclerosis in mice lacking apoE. Evaluation of lesional development and progression. *Arterioscler Thromb Vasc Biol* 1994;**14**:141–7.
103. Ishibashi S, Brown MS, Goldstein JL, Gerard RD, Hammer RE, Herz J. Hypercholesterolemia in low density lipoprotein receptor knockout mice and its reversal by adenovirus-mediated gene delivery. *J Clin Invest* 1993;**92**:883–93.
104. Ishibashi S, Goldstein JL, Brown MS, Herz J, Burns DK. Massive xanthomatosis and atherosclerosis in cholesterol-fed low density lipoprotein receptor-negative mice. *J Clin Invest* 1994;**93**:1885–93.
105. Bellosta S, Mahley RW, Sanan DA, Murata J, Newland DL, Taylor JM, et al. Macrophage-specific expression of human apolipoprotein E reduces atherosclerosis in hypercholesterolemic apolipoprotein E-null mice. *J Clin Invest* 1995;**96**:2170–9.
106. Schreyer SA, Vick C, Lystig TC, Mystkowski P, LeBoeuf RC. LDL receptor but not apolipoprotein E deficiency increases diet-induced obesity and diabetes in mice. *Am J Physiol Endocrinol Metab* 2002;**282**:E207–14.
107. Tian J, Pei H, Sanders JM, Angle JF, Sarembock IJ, Matsumoto AH, et al. Hyperlipidemia is a major determinant of neointimal formation in LDL receptor-deficient mice. *Biochem Biophys Res Commun* 2006;**345**:1004–9.
108. Shi W, Pei H, Fischer JJ, James JC, Angle JF, Matsumoto AH, et al. Neointimal formation in two apolipoprotein E-deficient mouse strains with different atherosclerosis susceptibility. *J Lipid Res* 2004;**45**:2008–14.
109. Van Vlijmen BJ, Van Den Maagdenberg AM, Gijbels MJ, Van der Boom H, HogenEsch H, Frants RR, et al. Diet-induced hyperlipoproteinemia and atherosclerosis in apolipoprotein E3-Leiden transgenic mice. *J Clin Invest* 1994;**93**:1403–10.

110. Williams H, Johnson JL, Carson KGS, Jackson CL. Characteristics of intact and ruptured atherosclerotic plaques in brachiocephalic arteries of apolipoprotein E-knockout mice. *Arterioscler Thromb Vasc Biol* 2002;**22**:788–92.
111. Johnson J, Carson K, Williams H, Karanam S, Newby A, Angelini G, et al. Plaque rupture after short periods of fat feeding in the apolipoprotein E-knockout mouse: model characterization and effects of pravastatin treatment. *Circulation* 2005;**111**:1422–30.
112. Jackson CL, Bennett MR, Biessen EAL, Johnson JL, Krams R. Assessment of unstable atherosclerosis in mice. *Arterioscler Thromb Vasc Biol* 2007;**27**:714–20.
113. Reddick RL, Zhang SH, Maeda N. Aortic atherosclerotic plaque injury in apolipoprotein E deficient mice. *Atherosclerosis* 1998;**140**:297–305.
114. Sasaki T, Kuzuya M, Nakamura K, Cheng XW, Shibata T, Sato K, et al. A simple method of plaque rupture induction in apolipoprotein E-deficient mice. *Arterioscler Thromb Vasc Biol* 2006;**26**:1304–9.
115. Eitzman DT, Westrick RJ, Xu Z, Tyson J, Ginsburg D. Hyperlipidemia promotes thrombosis after injury to atherosclerotic vessels in apolipoprotein E–deficient mice. *Arterioscler Thromb Vasc Biol* 2000;**20**:1831–4.
116. Gough PJ, Gomez IG, Wille PT, Raines EW. Macrophage expression of active MMP-9 induces acute plaque disruption in apoE-deficient mice. *J Clin Invest* 2006;**116**:59–69.
117. von der Thüsen JH, van Vlijmen BJM, Hoeben RC, Kockx MM, Havekes LM, van Berkel TJC, et al. Induction of atherosclerotic plaque rupture in apolipoprotein $E^{-/-}$ mice after adenovirus-mediated transfer of p53. *Circulation* 2002;**105**:2064–70.
118. Baker HJ, Lindsey JR, Weisbroth SH. *The laboratory rat.* New York; London: Academic Press; 1979.
119. Wexler BC. Spontaneous arteriosclerosis in repeatedly bred male and female rats. *J Atheroscler Res* 1964;**4**:57–80.
120. Wexler BC, Antony CD, Kittinger GW. Serum lipoprotein and lipid changes in arteriosclerotic breeder rats. *J Atheroscler Res* 1964;**4**:131–43.
121. Koletsky S. Obese spontaneously hypertensive rats—a model for study of atherosclerosis. *Exp Mol Pathol* 1973;**19**:53–60.
122. Dolphin PJ. Serum and hepatic nascent lipoproteins in normal and hypercholesterolemic rats. *J Lipid Res* 1981;**22**:971–89.
123. Joris I, Zand T, Nunnari JJ, Krolikowski FJ, Majno G. Studies on the pathogenesis of atherosclerosis. I. Adhesion and emigration of mononuclear cells in the aorta of hypercholesterolemic rats. *Am J Pathol* 1983;**113**:341–58.
124. Koletsky S. Pathologic findings and laboratory data in a new strain of obese hypertensive rats. *Am J Pathol* 1975;**80**:129–42.
125. St. John LC, Bell FP. Arterial lipid biochemistry in the spontaneously hyperlipidemic Zucker rat and its similarity to early atherogenesis. *Atherosclerosis* 1991;**86**:139–44.
126. Russell JC, Amy RM. Early atherosclerotic lesions in a susceptible rat model. The LA/N-corpulent rat. *Atherosclerosis* 1986;**60**:119–29.
127. Vance JE, Russell JC. Hypersecretion of VLDL, but not HDL, by hepatocytes from the JCR: LA-corpulent rat. *J Lipid Res* 1990;**31**:1491–501.
128. Russell JC, Shillabeer G, Bar-Tana J, Lau DC, Richardson M, Wenzel LM, et al. Development of insulin resistance in the JCR:LA-cp rat: role of triacylglycerols and effects of MEDICA 16. *Diabetes* 1998;**47**:770–8.
129. Russell JC, Graham SE, Dolphin PJ. Glucose tolerance and insulin resistance in the JCR: LA-corpulent rat: effect of miglitol (Bay m1099)°1. *Metabolism* 1999;**48**:701–6.
130. Russell JC, Graham SE, Richardson M. Cardiovascular disease in the JCR: LA-cp rat. *Mol Cell Biochem* 1998;**188**:113–26.

131. O'Brien SF, Russell JC, Davidge ST. Vascular wall dysfunction in JCR: LA-cp rats: effects of age and insulin resistance. *Am J Physiol Cell Physiol* 1999;**277**:C987–93.
132. Richardson M, Schmidt AM, Graham SE, Achen B, DeReske M, Russell JC. Vasculopathy and insulin resistance in the JCR: LA-cp rat. *Atherosclerosis* 1998;**138**:135–46.
133. Russell JC, Proctor SD. Small animal models of cardiovascular disease: tools for the study of the roles of metabolic syndrome, dyslipidemia, and atherosclerosis. *Cardiovasc Pathol* 2006;**15**:318–30.
134. Jacob HJ, Lazar J, Dwinell MR, Moreno C, Geurts AM. Gene targeting in the rat: advances and opportunities. *Trends Genet* 2010;**26**:510–8.
135. Herrera VLM, Makrides SC, Xie HX, Adari H, Krauss RM, Ryan US, et al. Spontaneous combined hyperlipidemia, coronary heart disease and decreased survival in Dahl salt-sensitive hypertensive rats transgenic for human cholesteryl ester transfer protein. *Nat Med* 1999;**5**:1383–9.
136. Herrera VLM, Tsikoudakis A, Didishvili T, Ponce LRB, Bagamasbad P, Gantz D, et al. Analysis of gender-specific atherosclerosis susceptibility in transgenic [hCETP]25DS rat model. *Atherosclerosis* 2004;**177**:9–18.
137. Fernández ML, Wilson TA, Conde K, Vergara-Jiménez M, Nicolosi RJ. Hamsters and guinea pigs differ in their plasma lipoprotein cholesterol distribution when fed diets varying in animal protein, soluble fiber, or cholesterol content. *J Nutr* 1999;**129**:1323–32.
138. Nikkari ST, Solakivi T, Jaakkola O. The hyperlipidemic hamster as an atherosclerosis model. *Artery* 1991;**18**:285–90.
139. Nistor A, Bulla A, Filip DA, Radu A. The hyperlipidemic hamster as a model of experimental atherosclerosis. *Atherosclerosis* 1987;**68**:159–73.
140. Kahlon TS, Chow FI, Irving DW, Sayre RN. Cholesterol response and foam cell formation in hamsters fed two levels of saturated fat and various levels of cholesterol. *Nutr Res* 1996;**16**:1353–68.
141. Kowala MC, Nunnari JJ, Durham SK, Nicolosi RJ. Doxazosin and cholestyramine similarly decrease fatty streak formation in the aortic arch of hyperlipidemic hamsters. *Atherosclerosis* 1991;**91**:35–49.
142. Dillard A, Matthan NR, Lichtenstein AH. Use of hamster as a model to study diet-induced atherosclerosis. *Nutr Metab* 2010;**7**:89.
143. Cos E, Ramjiganesh T, Roy S, Yoganathan S, Nicolosi RJ, Fernandez ML. Soluble fiber and soybean protein reduce atherosclerotic lesions in guinea pigs. Sex and hormonal status determine lesion extension. *Lipids* 2001;**36**:1209–16.

Genetic Animal Models of Cerebral Vasculopathies

Jeong Hyun Lee,* Brian J. Bacskai,† and Cenk Ayata*,‡

*Department of Radiology, Neurovascular Research Laboratory, Massachusetts General Hospital, Harvard Medical School, Charlestown, Massachusetts, USA

†Department of Neurology, MassGeneral Institute for Neurodegenerative Disease, Massachusetts General Hospital, Harvard Medical School, Charlestown, Massachusetts, USA

‡Stroke Service and Neuroscience Intensive Care Unit, Department of Neurology, Massachusetts General Hospital, Harvard Medical School, Charlestown, Massachusetts, USA

Cerebral amyloid angiopathy (CAA) and cerebral autosomal dominant arteriopathy with subcortical infarcts and leukoencephalopathy (CADASIL) are genetic cerebrovasculopathies associated with neurodegeneration and vascular cognitive impairment. Linked to autosomal dominant mutations in diverse genes that encode cell-surface receptors (i.e., amyloid precursor protein in CAA and NOTCH3 in CADASIL), both diseases are associated with accumulation of abnormal material around cerebral vessels, such as amyloid in CAA or granular osmiophilic material in CADASIL. Both CAA and CADASIL share clinical features of white matter degeneration and infarcts, and vascular dementia in the human adult; microbleeds occur in both CADASIL and CAA, but large intracerebral hemorrhages are more

Progress in Molecular Biology and Translational Science, Vol. 105
DOI: 10.1016/B978-0-12-394596-9.00002-0

characteristic for the latter. While the mechanisms are poorly understood, wall thickening, luminal narrowing, and eventual loss of vascular smooth muscle cells are overlapping pathologies involving leptomeningeal, and pial or penetrating small arteries and arterioles in CAA and CADASIL. Dysregulation of cerebral blood flow and eventual hypoperfusion are believed to be the key pathophysiological steps in neurodegeneration and cognitive impairment. Although animal models expressing CAA or CADASIL mutations have partially reproduced the human pathology, there has been marked heterogeneity in the phenotypic spectrum, possibly due to genetic background differences among mouse models, and obvious species differences between mouse and man. Here, we provide an overview of animal models of CAA and CADASIL and the insight on molecular and physiological mechanisms of disease gained from these models.

I. Cerebral Amyloid Angiopathy

A. Clinical Overview

Cerebral amyloid angiopathy (CAA), also known as congophilic angiopathy, is characterized by progressive mural deposition of amyloid fibrils in small to medium-sized arteries and arterioles predominantly located in the cerebral cortex and leptomeninges, and less frequently in the capillaries and veins of the central nervous system, often resulting in subsequent degenerative cerebrovascular changes.[1,2] It is generally not associated with systemic amyloidosis or common vascular risk factors.[2] The prevalence of CAA increases with age, so that 10–40% of elderly without Alzheimer's disease (AD) have some degree of CAA; in the presence of AD, this number reaches 80% or more.[3]

CAA is caused by patchy deposition of amyloid β (Aβ) protein around the smooth muscle cells in tunica media and in the adventitia. Progressive aggregation and accumulation of Aβ causes vessel wall thickening and luminal narrowing, and reduced wall compliance. In severe CAA, Aβ deposits replace the degenerating vascular smooth muscle cells (VSMCs) and become confluent, leading to fibrinoid necrosis, double-barrel lumen, microaneurysmal dilatations, and vessel rupture.[1–4]

While early CAA is mostly asymptomatic, progressive pathological changes are believed to cause cerebrovascular dysregulation and hypoperfusion. The most frequent and severe clinical manifestation of advanced CAA is rupture of weakened arterial wall causing spontaneous (i.e., nontraumatic) recurrent lobar intracerebral hemorrhage (ICH). Often in nonhypertensive elderly, approximately 5–20% of all ICHs are related to CAA.[3] Of course, the risk and severity of ICH are increased by the use of anticoagulant or antiplatelet agents.[2] CAA is also associated with ischemic infarction and white matter lesions presumably because of chronic hypoperfusion.[2,5]

CAA is associated with progressive vascular cognitive impairment and dementia in 10–30% of patients.[6,7] Although cerebrovascular dysfunction (e.g., hypoperfusion, impaired autoregulation) and secondary cumulative parenchymal lesions (e.g., recurrent ICH, ischemic white matter injury) are believed to contribute to progressive cognitive impairment, frequent coexistence of AD has made it difficult to substantiate this assertion.[6,8–10] It is possible that CAA accelerates the progression of AD pathology and related cognitive impairment; however, it is not clear whether early molecular abnormalities or elevated soluble Aβ levels in young individuals destined to develop CAA interfere with normal vessel function or accelerate AD, before actual vascular amyloid deposition takes place. Currently, there is no effective intervention to prevent or reverse CAA and related complications.

B. Molecular Biology

Amyloid is composed of highly insoluble proteinaceous fibrils in β-pleated sheet configuration that exhibit yellow-green fluorescence after thioflavin S or T staining or an apple-green birefringence under polarized light after Congo-red staining.[1] Besides the cerebral vessel walls in CAA, amyloid fibrils are deposited in the brain parenchyma as dense core, mature senile plaques in AD. Although several amyloidogenic proteins have been identified, Aβ is the predominant form of amyloid in sporadic and rare hereditary forms of CAA, as well as in AD and Down's syndrome.[2] Aβ peptides, ~4 kDa hydrophobic peptide of 39–43 amino acids long, are derived from Aβ precursor protein (APP) by sequential proteolytic cleavages, first at the N-terminus by either α- or β-secretase and then at the C-terminus by γ-secretase. Cleavage by α-secretase yields the nonamyloidogenic soluble 3-kDa peptide (p3) that does not form amyloid fibrils, whereas cleavage by β-secretase yields the amyloidogenic form $A\beta_{1-40}$ or the longer, more fibrillogenic $A\beta_{1-42}$. In normal brain, nonamyloidogenic α-secretase pathway predominates. When Aβ homeostasis is disturbed and production exceeds elimination, Aβ monomers polymerize to soluble oligomers or insoluble fibrils and are deposited in the brain parenchyma and the vessel wall.

APP is an evolutionary conserved type 1 transmembrane glycoprotein with ubiquitous expression in the central nervous system and peripheral tissues. It occurs in three most common isoforms, APP695, APP751, and APP770, which are created by tissue-specific alternative splicing of *APP* gene on chromosome 21. Although the physiological functions of APP and its cleavage derivatives still remain elusive, they may include regulation of G-protein coupling, Ca^{2+} homeostasis, neurite outgrowth, cell adhesion, synaptogenesis, and synaptic transmission.[11–16]

1. Sporadic CAA

It is the most common form present in approximately 10–40% of elderly, over 80% of patients with AD, and 100% of patients with Down's syndrome.[3,17] Sporadic CAA type 1 is characterized by Aβ deposition in cerebral capillaries with or without larger vessel involvement. Sporadic CAA type 2 is more common and affects cortical and leptomeningeal arteries, arterioles, and rarely veins, but lacks capillary involvement.[18] Capillary Aβ deposition in CAA type 1 is strongly associated with high frequency of apolipoprotein E (ApoE) ε4 allele, and characterized by perivascular inflammation, phosphorylated tau protein accumulation, neutrophil-associated dyshoric angiopathy, and hypoperfusion induced by capillary occlusion.[18,19]

2. Hereditary CAA

It is rare and shows autosomal dominant inheritance with younger age of onset and generally more severe clinical manifestations. Missense mutations within the Aβ-encoding region in the *APP* gene have been linked to severe CAA, including Dutch, Italian, Arctic, Flemish, and Iowa types (Table I).[20,21,24,25,28–30] With the exception of the Flemish mutation, which increases total Aβ production by interfering with α-secretase cleavage,

TABLE I
Hereditary Forms of Severe Aβ CAA Associated with Missense Mutations Within the Aβ Region of APP

	Mutation	Amino acid substitution	DP	SP	References
HCHWA-Dutch	APP E693Q	Aβ residue 22 Glu → Gln	Present	Absent	20–23
HCHWA-Italian	APP E693K	Aβ residue 22 Glu → Lys	Abundant	Absent	24
Flemish mutation	APP A692G	Aβ residue 21 Ala → Gly	Present	Abundant	25–27
Iowa mutation	APP D694N	Aβ residue 23 Asp → Asn	Abundant	Rare	28,29
Arctic mutation	APP E693G	Aβ residue 22 Glu → Gly	Present	Abundant	30,31

HCHWA, hereditary cerebral hemorrhage with amyloidosis; NFT, intracellular neurofibrillary tangles composed of hyperphosphorylated tau proteins; DP, diffuse plaques; SP, mature senile plaques.

mutations within the Aβ-encoding region primarily affect the biochemical properties of Aβ.[32] Dutch, Italian, Iowa, and Arctic mutant Aβ peptides show increased tendency to aggregate into protofibrils and/or fibrils, *in vitro*; interestingly, Arctic mutation is also associated with decreased plasma $A\beta_{1-40}$ and $A\beta_{1-42}$ levels.[30,33–36] Clinically, recurrent lobar ICH is characteristic for Dutch, Italian, and Flemish mutations, and one out of two families identified to date with Iowa mutation. In contrast, Arctic mutation is not associated with ICH. Additionally, leukoencephalopathy, ischemic strokes, epilepsy, recurrent headaches, and dementia that can be AD-like, early-onset, and progressive have been described in these families.

Other APP mutations flanking the secretase cleavage sites also exist that enhance total Aβ production and lead to early-onset AD, such as Indiana and London (γ-secretase), and Swedish double (β-secretase) mutations; however, these mutations do not cause excessive CAA disproportionate to patients' age.[37–41] In addition, increased dosage of APP, including French families with APP duplication and early-onset familial AD, and trisomy 21 in Down's syndrome are associated with severe CAA.[17,42] Abnormal APP processing, such as mutations in presenilin-1 or presenilin-2 catalytic subunit of γ-secretase, also leads to increased Aβ production and $A\beta_{1-42}/A\beta_{1-40}$ ratio, and severe CAA.[43–47] Moreover, hereditary non-Aβ forms of CAA also exist in young and middle-aged patients, such as autosomal dominant hereditary cerebral hemorrhage with amyloidosis of Icelandic type associated with mutations in cystatin C (ACys C) leading to severe, early-onset hemorrhage, cognitive decline, and dementia; Gerstmann–Sträussler–Scheinker syndrome associated with mutations in prion protein (PrP^{Sc}); leptomeningeal amyloidosis associated with mutations in transthyretin (TTR); familial amyloidosis of Finnish type associated with mutations in gelsolin (AGel); familial British dementia associated with mutations in amyloid-British protein (ABri); and familial Danish dementia associated with mutations in amyloid-Danish protein (ADan).[48–53]

C. Animal Models for CAA

Spontaneous sporadic Aβ-CAA has been reported in aged dogs and non-human primates.[54] However, their use as experimental models is not practical. Because spontaneous Aβ-CAA has not been observed naturally in rodents, mouse models have been based on transgenic expression of APP harboring mutations associated with rare hereditary forms of CAA, alone or in combination with PS1, PS2, or ApoE mutations that alter Aβ production; all models to date express endogenous APP at normal levels (Table II).

These mutant mouse models have reproduced various levels of progressive and age-dependent Aβ-CAA.[82] *PDAPP* mouse was the first successful APP-based transgenic mouse model that showed neuropathological hallmarks of typical early-onset familial AD; however, CAA was not a prominent feature in

TABLE II
COMMON TRANSGENIC MUTANT APP MOUSE MODELS

APP models	PDAPP	Tg2576	APP23	APPDutch	APP/Ld2	TgCRND8	TgSwDI
Background strain	C57BL/6-DBA/2-Swiss Webster	C57BL/6-SJL	C57BL/6xDBA2	C57BL/6J	FVB/N	C3H/He-C57BL/6	C57BL/6
Mutant human APP	APP770 V717F (Indiana)	APP695 K670N/M671L (Swedish)	APP751 K670N/M671L (Swedish)	APP751 E693Q (Dutch)	APP695 V717I (London)	APP695 K670N/M671L/V717F (Swedish/Indiana)	APP770 E693Q/D694N/K670N/M671L (Dutch/Iowa/Swedish)
Promoter (CNS neurons)	PDGF-β	Hamster PrP	Murine Thy-1	Murine Thy-1	Murine Thy-1	Hamster PrP	Murine Thy-1
Aβ species	Wild type	Wild type	Wild type	Mutant	Wild type	Wild type	Mutant
Expression level of mutant APP versus endogenous mouse APP	10-fold	5.5-fold	7-fold	5-fold	2–5 fold	5-fold	0.5-fold
Predominant form of Aβ	$A\beta_{1-42}$	$A\beta_{1-40}$	$A\beta_{1-40}$	$A\beta_{1-40}$	$A\beta_{1-40}$	$A\beta_{1-42}$	$A\beta_{1-40}$
Severity of parenchymal Aβ deposits	++	++	+++	–	++	+++	+
Senile plaques	+	+	+	–	+	+	–
Diffuse plaques	+	+	+	Rare	+	+	+
Onset of parenchymal Aβ deposits	6–9 months	7–8 months	6 months	–	10–12 months	2–3 months	3 months
Dystrophic neurites	+	+	+	N.D.	+	+	N.D.
Hyperphosphorylated tau pathology	+	+	+	N.D.	+	+	–
NFT	–	–	–	N.D.	+	–	–
Synaptic loss	+	+	+	N.D.	N.D.	+	N.D.
Severity of CAA	+	++	+++	+++	++	++	+++
Onset of CAA	21–25 months	9 months	8 months	22–25 months	15 months	5 months	6 months
VSMC loss	+	+	+	+	+	N.D.	+
Hemorrhage	+	+	+	+	–	N.D.	Occasional
Cognitive deficits (onset)	+ (4 months)	+ (6–9 months)	+ (3 months)	N.D.	+ (3 months)	+ (11 weeks)	+ (3 months, homozygotes)
References	55–61	39,55,62–67	68–74	22	75,76	44,77,78	79–81

PDGF-β, platelet-derived growth factor-β; NFT, neurofibrillary tangles; N.D., not determined.
+++, severe; ++, moderate; +, mild; –, none.

this model. *APP23* is the first transgenic model that developed significant CAA, associated with microaneurysms and vasculitis. Unlike other transgenic models, APP23 mice also develop selective neuronal loss in the hippocampal CA1 region at 14–18 months of age. *APPDutch* mice closely mimic human HCHWA-D to develop late onset of vascular Aβ deposition. *APP/Ld2* mouse additionally develops vessel dilatation and microaneurysm formation, but interestingly, hemorrhages are not common. In most mouse models, CAA is detected primarily in leptomeningeal, and occasionally in neocortical, hippocampal, and thalamic arteries and arterioles; VSMC degeneration and microhemorrhages are also common in aged animals. In contrast, TgSwDI mice develop CAA predominantly in parenchymal microvasculature and capillaries. Interestingly, TgSwDI mice crossbred with nitric oxide synthase 2 (iNOS) knockout mice (TgSwDI/NOS2$^{-/-}$) develop not only the pathology of TgSwDI mice but also extensive tau pathology, significant neuronal loss in the hippocampus, and severe behavioral impairments.[83] Therefore, the phenotype of this mouse model better approximates CAA type 1, though the complexity of the model limits its utility.

In addition to CAA, these mutant models variably develop parenchymal diffuse and senile plaques most commonly in the neocortex, hippocampus, and thalamus, and in some models, in striatum, cerebellum, white matter, and brain stem. Parenchymal plaques are associated with dystrophic neurites and synaptic loss, reactive astrocytosis, and an inflammatory response characterized by microgliosis and release of inflammatory mediators. Age-dependent cognitive deficits reminiscent of AD have also been demonstrated in some mutant models (e.g., impaired spatial learning and memory) associated with impaired synaptic transmission and plasticity, and long-term potentiation.[44,56,62,66,73,75]

Recently, two more transgenic mouse models expressing human APP with Arctic (E693G) and Swedish (K670M/N671L) mutations have been developed. APP_{ArcSwe} mice express the mutant transgene approximately threefold higher than endogenous APP under the control of murine Thy-1 promoter, and develop neuronal Aβ accumulation at 2 months and senile plaques at 5–6 months.[84] The arcAβ mice express human APP695-containing Arctic and Swedish mutations sixfold higher than endogenous APP under the control of PrP promoter, and develop senile plaques and CAA starting at 9–15 months of age.[85,86] Overall, these mutant models are less well characterized.

D. Insights into Pathophysiology from Transgenic Models

1. Origin of Aβ in CAA

The vascular hypothesis proposes that Aβ deposited in CAA is locally produced by cells within the vessel wall. APP expression and Aβ production and deposition have been demonstrated in proliferating and degenerating

VSMCs,[87–90] pericytes,[91] and endothelial cells, as well as in isolated cerebral microvessels and meningeal vessels.[92,93] Arguments against the vascular hypothesis are that Aβ deposits are detected in capillaries which lack VSMCs, and small arteries and arterioles with fewer VSMCs are more severely affected by CAA than larger arteries. *The systemic hypothesis* proposes that cerebral Aβ from blood enters the vessel wall across the blood–brain barrier by the receptor for advanced glycation end-products (RAGE)-mediated transport. This hypothesis is supported by the observation that APP is ubiquitously expressed throughout the body, and that Aβ is detected in plasma. However, transgenic mice expressing high plasma levels of soluble Aβ fail to show cerebrovascular amyloid deposition,[94] aged APP/Ld2 mice do not have detectable Aβ in plasma despite high levels of Aβ in cerebrospinal fluid,[76] and initial Aβ deposition occurs in the abluminal basement membrane of blood vessels,[95] arguing against a systemic origin for deposited Aβ. *The drainage hypothesis* proposes that Aβ originating exclusively from neurons is trapped and deposited in the interstitial fluid perivascular drainage pathways around parenchymal capillaries and arteries, and leptomeningeal vessels, on its way into cervical lymph nodes and systemic venous circulation,[96] by which it is cleared under normal physiological conditions. According to this hypothesis, accumulation of insoluble Aβ in the basement membranes of capillary and arterial walls blocks the perivascular spaces and further impedes the drainage.[96–98] This hypothesis is supported by data showing that: (1) soluble tracers injected into the extracellular space are drained along basement membranes of capillaries and arteries in adult mouse brain[99]; (2) APP23 mice with neuron-specific expression of mutant APP continues to develop CAA on an APP null background[74]; (3) both parenchymal and vascular Aβ deposition are effectively prevented when APP/Ld2 mice are crossed with conditional *PS1* knockout mice, presumably because of reduced Aβ production by neuron-specific PS1 deletion[100]; and (4) altered interstitial fluid drainage along basement membranes was demonstrated after intracerebral injection of dextran in both aged wild-type and Tg2576 mice.[101] Therefore, it is plausible that age-related impairment in perivascular drainage pathways impedes clearance and promotes accumulation and deposition of Aβ derived from brain parenchyma.

2. Different Roles for $A\beta_{1-42}$ and $A\beta_{1-40}$ in Amyloid Deposition

$A\beta_{1-40}$ is shorter and more soluble than $A\beta_{1-42}$, and more likely to reach the blood vessels to be deposited in the vascular wall as Aβ-CAA. In contrast, $A\beta_{1-42}$ is more prone to aggregate as fibrils, and although it represents less than 20% of total secreted Aβ, it constitutes the predominant form of Aβ in senile plaques and capillary CAA.[102–104] Nevertheless, several lines of evidence suggest that $A\beta_{1-42}$ is also essential for CAA, by serving as the seed on which $A\beta_{1-40}$ subsequently aggregates.[22,76] For example, transgenic overexpression of $A\beta_{1-42}$

alone without human APP overexpression (BRI-Aβ42) is enough to develop both CAA and senile plaques, while overexpression of $A\beta_{1-40}$ alone (BRI-Aβ40) is not sufficient to generate overt amyloid deposits.[105] APPDutch mice with significantly higher $A\beta_{1-40}/A\beta_{1-42}$ ratio generate extensive vascular deposition; however, when $A\beta_{1-42}$ production is increased and $A\beta_{1-40}/A\beta_{1-42}$ ratio reversed (e.g., double APPDutch/PS1 G384A transgenic mice),[106] CAA is diminished while parenchymal Aβ plaques are increased.[22] Moreover, crossbreeding of BRI-Aβ42 mice with predominantly $A\beta_{1-40}$-secreting Tg2576 mice leads to a marked increase in Aβ deposition.[105] In contrast, BRI-Aβ40/Tg2576 transgenic mice display decreased Aβ deposition.[107] And lastly, immunization of TgCRND8 against $A\beta_{1-42}$ significantly reduces plaque load, cognitive impairments, and CAA.[108,109] These data support an essential role for $A\beta_{1-42}$ in initiating amyloid deposition, whereas $A\beta_{1-40}$ can be amyloidogenic or antiamyloidogenic depending on the critical level of $A\beta_{1-42}$.

3. Clearance of Aβ

Amyloid deposition appears to be a result of imbalance between Aβ production and clearance. In hereditary CAA, excessive or aberrant production of Aβ by missense mutations in the CAA-related genes, while in sporadic CAA, impaired Aβ clearance could be responsible.[98] In addition to the perivascular drainage pathways, Aβ is eliminated from the brain by proteolytic degradation by enzymes including neprilysin,[110] insulin-degrading enzyme (IDE, insulysin),[111] and endothelin-converting enzyme.[112] Neprilysin and IDE are zinc metalloendopeptidases expressed in both neurons and vasculature. Decreased expression and activity of neprilysin and IDE in blood vessels in association with ApoE e4 allele contribute to the development of CAA, a strong risk factor for CAA.[113,114] Neprilysin degrades both monomeric and oligomeric extracellular Aβ, whereas IDE degrades only monomeric Aβ.[115] Overexpression of neprilysin or IDE in neurons significantly reduces Aβ level and amyloid plaque burden as well as gliosis in mutant APP transgenic mice.[116] Further, loss of enzyme activity in neprilysin- or IDE-deficient mice significantly increases production and accumulation of endogenous Aβ peptides.[117,118]

Brain Aβ levels are also regulated by receptor-mediated bidirectional transport to and from circulating blood by low density lipoprotein receptor-related protein-1 (LRP-1), and as mentioned above, RAGE.[119–121] LRP-1 mediates efflux of Aβ from the interstitial space into blood via transcytosis, either alone or in complex with ApoE or α_2-macroglobulin, and thus serves as an additional mechanism for Aβ clearance. In addition, soluble LRP-1, a major Aβ-binding protein in plasma, serves as a sink by sequestering plasma Aβ and blocking its transport across the blood–brain barrier back into the brain. TgSwDI mutant Aβ shows reduced affinity for LRP-1, resulting in diminished LRP-1-mediated efflux of Aβ peptides across the blood–brain barrier; as a result, TgSwDI mice develop robust, early-onset CAA, and extensive diffuse plaques, despite significantly lower transgene expression and Aβ production in the brain compared to other mouse models with higher expression such as Tg2576.[80,119]

4. Risk Factors

The frequency and severity of CAA increase with age, but are not correlated with common vascular risk factors such as hypertension, hyperlipidemia, and diabetes mellitus.[2] Several genetic risk factors (alleles, polymorphisms, mutations) are associated with increased risk of CAA, such as ApoE, α1-antichymotrypsin,[122,123] neprilysin,[124] transforming growth factor-β1,[125] and LRP-1.[126] ApoE is a polymorphic and multifunctional glycoprotein expressed at high levels in the central nervous system, and plays a pivotal role in lipid metabolism, synaptic activity, and axon growth.[127] The interaction between ApoE and Aβ peptides can influence Aβ fibril formation. Among the three major alleles of ApoE,[128] ε4 increases the risk of developing both late onset AD and sporadic CAA by enhancing the rate of amyloid fibril formation and deposition of Aβ in vessels and senile plaques,[129–131] whereas ε2 promotes the rupture of amyloid-laden blood vessels,[132–134] although it appears to reduce the risk of AD as well.[135] Deletion of *ApoE* gene in PDAPP and Tg2576 mice significantly reduces parenchymal and vascular Aβ deposition[57,136] and microhemorrhages,[55] supporting a role for murine ApoE in pathological formation of senile plaques and CAA as well.

5. Vascular Dysfunction

Patients with CAA develop diminished visual evoked flow velocity response compared to controls, suggesting that cerebrovascular structural changes are associated with functional vasomotor impairment as well[10] While established CAA (i.e., Aβ deposition around blood vessels) would be expected to disrupt vascular function, animal studies yielded conflicting results on whether soluble Aβ contributes to vascular dysfunction before actual deposition takes place. In some studies, elevated soluble Aβ levels alone in the absence of vascular Aβ deposition disrupted endothelium-dependent relaxation, cerebrovascular autoregulation, and neurovascular coupling.[137–140] In another study, 18–19-month-old Tg2576 mice with extensive CAA showed impaired vasodilator and constrictor responses to physiological and pharmacological stimuli, whereas 9-month-old mice with elevated soluble Aβ without CAA had normal vascular reactivity.[64] Recent data from the same mouse model did find impaired vascular reactivity in the absence of deposited Aβ.[67] Yet in two other studies in APP23 and arcAβ mouse models, vasodilator responses were impaired when there was severe CAA at 16–24, but not at 2–7 months of age.[85,86,141] Interestingly, APP/Ld2 mouse reportedly had normal hypercapnic hyperemia even in the presence of extensive CAA at 20–24 months of age.[76] Although it is difficult to reconcile these conflicting data, a common methodological overlap among the studies that show impaired vascular reactivity in the absence of deposited Aβ is their invasive nature involving cranial windows. Indeed, studies that found normal vasoreactivity performed measurements noninvasively, that is, through intact cranium without breaching dura, using laser speckle flowmetry or functional MRI. Therefore, it is possible that soluble Aβ sensitizes cerebral vessels to heat, trauma, and direct exposure to room air during the craniotomy procedure, reduced brain temperature under the window during the measurements, and additional inflammatory and reactive changes in case of chronic window preparations. As such, impaired vasoreactivity has not yet been shown using noninvasive methods in young mutant mice in the absence of CAA. Similarly, clinical evidence is also lacking to support cerebrovascular dysfunction in young asymptomatic individuals with elevated levels of soluble Aβ who are destined to develop AD or CAA.

II. Cerebral Autosomal Dominant Arteriopathy with Subcortical Infarcts and Leukoencephalopathy (CADASIL)

A. Clinical Overview

CADASIL is the most common monogenic inherited form of degenerative small vessel disease (SVD). It is linked to mutations in the *NOTCH3* gene.[142] Since its genetic definition in 1990s, CADASIL has emerged as an important cause of stroke and progressive pure vascular dementia in young or middle-aged adults.[142–144] Migraine headache often with severe aura is generally the first CADASIL symptom present in 20–40% of patients, five times greater than the general population. Progressive white matter degeneration (leukoaraiosis) and ischemic strokes are present in 60–85%. Indeed, $> 10\%$ of patients who suffer from lacunar infarcts and leukoaraiosis under age 50 may carry a CADASIL mutation.[145] Leukoaraiosis is the earliest sign of CADASIL preceding the onset of strokes and cognitive symptoms by 10–15 years. Recurrent cerebral ischemic events are almost exclusively lacunar infarcts involving subcortical white matter, deep gray matter, and brain stem, and begin at an average within the fifth decade without conspicuous vascular risk factors. Presence of vascular risk factors such as hypertension, hypercholesterolemia, and smoking can exacerbate disease progression[146] Cognitive impairment and dementia correlate with the extent of cumulative subcortical pathology, in particular, the lacunar infarct load and cerebellar atrophy.[147,148] Early vascular cognitive impairment progresses into subcortical vascular dementia without AD-related pathology. Additional clinical manifestations include mental disorders, mood disturbances, seizures, and apathy.[142–144]

The pathological hallmark of CADASIL is the pathognomonic accumulation of granular osmiophilic material (GOM) within the tunica media in the vicinity to the VSMC membranes.[149] NOTCH3 extracellular domain also accumulates on degenerating VSMCs, apparently because of impaired clearance of the receptor from VSMC and pericyte membranes. Eventual loss of VSMCs is followed by adventitial fibrosis, thickening of arterial walls, narrowing of arterial lumen, and markedly enlarged perivascular spaces.[150,151] As a result, long, small, and medium-sized penetrating arteries supplying subcortical structures are stenosed or obliterated leading to lacunar infarcts.[152]

In adult brain, NOTCH3 plays important roles in postnatal maturation and specification of VSMCs, regulation of VSMC growth and apoptosis, response to vascular injury, and regulation of actin cytoskeleton in response to mechanical stretching of the vessel wall by intraluminal pressure.[153,154] Hemodynamic studies suggest that resting cerebral blood flow (CBF) is reduced even at the presymptomatic stage, with complete penetrance by the end of fourth decade. This is followed by an age-dependent impairment in dilatory reserve and increase in oxygen extraction, suggesting metabolic supply–demand mismatch.[155–159] In addition, endothelial morphological and functional changes have been demonstrated in systemic vessels.[160–162] In most imaging studies, hypoperfusion was particularly prominent in regions that showed leukoaraiosis, and was comparable to that observed in leukoaraiosis of other etiologies. Hence, it has been difficult to ascertain whether hypoperfusion is the cause or consequence of leukoaraiosis in CADASIL. Although large ICHs typical for hypertensive patients are rare in CADASIL, microhemorrhages in cortical gray matter are not uncommon.[163]

B. Molecular Biology

Most CADASIL patients carry one of more than 170 highly stereotyped mutations in the *NOTCH3* gene identified to date, with autosomal dominant inheritance; *de novo* mutations (e.g., Arg182 > Cys or Cys128 > Gly) with no familial history can also occur albeit rarely.[149,164,165] *NOTCH3* encodes a transmembrane receptor which, in adult brain, is expressed predominantly in VSMCs and pericytes.[166,167] NOTCH3 receptor belongs to Notch receptor superfamily (*Drosophila Notch homologues 1 to 4*), which participates in an evolutionarily conserved intercellular signaling pathway that modulates context-dependent cell-fate determination during embryonic development.[168] It is a heterodimer with a large extracellular domain consisting of 34 tandem epidermal growth factor (EGF)-like repeats, each including six conserved cysteine residues. This extracellular domain is noncovalently linked to the intracellular domain containing six tandem ankyrin repeats.[143] Binding of ligands such as the canonical DSL (Delta/Serrate/LAG-2) to the extracellular domain activate sequential proteolytic cleavages that shed the extracellular domain and allow translocation of the intracellular domain to the nucleus for transcriptional regulation.[169]

The vast majority of pathogenic CADASIL mutations are missense mutations within the extracellular domain of NOTCH3 receptor that cause either a gain or a loss of cysteine residue resulting in an odd number of cysteins.[170,171] The unpaired cysteine residues lead to conformational changes of NOTCH3 extracellular domain, causing homo/heterodimerization or multimerization.[172–174]

CADASIL mutations do not appear to compromise NOTCH3 ligand binding and canonical downstream signaling, except for those in the predicted putative ligand-binding site (e.g., C428S and C455R in EGF-like repeats 10–11) that disrupt ligand binding and NOTCH3 signaling.[175,176] It has been difficult to characterize genotype–phenotype correlations, because marked phenotypic heterogeneity exists among CADASIL families and even among the affected members of a single family.[177,178] It is possible that the presence of common vascular risk factors or polymorphisms in other genes that modulate NOTCH3 signaling modify the disease progression.[146] CADASIL families carrying a mutation at the ligand-binding site revealed very early onset of strokes and severe white matter hyperintensities detected on MRI serve as an indicator of future risk of stroke, although, paradoxically, progression of disability and dementia appeared to be slower.[179,180] In addition, two other missense mutations (C117F and C174T) are associated with shorter survival, and a splice site mutation causing small in-frame deletion was associated with a phenotype dominated by high prevalence of migraine with aura.[181,182] Therefore, evidence does support distinct genotype–phenotype correlations.

C. Animal Models of CADASIL

1. *Notch3* Knockout Mice

Despite not being a CADASIL model, these mice have provided valuable insight into *Notch3* in cerebrovascular development and homeostasis. *Notch3* knockouts have been viable and fertile and did not show anatomic developmental abnormalities in circle of Willis or larger conductance arteries.[153,183–186] Ultrastructurally, however, one homozygous knockout model showed prominent structural defects in small systemic and cerebral resistance arteries, including enlarged diameters, thinner media resembling veins, and alteration of the shape and size of VSMCs, consistent with impaired VSMC differentiation and maturation.[153] Interestingly, such ultrastructural abnormalities were absent in another knockout model when examined in brain vessels or aorta, although canonical Notch3 signal transduction as well as the transcription of genes associated with muscle development and contraction were downregulated.[183] Knockout mice did show impaired cerebrovascular reactivity and resistance, diminished pressure-induced myogenic tone, augmented flow-mediated (shear-induced) dilation, and more severe breakthrough hyperemia during hypertensive transients, but absolute resting CBF and agonist- or KCl-induced vasomotor responses appeared to be normal at 10–12 weeks of age.[153,183,184] Despite these structural and functional defects, *Notch3* knockout mice did not develop CADASIL pathology such as GOM deposits, leukoaraiosis, or spontaneous lacunar strokes,[153] suggesting that CADASIL is not simply a hypomorphic (i.e., loss-of-function) phenotype. Nevertheless, homozygous *Notch3* knockout mice developed more than twofold larger infarcts compared to both wild type and heterozygous knockout after transient middle cerebral artery occlusion by an intraluminal filament.[183] Larger perfusion defects in the knockout suggested impaired autoregulation and collateral flow as one mechanism. Moreover, conditional expression of wild-type human *Notch3* in VSMC rescued the stroke phenotype in the knockout[183] clearly linking Notch3 signaling to a stroke phenotype.

2. CADASIL Mutant Mice

To date, several transgenic mouse models expressing common CADASIL mutations have been developed and characterized (Table III). Models differed in transgenic strategy and expression level, expression of endogenous *Notch3*, and the effects of mutations on Notch3 function, and variably showed age-dependent ultrastructural abnormalities in cerebral and systemic resistance arteries characteristic for CADASIL, including accumulation of GOM and NOTCH3 ectodomain. Evidence for VSMC degeneration was observed in systemic but not cerebral arteries. The rate of NOTCH3 deposition in VSMC membranes appeared to be related to the expression level of the mutant transgene in different mouse lines, despite the fact that transgenic overexpression of wild-type Notch3 does not produce vascular pathology.

a. R90C Mice. *R90C mice* express an archetypal human *NOTCH3* mutation associated with early onset of symptoms in CADASIL patients.[190,194] Subtle changes in tail arteries, including disruption of VSMC anchorage to adjacent cells and matrix, precede the accumulation of NOTCH3 ectodomain by several months in this model. Arterial VSMC loss is detectable in tail arteries at 10 months. Despite these alterations, R90C mice do not develop parenchymal lesions. Importantly, R90C mutation retains NOTCH3 receptor function (i.e., NOTCH3/RBP-Jκ activity) at normal levels with no dominant-negative activity at 1 or 12 months of age.[189] Moreover, vascular dysfunction starts even

TABLE III
CADASIL MUTANT MOUSE MODELS[a]

Genetics	Mutant transgene	R90C human *NOTCH3*	C428S human *NOTCH3*	R169C rat *Notch3*	R142C mouse *Notch3*	R1031C human *NOTCH3*	C455R human *NOTCH3*
	Promoter	Murine Sm22α	Sm22α	PAC	Knockin	Conditional knockin into ROSA26	Conditional knockin into ROSA26
	Background	C57BL/6 J	C57BL/6 J	FVB/N	C57BL/6 J SV129	C57BL/6	C57BL/6
	Expression level (% of endogenous Notch3 mRNA)	85%	50–150%	200–400%	100%	N.D.	N.D.
	Expression distribution	Arteries	Arteries	Arteries, capillaries, glia	Endogenous pattern	VSMCs	VSMCs
	Endogenous *Notch3*	+ or −	+ or −	+	−	−	−
	Mutation in ligand-binding domain	−	+	−	−	−	+
	Downstream signaling	↔	↓[b]	↔	↔	↓	↓↓
Histopathology and ultrastructure	Onset of GOM and *NOTCH3* ectodomain deposits	17–20 months	8 months	2–5 months	Absent	12 months	6 months
	VSMC abnormalities (see text for details)	+	+	+	−	+	+
	Capillary density	N.D.	N.D.	↓	↔	N.D.	N.D.
	Cerebral parenchymal pathology	−	−	+	−	N.D.	N.D.

(*Continues*)

TABLE III (*Continued*)

Cerebrovascular function	Resting CBF	N.D.	N.D.	↓	N.D.	N.D.	N.D.
	Breakthrough hyperemia during hypertensive transients	↓	N.D.	↓	N.D.	N.D.	N.D.
	Upper BP limit of CBF autoregulation	↑	N.D.	N.D.	N.D.	N.D.	N.D.
	Lower BP limit of CBF autoregulation	↑	N.D.	↑	N.D.	N.D.	N.D.
	Pressure-induced (myogenic) contractions	↑	N.D.	↓	N.D.	N.D.	N.D.
	Flow-mediated (shear stress) dilations	↓	N.D.	N.D.	N.D.	N.D.	N.D.
	Hypercapnic and acetazolamide-induced hyperemia	↓	N.D.	↔ [c]	N.D.	N.D.	N.D.
	Functional hyperemia (neurovascular coupling)	N.D.	N.D.	↓	N.D.	N.D.	N.D.
	KCl- and agonist-mediated contractions	↔	N.D.	↔	N.D.	N.D.	N.D.
	Agonist-mediated dilations	↔	N.D.	↔	N.D.	N.D.	N.D.
	Infarct volume after focal ischemia	N.D.	N.D.	N.D.	N.D.	↔	↑
References		187–190	180	167	191	192	192

N.D., not determined; PAC, P1-derived artificial chromosome; GOM, granular osmiophilic material; VSMC, vascular smooth muscle cell; CBF, cerebral blood flow; BP, blood pressure; +, present; −, absent; ↓, diminished; ↑, increased; ↔, unchanged; SM22α promoter, smooth muscle cell-specific.

[a]Modified from Ref.193.

[b]Normal when expressed on wild-type *Notch3* background.

[c]Normal at 6 months, but older mice were not tested.

before the onset of structural defects, and does not appear to worsen with age suggesting that it is not dependent upon GOM and NOTCH3 ectodomain deposits or VSMC degeneration.[187,188]

b. C428S Mice. *C428S mice* express a CADASIL mutation located in the putative ligand-binding domain (EGF-like repeat 10).[180] Similar to R90C mice, these mice also develop characteristic vascular pathology including arterial accumulation of GOM and Notch3 ectodomain upon aging. Unlike the R90C mutant, however, C428S mutation disrupts normal Notch3 receptor function with mild dominant-negative activity detectable on a heterozygous Notch3 knockout background.

c. R169C Mice. *R169C mice* overexpress CADASIL R169C point mutation [p.Arg169 > Cys] via rat P1-derived artificial chromosome (PAC) yielding up to fourfold higher mutant Notch3 expression compared to endogenous Notch3.[167] Characteristic pathological features appear earlier in this mutant model, such as Notch3 ectodomain aggregations at 1–2 months of age, and GOM deposits around 5 months of age. Interestingly, VSCM loss was absent in R169C mutants even at 20 months. Instead, capillary length progressively declined in white but not gray matter starting in corpus callosum as early as 5 months, and decreasing to < 50% of wild-type *Notch3* transgenic mice at 20 months. As a result, aged R169C mice showed significant white matter degeneration at 18–20 months of age. There was marked astrogliosis in white matter by 12 months, but lacunar infarcts were not observed at any age. The onset of vascular dysfunction precedes conspicuous vascular pathology in R169C mutants. Significant reduction of resting CBF is seen in gray matter at 11–12 months of age, and reached 10% and 20% lower than wild type in white and gray matter, respectively, at 18–20 months of age.

d. R142C Mice. *R142C mice* express one of the most prevalent CADASIL mutations.[191] R142C mutation has been shown to affect intracellular trafficking and localization of the receptor, *in vitro*, but does not disrupt Notch3 signaling.[195] Yet the R142C knockin model showed normal Notch3 receptor expression, cleavage, intracellular trafficking, and signaling, and no ultrastructural, histopathologic, or anatomical abnormalities either heterozygous or homozygous mice up to 20 months of age; vasomotor function has not been studied. The reasons for failure to express CADASIL phenotype in this knockin model are not clear.

e. R1031C and C455R. *R1031C and C455R* are conditional knockin mouse models expressing mutant human *NOTCH3* identified in two large Colombian CADASIL families.[192] The C455R mutation is in the putative ligand-binding site and leads to earlier onset of stroke by more than two decades compared to individuals with the R1031C mutation.[179] Expression of either mutation on a *Notch3* knockout background reproduced some of the histopathological features of CADASIL, C455R showing earlier age of onset than R1031C. Interestingly, however, the R1031C mutation rescued the experimental stroke phenotype in the knockout to a similar extent as with the wild-type NOTCH3 transgenic mice, whereas the C455R mutant NOTCH3 not only failed to rescue the phenotype but also tended to worsen it.[183,192] These findings strongly support distinct genotype–phenotype associations, and suggest an age-dependent hypomorphic mechanism for CADASIL, at least when interrogated using an experimental stroke model. Importantly, proteomic analysis of GOM deposits in R1031C mutant revealed increased levels of a number of proteins, including clusterin and endostatin[192]; the pathophysiological importance of this finding remains to be established.

D. Insights into Pathophysiology from Transgenic Models

1. Hypomorphic Phenotype

Notch signaling is known to be gene-dosage sensitive. Therefore, CADASIL mutations may disrupt NOTCH3 receptor function and downstream transcriptional signaling as the disease mechanism. Supporting this notion, Notch3 knockout mice show a more severe experimental stroke phenotype, and both Notch3 knockout and R90C transgenic mice display increased susceptibility to cortical spreading depression compared as a common determinant for severe migraine with aura phenotype in CADASIL.[196] However, evidence argues against a simple hypomorphic phenotype in CADASIL, which is, by definition, an autosomal dominant disease. First, despite impaired signaling, Notch3 knockout mice do not develop characteristic CADASIL pathology. Second, abnormalities in cerebrovascular reactivity in the knockout tend to be in the opposite direction compared to CADASIL mutants. Third, except for those in the ligand-binding site, most CADASIL mutations do not interfere with the downstream canonical NOTCH3 signaling; indeed, most mutant models reproduce the same ultrastructural features of CADASIL regardless of the functional status of the receptor. Fourth, R90C mutant NOTCH3 restores signal transduction in *Notch3* knockout mice despite the presence of GOMs and NOTCH3 ectodomain.[189] Nevertheless, loss of receptor function may be a modulator of the clinical phenotype associated with specific mutations such as C455R.

2. Hypermorphic Phenotype

There is no clear evidence that common CADASIL mutations increase downstream NOTCH3 signaling.[189] A novel heterozygous missense mutation (L1515P) in the extracellular heterodimerization domain has been described that does not alter the number of cysteines, but ligand independently leads to constitutively increased canonical Notch3 signaling by approximately ninefold. This mutation causes SVD with early-onset lacunar strokes and leukoaraiosis in patients, but does not show accumulation of GOM or NOTCH3 ectodomain, and therefore, distinct from CADASIL.[197]

3. Neomorphic Phenotype

Alternatively, the histopathological features of CADASIL, including accumulation of GOMs and NOTCH3 ectodomain might suggest a neomorphic phenotype (i.e., gain of a novel, toxic function), resembling CAA.[2] However, it is not known whether GOM or NOTCH3 aggregates are causally related to vascular dysfunction, leukoaraiosis, and lacunar infarcts, or whether they are innocent biomarkers, and degeneration of VSMC is detected before accumulation of GOM and NOTCH3 ectodomain in the R90C mutant. Animal models also addressed the issue of dominant-negative effect (i.e., mutant NOTCH3 antagonizing wild-type NOTCH3) with mixed results. For example, heterozygous Notch3 knockout mice have one functional copy of endogenous *Notch3*, which, alone, is not associated with any structural abnormalities.[189] However, when heterozygous knockout was crossed with C428S transgenic mice (i.e., presumed loss of function mutation), downstream Notch3 signaling was strongly reduced compared to heterozygous knockout.[180] In contrast, transgenic R90C mutant receptor restored Notch3 signaling in the knockout to normal level with no dominant-negative effect.[189] It should be noted, however, that the level of transgene expression was clearly higher in the C428S mutant compared to the R90C. Overall, the data from mutant mouse models suggest that one or all of these mechanisms may contribute to or modulate the phenotype, possibly explaining some of the clinical heterogeneity in CADASIL.

III. Concluding Remarks

Accumulation of abnormal folding proteins is associated with many vascular disorders. However, the exact mechanisms of how these deposits are formed and induce cellular dysfunction remain to be elucidated. In two types of SVD, CAA and CADASIL, Aβ and GOM/NOTCH3 ectodomain have been identified, respectively, as disease-relevant deposited materials with characteristically complex features. Several genetically altered animal models of CAA and

CADASIL exist that display many histopathological and clinical features of the diseases. As helpful as they are, however, mutant mouse models have serious caveats that limit their ability to fully reproduce the human condition. These include caveats associated with the genetic modeling (e.g., expression level of the mutant gene, presence or absence of endogenous gene) and those that are species specific (e.g., shorter life span, smaller brain size, neuroanatomical and cerebrovascular differences, limited volume of white matter). Absence in animal models of key features of human CAA and CADASIL, such as vascular cognitive impairment and lacunar strokes, makes extrapolation and translation of animal data to human therapy a challenge. Nevertheless, in the absence of animal models spontaneously developing the natural disease, genetically manipulated animals provide opportunities to explore the molecular and physiological pathogenesis of CAA and CADASIL, and will undoubtedly lead to diagnostic, therapeutic, and prognostic advances.

Acknowledgments

This work is supported by the National Institute of Health (NS055104, NS061505, AG021084), Harvard Catalyst—The Harvard Clinical and Translational Science Center (NIH Award #UL1 RR 025758 and financial contributions from Harvard University and its affiliated academic health care centers), and the National Research Foundation of Korea Grant funded by the Korean Government (NRF-2009-352-E00040).

References

1. Vinters HV. Cerebral amyloid angiopathy. A critical review. *Stroke* 1987;**18**:311–24.
2. Yamada M. Cerebral amyloid angiopathy: an overview. *Neuropathology* 2000;**20**:8–22.
3. Jellinger KA. Alzheimer disease and cerebrovascular pathology: an update. *J Neural Transm* 2002;**109**:813–36.
4. Kawai M, Kalaria RN, Cras P, Siedlak SL, Velasco ME, Shelton ER, et al. Degeneration of vascular muscle cells in cerebral amyloid angiopathy of Alzheimer disease. *Brain Res* 1993;**623**:142–6.
5. Gray F, Dubas F, Roullet E, Escourolle R. Leukoencephalopathy in diffuse hemorrhagic cerebral amyloid angiopathy. *Ann Neurol* 1985;**18**:54–9.
6. Attems J, Quass M, Jellinger KA, Lintner F. Topographical distribution of cerebral amyloid angiopathy and its effect on cognitive decline are influenced by Alzheimer disease pathology. *J Neurol Sci* 2007;**257**:49–55.
7. Greenberg SM, Vonsattel JP, Stakes JW, Gruber M, Finklestein SP. The clinical spectrum of cerebral amyloid angiopathy: presentations without lobar hemorrhage. *Neurology* 1993;**43**:2073–9.
8. Arvanitakis Z, Leurgans SE, Wang Z, Wilson RS, Bennett DA, Schneider JA. Cerebral amyloid angiopathy pathology and cognitive domains in older persons. *Ann Neurol* 2011;**69**:320–7.
9. Iadecola C. Neurovascular regulation in the normal brain and in Alzheimer's disease. *Nat Rev Neurosci* 2004;**5**:347–60.

10. Smith EE, Vijayappa M, Lima F, Delgado P, Wendell L, Rosand J, et al. Impaired visual evoked flow velocity response in cerebral amyloid angiopathy. *Neurology* 2008;**71**:1424–30.
11. Linde CI, Baryshnikov SG, Mazzocco-Spezzia A, Golovina VA. Dysregulation of Ca2+ signaling in astrocytes from mice lacking amyloid precursor protein. *Am J Physiol Cell Physiol* 2011;**300**:C1502–12.
12. Milward EA, Papadopoulos R, Fuller SJ, Moir RD, Small D, Beyreuther K, et al. The amyloid protein precursor of Alzheimer's disease is a mediator of the effects of nerve growth factor on neurite outgrowth. *Neuron* 1992;**9**:129–37.
13. Okamoto T, Takeda S, Murayama Y, Ogata E, Nishimoto I. Ligand-dependent G protein coupling function of amyloid transmembrane precursor. *J Biol Chem* 1995;**270**:4205–8.
14. Priller C, Bauer T, Mitteregger G, Krebs B, Kretzschmar HA, Herms J. Synapse formation and function is modulated by the amyloid precursor protein. *J Neurosci* 2006;**26**:7212–21.
15. Schubert D, Jin LW, Saitoh T, Cole G. The regulation of amyloid beta protein precursor secretion and its modulatory role in cell adhesion. *Neuron* 1989;**3**:689–94.
16. Wang Z, Wang B, Yang L, Guo Q, Aithmitti N, Songyang Z, et al. Presynaptic and postsynaptic interaction of the amyloid precursor protein promotes peripheral and central synaptogenesis. *J Neurosci* 2009;**29**:10788–801.
17. Belza MG, Urich H. Cerebral amyloid angiopathy in Down's syndrome. *Clin Neuropathol* 1986;**5**:257–60.
18. Thal DR, Ghebremedhin E, Rub U, Yamaguchi H, Del Tredici K, Braak H. Two types of sporadic cerebral amyloid angiopathy. *J Neuropathol Exp Neurol* 2002;**61**:282–93.
19. Oshima K, Akiyama H, Tsuchiya K, Kondo H, Haga C, Shimomura Y, et al. Relative paucity of tau accumulation in the small areas with abundant Abeta42-positive capillary amyloid angiopathy within a given cortical region in the brain of patients with Alzheimer pathology. *Acta Neuropathol* 2006;**111**:510–8.
20. Levy E, Carman MD, Fernandez-Madrid IJ, Power MD, Lieberburg I, van Duinen SG, et al. Mutation of the Alzheimer's disease amyloid gene in hereditary cerebral hemorrhage, Dutch type. *Science* 1990;**248**:1124–6.
21. Van Broeckhoven C, Haan J, Bakker E, Hardy JA, Van Hul W, Wehnert A, et al. Amyloid beta protein precursor gene and hereditary cerebral hemorrhage with amyloidosis (Dutch). *Science* 1990;**248**:1120–2.
22. Herzig MC, Winkler DT, Burgermeister P, Pfeifer M, Kohler E, Schmidt SD, et al. Abeta is targeted to the vasculature in a mouse model of hereditary cerebral hemorrhage with amyloidosis. *Nat Neurosci* 2004;**7**:954–60.
23. Maat-Schieman ML, Yamaguchi H, van Duinen SG, Natte R, Roos RA. Age-related plaque morphology and C-terminal heterogeneity of amyloid beta in Dutch-type hereditary cerebral hemorrhage with amyloidosis. *Acta Neuropathol* 2000;**99**:409–19.
24. Bugiani O, Giaccone G, Rossi G, Mangieri M, Capobianco R, Morbin M, et al. Hereditary cerebral hemorrhage with amyloidosis associated with the E693K mutation of APP. *Arch Neurol* 2010;**67**:987–95.
25. Hendriks L, van Duijn CM, Cras P, Cruts M, Van Hul W, van Harskamp F, et al. Presenile dementia and cerebral haemorrhage linked to a mutation at codon 692 of the beta-amyloid precursor protein gene. *Nat Genet* 1992;**1**:218–21.
26. Kumar-Singh S, Cras P, Wang R, Kros JM, van Swieten J, Lubke U, et al. Dense-core senile plaques in the Flemish variant of Alzheimer's disease are vasocentric. *Am J Pathol* 2002;**161**:507–20.
27. Roks G, Van Harskamp F, De Koning I, Cruts M, De Jonghe C, Kumar-Singh S, et al. Presentation of amyloidosis in carriers of the codon 692 mutation in the amyloid precursor protein gene (APP692). *Brain* 2000;**123**(Pt 10):2130–40.

28. Grabowski TJ, Cho HS, Vonsattel JP, Rebeck GW, Greenberg SM. Novel amyloid precursor protein mutation in an Iowa family with dementia and severe cerebral amyloid angiopathy. *Ann Neurol* 2001;**49**:697–705.
29. Greenberg SM, Shin Y, Grabowski TJ, Cooper GE, Rebeck GW, Iglesias S, et al. Hemorrhagic stroke associated with the Iowa amyloid precursor protein mutation. *Neurology* 2003;**60**:1020–2.
30. Nilsberth C, Westlind-Danielsson A, Eckman CB, Condron MM, Axelman K, Forsell C, et al. The 'Arctic' APP mutation (E693G) causes Alzheimer's disease by enhanced Abeta protofibril formation. *Nat Neurosci* 2001;**4**:887–93.
31. Basun H, Bogdanovic N, Ingelsson M, Almkvist O, Naslund J, Axelman K, et al. Clinical and neuropathological features of the arctic APP gene mutation causing early-onset Alzheimer disease. *Arch Neurol* 2008;**65**:499–505.
32. Zhang-Nunes SX, Maat-Schieman ML, van Duinen SG, Roos RA, Frosch MP, Greenberg SM. The cerebral beta-amyloid angiopathies: hereditary and sporadic. *Brain Pathol* 2006;**16**:30–9.
33. Davis J, Van Nostrand WE. Enhanced pathologic properties of Dutch-type mutant amyloid beta-protein. *Proc Natl Acad Sci USA* 1996;**93**:2996–3000.
34. Johansson AS, Berglind-Dehlin F, Karlsson G, Edwards K, Gellerfors P, Lannfelt L. Physiochemical characterization of the Alzheimer's disease-related peptides A beta 1-42Arctic and A beta 1-42wt. *FEBS J* 2006;**273**:2618–30.
35. Melchor JP, McVoy L, Van Nostrand WE. Charge alterations of E22 enhance the pathogenic properties of the amyloid beta-protein. *J Neurochem* 2000;**74**:2209–12.
36. Van Nostrand WE, Melchor JP, Cho HS, Greenberg SM, Rebeck GW. Pathogenic effects of D23N Iowa mutant amyloid beta-protein. *J Biol Chem* 2001;**276**:32860–6.
37. Citron M, Oltersdorf T, Haass C, McConlogue L, Hung AY, Seubert P, et al. Mutation of the beta-amyloid precursor protein in familial Alzheimer's disease increases beta-protein production. *Nature* 1992;**360**:672–4.
38. Goate A, Chartier-Harlin MC, Mullan M, Brown J, Crawford F, Fidani L, et al. Segregation of a missense mutation in the amyloid precursor protein gene with familial Alzheimer's disease. *Nature* 1991;**349**:704–6.
39. Mullan M, Crawford F, Axelman K, Houlden H, Lilius L, Winblad B, et al. A pathogenic mutation for probable Alzheimer's disease in the APP gene at the N-terminus of beta-amyloid. *Nat Genet* 1992;**1**:345–7.
40. Murrell J, Farlow M, Ghetti B, Benson MD. A mutation in the amyloid precursor protein associated with hereditary Alzheimer's disease. *Science* 1991;**254**:97–9.
41. Rensink AA, de Waal RM, Kremer B, Verbeek MM. Pathogenesis of cerebral amyloid angiopathy. *Brain Res Brain Res Rev* 2003;**43**:207–23.
42. Cabrejo L, Guyant-Marechal L, Laquerriere A, Vercelletto M, De la Fourniere F, Thomas-Anterion C, et al. Phenotype associated with APP duplication in five families. *Brain* 2006;**129**:2966–76.
43. Hardy J, Selkoe DJ. The amyloid hypothesis of Alzheimer's disease: progress and problems on the road to therapeutics. *Science* 2002;**297**:353–6.
44. Chishti MA, Yang DS, Janus C, Phinney AL, Horne P, Pearson J, et al. Early-onset amyloid deposition and cognitive deficits in transgenic mice expressing a double mutant form of amyloid precursor protein 695. *J Biol Chem* 2001;**276**:21562–70.
45. Mann DM, Pickering-Brown SM, Takeuchi A, Iwatsubo T. Amyloid angiopathy and variability in amyloid beta deposition is determined by mutation position in presenilin-1-linked Alzheimer's disease. *Am J Pathol* 2001;**158**:2165–75.
46. Nochlin D, Bird TD, Nemens EJ, Ball MJ, Sumi SM. Amyloid angiopathy in a Volga German family with Alzheimer's disease and a presenilin-2 mutation (N141I). *Ann Neurol* 1998;**43**:131–5.

47. Steiner H, Revesz T, Neumann M, Romig H, Grim MG, Pesold B, et al. A pathogenic presenilin-1 deletion causes abberrant Abeta 42 production in the absence of congophilic amyloid plaques. *J Biol Chem* 2001;**276**:7233–9.
48. Ghetti B, Piccardo P, Spillantini MG, Ichimiya Y, Porro M, Perini F, et al. Vascular variant of prion protein cerebral amyloidosis with tau-positive neurofibrillary tangles: the phenotype of the stop codon 145 mutation in PRNP. *Proc Natl Acad Sci USA* 1996;**93**:744–8.
49. Levy E, Haltia M, Fernandez-Madrid I, Koivunen O, Ghiso J, Prelli F, et al. Mutation in gelsolin gene in Finnish hereditary amyloidosis. *J Exp Med* 1990;**172**:1865–7.
50. Palsdottir A, Snorradottir AO, Thorsteinsson L. Hereditary cystatin C amyloid angiopathy: genetic, clinical, and pathological aspects. *Brain Pathol* 2006;**16**:55–9.
51. Petersen RB, Goren H, Cohen M, Richardson SL, Tresser N, Lynn A, et al. Transthyretin amyloidosis: a new mutation associated with dementia. *Ann Neurol* 1997;**41**:307–13.
52. Vidal R, Frangione B, Rostagno A, Mead S, Revesz T, Plant G, et al. A stop-codon mutation in the BRI gene associated with familial British dementia. *Nature* 1999;**399**:776–81.
53. Vidal R, Revesz T, Rostagno A, Kim E, Holton JL, Bek T, et al. A decamer duplication in the 3' region of the BRI gene originates an amyloid peptide that is associated with dementia in a Danish kindred. *Proc Natl Acad Sci USA* 2000;**97**:4920–5.
54. Walker LC. Animal models of cerebral beta-amyloid angiopathy. *Brain Res Brain Res Rev* 1997;**25**:70–84.
55. Fryer JD, Taylor JW, DeMattos RB, Bales KR, Paul SM, Parsadanian M, et al. Apolipoprotein E markedly facilitates age-dependent cerebral amyloid angiopathy and spontaneous hemorrhage in amyloid precursor protein transgenic mice. *J Neurosci* 2003;**23**:7889–96.
56. Hartman RE, Izumi Y, Bales KR, Paul SM, Wozniak DF, Holtzman DM. Treatment with an amyloid-beta antibody ameliorates plaque load, learning deficits, and hippocampal long-term potentiation in a mouse model of Alzheimer's disease. *J Neurosci* 2005;**25**:6213–20.
57. Bales KR, Verina T, Dodel RC, Du Y, Altstiel L, Bender M, et al. Lack of apolipoprotein E dramatically reduces amyloid beta-peptide deposition. *Nat Genet* 1997;**17**:263–4.
58. Racke MM, Boone LI, Hepburn DL, Parsadainian M, Bryan MT, Ness DK, et al. Exacerbation of cerebral amyloid angiopathy-associated microhemorrhage in amyloid precursor protein transgenic mice by immunotherapy is dependent on antibody recognition of deposited forms of amyloid beta. *J Neurosci* 2005;**25**:629–36.
59. Masliah E, Sisk A, Mallory M, Mucke L, Schenk D, Games D. Comparison of neurodegenerative pathology in transgenic mice overexpressing V717F beta-amyloid precursor protein and Alzheimer's disease. *J Neurosci* 1996;**16**:5795–811.
60. Irizarry MC, Soriano F, McNamara M, Page KJ, Schenk D, Games D, et al. Abeta deposition is associated with neuropil changes, but not with overt neuronal loss in the human amyloid precursor protein V717F (PDAPP) transgenic mouse. *J Neurosci* 1997;**17**:7053–9.
61. Games D, Adams D, Alessandrini R, Barbour R, Berthelette P, Blackwell C, et al. Alzheimer-type neuropathology in transgenic mice overexpressing V717F beta-amyloid precursor protein. *Nature* 1995;**373**:523–7.
62. Westerman MA, Cooper-Blacketer D, Mariash A, Kotilinek L, Kawarabayashi T, Younkin LH, et al. The relationship between Abeta and memory in the Tg2576 mouse model of Alzheimer's disease. *J Neurosci* 2002;**22**:1858–67.
63. Jacobsen JS, Wu CC, Redwine JM, Comery TA, Arias R, Bowlby M, et al. Early-onset behavioral and synaptic deficits in a mouse model of Alzheimer's disease. *Proc Natl Acad Sci USA* 2006;**103**:5161–6.
64. Shin HK, Jones PB, Garcia-Alloza M, Borrelli L, Greenberg SM, Bacskai BJ, et al. Age-dependent cerebrovascular dysfunction in a transgenic mouse model of cerebral amyloid angiopathy. *Brain* 2007;**130**:2310–9.

65. Irizarry MC, McNamara M, Fedorchak K, Hsiao K, Hyman BT. APPSw transgenic mice develop age-related A beta deposits and neuropil abnormalities, but no neuronal loss in CA1. *J Neuropathol Exp Neurol* 1997;**56**:965–73.
66. Hsiao K, Chapman P, Nilsen S, Eckman C, Harigaya Y, Younkin S, et al. Correlative memory deficits, Abeta elevation, and amyloid plaques in transgenic mice. *Science* 1996;**274**:99–102.
67. Han BH, Zhou ML, Abousaleh F, Brendza RP, Dietrich HH, Koenigsknecht-Talboo J, et al. Cerebrovascular dysfunction in amyloid precursor protein transgenic mice: contribution of soluble and insoluble amyloid-beta peptide, partial restoration via gamma-secretase inhibition. *J Neurosci* 2008;**28**:13542–50.
68. Pfeifer M, Boncristiano S, Bondolfi L, Stalder A, Deller T, Staufenbiel M, et al. Cerebral hemorrhage after passive anti-Abeta immunotherapy. *Science* 2002;**298**:1379.
69. Winkler DT, Bondolfi L, Herzig MC, Jann L, Calhoun ME, Wiederhold KH, et al. Spontaneous hemorrhagic stroke in a mouse model of cerebral amyloid angiopathy. *J Neurosci* 2001;**21**:1619–27.
70. Winkler DT, Biedermann L, Tolnay M, Allegrini PR, Staufenbiel M, Wiessner C, et al. Thrombolysis induces cerebral hemorrhage in a mouse model of cerebral amyloid angiopathy. *Ann Neurol* 2002;**51**:790–3.
71. Van Dam D, D'Hooge R, Staufenbiel M, Van Ginneken C, Van Meir F, De Deyn PP. Age-dependent cognitive decline in the APP23 model precedes amyloid deposition. *Eur J Neurosci* 2003;**17**:388–96.
72. Sturchler-Pierrat C, Abramowski D, Duke M, Wiederhold KH, Mistl C, Rothacher S, et al. Two amyloid precursor protein transgenic mouse models with Alzheimer disease-like pathology. *Proc Natl Acad Sci USA* 1997;**94**:13287–92.
73. Kelly PH, Bondolfi L, Hunziker D, Schlecht HP, Carver K, Maguire E, et al. Progressive age-related impairment of cognitive behavior in APP23 transgenic mice. *Neurobiol Aging* 2003;**24**:365–78.
74. Calhoun ME, Burgermeister P, Phinney AL, Stalder M, Tolnay M, Wiederhold KH, et al. Neuronal overexpression of mutant amyloid precursor protein results in prominent deposition of cerebrovascular amyloid. *Proc Natl Acad Sci USA* 1999;**96**:14088–93.
75. Moechars D, Dewachter I, Lorent K, Reverse D, Baekelandt V, Naidu A, et al. Early phenotypic changes in transgenic mice that overexpress different mutants of amyloid precursor protein in brain. *J Biol Chem* 1999;**274**:6483–92.
76. Van Dorpe J, Smeijers L, Dewachter I, Nuyens D, Spittaels K, Van Den Haute C, et al. Prominent cerebral amyloid angiopathy in transgenic mice overexpressing the london mutant of human APP in neurons. *Am J Pathol* 2000;**157**:1283–98.
77. Dudal S, Krzywkowski P, Paquette J, Morissette C, Lacombe D, Tremblay P, et al. Inflammation occurs early during the Abeta deposition process in TgCRND8 mice. *Neurobiol Aging* 2004;**25**:861–71.
78. Hawkes CA, McLaurin J. Selective targeting of perivascular macrophages for clearance of beta-amyloid in cerebral amyloid angiopathy. *Proc Natl Acad Sci USA* 2009;**106**:1261–6.
79. Davis J, Xu F, Deane R, Romanov G, Previti ML, Zeigler K, et al. Early-onset and robust cerebral microvascular accumulation of amyloid beta-protein in transgenic mice expressing low levels of a vasculotropic Dutch/Iowa mutant form of amyloid beta-protein precursor. *J Biol Chem* 2004;**279**:20296–306.
80. Miao J, Xu F, Davis J, Otte-Holler I, Verbeek MM, Van Nostrand WE. Cerebral microvascular amyloid beta protein deposition induces vascular degeneration and neuroinflammation in transgenic mice expressing human vasculotropic mutant amyloid beta precursor protein. *Am J Pathol* 2005;**167**:505–15.

81. Xu F, Grande AM, Robinson JK, Previti ML, Vasek M, Davis J, et al. Early-onset subicular microvascular amyloid and neuroinflammation correlate with behavioral deficits in vasculotropic mutant amyloid beta-protein precursor transgenic mice. *Neuroscience* 2007;**146**:98–107.
82. Herzig MC, Van Nostrand WE, Jucker M. Mechanism of cerebral beta-amyloid angiopathy: murine and cellular models. *Brain Pathol* 2006;**16**:40–54.
83. Wilcock DM, Lewis MR, Van Nostrand WE, Davis J, Previti ML, Gharkholonarehe N, et al. Progression of amyloid pathology to Alzheimer's disease pathology in an amyloid precursor protein transgenic mouse model by removal of nitric oxide synthase 2. *J Neurosci* 2008;**28**:1537–45.
84. Lord A, Kalimo H, Eckman C, Zhang XQ, Lannfelt L, Nilsson LN. The Arctic Alzheimer mutation facilitates early intraneuronal Abeta aggregation and senile plaque formation in transgenic mice. *Neurobiol Aging* 2006;**27**:67–77.
85. Knobloch M, Konietzko U, Krebs DC, Nitsch RM. Intracellular Abeta and cognitive deficits precede beta-amyloid deposition in transgenic arcAbeta mice. *Neurobiol Aging* 2007;**28**:1297–306.
86. Princz-Kranz FL, Mueggler T, Knobloch M, Nitsch RM, Rudin M. Vascular response to acetazolamide decreases as a function of age in the arcA beta mouse model of cerebral amyloidosis. *Neurobiol Dis* 2010;**40**:284–92.
87. Davis-Salinas J, Saporito-Irwin SM, Cotman CW, Van Nostrand WE. Amyloid beta-protein induces its own production in cultured degenerating cerebrovascular smooth muscle cells. *J Neurochem* 1995;**65**:931–4.
88. Frackowiak J, Miller DL, Potempska A, Sukontasup T, Mazur-Kolecka B. Secretion and accumulation of Abeta by brain vascular smooth muscle cells from AbetaPP-Swedish transgenic mice. *J Neuropathol Exp Neurol* 2003;**62**:685–96.
89. Frackowiak J, Potempska A, LeVine H, Haske T, Dickson D, Mazur-Kolecka B. Extracellular deposits of A beta produced in cultures of Alzheimer disease brain vascular smooth muscle cells. *J Neuropathol Exp Neurol* 2005;**64**:82–90.
90. Wisniewski HM, Frackowiak J, Mazur-Kolecka B. *In vitro* production of beta-amyloid in smooth muscle cells isolated from amyloid angiopathy-affected vessels. *Neurosci Lett* 1995;**183**:120–3.
91. Verbeek MM, de Waal RM, Schipper JJ, Van Nostrand WE. Rapid degeneration of cultured human brain pericytes by amyloid beta protein. *J Neurochem* 1997;**68**:1135–41.
92. Kalaria RN, Premkumar DR, Pax AB, Cohen DL, Lieberburg I. Production and increased detection of amyloid beta protein and amyloidogenic fragments in brain microvessels, meningeal vessels and choroid plexus in Alzheimer's disease. *Brain Res Mol Brain Res* 1996;**35**:58–68.
93. Natte R, de Boer WI, Maat-Schieman ML, Baelde HJ, Vinters HV, Roos RA, et al. Amyloid beta precursor protein-mRNA is expressed throughout cerebral vessel walls. *Brain Res* 1999;**828**:179–83.
94. Fukuchi K, Ho L, Younkin SG, Kunkel DD, Ogburn CE, LeBoeuf RC, et al. High levels of circulating beta-amyloid peptide do not cause cerebral beta-amyloidosis in transgenic mice. *Am J Pathol* 1996;**149**:219–27.
95. Yamaguchi H, Yamazaki T, Lemere CA, Frosch MP, Selkoe DJ. Beta amyloid is focally deposited within the outer basement membrane in the amyloid angiopathy of Alzheimer's disease. An immunoelectron microscopic study. *Am J Pathol* 1992;**141**:249–59.
96. Weller RO, Massey A, Newman TA, Hutchings M, Kuo YM, Roher AE. Cerebral amyloid angiopathy: amyloid beta accumulates in putative interstitial fluid drainage pathways in Alzheimer's disease. *Am J Pathol* 1998;**153**:725–33.

97. Burgermeister P, Calhoun ME, Winkler DT, Jucker M. Mechanisms of cerebrovascular amyloid deposition. Lessons from mouse models. *Ann N Y Acad Sci* 2000;**903**:307–16.
98. Weller RO, Subash M, Preston SD, Mazanti I, Carare RO. Perivascular drainage of amyloid-beta peptides from the brain and its failure in cerebral amyloid angiopathy and Alzheimer's disease. *Brain Pathol* 2008;**18**:253–66.
99. Carare RO, Bernardes-Silva M, Newman TA, Page AM, Nicoll JA, Perry VH, et al. Solutes, but not cells, drain from the brain parenchyma along basement membranes of capillaries and arteries: significance for cerebral amyloid angiopathy and neuroimmunology. *Neuropathol Appl Neurobiol* 2008;**34**:131–44.
100. Dewachter I, Reverse D, Caluwaerts N, Ris L, Kuiperi C, Van den Haute C, et al. Neuronal deficiency of presenilin 1 inhibits amyloid plaque formation and corrects hippocampal long-term potentiation but not a cognitive defect of amyloid precursor protein [V717I] transgenic mice. *J Neurosci* 2002;**22**:3445–53.
101. Hawkes CA, Hartig W, Kacza J, Schliebs R, Weller RO, Nicoll JA, et al. Perivascular drainage of solutes is impaired in the ageing mouse brain and in the presence of cerebral amyloid angiopathy. *Acta Neuropathol* 2011;**121**:431–43.
102. Attems J, Lintner F, Jellinger KA. Amyloid beta peptide 1–42 highly correlates with capillary cerebral amyloid angiopathy and Alzheimer disease pathology. *Acta Neuropathol* 2004;**107**:283–91.
103. Iwatsubo T, Odaka A, Suzuki N, Mizusawa H, Nukina N, Ihara Y. Visualization of A beta 42 (43) and A beta 40 in senile plaques with end-specific A beta monoclonals: evidence that an initially deposited species is A beta 42(43). *Neuron* 1994;**13**:45–53.
104. Suzuki N, Iwatsubo T, Odaka A, Ishibashi Y, Kitada C, Ihara Y. High tissue content of soluble beta 1–40 is linked to cerebral amyloid angiopathy. *Am J Pathol* 1994;**145**:452–60.
105. McGowan E, Pickford F, Kim J, Onstead L, Eriksen J, Yu C, et al. Abeta42 is essential for parenchymal and vascular amyloid deposition in mice. *Neuron* 2005;**47**:191–9.
106. De Jonghe C, Cras P, Vanderstichele H, Cruts M, Vanderhoeven I, Smouts I, et al. Evidence that Abeta42 plasma levels in presenilin-1 mutation carriers do not allow for prediction of their clinical phenotype. *Neurobiol Dis* 1999;**6**:280–7.
107. Kim J, Onstead L, Randle S, Price R, Smithson L, Zwizinski C, et al. Abeta40 inhibits amyloid deposition *in vivo*. *J Neurosci* 2007;**27**:627–33.
108. Janus C, Pearson J, McLaurin J, Mathews PM, Jiang Y, Schmidt SD, et al. A beta peptide immunization reduces behavioural impairment and plaques in a model of Alzheimer's disease. *Nature* 2000;**408**:979–82.
109. DaSilva KA, Brown ME, McLaurin J. Reduced oligomeric and vascular amyloid-beta following immunization of TgCRND8 mice with an Alzheimer's DNA vaccine. *Vaccine* 2009;**27**:1365–76.
110. Iwata N, Tsubuki S, Takaki Y, Watanabe K, Sekiguchi M, Hosoki E, et al. Identification of the major Abeta1-42-degrading catabolic pathway in brain parenchyma: suppression leads to biochemical and pathological deposition. *Nat Med* 2000;**6**:143–50.
111. Kurochkin IV, Goto S. Alzheimer's beta-amyloid peptide specifically interacts with and is degraded by insulin degrading enzyme. *FEBS Lett* 1994;**345**:33–7.
112. Eckman EA, Reed DK, Eckman CB. Degradation of the Alzheimer's amyloid beta peptide by endothelin-converting enzyme. *J Biol Chem* 2001;**276**:24540–8.
113. Cook DG, Leverenz JB, McMillan PJ, Kulstad JJ, Ericksen S, Roth RA, et al. Reduced hippocampal insulin-degrading enzyme in late-onset Alzheimer's disease is associated with the apolipoprotein E-epsilon4 allele. *Am J Pathol* 2003;**162**:313–9.
114. Miners JS, Van Helmond Z, Chalmers K, Wilcock G, Love S, Kehoe PG. Decreased expression and activity of neprilysin in Alzheimer disease are associated with cerebral amyloid angiopathy. *J Neuropathol Exp Neurol* 2006;**65**:1012–21.

115. Kanemitsu H, Tomiyama T, Mori H. Human neprilysin is capable of degrading amyloid beta peptide not only in the monomeric form but also the pathological oligomeric form. *Neurosci Lett* 2003;**350**:113–6.
116. Leissring MA, Farris W, Chang AY, Walsh DM, Wu X, Sun X, et al. Enhanced proteolysis of beta-amyloid in APP transgenic mice prevents plaque formation, secondary pathology, and premature death. *Neuron* 2003;**40**:1087–93.
117. Iwata N, Tsubuki S, Takaki Y, Shirotani K, Lu B, Gerard NP, et al. Metabolic regulation of brain Abeta by neprilysin. *Science* 2001;**292**:1550–2.
118. Farris W, Mansourian S, Chang Y, Lindsley L, Eckman EA, Frosch MP, et al. Insulin-degrading enzyme regulates the levels of insulin, amyloid beta-protein, and the beta-amyloid precursor protein intracellular domain in vivo. *Proc Natl Acad Sci USA* 2003;**100**:4162–7.
119. Deane R, Wu Z, Sagare A, Davis J, Du Yan S, Hamm K, et al. LRP/amyloid beta peptide interaction mediates differential brain efflux of Abeta isoforms. *Neuron* 2004;**43**:333–44.
120. Shibata M, Yamada S, Kumar SR, Calero M, Bading J, Frangione B, et al. Clearance of Alzheimer's amyloid-ss(1–40) peptide from brain by LDL receptor-related protein-1 at the blood–brain barrier. *J Clin Invest* 2000;**106**:1489–99.
121. Deane R, Du Yan S, Submamaryan RK, LaRue B, Jovanovic S, Hogg E, et al. RAGE mediates amyloid-beta peptide transport across the blood–brain barrier and accumulation in brain. *Nat Med* 2003;**9**:907–13.
122. Durany N, Ravid R, Riederer P, Cruz-Sanchez FF. Increased frequency of the alpha-1-antichymotrypsin T allele in cerebral amyloid angiopathy. *Neuropathology* 2000;**20**:184–9.
123. Yamada M, Sodeyama N, Itoh Y, Suematsu N, Otomo E, Matsushita M, et al. Association of alpha1-antichymotrypsin polymorphism with cerebral amyloid angiopathy. *Ann Neurol* 1998;**44**:129–31.
124. Yamada M, Sodeyama N, Itoh Y, Takahashi A, Otomo E, Matsushita M, et al. Association of neprilysin polymorphism with cerebral amyloid angiopathy. *J Neurol Neurosurg Psychiatry* 2003;**74**:749–51.
125. Hamaguchi T, Okino S, Sodeyama N, Itoh Y, Takahashi A, Otomo E, et al. Association of a polymorphism of the transforming growth factor-beta1 gene with cerebral amyloid angiopathy. *J Neurol Neurosurg Psychiatry* 2005;**76**:696–9.
126. Christoforidis M, Schober R, Krohn K. Genetic-morphologic association study: association between the low density lipoprotein-receptor related protein (LRP) and cerebral amyloid angiopathy. *Neuropathol Appl Neurobiol* 2005;**31**:11–9.
127. Verghese PB, Castellano JM, Holtzman DM. Apolipoprotein E in Alzheimer's disease and other neurological disorders. *Lancet Neurol* 2011;**10**:241–52.
128. Rubinsztein DC. Apolipoprotein E: a review of its roles in lipoprotein metabolism, neuronal growth and repair and as a risk factor for Alzheimer's disease. *Psychol Med* 1995;**25**:223–9.
129. Corder EH, Saunders AM, Strittmatter WJ, Schmechel DE, Gaskell PC, Small GW, et al. Gene dose of apolipoprotein E type 4 allele and the risk of Alzheimer's disease in late onset families. *Science* 1993;**261**:921–3.
130. Greenberg SM, Rebeck GW, Vonsattel JP, Gomez-Isla T, Hyman BT. Apolipoprotein E epsilon 4 and cerebral hemorrhage associated with amyloid angiopathy. *Ann Neurol* 1995;**38**:254–9.
131. Premkumar DR, Cohen DL, Hedera P, Friedland RP, Kalaria RN. Apolipoprotein E-epsilon4 alleles in cerebral amyloid angiopathy and cerebrovascular pathology associated with Alzheimer's disease. *Am J Pathol* 1996;**148**:2083–95.
132. Greenberg SM, Vonsattel JP, Segal AZ, Chiu RI, Clatworthy AE, Liao A, et al. Association of apolipoprotein E epsilon2 and vasculopathy in cerebral amyloid angiopathy. *Neurology* 1998;**50**:961–5.

133. McCarron MO, Nicoll JA. Apolipoprotein E genotype and cerebral amyloid angiopathy-related hemorrhage. *Ann N Y Acad Sci* 2000;**903**:176–9.
134. Nicoll JA, Burnett C, Love S, Graham DI, Dewar D, Ironside JW, et al. High frequency of apolipoprotein E epsilon 2 allele in hemorrhage due to cerebral amyloid angiopathy. *Ann Neurol* 1997;**41**:716–21.
135. Corder EH, Saunders AM, Risch NJ, Strittmatter WJ, Schmechel DE, Gaskell Jr. PC, et al. Protective effect of apolipoprotein E type 2 allele for late onset Alzheimer disease. *Nat Genet* 1994;**7**:180–4.
136. Holtzman DM, Fagan AM, Mackey B, Tenkova T, Sartorius L, Paul SM, et al. Apolipoprotein E facilitates neuritic and cerebrovascular plaque formation in an Alzheimer's disease model. *Ann Neurol* 2000;**47**:739–47.
137. Iadecola C, Zhang F, Niwa K, Eckman C, Turner SK, Fischer E, et al. SOD1 rescues cerebral endothelial dysfunction in mice overexpressing amyloid precursor protein. *Nat Neurosci* 1999;**2**:157–61.
138. Niwa K, Carlson GA, Iadecola C. Exogenous A beta1-40 reproduces cerebrovascular alterations resulting from amyloid precursor protein overexpression in mice. *J Cereb Blood Flow Metab* 2000;**20**:1659–68.
139. Niwa K, Kazama K, Younkin L, Younkin SG, Carlson GA, Iadecola C. Cerebrovascular autoregulation is profoundly impaired in mice overexpressing amyloid precursor protein. *Am J Physiol Heart Circ Physiol* 2002;**283**:H315–23.
140. Park L, Anrather J, Forster C, Kazama K, Carlson GA, Iadecola C. Abeta-induced vascular oxidative stress and attenuation of functional hyperemia in mouse somatosensory cortex. *J Cereb Blood Flow Metab* 2004;**24**:334–42.
141. Mueggler T, Sturchler-Pierrat C, Baumann D, Rausch M, Staufenbiel M, Rudin M. Compromised hemodynamic response in amyloid precursor protein transgenic mice. *J Neurosci* 2002;**22**:7218–24.
142. Joutel A, Corpechot C, Ducros A, Vahedi K, Chabriat H, Mouton P, et al. Notch3 mutations in CADASIL, a hereditary adult-onset condition causing stroke and dementia. *Nature* 1996;**383**:707–10.
143. Chabriat H, Joutel A, Dichgans M, Tournier-Lasserve E, Bousser MG. Cadasil. *Lancet Neurol* 2009;**8**:643–53.
144. Tournier-Lasserve E, Joutel A, Melki J, Weissenbach J, Lathrop GM, Chabriat H, et al. Cerebral autosomal dominant arteriopathy with subcortical infarcts and leukoencephalopathy maps to chromosome 19q12. *Nat Genet* 1993;**3**:256–9.
145. Dong Y, Hassan A, Zhang Z, Huber D, Dalageorgou C, Markus HS. Yield of screening for CADASIL mutations in lacunar stroke and leukoaraiosis. *Stroke* 2003;**34**:203–5.
146. Singhal S, Bevan S, Barrick T, Rich P, Markus HS. The influence of genetic and cardiovascular risk factors on the CADASIL phenotype. *Brain* 2004;**127**:2031–8.
147. Liem MK, van der Grond J, Haan J, van den Boom R, Ferrari MD, Knaap YM, et al. Lacunar infarcts are the main correlate with cognitive dysfunction in CADASIL. *Stroke* 2007;**38**:923–8.
148. Viswanathan A, Gschwendtner A, Guichard JP, Buffon F, Cumurciuc R, O'Sullivan M, et al. Lacunar lesions are independently associated with disability and cognitive impairment in CADASIL. *Neurology* 2007;**69**:172–9.
149. Tikka S, Mykkanen K, Ruchoux MM, Bergholm R, Junna M, Poyhonen M, et al. Congruence between NOTCH3 mutations and GOM in 131 CADASIL patients. *Brain* 2009;**132**:933–9.
150. Okeda R, Arima K, Kawai M. Arterial changes in cerebral autosomal dominant arteriopathy with subcortical infarcts and leukoencephalopathy (CADASIL) in relation to pathogenesis of diffuse myelin loss of cerebral white matter: examination of cerebral medullary arteries by reconstruction of serial sections of an autopsy case. *Stroke* 2002;**33**:2565–9.

151. Yamamoto Y, Ihara M, Tham C, Low RW, Slade JY, Moss T, et al. Neuropathological correlates of temporal pole white matter hyperintensities in CADASIL. *Stroke* 2009;**40**:2004–11.
152. Miao Q, Paloneva T, Tuominen S, Poyhonen M, Tuisku S, Viitanen M, et al. Fibrosis and stenosis of the long penetrating cerebral arteries: the cause of the white matter pathology in cerebral autosomal dominant arteriopathy with subcortical infarcts and leukoencephalopathy. *Brain Pathol* 2004;**14**:358–64.
153. Domenga V, Fardoux P, Lacombe P, Monet M, Maciazek J, Krebs LT, et al. Notch3 is required for arterial identity and maturation of vascular smooth muscle cells. *Genes Dev* 2004;**18**:2730–5.
154. Wang T, Baron M, Trump D. An overview of Notch3 function in vascular smooth muscle cells. *Prog Biophys Mol Biol* 2008;**96**:499–509.
155. Bruening R, Dichgans M, Berchtenbreiter C, Yousry T, Seelos KC, Wu RH, et al. Cerebral autosomal dominant arteriopathy with subcortical infarcts and leukoencephalopathy: decrease in regional cerebral blood volume in hyperintense subcortical lesions inversely correlates with disability and cognitive performance. *AJNR Am J Neuroradiol* 2001;**22**:1268–74.
156. Chabriat H, Pappata S, Ostergaard L, Clark CA, Pachot-Clouard M, Vahedi K, et al. Cerebral hemodynamics in CADASIL before and after acetazolamide challenge assessed with MRI bolus tracking. *Stroke* 2000;**31**:1904–12.
157. Pfefferkorn T, von Stuckrad-Barre S, Herzog J, Gasser T, Hamann GF, Dichgans M. Reduced cerebrovascular CO(2) reactivity in CADASIL: a transcranial Doppler sonography study. *Stroke* 2001;**32**:17–21.
158. Tuominen S, Miao Q, Kurki T, Tuisku S, Poyhonen M, Kalimo H, et al. Positron emission tomography examination of cerebral blood flow and glucose metabolism in young CADASIL patients. *Stroke* 2004;**35**:1063–7.
159. van den Boom R, Lesnik Oberstein SA, Spilt A, Behloul F, Ferrari MD, Haan J, et al. Cerebral hemodynamics and white matter hyperintensities in CADASIL. *J Cereb Blood Flow Metab* 2003;**23**:599–604.
160. Peters N, Freilinger T, Opherk C, Pfefferkorn T, Dichgans M. Enhanced L-arginine-induced vasoreactivity suggests endothelial dysfunction in CADASIL. *J Neurol* 2008;**255**:1203–8.
161. Ruchoux MM, Maurage CA. Endothelial changes in muscle and skin biopsies in patients with CADASIL. *Neuropathol Appl Neurobiol* 1998;**24**:60–5.
162. Stenborg A, Kalimo H, Viitanen M, Terent A, Lind L. Impaired endothelial function of forearm resistance arteries in CADASIL patients. *Stroke* 2007;**38**:2692–7.
163. Choi JC, Kang SY, Kang JH, Park JK. Intracerebral hemorrhages in CADASIL. *Neurology* 2006;**67**:2042–4.
164. Joutel A, Dodick DD, Parisi JE, Cecillon M, Tournier-Lasserve E, Bousser MG. *De novo* mutation in the Notch3 gene causing CADASIL. *Ann Neurol* 2000;**47**:388–91.
165. Coto E, Menendez M, Navarro R, Garcia-Castro M, Alvarez V. A new *de novo* Notch3 mutation causing CADASIL. *Eur J Neurol* 2006;**13**:628–31.
166. Joutel A, Andreux F, Gaulis S, Domenga V, Cecillon M, Battail N, et al. The ectodomain of the Notch3 receptor accumulates within the cerebrovasculature of CADASIL patients. *J Clin Invest* 2000;**105**:597–605.
167. Joutel A, Monet-Lepretre M, Gosele C, Baron-Menguy C, Hammes A, Schmidt S, et al. Cerebrovascular dysfunction and microcirculation rarefaction precede white matter lesions in a mouse genetic model of cerebral ischemic small vessel disease. *J Clin Invest* 2010;**120**:433–45.
168. Artavanis-Tsakonas S, Rand MD, Lake RJ. Notch signaling: cell fate control and signal integration in development. *Science* 1999;**284**:770–6.
169. D'Souza B, Miyamoto A, Weinmaster G. The many facets of Notch ligands. *Oncogene* 2008;**27**:5148–67.

170. Joutel A, Vahedi K, Corpechot C, Troesch A, Chabriat H, Vayssiere C, et al. Strong clustering and stereotyped nature of Notch3 mutations in CADASIL patients. *Lancet* 1997;**350**:1511–5.
171. Peters N, Opherk C, Bergmann T, Castro M, Herzog J, Dichgans M. Spectrum of mutations in biopsy-proven CADASIL: implications for diagnostic strategies. *Arch Neurol* 2005;**62**:1091–4.
172. Dichgans M, Ludwig H, Muller-Hocker J, Messerschmidt A, Gasser T. Small in-frame deletions and missense mutations in CADASIL: 3D models predict misfolding of Notch3 EGF-like repeat domains. *Eur J Hum Genet* 2000;**8**:280–5.
173. Donahue CP, Kosik KS. Distribution pattern of Notch3 mutations suggests a gain-of-function mechanism for CADASIL. *Genomics* 2004;**83**:59–65.
174. Opherk C, Duering M, Peters N, Karpinska A, Rosner S, Schneider E, et al. CADASIL mutations enhance spontaneous multimerization of NOTCH3. *Hum Mol Genet* 2009;**18**:2761–7.
175. Joutel A, Monet M, Domenga V, Riant F, Tournier-Lasserve E. Pathogenic mutations associated with cerebral autosomal dominant arteriopathy with subcortical infarcts and leukoencephalopathy differently affect Jagged1 binding and Notch3 activity via the RBP/JK signaling Pathway. *Am J Hum Genet* 2004;**74**:338–47.
176. Peters N, Opherk C, Zacherle S, Capell A, Gempel P, Dichgans M. CADASIL-associated Notch3 mutations have differential effects both on ligand binding and ligand-induced Notch3 receptor signaling through RBP-Jk. *Exp Cell Res* 2004;**299**:454–64.
177. Dichgans M, Filippi M, Bruning R, Iannucci G, Berchtenbreiter C, Minicucci L, et al. Quantitative MRI in CADASIL: correlation with disability and cognitive performance. *Neurology* 1999;**52**:1361–7.
178. Dichgans M, Mayer M, Uttner I, Bruning R, Muller-Hocker J, Rungger G, et al. The phenotypic spectrum of CADASIL: clinical findings in 102 cases. *Ann Neurol* 1998;**44**:731–9.
179. Arboleda-Velasquez JF, Lopera F, Lopez E, Frosch MP, Sepulveda-Falla D, Gutierrez JE, et al. C455R notch3 mutation in a Colombian CADASIL kindred with early onset of stroke. *Neurology* 2002;**59**:277–9.
180. Monet-Lepretre M, Bardot B, Lemaire B, Domenga V, Godin O, Dichgans M, et al. Distinct phenotypic and functional features of CADASIL mutations in the Notch3 ligand binding domain. *Brain* 2009;**132**:1601–12.
181. Joutel A, Chabriat H, Vahedi K, Domenga V, Vayssiere C, Ruchoux MM, et al. Splice site mutation causing a seven amino acid Notch3 in-frame deletion in CADASIL. *Neurology* 2000;**54**:1874–5.
182. Opherk C, Peters N, Herzog J, Luedtke R, Dichgans M. Long-term prognosis and causes of death in CADASIL: a retrospective study in 411 patients. *Brain* 2004;**127**:2533–9.
183. Arboleda-Velasquez JF, Zhou Z, Shin HK, Louvi A, Kim HH, Savitz SI, et al. Linking Notch signaling to ischemic stroke. *Proc Natl Acad Sci USA* 2008;**105**:4856–61.
184. Belin de Chantemele EJ, Retailleau K, Pinaud F, Vessieres E, Bocquet A, Guihot AL, et al. Notch3 is a major regulator of vascular tone in cerebral and tail resistance arteries. *Arterioscler Thromb Vasc Biol* 2008;**28**:2216–24.
185. Kitamoto T, Takahashi K, Takimoto H, Tomizuka K, Hayasaka M, Tabira T, et al. Functional redundancy of the Notch gene family during mouse embryogenesis: analysis of Notch gene expression in Notch3-deficient mice. *Biochem Biophys Res Commun* 2005;**331**:1154–62.
186. Krebs LT, Xue Y, Norton CR, Sundberg JP, Beatus P, Lendahl U, et al. Characterization of Notch3-deficient mice: normal embryonic development and absence of genetic interactions with a Notch1 mutation. *Genesis* 2003;**37**:139–43.
187. Dubroca C, Lacombe P, Domenga V, Maciazek J, Levy B, Tournier-Lasserve E, et al. Impaired vascular mechanotransduction in a transgenic mouse model of CADASIL arteriopathy. *Stroke* 2005;**36**:113–7.

188. Lacombe P, Oligo C, Domenga V, Tournier-Lasserve E, Joutel A. Impaired cerebral vasoreactivity in a transgenic mouse model of cerebral autosomal dominant arteriopathy with subcortical infarcts and leukoencephalopathy arteriopathy. *Stroke* 2005;**36**:1053–8.
189. Monet M, Domenga V, Lemaire B, Souilhol C, Langa F, Babinet C, et al. The archetypal R90C CADASIL-NOTCH3 mutation retains NOTCH3 function *in vivo*. *Hum Mol Genet* 2007;**16**:982–92.
190. Ruchoux MM, Domenga V, Brulin P, Maciazek J, Limol S, Tournier-Lasserve E, et al. Transgenic mice expressing mutant Notch3 develop vascular alterations characteristic of cerebral autosomal dominant arteriopathy with subcortical infarcts and leukoencephalopathy. *Am J Pathol* 2003;**162**:329–42.
191. Lundkvist J, Zhu S, Hansson EM, Schweinhardt P, Miao Q, Beatus P, et al. Mice carrying a R142C Notch 3 knock-in mutation do not develop a CADASIL-like phenotype. *Genesis* 2005;**41**:13–22.
192. Arboleda-Velasquez JF, Manent J, Lee JH, Tikka S, Ospina C, Vanderburg CR, et al. PNAS Plus: hypomorphic Notch 3 alleles link Notch signaling to ischemic cerebral small-vessel disease. *Proc Natl Acad Sci USA* 2011;**108**:E128–35.
193. Ayata C. CADASIL: experimental insights from animal models. *Stroke* 2010;**41**:S129–34.
194. Utku U, Celik Y, Uyguner O, Yuksel-Apak M, Wollnik B. CADASIL syndrome in a large Turkish kindred caused by the R90C mutation in the Notch3 receptor. *Eur J Neurol* 2002;**9**:23–8.
195. Karlstrom H, Beatus P, Dannaeus K, Chapman G, Lendahl U, Lundkvist J. A CADASIL-mutated Notch 3 receptor exhibits impaired intracellular trafficking and maturation but normal ligand-induced signaling. *Proc Natl Acad Sci USA* 2002;**99**:17119–24.
196. Eikermann-Haerter K, Yuzawa I, Dilekoz E, Joutel A, Moskowitz MA, Ayata C. Cerebral autosomal dominant arteriopathy with subcortical infarcts and leukoencephalopathy syndrome mutations increase susceptibility to spreading depression. *Ann Neurol* 2011;**69**:413–8.
197. Fouillade C, Chabriat H, Riant F, Mine M, Arnoud M, Magy L, et al. Activating NOTCH3 mutation in a patient with small-vessel-disease of the brain. *Hum Mutat* 2008;**29**:452.

Experimental Models of Seizures and Epilepsies

Yogendra H. Raol* and
Amy R. Brooks-Kayal*,†

*Division of Neurology, Department of Pediatrics, School of Medicine, University of Colorado Denver, Aurora, Colorado, USA

†Children's Hospital Colorado, Department of Neurology, Aurora, Colorado, USA

Epilepsy is one of the most common neurological conditions that affect people of all ages. Epilepsy is characterized by occurrence of spontaneous recurrent seizures. Currently available drugs are ineffective in controlling seizures in approximately one-third of patients with epilepsy. Moreover, these drugs are associated with adverse effects, and none of them are effective in preventing development of epilepsy following an insult or injury. To develop an effective therapeutic strategy that can interfere with the process of development of epilepsy (epileptogenesis), it is crucial to study the changes that occur in the brain after an injury and before epilepsy develops. It is not possible to determine these changes in human tissue for obvious ethical reasons. Over the years, experimental models of epilepsies have contributed immensely in improving our understanding of mechanism of epileptogenesis as well as of seizure generation. There are many models that replicate at least some of the characteristics of human epilepsy. Each model has its advantages and disadvantages, and the investigator should be aware of this before selecting a specific model for his/her studies. Availability of a good animal

Progress in Molecular Biology
and Translational Science, Vol. 105
DOI: 10.1016/B978-0-12-394596-9.00003-2

1877-1173/12 $35.00

model is a key to the development of an effective treatment. Unfortunately, there are many epilepsy syndromes, specifically pediatric, which still lack a valid animal model. It is vital that more research is done to develop animal models for such syndromes.

I. Introduction

Epilepsy is a brain disorder characterized by occurrence of more than one epileptic seizure with a continuing predisposition to generate more epileptic seizures associated with neurobiologic, cognitive, psychological, and social disturbances.[1] The epileptic seizure is an event consisting of a sudden and transient occurrence of signs and/or symptoms due to abnormal excessive or synchronous neuronal activity in the brain.[1] An epileptic seizure can be as subtle as a momentary lapse of attention to very obvious involving violent and involuntary contractions of muscles. Epilepsy is one of the most common neurological disorders and affects approximately 50 million people worldwide. It is estimated that 1 in 10 people will experience an epileptic seizure during their lifetime. Almost 500 new cases are diagnosed every day and approximately 50,000 people die due to seizure-related causes every year in the USA alone. It is an economic burden on individuals and society because of increased health care cost, loss of work, and wages. The annual cost of epilepsy has been estimated at $12.5 billion.[2] In spite of about 20 medications available for the treatment of epileptic seizures, approximately 30% patients remain resistant to treatment by drugs. Further, all of the available anticonvulsant drugs have multiple adverse effects. There is also a lack of treatment that can prevent epileptogenesis, the process by which a normal brain develops epilepsy following an injury or an insult.

Since the discovery of the anticonvulsant activity of phenytoin in an animal model of epilepsy in 1937 by Tracy Putnam and H. Houston Merritt,[3] animal models have played a crucial role in our understanding of the basic mechanisms of epilepsy and in development of novel antiepileptic treatments. Epilepsy is not a single disorder but consists of multiple heterogeneous syndromes with many etiologies. Therefore, there exist more than 100 *in vitro* and *in vivo* models that mimic different types of epileptic seizures or syndrome and/or specific aspects of the disease. As it is not possible to explain all the available models in a single chapter, only the animal models that are most commonly used for studying a particular type of epileptic seizure or epilepsy syndrome will be described. Toward the end of the chapter, we introduce some of the newer and emerging models.

II. Classification of Epileptic Seizures

According to the classification developed by the International League Against Epilepsy, epileptic seizures are divided into three categories: focal onset seizures, generalized onset seizures, and seizures of unknown origin.[4] Focal onset epileptic seizures originate from a small area that is limited to one hemisphere, whereas generalized onset seizures originate simultaneously from both hemispheres. Epileptic seizures that do not fit into either of the category are termed as of unknown origin.[4]

A. Focal Onset Seizures

Manifestation of focal onset seizure symptoms is determined by the area from which the seizure originates and may consist of sensory, autonomic, motor, or psychic symptoms. Focal onset seizures can be broadly classified into the following three subcategories.

Neocortical seizures: Neocortical seizures manifest differently, depending on the exact location of its origin and pattern of its spread. For example, a seizure that originates from primary sensory visual cortex may involve visual hallucinations such as flickering lights.

Temporal lobe seizures: The majority of temporal lobe seizures originate in the mesial structures, mainly from the hippocampus. Typical mesial temporal lobe seizures begin with aura of rising epigastric sensation followed by automatisms, dystonia of hand contralateral to the side of seizure origin and postictal (ictal is period during seizure, postictal is after seizure) confusion. Automatisms frequently consist of chewing, lip smacking, and fumbling with the hands.

Multifactorial: Seizures can simultaneously originate from two independent foci such as neocortex and mesial temporal lobe.[5] For example, in patients suffering from posttraumatic epilepsy, it has been shown that seizures can originate from mesial temporal lobe as well as neocortex.

B. Generalized Onset Seizures

Based on the clinical seizure manifestation, generalized seizures can be broadly classified into the following six categories.

Absence seizures: Absence seizures were previously referred to as *petit mal* seizures. Absence seizures occur most commonly in children and involve loss of consciousness, eye blinking, staring, and other facial movements. The seizures last for a few seconds to a minute but can occur multiple times in a day. The main electroencephalographic (EEG) characteristic of absence seizure is 3-Hz generalized spike-and-slow-wave complexes.

Myoclonic seizure types: Myoclonic seizures consist of sudden, brief, arrhythmic, quick muscle jerks. These may involve only one or both sides of the body. Consciousness is not usually impaired. Myoclonic seizures are typically associated with generalized spike-and-wave discharges on the EEG.

Clonic seizures: Clonic seizures are very similar to myoclonic seizures. However, unlike myoclonic seizures, the clonic seizure consists of rhythmic muscle jerks and the consciousness is usually impaired during seizures. They occur most commonly in neonates and young children. During clonic seizures, EEG usually shows fast activity (≥ 10 Hz) and slow waves and occasional spike-wave patterns.

Tonic seizures: During tonic seizures, there is a sudden increase in muscle tone resulting in tonic seizure consists of tonic extension or flexion of head, trunk, and/or extremities for several seconds. The EEG shows bilateral synchronous medium to high-voltage fast activity (10–25 Hz). Tonic seizures are also most common in childhood.

Tonic–clonic seizures: Generalized tonic–clonic seizures were previously referred to as *grand-mal* seizures and consist of tonic extension of muscles followed by clonic rhythmic movements and postictal confusion. The EEG shows generalized rhythmic discharges decreasing in frequency and increasing in amplitude during the tonic phase and slow waves during the clonic phase. The seizure may last for 1–2 min.

Atonic seizures: These seizures are manifested as brief loss of postural tone, often resulting in falls and injuries. The EEG usually shows rhythmic 1–2 Hz spike-and-wave complexes or more rapid, irregular multifocal spike-and-wave activity.

C. Seizures of Unknown Origin

Seizures whose origin is uncertain and less clear are termed as seizures of unknown origin.

Epileptic spasms: A spasm is a sudden extension and/or flexion of mainly proximal and truncal muscles.[6] Epileptic spasms last longer than myoclonic seizures, but are shorter than tonic seizures.[6] In infants, EEG typically shows disorganized high-amplitude waves with irregular spikes in the background and attenuation (electrodecrement) during the spasms.

III. Classification of Epileptic Syndromes

An epileptic syndrome is a clinical condition that is defined by a group of signs and symptoms such as type of epileptic seizure, age of onset of epileptic seizure, EEG pattern, and family history. For example, benign familial neonatal epilepsy (BFNE) is an idiopathic generalized epilepsy where the seizures appear during the first few days of life and stop spontaneously by 2–3 months of life. The clinical seizures are generally brief clonic movements lasting for

1–2 min. BFNE is caused by mutations in genes that encode for potassium channel subunits KCNQ2 and KCNQ3.[7] Mutation in KCNQ2/3 subunits results in dysfunction of M-channels (a type of potassium channel) that plays an important role in preventing hyperexcitability in the brain, specifically in early-life when γ-amino-butyric acid (GABA)-mediated inhibition is not fully developed.[8–10] Most of the neonates with BFNE go on to develop well, however, about 10–15% of the neonates with BFNE develop epilepsy (spontaneous seizures) in latter life.[11] A recently modified classification of epileptic syndromes according to the age at onset of syndrome is given in Table I.[4]

IV. Models for Seizures and Epilepsies

There are many models, *in vitro* and *in vivo*, that replicate some of the characteristics of the epilepsy and/or seizures in human. A description of the most commonly used models and the characteristics of the human condition it best replicates are given in Table II. Some of the more popular models from Table II are described in more detail in the text of the chapter.

A. The Perfect Animal Model

A perfect model for human seizures or epilepsy should satisfy the following criteria[26,27]:

Age of onset: It is now well understood that significant differences exist, in terms of anatomy, physiology, and neurochemistry, between an immature and an adult brain. These differences affect how a brain responds to an injury[28] and a treatment.[29,30] The combination of increased excitation and underdeveloped inhibition in developing brain results in increased neuronal activity that is important for synaptogenesis and brain development, but it also makes immature brain susceptible to hyperexcitability (for review, see Refs. 31,32). The immature brain has fewer $GABA_A$ receptors (a subtype of GABA receptor),[33–35] smaller GABA-mediated currents,[36] is less sensitive to benzodiazepine augmentation,[37,38] and due to developmental differences in the chloride gradient GABA is excitatory in the immature brain.[39] Clinical and animal studies suggest that the developing brain is more susceptible to seizures than the mature brain. The incidence of seizures is highest during the first year of life,[40] and the amount of chemoconvulsant required to induce seizures in immature animals is much lower than that required for induction of seizures in adult animals.[41,42] Similarly, phenobarbital, an antiepileptic drug (AED) that mediates its action by augmenting $GABA_A$ receptor activity, also shows age-dependent efficacy in treating seizures. Phenobarbital is very effective in stopping seizures in adult patients but has been poorly effective in treating neonatal seizures.[29,30] Moreover, treatment of developing brain with AEDs like phenobarbital may also be associated

TABLE I
EPILEPSY SYNDROMES

Neonatal period
Benign familial neonatal epilepsy (BFNE)
Early myoclonic encephalopathy (EME)
Ohtahara syndrome
Infancy
Epilepsy of infancy with migrating focal seizures
West syndrome
Myoclonic epilepsy in infancy (MEI)
Benign infantile epilepsy
Benign familial infantile epilepsy
Dravet syndrome
Myoclonic encephalopathy in nonprogressive disorders
Childhood
Febrile seizures plus (FS^{+}) (can start in infancy)
Panayiotopoulos syndrome
Epilepsy with myoclonic atonic (previously astatic) seizures
Benign epilepsy with centrotemporal spikes (BECTS)
Autosomal-dominant nocturnal frontal lobe epilepsy (ADNFLE)
Late onset childhood occipital epilepsy (Gastaut type)
Epilepsy with myoclonic absences
Lennox–Gastaut syndrome
Epileptic encephalopathy with continuous spike-and-wave during sleep (CSWS)
Landau–Kleffner syndrome (LKS)
Childhood absence epilepsy (CAE)
Adolescence–Adult
Juvenile absence epilepsy (JAE)
Juvenile myoclonic epilepsy (JME)
Epilepsy with generalized tonic–clonic seizures alone
Progressive myoclonus epilepsies (PME)
Autosomal-dominant epilepsy with auditory features (ADEAF)
Other familial temporal lobe epilepsies
Less specific age relationship
Familial focal epilepsy with variable foci (childhood to adult)
Reflex epilepsies
Distinctive constellations
Mesial temporal lobe epilepsy with hippocampal sclerosis (MTLE with HS)
Rasmussen syndrome
Gelastic seizures with hypothalamic hamartoma
Hemiconvulsion–hemiplegia–epilepsy
Epilepsies that do not fit into any of these diagnostic categories can be distinguished first on the basis of the presence or absence of a known structural or metabolic condition (presumed cause) and then on the basis of the primary mode of seizure onset (generalized vs. focal)
Epilepsies attributed to and organized by structural-metabolic causes
Malformations of cortical development (hemimegalencephaly, heterotopias, etc.)
Neurocutaneous syndromes (tuberous sclerosis complex, Sturge–Weber, etc.)

(*Continues*)

TABLE I (*Continued*)

Tumor
Infection
Trauma
Angioma
Perinatal insults
Stroke, etc.
Epilepsies of unknown cause
Conditions with epileptic seizures that are traditionally not diagnosed as a form of epilepsy per se
Benign neonatal seizures (BNS)
Febrile seizures (FS)

Printed with permission from Ref. 4.

with side effects such as reduced IQ.[43] Therefore, to identify the most effective therapeutic intervention strategy, it is imperative to target age-specific mechanisms and test new therapies in age-specific disease models.

Etiology: The underlying cause of the seizure/epilepsy, which can be genetic or acquired after infection, injury, prolonged seizure with fever, etc., should be similar to that of human condition that the model is replicating. Many of the epilepsy syndromes can result from multiple etiologies and ideally a model should incorporate such etiologies.

Seizure phenotype and EEG characteristics: A good model will reproduce seizure phenotype of human condition. The seizure phenotype may depend on the area involved in generation of seizure. For example, seizures that involve temporal lobe may begin with an aura or feeling of epigastric rising, followed by motor seizures such as staring and facial automatisms. The other important component that defines a seizure or an epilepsy syndrome is EEG abnormalities. For example, absence seizures are defined by a 3-Hz spike-wave pattern on the EEG, and therefore, a good animal model of absence seizures would show similar EEG characteristics.

Long-term consequences: An insult or an injury such as hypoxia–ischemia (HI) not only causes acute seizures, but it can also lead to other long-term deleterious consequences. Approximately two-thirds of human infants that experience HI in early-life develop epilepsy in later-life, and the majority of adult patients that had even moderate neonatal HI suffer from cognitive impairment.[44,45] An animal model of epilepsy that replicates both acute and long-term effects of injury, for example, can be useful in identifying a drug that is an anticonvulsant (stops acute seizures) as well as antiepileptogenic (prevents development of later spontaneous seizures, i.e., epilepsy).

Currently, majority of the animal models of epilepsy replicate only some of the above-mentioned criteria. Selection of a model, therefore, should depend on the specific question an investigator wishes to address.

TABLE II

Examples of Commonly used Models of Seizures and Epilepsies

Animal model	Human condition	Common use	Reference (review)
Pilocarpine	Focal (temporal lobe)	Mechanism of epileptogenesis, cognitive deficits and epilepsy, therapy development	12
Kainic acid	Focal (temporal lobe)	Mechanism of epileptogenesis, cognitive deficits and epilepsy, therapy development	13
Pentylenetetrazol	Generalized seizures	Seizure mechanism, drug screening	14
Flurothyl	Multiple acute seizures, childhood epilepsies	Seizure-induced cognitive deficits	15
Focal application of penicillin, picrotoxin	Focal (neocortical) seizures	Mechanism of seizure generation and spread	16
Tetanus toxin	Focal seizures	Mechanism of seizure generation and spread, mechanism of epileptogenesis	16
Tetrodotoxin	Infantile spasms	Mechanism of long-term consequences of seizures, therapy development	17
Doxorubicin + lipopolysaccharides + *p*-chlorophenylamine	Infantile spasms	Mechanism of long-term consequences of seizures, therapies development	18
Electrical and chemical kindling	Focal (temporal lobe) epilepsy	Mechanism of epileptogenesis, anticonvulsant and antiepileptogenic drug screening, cognitive deficits	19
Maximal electroshock (MES)	Generalized seizures	Drug screening	20
Trauma	Focal epilepsy	Mechanism of epileptogenesis, cognitive deficits and cell death, therapy development	21
Hypoxia/ischemia	Hypoxic–ischemic encephalopathy	Mechanism of epileptogenesis, cognitive deficits and cell death, therapy development	22
Temperature	Febrile seizures	Mechanism of long-term consequences of seizures, therapy development	23
Genetic models	Mutations-linked epilepsies	Mechanism of seizures and epilepsy	24
In vitro models	Seizures	Mechanism of seizures, preliminary drug screening	25

B. Chemical Models

Pilocarpine model: The pilocarpine model of epilepsy in rodents is a commonly used model to study temporal lobe epilepsy (TLE). TLE is the most common type of epilepsy in humans and is characterized by an initial event of an injury (early-life febrile seizures, trauma), a latent period followed by appearance of spontaneous recurrent seizures (epilepsy) and mesial temporal lobe sclerosis.[46] Pilocarpine is an acetylcholine receptor agonist that activates muscarinic receptors.[47] It has been suggested that M1 muscarinic receptors are involved in the initiation of the seizures, whereas *N*-methyl-D-aspartic acid (NMDA) glutamate receptors are involved in the maintenance of seizures. The typical protocol involves intraperitoneal (i.p.) administration of 1 mg/kg of scopolamine methylnitrate 30 min before i.p. injection of 350–385 mg/kg pilocarpine. The scopolamine methylnitrate is given to reduce the peripheral effects of cholinergic action. Approximately 45 min after pilocarpine injection, the rats develop electrographic and behavioral seizures that can last for several hours. The duration of status epilepticus can be controlled by benzodiazepine or phenobarbital treatment. After 1–2 weeks of latent period, the majority of the adult rats treated with pilocarpine develop spontaneous recurrent seizures[48–50] (Fig. 1). The latency to develop spontaneous recurrent seizures and the percent of animals developing spontaneous seizures following pilocarpine injection depends on the age of the animal at the treatment. For example, only 25% of the rats that experience pilocarpine-induced prolonged seizures at postnatal-day 12 (P12) developed spontaneous seizures,[51] whereas 67% of rats treated with pilocarpine at P20 develop spontaneous seizures.[52] Adult rats treated with pilocarpine display hippocampal pathology similar to mesial temporal sclerosis found in many TLE patients. Significant cell loss in various regions of hippocampus and mossy fiber sprouting in the inner molecular layer of the dentate gyrus has been observed in the rats following pilocarpine-induced status epilepticus (Fig. 1).

Other variations in the pilocarpine treatment protocol and the route of pilocarpine administration have also been used. Some laboratories administer pilocarpine directly into the hippocampus or into the cerebral ventricle.[53] Treatment with lithium (3 mEq/kg, i.p.) 14–24 h prior to pilocarpine treatment is also a commonly used protocol for the induction of status epilepticus.[51,52,54] Pretreatment with lithium reduces amount of pilocarpine (30–60 mg/kg, i.p.) required to induce prolong seizures. Instead of a single high dose of pilocarpine, repeated low doses of pilocarpine can also be given to rats or mice to induce status epilepticus.[55]

Kainic acid model: Kainic acid, which was first isolated from seaweed *Digenea Simplex*, is a glutamate analogue and an agonist for α-amino-3-hydroxy-5-methyl-4-isoxazole propionic acid (AMPA) and kainate glutamate receptors. Kainic acid can be administered systemically, intracerebroventricularly or directly into a

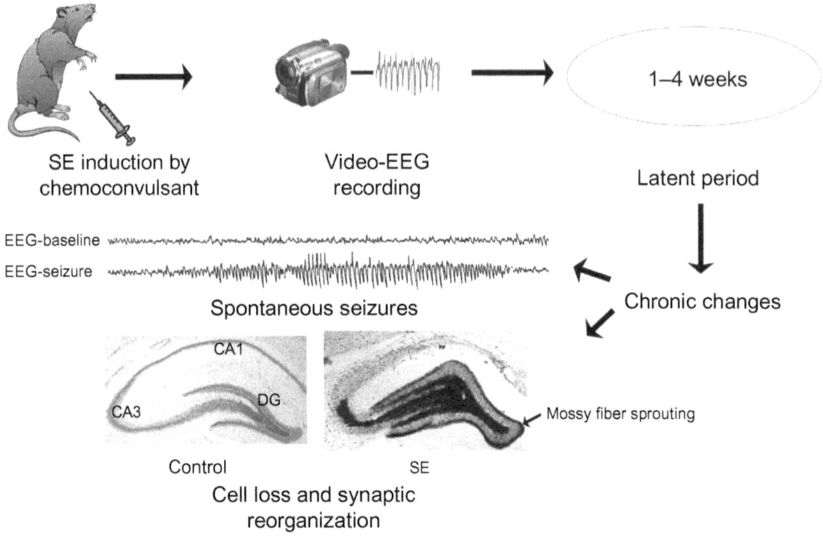

FIG. 1. Schematic diagram depicting the typical protocol to study TLE using chemoconvulsant-induced status epilepticus (SE) model. SE is induced in a rat or mouse by a chemoconvulsant such as kainic acid or pilocarpine. The animal is implanted with intracranial electrodes before or after SE induction for video-EEG monitoring to determine occurrence of spontaneous recurrent seizures (epilepsy) following SE. Approximately 1–4 weeks after kainic acid or pilocarpine-induced SE, animals demonstrate spontaneous seizures (percentage of animals developing spontaneous seizures depends on chemoconvulsant used, severity of SE, and age at SE induction). The "latent period" between SE and onset of spontaneous seizures has been actively studied by epilepsy researchers in order to identify changes that occur after SE, which may play a crucial role in epilepsy development. Cell loss in multiple brain regions including the hippocampus (as seen in CA1 and CA3 areas of hippocampal sections obtained from animal that had experienced SE) and synaptic reorganization of mossy fibers in the dentate gyrus (DG) of the hippocampus, similar to what has been observed in patients with TLE, occurs in many animals that have experienced SE.

specific brain region to induce status epilepticus. A single high-dose (10–18 mg/kg) kainic acid injection induces stereotypical behavioral seizures that start with staring spells, followed by head nodding and wet-dog shakes. Approximately 1 h after kainic acid injection, rats demonstrate recurrent limbic motor seizures such as mastication, facial clonus, forelimb tremor, rearing, and falling before entering into status epilepticus that can last for hours.[56] Instead of a single high dose, many laboratories inject multiple smaller doses of kainic acid (5 mg/kg) until an animal develops status epilepticus.[57] This approach significantly reduces animal mortality during status.

Similar to the pilocarpine model, the kainic acid model also mimics many important characteristics of human TLE. After a latent period of approximately 1 week, rats treated with repetitive low dose of kainic acid develop spontaneous recurrent seizures.[58] The spontaneous recurrent seizure phenotype range from nonconvulsive electrographic seizure to very obvious convulsive seizures characterized by forelimb clonus, rearing, and loss of righting reflex.[58] Kainic acid-treated animals also demonstrate hippocampal sclerosis and mossy fiber sprouting in dentate gyrus region of hippocampus.[59,60] However, one of the criticism of the chemoconvulsant models of TLE (kainic acid and pilocarpine) is that the cell death observed in these models is much more severe and extensive than that typically observed in human TLE patients.

Flurothyl model: Flurothyl (2,2,2-trifluoroethyl ether) is a volatile liquid and its vapor is a potent convulsant. Flurothyl is a noncompetitive antagonist at $GABA_A$ receptors[61] and is most commonly used as a model for recurrent neonatal generalized tonic–clonic seizures.[62] However, flurothyl can also be used for seizure induction in adult rats as well as mice. To induce seizures in rat pups, an animal is placed in an airtight plastic cylindrical chamber within a fumehood. Using a syringe pump, flurothyl is then injected into the chamber at the rate of 50 μl/min. After an average of 2.22 min of flurothyl exposure P10 rat pups develop forelimb and hindlimb tonic extension.[30] Seizures stop almost immediately when the pup is removed from the chamber and is returned to room air. Immature rats that experience recurrent flurothyl-induced seizures do not have cell loss[63] and do not develop spontaneous seizures in later-life[64]; however, they have higher susceptibility to develop seizures[64] and have memory impairment.[62]

C. Electrical Stimulation Models

Kindling: Kindling is a process wherein low-intensity electrical stimulation (or a low-dose chemical convulsant such as pentylenetetrazol) is repeatedly delivered to a brain region until an epileptic seizure is triggered. To electrically kindle an animal, first a stimulating electrode is implanted stereotaxically in the region of interest. The amygdala, hippocampus, perirhinal cortex, and piriform cortex are some of the most commonly kindled areas. The rate of kindling varies between the brain regions and among different rat strains.[65,66] Similarly, the rate of development of kindling and the behavioral manifestations of seizures varies according to the age of animal.[67] Typically, rats are stimulated with a 1-s train of 60 Hz sine or biphasic square-wave pulses.[68] A stimulus intensity that evokes an after-discharge of predefined characteristics (e.g., discharge that lasts for 8 s or more and the EEG amplitude that is three times the prestimulation baseline), usually referred as after-discharge threshold, is first established. The animal is then stimulated once a day, everyday at the after-discharge threshold until it develops stage 5 seizures. The development of behavioral seizures by kindling was classified into following five stages by Ronald J. Racine[65]: (1) mouth and facial clonus, (2) head nodding, (3) forelimb clonus, (4) rearing, and (5) rearing and falling. Kindling-induced seizures, which originate from the site of stimulation and then spread to the other areas with repeated stimulation, replicate the clinical phenotype of partial seizures with secondary generalization.

Kindling is one of the commonly used models to study mechanisms of TLE. The major advantage of the kindling is that a specific brain area can be stimulated in a controlled fashion to study its role in seizure generation and epileptogenesis. The kindling model is also used for preclinical investigation of efficacy of AEDs. To test the anticonvulsant effect, the drug is given to fully kindled rats and its efficacy to block the seizures is measured. To find out if the drug can stop the process of development of epilepsy (antiepileptogenic property), the drug is given during the kindling process (before each stimulation) and its efficacy to suppress the kindling is determined.[69]

The kindling model also has certain limitations and disadvantages as compared to other TLE models such as chemoconvulsant-induced status epilepticus. The process of kindling is laborious and time consuming, requiring precise electrode implantation and electrical stimulation over many days. However, the major weakness of the model is that a large number of stimulations and over 90 stage 5 seizures are required before 40% of stimulated rats develop spontaneous seizures.[70] Similarly, visually apparent cell death in hippocampus is observed only after 30 stage 5 seizures,[71] and the extent of hippocampal damage is not similar to that observed in TLE patients.

Self-sustaining status epilepticus (SSSE): Repetitive electrical stimulation of the limbic structures or of hippocampal circuitry such as perforant pathway can induce status epilepticus in rodents that continues even after stimulation is withdrawn, that is, the seizures become self-sustaining.[72] A common protocol to induce SSSE in rats involves stimulation of the area of interest of a freely moving rat every 0.5 s for 20–30 min with 100-ms trains of 1-ms biphasic square-wave pulses of 400 μA and 60 Hz.[73] SSSE is used as a model for human mesial temporal lobe epilepsy (MTLE). SSSE induces neuronal loss and synaptic reorganization in hippocampus.[74] After a latent period of approximately 3 weeks, the majority of rats develop spontaneous seizures.[74] The disadvantage of the model, compared to chemoconvulsant models of status epilepticus, is that the electrode implantation is cumbersome and labor intensive.

Maximal Electroshock Seizures (MES): MES are a model of generalized tonic–clonic seizures. MES are induced by delivering electroshock through corneal electrodes (most commonly used), transcranial electrodes, or ear clips. For the corneal stimulation, the eyes are anesthetized by local application of 0.5% tetracaine before the electrodes are inserted on to the cornea. The eyes are then moistened with a drop of physiologic saline and 0.2-s series of alternating current with 60-Hz frequency and intensity of 50 mA for mice and 150 mA for rats is applied to the cornea through the electrode.[75] The stimulation induces seizures consisting of tonic extension of forelimbs and hindlimbs.

Because of its ease of use, MES testing is routinely used for screening of potential AEDs. The anticonvulsant activity of phenytoin was identified using the MES model.[3] The duration of tonic maximal extension of the hind leg is usually measured to determine the efficacy of the drug.

D. Genetic Models

Mutation in many genes can cause epilepsy. Dysfunction of ion channels, receptors, enzymes, and transporters caused by mutations in their genes has been reported to cause epilepsy (for review, see Ref. 76). Mutation in KCNQ2/3 subunits of voltage-activated potassium channel causes BFNE, a type of neonatal epilepsy syndrome,[7] (for review, see Ref. 77). Newborns with BFNE experience seizures that remit spontaneously in majority of cases; however, 10–15% of the neonates go on to develop epilepsy in later-life.[11] KCNQ2 and KCNQ3 homozygous mutant mice develop spontaneous seizures latter in life[9,78] without significant neuronal death and mossy fiber sprouting in the hippocampus.[78] Mutation in SCN1A, SCN2A, or SCN1B subunit of sodium channels causes generalized epilepsy with febrile seizures plus, a type of childhood epilepsy. Mutation in SCN1A has been also identified in the patients with severe myoclonic epilepsy of infancy (SMEI or Dravet syndrome) (for review, see Ref. 79). Dravet syndrome is characterized by tonic–clonic seizures in infancy followed by myoclonic, tonic–clonic, absence, and simple partial seizures in later-life. Interestingly, SMEI has been reported to arise from loss-of-function as well as gain-of-function of sodium channels. Several mouse models of SMEI are available that recapitulate many characteristics of Dravet syndrome including seizures during infancy and epilepsy in later-life (for review, see Ref. 79). Variations in the seizure severity, clinical presentation of the seizures, and long-term effects of the epilepsy syndrome are observed in the individuals with same mutation. Genetic or environmental modifiers are hypothesized to be the cause of these variations. Double mutant mice carrying mutated scn1a/scn2a and kcnq2 gene experience more severe seizures with early onset than the mice carrying only mutated sodium channel subunit.[80,81] This suggests that genetic background greatly influences the phenotypic outcome of the mutation. Mutations in the proteins that are involved in the migration of neurons during development are also associated with epilepsy in both mice and humans. For example, mutation in aristaless-related homebox (ARX) gene is linked to developmental epilepsy. ARX mutant mice exhibit behavioral and electrographic early-life seizures resembling the infantile spasms (West syndrome) seen in human patients with the mutation.[82,83]

In many mouse strains, genetic mutations that lead to epilepsy occur spontaneously. These models have contributed immensely to our basic understanding of the pathophysiology of epilepsy. For example, *tottering*, *lethargic*, *ducky*, and *stargazer* mice all show absence-type epilepsy and have mutations in calcium ion channels (for review, see Ref. 84). In tottering mice, mutation in a voltage-gated calcium channel leads to reduced calcium currents[85] in neurons that affect the release of neurotransmitters. In fact, in tottering mice, a reduced presynaptic release of glutamate has been reported.[86] It has been suggested that reduced glutamate release in thalamus may increase GABA-mediated inhibition leading to synchronized firing of thalamocortical circuit (and hence spike-wave on EEG and absence seizures). Moreover, reduction in calcium currents, and therefore reduced neurotransmitter release, may affect the formation of proper synaptic connectivity and brain development (for review, see Ref. 84). Some rat strains also show hereditary absence-type seizures. The genetic absence epilepsy rats from Strasbourg (GAERS) and WAG/Rij (Wistar Albino Glaxo strain inbred in the UK and then kept in Rijswijk, The Netherlands) strain of rats spontaneously develop absence epilepsy (for review, see Ref. 87). The EEG recorded from the cortex of these rats shows spike-and-wave discharges that behaviorally correlate with immobility (freezing) (for review, see Ref. 87). Drugs that are effective against human absence seizures, such as ethosuximide and valporate, are also effective in blocking seizures in GAERS and WAG/Rij rats.[88,89] Therefore, these rats are very useful for testing efficacy of new drugs to treat absence seizures. Absence epilepsy in humans is a childhood disease that disappears around puberty, whereas in GAERS and WAG/Rij rats, seizures first appear in early adulthood and remain for the rest of life.

E. Developmental Models

Many of the chemical convulsants that are used for inducing seizures in adult animals, as described above, can be used at reduced dose to induce seizures in young developing animals.[30,51,52,54,90] However, we would like to describe in more detail some unique models that closely replicate the clinical situation that leads to epilepsy during development.

Febrile seizure model: Febrile seizures, which occur during fever, affect 2–5% of children between age 6 months and 5 years[91] and is the most common type of seizures in infants and young children. Retrospective studies suggest that most patients with TLE have experienced prolonged febrile seizures in early-life.[92,93] Animal models of febrile seizures involve slowly increasing the core body temperature of mice or rats until they develop convulsions.[94–96] For example, in one of the widely used animal models of febrile seizures, a P10 rat pup is put in a 3-l jar, and using a commercially available hair-dryer, a steady stream of moderately hot air is blown over the rat until seizure onset.[96] At approximately 41 °C of body temperature (measured via rectal probe), almost all of the rat pups developed stereotyped behavioral seizures consisting of facial automatisms, wet-dog shakes, and tonic flexion of the body.[96] The EEG recording taken during experimental febrile seizures suggest that the seizures originate from the amygdala or hippocampus.[96] The seizures cause a transient neuronal injury in limbic structures but do not cause acute or chronic neuronal death.[97,98] However, 35% adults rats that had febrile seizures at P10 develop spontaneous seizures.[99]

HI seizure model: The incidence of seizures is highest in the neonatal period,[40,100] and HI is one of the most common causes of neonatal seizures.[40] Neonatal HI increases the risk of epilepsy development in later-life by five times[101] and causes chronic behavioral and cognitive deficits.[102] The majority of adult patients that had even moderate neonatal HI suffer from cognitive impairment.[45] There are a number of animal models that reproduce important features of human hypoxic–ischemic encephalopathy (HIE) (for review, see Ref. 103). The two most commonly used neonatal rat models are described. In one of the neonatal HI model, P10–P12 rat pups are exposed to graded oxygen concentrations of 7% for 8 min, 6% for 4 min, 5% for 2 min, and 4% for 1 min.[28] This technique reliably induces neonatal seizures that are characterized by myoclonic jerks, automatisms, and tonic–clonic head and limb movements.[28] The behavioral seizures are associated with epileptiform EEG activity that persists for days.[28] The adult rats that were exposed to hypoxia at P10 had hyperexcitable hippocampus,[104] synaptic reorganization in CA3 area of hippocampus,[105] and spontaneous recurrent seizures.[105] However, no acute or chronic cell death has been reported in this model.[105–107]

A longer exposure to hypoxia along with carotid ligation (ischemia) is required to induce brain injury similar to what has been observed in patients with HIE. The commonly used protocol involves ligation of the right carotid artery of P7 rats with absorbable sutures and 2 h later exposure to 8% O_2 for 2 h.[108,109] During the hypoxic exposure, the pup becomes hypocapnic because of hyperventilation with no change in systemic pH.[110] Cerebral blood flow reduces by 40–60% of control values in the hemisphere ipsilateral to the ligation during hypoxia and returns to control values immediately upon return to normoxic condition.[110] A combination of carotid ligation and prolonged hypoxia (more than 90 min) is required to produce infarction, neuropathological damage,[111,112] and development of epilepsy in later-life.[113] This model exhibits many of the salient features of human HI such as extent of brain injury,[109,114,115] development of epilepsy in some but not all animals,[109,113,116] and cognitive deficits.[45,117,118] In contrast to the graded hypoxia model, the development of spontaneous seizures in adult life following HI at P7 correlated with cell death and mossy fiber sprouting in the hippocampus.[109]

Infantile spasm/West syndrome models: West syndrome or infantile spasms most commonly occurs in infancy between the ages of 4 and 9 months. It is a devastating disease that affects 1 person per 2000–6000 live births. West syndrome is characterized by epileptic spasms, a unique interictal EEG pattern called hypsarrhythmia and poor developmental outcome. The seizures consist of sudden and brief flexion or extension of the body and extremities. The EEG during seizure shows sudden voltage attenuation called electrodecremental response. The EEG background is known as hypsarrhythmia and consists of high-amplitude slow waves intermixed with multifocal spikes. There are multiple etiologies, such as HIE, intraventricular hemorrhage, tuberous sclerosis, or mutation in ARX gene, which can be responsible for infantile spasm syndrome.

Several animal models of infantile spasms are available that satisfy at least some of the prominent characteristics of the disease.[82,83,119–125] Most of these models have been developed very recently.[82,83,125] A brief description of some of the older models is provided below. The newer models are described latter in the chapter under the Section III.G. One of the earliest models of infantile spasms involved injection of corticotropin-releasing hormone (CRH) directly into the brain of rat pups.[119] The administration of CRH caused severe seizures. However, the seizures were not spasm like and resembled limbic seizures. The CRH model is based on the observation that infantile spasms respond positively to the adrenocorticotropic hormone (ACTH) treatment and the fact that ACTH suppresses CRH production. Another model of infantile spasms involves intraperitoneal injection of NMDA, a glutamate receptor agonist, to P15 rat pups that were previously exposed to prenatal betamethasone.[123] The acute seizures in this model were similar to human infantile spasms. However, the spasms never evolved into limbic seizures and animals never developed spontaneous seizures following acute insult. Slow infusion of the sodium channel blocker, tetrodotoxin (TTX), into the neocortex or hippocampus caused brief spasm-like seizures.[124] The infusion began at P10–P12 and the seizures first appeared after a minimum of 10 days of infusion. The seizures consisted of flexion or extension of the trunk. The EEG showed high-amplitude slow wave followed by an electrodecrement with superimposed fast activity during the seizures. The criticism of the TTX model is that spasms appear at the older age, that is, lacks age specificity of the infantile spasm syndrome.

F. Trauma Models

Traumatic brain injury (TBI), which annually affects about 1.5 million Americans, is a common cause of acquired epilepsy (for review, see Ref. 126). Incidence of epilepsy following TBI directly correlates with the severity of injury, and the TBI patients with the penetrating head injury are most likely to develop epilepsy[127] (for review, see Ref. 128). Also, risk of epilepsy development is higher in older TBI patients than the younger patients[127] (for review, see Ref. 128). Clinical studies suggest that multiple brain regions, such as hippocampus, thalamus, and basal forebrain, undergo neurodegeneration following TBI (for review, see Ref. 126).

Several methods such as cortical undercut (for review, see Ref. 129), weight drop,[130] controlled cortical impact (CCI) (for review, see Ref. 131), and fluid percussion (for review, see Ref. 126) have been employed to induce TBI in animals. In lateral fluid percussion injury (FPI) model, an injury is induced by forced application of saline over a small exposed surface of the cortex.[132] The force of the impact can be adjusted to acquire the desired severity of injury. Long-term video-EEG monitoring in severely injured adult rats revealed that 50% of the injured rats develop epilepsy.[126] Histological analyses reveal injury to the cortex as well as hippocampus.[126,133] One of the limitations of the FPI model is that mildly to moderately injured rats do not develop epilepsy. Further, the latency to spontaneous seizures in the severely injured rats can vary between 2 weeks and 11 months.[126,133]

CCI is a model for closed head injury and has been modified from the original weight drop model to precisely control extent of injury. A craniotomy is first performed on anesthetized animals to expose the cortex. An air-driven piston, which is controlled by a computer, then compresses the cortex to induce injury. The velocity of the piston and the depth of cortical penetration determine the extent of injury. Cell loss in neocortex and hippocampus, and mossy fiber sprouting in the inner molecular layer of the dentate gyrus has been observed following CCI-induced injury in rats and mice.[134] Similar to what has been observed in the clinical studies regarding correlation between severity of trauma and incidence of posttrauma epilepsy, the percent of mice developing epilepsy following CCI-induced trauma depends on the severity of injury. Only 20% of mice with mild injury developed spontaneous seizures, whereas 36% of mice with severe injury became epileptic.[134] The latency to spontaneous seizures was approximately 60 days.

G. Emerging Epilepsy Models

A good animal model is the most important tool required in a fight against a disease. There are a number of epilepsy syndromes, such as Lennox–Gaustaut syndrome or Sturge–Weber syndrome, which lack an appropriate animal model. In recent years, a number of new promising models of various epilepsy syndromes have emerged. These models will be instrumental in unraveling the basic mechanism of the underlying pathology of the disease. Increased understanding of the disease will be helpful in designing new therapeutic strategies. Moreover, a valid animal model is critical to test new therapies.

Infantile spasm/West syndrome: In recent years, three promising new animal models of infantile spasms have been proposed.[82,83,125] One model, termed as "multiple-hit model," involves intracerebral injection of doxorubicin and lipopolysaccahride at P3 and *p*-chlorophenylalanine at P5 in rats.[125] Doxorubicin was used because it causes diffuse brain injury in forebrain and brainstem.[125] Lipopolysaccahride was used because it activates inflammatory cascades and can cause hypomyelination, white matter rarefaction, and necrosis.[125] *p*-Chlorophenylamine, a serotonin depleting agent, was injected because depletion of serotonin is considered as one mechanism that may lead to infantile spasm syndrome.[125] Flexion spasms with an EEG correlate were observed in the treated animals during P4–P13. Interictally, slow waves discharges and spikes were observed in the treated pups during P7–P13. However, the spasms could not be suppressed by ACTH treatment, currently the first-line therapy for spasms in human infants.

Mutation in the ARX gene has been linked to multiple disorders including infantile spasms, Ohtahara syndrome, X-linked lissencephaly with abnormal genitalia, and intellectual disability (for review, see Ref. 135). ARX is a transcription factor that has been suggested to be involved in nonradial cell migration of interneurons from the subpallial ganglionic eminence into the cortex during development (for review, see Ref. 135). A recent study by Marsh and colleagues shows that ARX knockout mice display some of the prominent characteristics of infantile spasm syndrome such as spasm-like seizures associated with electrodecrement in EEG.[82] ARX knockout mice also develop convulsive spontaneous seizures early in life (P14–P17) resembling partial seizures in humans with the West syndrome.[82] However, the evolution of the spasm-like seizure phenotype in the ARX knockout mice appears at latter age (in adult mice) than what has been observed in humans (during infancy). The authors suggest that the sequence of development of epilepsy phenotype in their model may actually resemble the human condition where the partial seizures often precede or occur concurrently with the onset of infantile spasms.[82] Price and colleagues also observed spasm-like movements beginning at P7 in ARX knock-in mice.[83] The EEG during the spasm-like movements showed slow-wave transient followed by electrodecrement. The mutant mice exhibited spontaneous seizure activity between the ages of 3.5 and 10 weeks that can be classified as limbic seizures. However, hypsarrhythmia in the EEG was not detected in the mutant mice.

Zebrafish as model for seizures: Epilepsy researchers around the world most commonly use rats or mice to study epilepsy. The obvious reasons are small size, faster growth, high breeding rate, and many similarities between rodent and human brain. However, for certain studies, a simpler preparation (cell cultures) or organism (flies, worm, fish) that are easy to manipulate, maintain, and grow is required. In recent years, use of zebrafish to study mechanisms of neurological disorders has increased.[136] Zebrafish (Danio rerio) is a simple vertebrate that is small and easy to maintain in large populations. It produces large clutches of embryos that are transparent (therefore easy to view fluorescent reporters) and undergo rapid development outside utero. It takes zebrafish 3–4 months after fertilization to reach full maturity.[137] To develop a zebrafish seizure model, Baraban and colleagues exposed 7 days postfertilization zebrafish larvae to various concentrations of pentylenetetrazole and studied its effects on zebrafish swimming behavior and electrical activity of the brain.[138] Exposure of zebrafish larvae to pentylenetetrazole resulted in a distinct seizure-like behaviors such as increase in swim activity which was followed by a rapid swimming in circles (near the wall of the Petri dish) and brief clonus-like activity leading to a loss of posture (i.e., fish falls to one side and remain immobile for 1–3 s).[138] An epileptiform-like electrographic activity was also observed from optic tectum (midbrain) of the fish during pentylenetetrazole exposure.[138] Similarly, intraperitoneal kainic acid injection also triggered behavioral seizures in adult (6–9-month-old) zebrafish.[139] Zebrafish have now been used to identify critical pathways involved in epilepsy,[140] to study effects of AED on seizure-induced learning deficits,[141] and to determine the genetic aspects of epilepsy.[137]

V. Summary

Animal models have played a very important role in advancing our understanding of epilepsy. With the help of animal models, scientists have started to unravel the basic mechanisms of epileptogenesis. Since the discovery of the anticonvulsant effect of phenytoin, animal models have played a crucial role in determining efficacy of drugs to treat specific seizure types and to prevent development of epilepsy. Developmental epilepsy models have allowed researchers to understand the differences in mechanisms of seizure generation and response to treatment between mature and immature brain. It is extremely difficult to design an animal model that recapitulates all the characteristics of specific epilepsy syndromes and none of the current models are perfect. It is therefore crucial that the investigator is aware of the limitations of the model that is being used in any study.

References

1. Fisher RS, van Emde Boas W, Blume W, Elger C, Genton P, Lee P, et al. Epileptic seizures and epilepsy: definitions proposed by the International League Against Epilepsy (ILAE) and the International Bureau for Epilepsy (IBE). *Epilepsia* 2005;**46**:470–2.
2. Begley CE, Famulari M, Annegers JF, Lairson DR, Reynolds TF, Coan S, et al. The cost of epilepsy in the United States: an estimate from population-based clinical and survey data. *Epilepsia* 2000;**41**:342–51.
3. Putnam TJ, Merritt HH. Experimental determination of the anticonvulsant properties of some phenyl derivatives. *Science* 1937;**85**:525–6.
4. Berg AT, Berkovic SF, Brodie MJ, Buchhalter J, Cross JH, van Emde Boas W, et al. Revised terminology and concepts for organization of seizures and epilepsies: report of the ILAE Commission on Classification and Terminology, 2005–2009. *Epilepsia* 2010;**51**:676–85.

5. Burgerman RS, Sperling MR, French JA, Saykin AJ, O'Connor MJ. Comparison of mesial versus neocortical onset temporal lobe seizures: neurodiagnostic findings and surgical outcome. *Epilepsia* 1995;**36**:662–70.
6. Blume WT, Luders HO, Mizrahi E, Tassinari C, van Emde Boas W, Engel Jr. J. Glossary of descriptive terminology for ictal semiology: report of the ILAE task force on classification and terminology. *Epilepsia* 2001;**42**:1212–8.
7. Singh NA, Charlier C, Stauffer D, DuPont BR, Leach RJ, Melis R, et al. A novel potassium channel gene, KCNQ2, is mutated in an inherited epilepsy of newborns. *Nat Genet* 1998;**18**:25–9.
8. Galanopoulou AS. GABA(A) receptors in normal development and seizures: friends or foes? *Curr Neuropharmacol* 2008;**6**:1–20.
9. Peters HC, Hu H, Pongs O, Storm JF, Isbrandt D. Conditional transgenic suppression of M channels in mouse brain reveals functions in neuronal excitability, resonance and behavior. *Nat Neurosci* 2005;**8**:51–60.
10. Pena F, Alavez-Perez N. Epileptiform activity induced by pharmacologic reduction of M-current in the developing hippocampus in vitro. *Epilepsia* 2006;**47**:47–54.
11. Ronen GM, Rosales TO, Connolly M, Anderson VE, Leppert M. Seizure characteristics in chromosome 20 benign familial neonatal convulsions. *Neurology* 1993;**43**:1355–60.
12. Curia G, Longo D, Biagini G, Jones RS, Avoli M. The pilocarpine model of temporal lobe epilepsy. *J Neurosci Methods* 2008;**172**:143–57.
13. Williams PA, Hellier JL, White AM, Staley KJ, Dudek FE. Development of spontaneous seizures after experimental status epilepticus: implications for understanding epileptogenesis. *Epilepsia* 2007;**48**(Suppl. 5):157–63.
14. Loscher W. Preclinical assessment of proconvulsant drug activity and its relevance for predicting adverse events in humans. *Eur J Pharmacol* 2009;**610**:1–11.
15. Velisek L, Veliskova J, Moshe SL. Developmental seizure models. *Ital J Neurol Sci* 1995;**16**:127–33.
16. Fisher RS. Animal models of the epilepsies. *Brain Res Brain Res Rev* 1989;**14**:245–78.
17. Stafstrom CE. Infantile spasms: a critical review of emerging animal models. *Epilepsy Curr* 2009;**9**:75–81.
18. Chudomelova L, Scantlebury MH, Raffo E, Coppola A, Betancourth D, Galanopoulou AS. Modeling new therapies for infantile spasms. *Epilepsia* 2010;**51**(Suppl. 3):27–33.
19. Morimoto K, Fahnestock M, Racine RJ. Kindling and status epilepticus models of epilepsy: rewiring the brain. *Prog Neurobiol* 2004;**73**:1–60.
20. Loscher W. Critical review of current animal models of seizures and epilepsy used in the discovery and development of new antiepileptic drugs. *Seizure* 2011;**20**:359–68.
21. Pitkanen A, Immonen RJ, Grohn OH, Kharatishvili I. From traumatic brain injury to posttraumatic epilepsy: what animal models tell us about the process and treatment options. *Epilepsia* 2009;**50**(Suppl. 2):21–9.
22. Jensen FE, Baram TZ. Developmental seizures induced by common early-life insults: short- and long-term effects on seizure susceptibility. *Ment Retard Dev Disabil Res Rev* 2000;**6**:253–7.
23. Bender RA, Dube C, Baram TZ. Febrile seizures and mechanisms of epileptogenesis: insights from an animal model. *Adv Exp Med Biol* 2004;**548**:213–25.
24. Avanzini G, Franceschetti S, Mantegazza M. Epileptogenic channelopathies: experimental models of human pathologies. *Epilepsia* 2007;**48**(Suppl. 2):51–64.
25. Wahab A, Albus K, Gabriel S, Heinemann U. In search of models of pharmacoresistant epilepsy. *Epilepsia* 2010;**51**(Suppl. 3):154–9.
26. Sarkisian MR. Overview of the current animal models for human seizure and epileptic disorders. *Epilepsy Behav* 2001;**2**:201–16.

27. Stafstrom CE, Moshe SL, Swann JW, Nehlig A, Jacobs MP, Schwartzkroin PA. Models of pediatric epilepsies: strategies and opportunities. *Epilepsia* 2006;**47**:1407–14.
28. Jensen FE, Applegate CD, Holtzman D, Belin TR, Burchfiel JL. Epileptogenic effect of hypoxia in the immature rodent brain. *Ann Neurol* 1991;**29**:629–37.
29. Painter MJ, Scher MS, Stein AD, Armatti S, Wang Z, Gardiner JC, et al. Phenobarbital compared with phenytoin for the treatment of neonatal seizures. *N Engl J Med* 1999;**341**:485–9.
30. Raol YH, Lapides DA, Keating JG, Brooks-Kayal AR, Cooper EC. A KCNQ channel opener for experimental neonatal seizures and status epilepticus. *Ann Neurol* 2009;**65**:326–36.
31. Holmes G. Epilepsy in the developing brain: lessons from the laboratory and clinic. *Epilepsia* 1997;**38**:12–30.
32. Rakhade SN, Jensen FE. Epileptogenesis in the immature brain: emerging mechanisms. *Nat Rev Neurol* 2009;**5**:380–91.
33. Swann J, Brady R, Martin D. Postnatal development of GABA-mediated synaptic inhibition in rat hippocampus. *Neuroscience* 1989;**28**:551–61.
34. Brooks-Kayal A, Pritchett D. Developmental changes in human $GABA_A$ receptor subunit composition. *Ann Neurol* 1993;**34**:687–93.
35. Brooks-Kayal A, Shumate M, Jin H, Rikhter T, Kelly M, Coulter D. GABA(A) receptor subunit expression predicts functional changes in hippocampal dentate granule cells during postnatal development. *J Neurochem* 2001;**77**:1266–78.
36. Brooks-Kayal AR, Pritchett DB. Developmental changes in human gamma-aminobutyric acidA receptor subunit composition. *Ann Neurol* 1993;**34**:687–93.
37. Gibbs J, Berkow G, Coulter D. GABAA receptor function in developing rat thalamic reticular neurons: whole-cell recordings of GABA-mediated currents and modulation by benzodiazepines. *J Neurophysiol* 1996;**76**:2568–79.
38. Kapur J, Macdonald RL. Postnatal development of hippocampal dentate granule cell gamma-aminobutyric acidA receptor pharmacological properties. *Mol Pharmacol* 1999;**55**:444–52.
39. Ganguly K, Schinder AF, Wong ST, Poo M. GABA itself promotes the developmental switch of neuronal GABAergic responses from excitation to inhibition. *Cell* 2001;**105**:521–32.
40. Hauser WA, Annegers JF, Kurland LT. Incidence of epilepsy and unprovoked seizures in Rochester, Minnesota: 1935–1984. *Epilepsia* 1993;**34**:453–68.
41. Cavalheiro E, Silva D, Turski W, Calderazzo-Filho L, Bortolotto Z, Turski L. The susceptibility of rats to pilocarpine-induced seizures is age-dependent. *Dev Brain Res* 1987;**37**:43–58.
42. Stafstrom CE, Thompson JL, Holmes GL. Kainic acid seizures in the developing brain: status epilepticus and spontaneous recurrent seizures. *Dev Brain Res* 1992;**65**:227–36.
43. Farwell JR, Lee YJ, Hirtz DG, Sulzbacher SI, Ellenberg JH, Nelson KB. Phenobarbital for febrile seizures—effects on intelligence and on seizure recurrence. *N Engl J Med* 1990;**322**:364–9.
44. Clancy RR, Legido A. Postnatal epilepsy after EEG-confirmed neonatal seizures. *Epilepsia* 1991;**32**:69–76.
45. Legido A, Clancy RR, Berman PH. Neurologic outcome after electroencephalographically proven neonatal seizures. *Pediatrics* 1991;**88**:583–96.
46. Luby M, Spencer DD, Kim JH, deLanerolle N, McCarthy G. Hippocampal MRI volumetrics and temporal lobe substrates in medial temporal lobe epilepsy. *Magn Reson Imaging* 1995;**13**:1065–71.
47. Hamilton SE, Loose MD, Qi M, Levey AI, Hille B, McKnight GS, et al. Disruption of the m1 receptor gene ablates muscarinic receptor-dependent M current regulation and seizure activity in mice. *Proc Natl Acad Sci USA* 1997;**94**:13311–6.
48. Arida RM, Scorza FA, Peres CA, Cavalheiro EA. The course of untreated seizures in the pilocarpine model of epilepsy. *Epilepsy Res* 1999;**34**:99–107.

49. Goffin K, Nissinen J, Van Laere K, Pitkanen A. Cyclicity of spontaneous recurrent seizures in pilocarpine model of temporal lobe epilepsy in rat. *Exp Neurol* 2007;**205**:501–5.
50. Raol YH, Lund IV, Bandyopadhyay S, Zhang G, Roberts DS, Wolfe JH, et al. Enhancing GABA(A) receptor alpha 1 subunit levels in hippocampal dentate gyrus inhibits epilepsy development in an animal model of temporal lobe epilepsy. *J Neurosci* 2006;**26**:11342–6.
51. Kubova H, Mares P, Suchomelova L, Brozek G, Druga R, Pitkanen A. Status epilepticus in immature rats leads to behavioural and cognitive impairment and epileptogenesis. *Eur J Neurosci* 2004;**19**:3255–65.
52. Raol YS, Budreck EC, Brooks-Kayal AR. Epilepsy after early-life seizures can be independent of hippocampal injury. *Ann Neurol* 2003;**53**:503–11.
53. Furtado MA, Castro OW, Del Vecchio F, de Oliveira JA, Garcia-Cairasco N. Study of spontaneous recurrent seizures and morphological alterations after status epilepticus induced by intrahippocampal injection of pilocarpine. *Epilepsy Behav* 2011;**20**:257–66.
54. Clifford DB, Olney JW, Maniotis A, Collins RC, Zorumski CF. The functional anatomy and pathology of lithium-pilocarpine and high-dose pilocarpine seizures. *Neuroscience* 1987;**23**:953–68.
55. Glien M, Brandt C, Potschka H, Voigt H, Ebert U, Loscher W. Repeated low-dose treatment of rats with pilocarpine: low mortality but high proportion of rats developing epilepsy. *Epilepsy Res* 2001;**46**:111–9.
56. Ben-Ari Y. Limbic seizure and brain damage produced by kainic acid: mechanisms and relevance to human temporal lobe epilepsy. *Neuroscience* 1985;**14**:375–403.
57. Hellier JL, Patrylo PR, Buckmaster PS, Dudek FE. Recurrent spontaneous motor seizures after repeated low-dose systemic treatment with kainate: assessment of a rat model of temporal lobe epilepsy. *Epilepsy Res* 1998;**31**:73–84.
58. Williams P, White A, Clark S, Ferraro D, Swiercz W, Staley K, et al. Development of spontaneous recurrent seizures after kainate-induced status epilepticus. *J Neurosci* 2009;**29**:2103–12.
59. Nadler JV, Perry BW, Cotman CW. Intraventricular kainic acid preferentially destroys hippocampal pyramidal cells. *Nature* 1978;**271**:676–7.
60. Tauck DL, Nadler JV. Evidence of functional mossy fiber sprouting in hippocampal formation of kainic acid-treated rats. *J Neurosci* 1985;**5**:1016–22.
61. Krasowski MD. Differential modulatory actions of the volatile convulsant flurothyl and its anesthetic isomer at inhibitory ligand-gated ion channels. *Neuropharmacology* 2000;**39**:1168–83.
62. Hoffmann AF, Zhao Q, Holmes GL. Cognitive impairment following status epilepticus and recurrent seizures during early development: support for the "two- hit hypothesis" *Epilepsy Behav* 2004;**5**:873–7.
63. Riviello P, de Rogalski Landrot I, Holmes GL. Lack of cell loss following recurrent neonatal seizures. *Brain Res Dev Brain Res* 2002;**135**:101–4.
64. Isaeva E, Isaev D, Savrasova A, Khazipov R, Holmes GL. Recurrent neonatal seizures result in long-term increases in neuronal network excitability in the rat neocortex. *Eur J Neurosci* 2010;**31**:1446–55.
65. Racine R. Modification of seizure activity by electrical stimulation, II: motor seizure. *Electroencephalogr Clin Neurophysiol* 1972;**32**:281–94.
66. Loscher W, Cramer S, Ebert U. Differences in kindling development in seven outbred and inbred rat strains. *Exp Neurol* 1998;**154**:551–9.
67. Kubova H, Moshe S. Experimental models of epilepsy in young animals. *J Child Neurol* 1994;**9**(Suppl. 1):S3–S11.
68. Goddard GV, McIntyre DC, Leech CK. A permanent change in brain function resulting from daily electrical stimulation. *Exp Neurol* 1969;**25**:295–330.

69. Loscher W. Animal models of epilepsy for the development of antiepileptogenic and disease-modifying drugs. A comparison of the pharmacology of kindling and post-status epilepticus models of temporal lobe epilepsy. *Epilepsy Res* 2002;**50**:105–23.
70. Sayin U, Osting S, Hagen J, Rutecki P, Sutula T. Spontaneous seizures and loss of axo-axonic and axo-somatic inhibition induced by repeated brief seizures in kindled rats. *J Neurosci* 2003;**23**:2759–68.
71. Cavazos JE, Sutula TP. Progressive neuronal loss induced by kindling: a possible mechanism for mossy fiber synaptic reorganization and hippocampal sclerosis. *Brain Res* 1990;**527**:1–6.
72. Lothman EW, Bertram EH, Bekenstein JW, Perlin JB. Self-sustaining limbic status epilepticus induced by 'continuous' hippocampal stimulation: electrographic and behavioral characteristics. *Epilepsy Res* 1989;**3**:107–19.
73. Nissinen J, Halonen T, Koivisto E, Pitkanen A. A new model of chronic temporal lobe epilepsy induced by electrical stimulation of the amygdala in rat. *Epilepsy Res* 2000;**38**.177–205.
74. Mazarati A, Bragin A, Baldwin R, Shin D, Wilson C, Sankar R, et al. Epileptogenesis after self-sustaining status epilepticus. *Epilepsia* 2002;**43**(Suppl. 5):74–80.
75. Mares P, Kubova H. Electrical stimulation-induced models of seizures. In: Pitkanen A, Schwartzkroin PA, Moshe SL, editors. *Models of seizures and epilepsy*. Burlington, MA: Elsevier; 2006. pp. 153–9.
76. Reid CA, Berkovic SF, Petrou S. Mechanisms of human inherited epilepsies. *Prog Neurobiol* 2009;**87**:41–57.
77. Rogawski MA. KCNQ2/KCNQ3 K+ channels and the molecular pathogenesis of epilepsy: implications for therapy. *Trends Neurosci* 2000;**23**:393–8.
78. Singh NA, Otto JF, Dahle EJ, Pappas C, Leslie JD, Vilaythong A, et al. Mouse models of human KCNQ2 and KCNQ3 mutations for benign familial neonatal convulsions show seizures and neuronal plasticity without synaptic reorganization. *J Physiol* 2008;**586**:3405–23.
79. Escayg A, Goldin AL. Sodium channel SCN1A and epilepsy: mutations and mechanisms. *Epilepsia* 2010;**51**:1650–8.
80. Kearney JA, Yang Y, Beyer B, Bergren SK, Claes L, Dejonghe P, et al. Severe epilepsy resulting from genetic interaction between Scn2a and Kcnq2. *Hum Mol Genet* 2006;**15**:1043–8.
81. Hawkins NA, Martin MS, Frankel WN, Kearney JA, Escayg A. Neuronal voltage-gated ion channels are genetic modifiers of generalized epilepsy with febrile seizures plus. *Neurobiol Dis* 2011;**41**:655–60.
82. Marsh E, Fulp C, Gomez E, Nasrallah I, Minarcik J, Sudi J, et al. Targeted loss of Arx results in a developmental epilepsy mouse model and recapitulates the human phenotype in heterozygous females. *Brain* 2009;**132**:1563–76.
83. Price MG, Yoo JW, Burgess DL, Deng F, Hrachovy RA, Frost Jr. JD, et al. A triplet repeat expansion genetic mouse model of infantile spasms syndrome, Arx(GCG)10+7, with interneuronopathy, spasms in infancy, persistent seizures, and adult cognitive and behavioral impairment. *J Neurosci* 2009;**29**:8752–63.
84. Noebels J. Spontaneous epileptic mutations in the mouse. In: Pitkanen A, Schwartzkroin PA, Moshe SL, editors. *Models of seizures and epilepsy*. Burlington, MA: Elsevier; 2006. pp. 223–32.
85. Wakamori M, Yamazaki K, Matsunodaira H, Teramoto T, Tanaka I, Niidome T, et al. Single tottering mutations responsible for the neuropathic phenotype of the P-type calcium channel. *J Biol Chem* 1998;**273**:34857–67.
86. Caddick SJ, Wang C, Fletcher CF, Jenkins NA, Copeland NG, Hosford DA. Excitatory but not inhibitory synaptic transmission is reduced in lethargic (Cacnb4(lh)) and tottering (Cacna1atg) mouse thalami. *J Neurophysiol* 1999;**81**:2066–74.

87. Coenen AM, Van Luijtelaar EL. Genetic animal models for absence epilepsy: a review of the WAG/Rij strain of rats. *Behav Genet* 2003;**33**:635–55.
88. van Rijn CM, Sun MS, Deckers CL, Edelbroek PM, Keyser A, Renier W, et al. Effects of the combination of valproate and ethosuximide on spike wave discharges in WAG/Rij rats. *Epilepsy Res* 2004;**59**:181–9.
89. Dedeurwaerdere S, van Raay L, Morris MJ, Reed RC, Hogan RE, O'Brien TJ. Fluctuating and constant valproate administration gives equivalent seizure control in rats with genetic and acquired epilepsy. *Seizure* 2011;**20**:72–9.
90. Albala B, Moshe S, Okada R. Kainic-acid induced seizures developmental study. *Dev Brain Res* 1984;**13**:139–48.
91. Hauser WA. The prevalence and incidence of convulsive disorders in children. *Epilepsia* 1994;**35**:S1–6.
92. Cendes F, Andermann F, Dubeau F, Gloor P, Evans A, Jones-Gotman M, et al. Early childhood prolonged febrile convulsions, atrophy and sclerosis of mesial structures, and temporal lobe epilepsy: an MRI volumetric study. *Neurology* 1993;**43**:1083–7.
93. French JA, Williamson PD, Thadani VM, Darcey TM, Mattson RH, Spencer SS, et al. Characteristics of medial temporal lobe epilepsy: I. Results of history and physical examination. *Ann Neurol* 1993;**34**:774–80.
94. Holtzman D, Obana K, Olson J. Hyperthermia-induced seizures in the rat pup: a model for febrile convulsions in children. *Science* 1981;**213**:1034–6.
95. Hjeresen DL, Diaz J. Ontogeny of susceptibility to experimental febrile seizures in rats. *Dev Psychobiol* 1988;**21**:261–75.
96. Baram TZ, Gerth A, Schultz L. Febrile seizures: an appropriate-aged model suitable for long-term studies. *Brain Res Dev Brain Res* 1997;**98**:265–70.
97. Toth Z, Yan XX, Haftoglou S, Ribak CE, Baram TZ. Seizure-induced neuronal injury: vulnerability to febrile seizures in an immature rat model. *J Neurosci* 1998;**18**:4285–94.
98. Bender RA, Dube C, Gonzalez-Vega R, Mina EW, Baram TZ. Mossy fiber plasticity and enhanced hippocampal excitability, without hippocampal cell loss or altered neurogenesis, in an animal model of prolonged febrile seizures. *Hippocampus* 2003;**13**:399–412.
99. Dube C, Richichi C, Bender RA, Chung G, Litt B, Baram TZ. Temporal lobe epilepsy after experimental prolonged febrile seizures: prospective analysis. *Brain* 2006;**129**:911–22.
100. Saliba RM, Annegers JF, Waller DK, Tyson JE, Mizrahi EM. Incidence of neonatal seizures in Harris County, Texas, 1992–1994. *Am J Epidemiol* 1999;**150**:763–9.
101. Bergamasco B, Benna P, Ferrero P, Gavinelli R. Neonatal hypoxia and epileptic risk: a clinical prospective study. *Epilepsia* 1984;**25**:131–6.
102. Glass HC, Glidden D, Jeremy RJ, Barkovich AJ, Ferriero DM, Miller SP. Clinical neonatal seizures are independently associated with outcome in infants at risk for hypoxic-ischemic brain injury. *J Pediatr* 2009;**155**:318–23.
103. Epsztein J, Ben-Ari Y, Represa A, Crepel V. Late-onset epileptogenesis and seizure genesis: lessons from models of cerebral ischemia. *Neuroscientist* 2008;**14**:78–90.
104. Jensen FE, Wang C, Stafstrom CE, Liu Z, Geary C, Stevens MC. Acute and chronic increases in excitability in rat hippocampal slices after perinatal hypoxia in vivo. *J Neurophysiol* 1998;**79**:73–81.
105. Rakhade SN, Klein PM, Huynh T, Hilario-Gomez C, Kosaras B, Rotenberg A, et al. Development of later life spontaneous seizures in a rodent model of hypoxia-induced neonatal seizures. *Epilepsia* 2011;**52**:753–65.
106. Jensen FE, Wang C. Hypoxia-induced hyperexcitability in vivo and in vitro in the immature hippocampus. *Epilepsy Res* 1996;**26**:131–40.
107. Owens Jr. J, Robbins CA, Wenzel HJ, Schwartzkroin PA. Acute and chronic effects of hypoxia on the developing hippocampus. *Ann Neurol* 1997;**41**:187–99.

108. Rice 3rd JE, Vannucci RC, Brierley JB. The influence of immaturity on hypoxic-ischemic brain damage in the rat. *Ann Neurol* 1981;**9**:131–41.
109. Kadam SD, Dudek FE. Neuropathogical features of a rat model for perinatal hypoxic-ischemic encephalopathy with associated epilepsy. *J Comp Neurol* 2007;**505**:716–37.
110. Vannucci SJ, Hagberg H. Hypoxia-ischemia in the immature brain. *J Exp Biol* 2004;**207**:3149–54.
111. Towfighi J, Zec N, Yager J, Housman C, Vannucci RC. Temporal evolution of neuropathologic changes in an immature rat model of cerebral hypoxia: a light microscopic study. *Acta Neuropathol* 1995;**90**:375–86.
112. Vannucci RC, Vannucci SJ. A model of perinatal hypoxic-ischemic brain damage. *Ann N Y Acad Sci* 1997;**835**:234–49.
113. Kadam SD, White AM, Staley KJ, Dudek FE. Continuous electroencephalographic monitoring with radio-telemetry in a rat model of perinatal hypoxia-ischemia reveals progressive post-stroke epilepsy. *J Neurosci* 2010;**30**:404–15.
114. Ferriero DM. Neonatal brain injury. *N Engl J Med* 2004;**351**:1985–95.
115. Hossain MA. Molecular mediators of hypoxic-ischemic injury and implications for epilepsy in the developing brain. *Epilepsy Behav* 2005;**7**:204–13.
116. Williams PA, Dou P, Dudek FE. Epilepsy and synaptic reorganization in a perinatal rat model of hypoxia-ischemia. *Epilepsia* 2004;**45**:1210–8.
117. Almli CR, Levy TJ, Han BH, Shah AR, Gidday JM, Holtzman DM. BDNF protects against spatial memory deficits following neonatal hypoxia- ischemia. *Exp Neurol* 2000;**166**:99–114.
118. McAuliffe JJ, Miles L, Vorhees CV. Adult neurological function following neonatal hypoxia-ischemia in a mouse model of the term neonate: water maze performance is dependent on separable cognitive and motor components. *Brain Res* 2006;**1118**:208–21.
119. Baram TZ, Schultz L. Corticotropin-releasing hormone is a rapid and potent convulsant in the infant rat. *Brain Res Dev Brain Res* 1991;**61**:97–101.
120. Mares P, Velisek L. N-methyl-D-aspartate (NMDA)-induced seizures in developing rats. *Brain Res Dev Brain Res* 1992;**65**:185–9.
121. Kabova R, Liptakova S, Slamberova R, Pometlova M, Velisek L. Age-specific N-methyl-D-aspartate-induced seizures: perspectives for the West syndrome model. *Epilepsia* 1999;**40**:1357–69.
122. Galvan CD, Hrachovy RA, Smith KL, Swann JW. Blockade of neuronal activity during hippocampal development produces a chronic focal epilepsy in the rat. *J Neurosci* 2000;**20**:2904–16.
123. Velisek L, Jehle K, Asche S, Veliskova J. Model of infantile spasms induced by N-methyl-D-aspartic acid in prenatally impaired brain. *Ann Neurol* 2007;**61**:109–19.
124. Lee CL, Frost Jr. JD, Swann JW, Hrachovy RA. A new animal model of infantile spasms with unprovoked persistent seizures. *Epilepsia* 2008;**49**:298–307.
125. Scantlebury MH, Galanopoulou AS, Chudomelova L, Raffo E, Betancourth D, Moshe SL. A model of symptomatic infantile spasms syndrome. *Neurobiol Dis* 2010;**37**:604–12.
126. Pitkanen A, McIntosh TK. Animal models of post-traumatic epilepsy. *J Neurotrauma* 2006;**23**:241–61.
127. Annegers JF, Grabow JD, Groover RV, Laws Jr. ER, Elveback LR, Kurland LT. Seizures after head trauma: a population study. *Neurology* 1980;**30**:683–9.
128. Frey LC. Epidemiology of posttraumatic epilepsy: a critical review. *Epilepsia* 2003;**44** (Suppl. 10):11–7.
129. Prince DA, Parada I, Scalise K, Graber K, Jin X, Shen F. Epilepsy following cortical injury: cellular and molecular mechanisms as targets for potential prophylaxis. *Epilepsia* 2009;**50** (Suppl. 2):30–40.

130. Golarai G, Greenwood AC, Feeney DM, Connor JA. Physiological and structural evidence for hippocampal involvement in persistent seizure susceptibility after traumatic brain injury. *J Neurosci* 2001;**21**:8523–37.
131. Prins ML, Hovda DA. Developing experimental models to address traumatic brain injury in children. *J Neurotrauma* 2003;**20**:123–37.
132. McIntosh TK, Vink R, Noble L, Yamakami I, Fernyak S, Soares H, et al. Traumatic brain injury in the rat: characterization of a lateral fluid-percussion model. *Neuroscience* 1989;**28**:233–44.
133. D'Ambrosio R, Fender JS, Fairbanks JP, Simon EA, Born DE, Doyle DL, et al. Progression from frontal-parietal to mesial-temporal epilepsy after fluid percussion injury in the rat. *Brain* 2005;**128**:174–88.
134. Hunt RF, Scheff SW, Smith BN. Posttraumatic epilepsy after controlled cortical impact injury in mice. *Exp Neurol* 2009;**215**:243–52.
135. Shoubridge C, Fullston T, Gecz J. ARX spectrum disorders: making inroads into the molecular pathology. *Hum Mutat* 2010;**31**:889–900.
136. Bandmann O, Burton EA. Genetic zebrafish models of neurodegenerative diseases. *Neurobiol Dis* 2010;**40**:58–65.
137. Hortopan GA, Dinday MT, Baraban SC. Zebrafish as a model for studying genetic aspects of epilepsy. *Dis Model Mech* 2010;**3**:144–8.
138. Baraban SC, Taylor MR, Castro PA, Baier H. Pentylenetetrazole induced changes in zebrafish behavior, neural activity and c-fos expression. *Neuroscience* 2005;**131**:759–68.
139. Alfaro JM, Ripoll-Gomez J, Burgos JS. Kainate administered to adult zebrafish causes seizures similar to those in rodent models. *Eur J Neurosci* 2011;**33**:1252–5.
140. Hortopan GA, Dinday MT, Baraban SC. Spontaneous seizures and altered gene expression in GABA signaling pathways in a mind bomb mutant zebrafish. *J Neurosci* 2010;**30**:13718–28.
141. Lee Y, Kim D, Kim YH, Lee H, Lee CJ. Improvement of pentylenetetrazol-induced learning deficits by valproic acid in the adult zebrafish. *Eur J Pharmacol* 2010;**643**:225–31.

Animal Models of Muscular Dystrophy

Rainer Ng,* Glen B. Banks,* John K. Hall,* Lindsey A. Muir,† Julian N. Ramos,† Jacqueline Wicki,* Guy L. Odom,* Patryk Konieczny,* Jane Seto,* Joel R. Chamberlain,‡ and Jeffrey S. Chamberlain*

**Division of Medical Genetics, Department of Neurology, University of Washington, Seattle, Washington, USA*

†Program in Cellular and Molecular Biology, Department of Neurology, University of Washington, Seattle, Washington, USA

‡Division of Medical Genetics, Department of Medicine, University of Washington, Seattle, Washington, USA

Progress in Molecular Biology and Translational Science, Vol. 105
DOI: 10.1016/B978-0-12-394596-9.00004-4

The muscular dystrophies (MDs) represent a diverse collection of inherited human disorders, which affect to varying degrees skeletal, cardiac, and sometimes smooth muscle (Emery, 2002[1]). To date, more than 50 different genes have been implicated as causing one or more types of MD (Bansal *et al.*, 2003[2]). In many cases, invaluable insights into disease mechanisms, structure and function of gene products, and approaches for therapeutic interventions have benefited from the study of animal models of the different MDs (Arnett *et al.*, 2009[3]). The large number of genes that are associated with MD and the tremendous number of animal models that have been developed preclude a complete discussion of each in the context of this review. However, we summarize here a number of the more commonly used models together with a mixture of different types of gene and MD, which serves to give a general overview of the value of animal models of MD for research and therapeutic development.

I. Dystrophin

Duchenne muscular dystrophy (DMD) and the milder Becker MD (BMD) are caused by dystrophin gene mutations.[4,5] DMD/BMD is thought to be the most common form of muscular dystrophy (MD), and the most common lethal genetic disorder of children.[6] Full-length dystrophin is expressed in all muscles, while shorter, truncated isoforms are expressed in many other tissues.[7] Dystrophin is part of a large complex of integral and peripheral proteins that link the γ-actin cytoskeleton to the extracellular matrix *via* the dystrophin–glycoprotein complex (DGC) (Fig. 1).[8] Studies of dystrophin and DMD/BMD have been greatly facilitated by the availability of numerous murine and canine models of DMD.[9]

A. Mouse Models for Duchenne Muscular Dystrophy

Mouse models have been used extensively to elucidate the pathogenic mechanism of DMD and have been indispensable in the development of therapeutic approaches. Many mouse models with mutations in the *DMD* gene have been identified. Among these models, the *mdx* mouse is the most commonly used to analyze dystrophin gene expression and function. The *mdx* mouse contains a premature stop codon in exon 23 that leads to loss of full-length dystrophin, although smaller isoforms are still expressed.[10–12] Although *mdx* skeletal muscle shows moderate signs of dystrophy, young mice exhibit little weakness and live 80% as long as controls, significantly more than that of

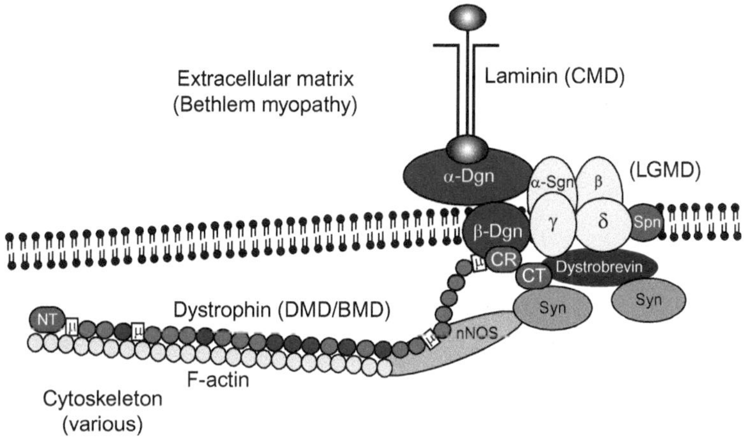

FIG. 1. The dystrophin–glycoprotein complex (DGC) contains many proteins that, when not expressed, lead to various forms of muscular dystrophy. Shown is the organization of the DGC in skeletal muscle. Dystrophin is defective in DMD and BMD; α-, β-, γ-, and δ- sarcoglycans (Sgn) are defective in four different types of LGMD; laminin α2 is defective in one form of congenital muscular dystrophy; collagen VI in the extracellular matrix is defective in Bethlem myopathy and Ullrich congenital muscular dystrophy; while numerous other cytosolic- and membrane-associated proteins lead to other MDs. Other abbreviations include Syn, syntrophin; SPN, sarcospan; Dgn, dystroglycan; nNOS, neuronal nitric oxide synthase. Dystrophin domains listed include the N-terminal actin-binding domain (NT), various hinges (H), a WW domain (W), a cysteine-rich domain (CR), and a C-terminal domain (CT). Orange and red balls in dystrophin represent spectrin-like repeats, orange ones are acidic, while blue ones are basic and contribute to actin binding. (See Color Insert.)

DMD patients.[13] This mild, yet progressive phenotype has been attributed in part to compensatory upregulation of utrophin.[14,15] Histological examination of *mdx* muscle during various stages of development reveals that muscle fiber necrosis and cellular infiltration begin at approximately 3 weeks of age. This is followed by a crises period that peaks at approximately 4–6 weeks of age and is characterized by the presence of extensive necrosis, regenerating muscle fibers with centrally located nuclei, and elevated levels of serum creatine kinase (CK).[10,16] After 12 weeks, the cycles of necrosis and regeneration begin to slow, although necrotic myofibers are present for the remainder of their lifespan. The fibrosis and infiltration of inflammatory cells in skeletal and cardiac muscle of the *mdx* are much milder than that observed in DMD patients.[17,18] In contrast, the *mdx* mouse diaphragm exhibits severe pathological changes and functional deficit comparable to that of DMD limb muscle.[19,20]

Four additional strains of *mdx* mice, $mdx^{2cv-5cv}$, have been generated with *N*-ethylnitrosourea chemical mutagenesis.[21] All these strains have point mutations that lead to loss of full-length dystrophin isoforms. The relative location of these mutations results in a series of *mdx* mouse mutants that vary in their expression of different dystrophin isoforms, such that these mice should be useful for analysis of the role of the various isoforms.[12] Despite their differences, all five *mdx* strains display essentially identical muscle pathology as *mdx* mice, although additional phenotypes have been observed.[21–23] In the mdx^{2cv} mouse, a single base change in the splice acceptor site of dystrophin intron 42 generates multiple transcripts due to aberrant splicing that eliminates the normal open reading frame. The mdx^{3cv} allele arises from a mutant splice acceptor site in intron 65,[12,21,22] leading to severe loss of most dystrophin isoforms, including Dp71. Of the five original *mdx* strains, only the mdx^{3cv} mice accumulate detectable levels of dystrophin-positive muscle fibers, although the expressed dystrophin lacks a key functional domain and is largely nonfunctional.[22–24] The mutation in the mdx^{4cv} mouse is a point mutation that creates a stop codon in exon 53, whereas in the mdx^{5cv} allele, a transversion mutation in exon 10 creates a new splice donor that generates a frame-shifting deletion in the mRNA. Both the mdx^{4cv} and mdx^{5cv} strains display a low background of reverent dystrophin-containing fibers, making them particularly useful strains in gene transfer studies exploring the feasibility of DMD therapy.[23,25,26]

In the $mdx^{\beta geo}$ mouse, a β-galactosidase/neomycin phosphotransferase (β*geo*) reporter gene is inserted 3′ to exon 63 of the dystrophin gene, affecting translation of dystrophin isoforms, including Dp71.[27] These mice develop a dilated esophagus, cardiac hypertrophy, and the same skeletal muscle pathology as *mdx* mice. The *lacZ* reporter gene of $mdx^{\beta geo}$ allows detection of dystrophin expression by staining for β-galactosidase activity.

Another mutant mouse known as *mdx*52 was generated by a homologous recombination technique.[28] In this model, exon 52 of the dystrophin gene is disrupted, and dystrophin along with shorter dystrophin isoforms (Dp140 and Dp260) are absent. This exon was targeted to replicate the most common out-of-frame deletion mutation that occurs in 13% of DMD patients.[29–31] Recently, antisense-mediated skipping of exon 51 using antisense oligonucleotides (AOs) to restore the reading frame of the dystrophin gene was found to be successful in some patients with DMD.[32–34] Exon 51 skipping using AO in *mdx*52 mice successfully converted an out-of-frame mutation into an in-frame mutation, resulting in amelioration of the dystrophic pathology and improved muscle function.[35]

Hanaoka and colleagues generated a new DMD mouse model known as DMD-null. In this model, the entire genomic region of the *DMD* gene on mouse chromosome X was deleted using a Cre-*loxP* recombination system.[36] These DMD-null males exhibited some behavioral abnormality, sterility, and pathological features very similar to those of the *mdx* mice. As the production of all dystrophin isoforms is prevented in the DMD-null mouse, this model system may be used to facilitate studies to clarify the function of individual dystrophin isoforms.

In efforts to make the *mdx* muscle phenotype more similar to that of patients, several additional mutations have been crossed onto the *mdx* background to generate double knockouts (DKOs). The most widely used is a dystrophin:utrophin DKO (*mdx:utrn*$^{-/-}$).[37,38] These mice display a severe phenotype and live on average only about 3 months. The severity of the phenotype supports the concept that utrophin upregulation in dystrophic muscles partially compensates for the absence of dystrophin. Further, the *mdx:utrn*$^{-/-}$ have proved useful in gene therapy studies, where the phenotype can be largely eliminated by muscle-specific expression of utrophin, mini-utrophin, or mini- or micro-dystrophin.[25,26,37–40] Two newer DKO strains have recently been generated that not only resulted in a several phonotype but also rely on mutations that appear to faithfully reproduce the human condition. The first is a cross between *mdx* and a mutation of cytidine monophosphate–sialic acid hydroxylase (Cmah).[41] Cmah is inactive in humans due to a mutation, and it prevents glycosylation with *N*-glycolylneuraminic acid. The second is an *mdx:mTR* cross that effectively lacks telomerase activity.[42] This cross becomes severe due to the impaired ability of myogenic stem cells to support ongoing muscle regeneration.

B. Canine Models for DMD

Clinical diagnosis of MD has been observed in the dog for several decades. Identifying a particular MD with broad symptoms (i.e., weakness and intolerance to exercise) has proven difficult. With the increasing availability of research tools, more human MDs have found a pathologically similar model in canines. Yet, the infrequency of these diagnoses and the difficulties in identifying and propagating the mutations by breeding have hindered multiple MD strains from being used as valuable research tools. One MD that has been successfully generated and utilized throughout the world is the X-linked dystrophin-deficient dog. Spontaneous mutations causing dystrophinopathy have been identified in several breeds of dog.[43–45] Multiple colonies of the golden retriever muscular dystrophy dog (GRMD) have been created and is the most extensively studied breed for this model, but it is not the only one.[9,46] Due to prior use in research and its smaller size, the GRMD mutation has been bred onto the beagle background.[47,48] Severe symptoms commonly appear at 6 months of age in the GRMD dog, but unlike the *mdx* mouse, the degree of severity and time of progression are quite variable. Pathology has been observed *in utero* and massive necrosis can result in neonatal death.[46] Compounded by the increased cost of maintaining a larger species, the GRMD model can be problematic. However, use of the GRMD model has gained much emphasis with its more clinically similar pathology than the *mdx* mouse model.

II. The Sarcoglycans

As a clinical group, the MDs were first appreciated in the mid-1800s with a detailed description of DMD.[49,50] Approximately a century following, another group of clinically comparable progressive MDs distinct from DMD/BMD muscular dystrophy emerged which were found to follow an autosomal recessive mode of inheritance. Given their phenotypic similarity, these diseases were originally termed Duchenne-like autosomal recessive MD or severe childhood autosomal recessive MD.[51] Several decades later, the discovery of mutations in the dystrophin gene as the genetic cause of DMD,[52] subsequent description of the DGC,[53] and demarcation of the striated muscle transmembrane sarcoglycan complex paved the way for initial characterization of the sarcoglycanopathies. Four genetically related MDs arose with pathological mutations in genes for alpha-sarcoglycan (α-SG),[54] beta-sarcoglycan (β-SG),[55] gamma-sarcoglycan (γ-SG),[56] and delta-sarcoglycan (δ-SG)[57] defining these distinct autosomal recessive MDs. Given that these disorders primarily affected proximal muscles around the scapular and pelvic girdles, and elevation of the genetic classification of limb-girdle muscular dystrophies (LGMD), there was the assignment of LGMD2C-F for the genes representing γ-SG, α-SG, β-SG, and δ-SG, respectively.

A. δ-Sarcoglycan and LGMD2F

The BIO14.6 hamster strain was established in 1962 and is recognized as the first animal model of sarcoglycanopathies.[58] However, it was not until after the discovery of the causative mutation in LGMD2F patients[59] that a large deletion in the 5′ portion of the *δ-SG* gene was identified.[60] The hamsters die prematurely (~ 8 months) from progressive myocardial necrosis and ultimately heart failure. Compared with skeletal muscle, damage to cardiac muscle occurs early (~ 5 weeks), consistent with a cardiomyopathic phenotype with decreased stroke volume, cardiac output, and increased ventricular hypertrophy.[61] As the original BIO14.6 strain was identified, there have been at least three additional strains developed: the J2-NK, the CHF147 (formerly VMX7.1), and the TO-2 strain. Differences, such as the expression profile of individual sarcoglycan isoforms, have been observed in the TO-2 as compared to the BIO14.6 model. For example, in the BIO14.6 muscles, residual protein levels of α-SG and γ-SG can be detected on the sarcolemma, while in the TO-2, there is a complete absence of these isoforms. Further, the TO-2 has a shorter lifespan with a tendency to develop a more severe cardiomyopathy displaying a necrotic mode of pathology, while the BIO14.6 model displays a hypertrophic cardiomyopathy.[62,63] Mice deficient in δ-SG have also been generated with subsequent complete loss of the entire SG complex,[64,65a] and as with the BIO14.6 hamster, these mice display classical histological findings of dystrophy inclusive of muscle degeneration, necrosis, inflammation, fibrosis, and reduced survival. However, muscle mass, peak twitch tension (P_1) and peak tetanic tension (P_0), and specific force (P_0/CSA) do not differ from wild-type muscle with the latter being in contrast to the BIO14.6 hamster[65b] suggesting species-specific variations in pathophysiology. Similar to the BIO14.6 hamster model, these mice develop cardiomyopathy with a relatively early onset (~8 weeks) with areas of fibrosis. By 32 weeks, a detrimental change in pathology is observed with concurrent reduction in ejection fraction, and evidence of right ventricular dilation suggestive of pulmonary dysfunction.[65c] Thus, similar to the hamster model, the δ-SG-null mouse reproduces many of the pathophysiological features of the human disease, providing a useful preclinical research model.

B. γ-Sarcoglycan and LGMD2C

Mice with a targeted deletion of γ-SG also exhibit progressive MD and cardiomyopathy with a subsequent destabilization of β-SG and δ-SG while α-SG is highly reduced.[66] Like the δ-SG-null hamster, cardiomyopathy in γ-SG-null mice may be dilated or hypertrophic.[62,66] Similar to findings in LGMD2C patients, membrane disruptions are a common feature of mice deficient for γ-SG as evidenced by Evans blue dye uptake and elevation of the muscle-specific enzyme M-CK in serum.[66,67] Importantly, in striated muscle, the expression of dystrophin, dystroglycan, and laminin appears unaltered with the loss of γ-SG when evaluated by immunostaining.[64] Thus, the mechanical linkage between the extracellular matrix and the intracellular cytoskeleton appears to be maintained in the absence of γ-SG. Further, γ-SG-deficient mice show normal resistance to mechanical strain and no evidence of contraction-induced injury after exercise when compared to its wild-type counterparts.[66] These results are similar to the BIO14.6 hamster and in contrast to muscles of α-SG, β-SG, and δ-SG-null mice (see below). Considering the sarcoglycans are secondarily reduced in dystrophin-deficient mice, this advocates that nonmechanical properties contribute to the pathophysiology of MD in genetically varied forms of disease.

C. β-Sarcoglycan and LGMD2E

Mice deficient for β-SG have also been generated which, like δ-SG-null mice, display MD with large areas of necrosis/fibrosis and cardiomyopathy.[68,69] As an indicator of acute myocardial necrosis preceding that of fibrotic lesion formation, serum levels of the cardiac-specific troponin-I have been found elevated in these mice.[65a,68,70] Additionally, like δ-SG-null mice, β-SG-null mice show severely reduced expression of the entire sarcoglycan complex inclusive of sarcospan, increased muscle mass, decreased specific force generation, myofiber cross-sectional area, and resistance to contraction-induced injury.[68,69,71] In support of the former, coimmunoprecipitation studies have shown a strong association between β-SG and δ-SG, proposed to form a core complex enabling delivery and retention of the sarcoglycan complex to the sarcolemma.[72] Correlative support of this hypothesis can be found in human patients with either *β-SG* or *δ-SG* mutations, where there is a strong reduction or complete absence of the entire sarcoglycan complex.[73–75]

D. α-Sarcoglycan and LGMD2D

Mice lacking α-SG have been generated by targeted disruption of the α-SG gene.[76] These mice show a complete loss of the sarcoglycan complex, demonstrate a progressive MD, and like LGMD2D patients present with an absence of cardiomyopathy.[76] Like the γ-, β-, and δ-SG-null mice, the α-SG-null mice demonstrate membrane permeability defects as evidenced by Evans blue dye uptake and elevated serum CK. Given this similarity across dystrophic models, it remains unclear whether or not changes in increased membrane permeability are causative for muscle degeneration or rather a result of it. A further interesting aspect of α-SG-null mice that differs from the γ-SG-null mice is an apparent decrease in expression of α-dystroglycan (αDG) and dystrophin with an inherent increase in expression of utrophin at the sarcolemma.[76] The molecular basis for these differences has not been established, though it is tempting to speculate that the residual amount of α-SG present in the γ-SG-null could be sufficient to allow normal levels of dystrophin and dystroglycan. In the α-SG-null mice, there are notable muscle-specific differences in contractile properties. For example, in hypertrophic slow-twitch soleus muscle, there is a demonstrated increase in maximal force values and relatively unchanged specific force relative to wild-type controls.[76] However, in fast-twitch extensor digitorum longus (EDL) muscles, maximal force values are unchanged while specific force is significantly lower than wild-type controls.[76] In diaphragm of α-SG-null mice, there is a significant reduction in maximal force, coupling between longitudinal and transverse properties, and muscle extensibility in the direction of muscle fibers and transverse to fibers.[77] Thus, as a whole, the sarcoglycan complex may also provide a complex mechanical function in the diaphragm by influencing the properties of muscle stiffness, muscle viscoelasticity, and modulation of the contractile properties.

E. Canine Models of Sarcoglycan Deficiency

Three breeds of dogs have also been reported with deficiencies in expression of the sarcoglycan complex in skeletal muscle.[78,79] Where the mutation lies in each of these cases has yet to be determined, but as discussed above for both rodents and humans, a deficiency in one sarcoglycan can destabilize the subcomplex made by all four of these isoforms.

III. Calcium and Dystrophic Pathology

A. Calcium Contributes to the Pathology of Dystrophic Skeletal Muscles

Calcium plays a dual role in muscle function, serving as a signaling molecule as well as a regulator of muscle contraction and relaxation.[80] To effectively regulate cytosolic [Ca^{2+}], muscle fibers contain a complex system of channels, pumps, and buffering proteins, each dedicated to mobilize, transport, or sequester Ca^{2+}.[80] In dystrophin-deficient muscle, all three components, that is, channels, pumps, and buffering proteins, appear to be dysfunctional. Ca^{2+} channels in the sarcolemma of *mdx* mice exhibit increased activity,[81,82] while Ca^{2+} buffering proteins[83,84] and ATPase pumps in the sarcoplasmic reticulum[85,86] demonstrate an impaired ability to sequester Ca^{2+}. Together, the compromised Ca^{2+}-cycling system in *mdx* muscles results in an elevated level of resting intracellular [Ca^{2+}].[87–89] Accumulation of Ca^{2+} is particularly deleterious to the survivability of muscle fibers, as Ca^{2+} activates a variety of free radical and enzymatic processes that target the cytoskeleton, sarcolemma, and mitochondria. Left unchecked, these processes may ultimately result in the necrosis of the muscle fiber.[90]

Investigators have gained valuable insight into the relationship between calcium and MD by developing several mouse models with altered calcium handling properties. One class of calcium handling proteins to be overexpressed is the transient receptor potential canonical (TRP) channels that facilitate the entry of Ca^{2+} into muscle fibers. Overexpression of TRP channels in wild-type mice causes skeletal muscles to exhibit increased fibrosis, immune cell infiltration, fatty tissue, and continuous cycles of muscle regeneration. This mouse model possesses many of the pathological characteristics of dystrophic mice and demonstrates that calcium misregulation alone is sufficient to trigger a MD-like phenotype. In a related set of experiments, investigators were able to partially correct the pathology of dystrophic *mdx* mice by overexpressing nonfunctional mutant TRP channels.[91,92] The rescue effect stems from the suppression of endogenous TRP channel activity elicited by the dominant-negative mutant TRP channels.

B. Dysferlin Gene Mutations

In addition to overactive calcium channels, compromised membrane-repair processes are also capable of eliciting MD. Dysferlin is a protein that augments membrane repair[2,93,94] and deficiencies have been linked to LGMD2B and Miyoshi myopathy.[95,96] In mouse models that lack dysferlin, skeletal muscles suffer from a progressive dystrophy and display altered morphology, muscle necrosis, macrophage infiltration, and fatty tissue replacement.[2,97] Without dysferlin, lesions in the membrane that may occur during contractile activity remain unrepaired, allowing calcium to rush into the muscle fiber. The uncontrolled influx of calcium into the cytosol is likely to overwhelm the endogenous calcium handling apparatus and trigger gradual muscle fiber necrosis. Studies on these mouse models suggest that imbalances in cytosolic calcium levels are sufficient to induce progressive MD, independent of the absence or presence of structural proteins such as dystrophin and the DGC. Other recent studies also suggest that dysferlin may be a component of T-tubules, but whether it regulates calcium flux there is not clear.[93,98]

C. Calpain 3 and LGMD2A

Mutations in calpain 3 gene cause LGMD2A, one of the most prevalent forms of the disease.[99–101] Calpain 3 is a calcium-dependent nonlysosomal protease involved primarily in regulating sarcomere formation and remodeling.[102] Additionally, calpain 3 serves structural functions as a component of the skeletal muscle triad[103] and the dysferlin complex.[104] Although calpain 3 has been found in many subcellular locations, the majority of it is anchored to titin, a giant sarcomeric protein that acts as a molecular ruler and a template for sarcomere assembly during muscle cell development.[105–107] As mutations in titin result in LGMD2J and a secondary deficiency of calpain 3,[108] mouse models for both forms of the disease will be discussed together.

Two calpain 3 knockout mouse models have been created. Richard *et al.* generated mice with disrupted proteolytic site of calpain 3 by substituting exons 2 and 3 with a neoR cassette using homologous recombination.[109] Skeletal muscles from the transgenic mice showed 80% decrease in calpain 3 mRNA and a complete absence of calpain 3 protein. Phenotypically, the muscles revealed regions of focal necrosis and apoptosis of nuclei associated with perturbation of NF-kB/IkBα as well as disruption of sarcolemmal integrity. Spencer's group took a different approach and generated calpain 3-deficient mice by utilizing a gene trap retroviral vector that introduced premature stop signals that completely abolished mRNA and protein expression.[110] The phenotype more closely resembled LGMD2A[111,112] as it showed muscle necrosis in combination with fiber atrophy. The studies utilizing these mice indicated that calpain 3 participates in remodeling of the sarcomere by promoting ubiquitination of its ligands.[102] Further, calpain 3-deficient mice were shown to have reduced calcium release in fibers due to perturbations in the triad-associated complex.[103]

Although most cases of LGMD2A result from mutations in the proteolytic region of calpain 3, pathogenic mutations are present along the entire sequence of calpain 3 gene. Mice expressing a proteolytically inactive mutant form (C129S) that retains its conformational properties enabled dissecting enzymatic and structural functions of calpain 3. Tagawa *et al.* generated transgenic mice expressing the mutated protein on the wild-type background.[113] Despite low expression levels, the aged mice showed signs of a mild myopathy, including reduced motor functions, as well as the presence of centrally nucleated nuclei, and lobulated and split fibers. In a more recent study,[114] analyses of C129S knock-in mice revealed progressive MD associated with inefficient redistribution of the mutant protein in stretched sarcomeres. Nevertheless, the mice showed a less severe phenotype than calpain 3 knockout mice, presumably due to preserved calpain 3 structural functions in the sarcoplasmic reticulum.[103,115] Another study showed that impaired titin–calpain 3 interaction results in reduced concentration of calpain 3 in the myofibrillar fraction along with its rapid degradation.[116] Here, generated transgenic mice expressed mutant calpains known to result in LGMD2A with preserved proteolytic function but mutated titin-binding site on the caveolin 3 knockout background.

D. Titin and LGMD2J

Mutations in the last 2 coding exons of titin lead to a mild late onset dominant tibial muscular dystrophy and a severe recessive LGMD2J. The knock-in mice carrying the most common mutation (FINmaj) recapitulate the human pathology, with heterozygotes showing a very mild phenotype and homozygotes developing a progressive MD.[117] This mouse model reveals also a secondary reduction in calpain 3 expression levels. Importantly, when the mice were bred onto the calpain 3-deficient background, the pathology of the FINmaj knock-in mice was attenuated, indicating that disrupted regulation of calpain 3 is responsible for the pathology of LGMD2J. It is also important to note that the pathology of another mouse model for titinopathy, severely affected MD with myositis (*mdm*), was not linked to deregulation of calpain 3.[118] The *Mdm* causing deletion of 83 amino acids in the N2A region of titin arose spontaneously in the Jackson laboratory in the early 1980s.[119] The mutation disrupts the calpain 3 binding site in the N2A region of titin. Despite that, *mdm*/calpain 3-deficient mice showed no change in the *mdm* pathology.

IV. α-Actinin

The α-actinins are a family of actin-binding proteins that are structurally related to dystrophin. Of the four members in the α-actinin family, only the sarcomeric isoforms α-actinin-2 and α-actinin-3 (Actn3) have been shown to bind dystrophin and other Z-disk associated proteins at the sarcomere,[120] while the cytoskeletal isoforms α-actinin-1 and α-actinin-4 are involved in cytoskeleton organization and cellular attachment to the extracellular matrix at focal adhesions.[121] As α-actinin-2 is the sole sarcomeric α-actinin isoform expressed in the heart, mutations in α-actinin-2 result in hypertrophic cardiomyopathy.[122]

In contrast, complete deficiency of α-actinin-3 occurs in 16% of the normal human population[123] and is associated with reduced muscle strength and power in elite athletes[124–127] and nonathletes.[128,129] Phenotypic analysis of the *Actn3* knockout mouse showed that absence of α-actinin-3 alters the structural, metabolic, and physiological characteristics of fast-twitch muscle fibers toward that of slow-twitch fibers,[130] suggesting that α-actinin-3 expression is essential for optimal rapid power generation in muscles of humans and mice. Recently, α-actinin-3 deficiency has also been associated with increased susceptibility to contraction-induced injury.[131] Work is ongoing to examine the influence of *ACTN3* genotype on muscle disease phenotypes by crossing the *Actn3* knockout mouse with mouse models of muscle disease, and to discover the therapeutic potential of α-actinin-3 overexpression by generation of *Actn3* transgenic models.

V. Fukutin-Related Protein and Dystroglycan Processing

Fukutin-related protein (FKRP) is a putative glycosyltransferase and is one of many that glycosylate αDG, allowing it to stabilize the sarcomere at the basal lamina *via* binding to laminin α2. Mutations in FKRP cause a broad spectrum of clinical phenotypes with variable disease severity. The most severe forms result in structural brain abnormalities (Walker–Warburg syndrome, WWS and muscle–eye–brain disease, MEB)[132] and an inability to walk (MDC1C).[133] FKRP mutations also result in LGMD2I, one of the most common forms of LGMD in childhood. LGMD2I itself is variable—one version has an early onset, Duchenne-like phenotype (severe LGMD2I) and the other a milder, slower progression (mild LGMD2I). Clinical severity has been correlated to depletion of glycosylated αDG and secondary reduction in laminin α2 as a consequence of reduced FKRP expression.[134,135]

The effect of FKRP downregulation has been explored in zebrafish[136] and mice[137] injected with AOs. While mouse models based on patient FKRP mutations that result in MEB (Y307N),[138] WWS (E310del), and MDC1C (P448L)[139] have also been generated, only those retaining the neomycin cassette succeeded in reducing *Fkrp* transcript levels and recapitulating the severity of the disease phenotypes.[138,139] Of the three mouse models, the E310del was embryonically lethal, FKRP-neo-Y307N resulted in perinatal lethality, while some FKRP-neo-P448L mutant mice survived to 10 months. The difference in lifespan of these models could be associated with the level of *Fkrp* mRNA/protein present. Homozygous E310del mice completely lack functional FKRP due to deletion of the C-terminus of FKRP that contains a consensus D × D motif commonly found in glycosyltransferases, suggesting that the presence of at least some FKRP activity is critical for embryonic development.[139] In contrast, the FKRP-neo-Y307N homozygotes expressed approximately 40% of *Fkrp* mRNA compared to controls,[138] while FKRP-neo-P448L homozygotes expressed approximately 55%.[139] This is consistent with results from knockdown of *Fkrp* transcript using RNA interference, which showed that the greater reductions of *Fkrp* mRNA (up to 75%) using dual shRNA cassettes induce overt dystrophic pathology, compared with the lower reductions from using single shRNA cassettes.[137] However, as shRNA expression by itself can cause toxicity, it remains unclear how much of the phenotype was FKRP specific.[140] Differences in disease severity between the latter two models could also be attributed to laminin α2 localization, which was reduced in FKRP-neo-Y307N mutant muscles but was normal in FKRP neo-P448L mutants. Interestingly, disease severity may not be correlated with the level of glycosylated αDG expressed, as some expression of glycosylated αDG was detected in FKRP-neo-Y307N muscles by Western blot but not in FKRP-neo-P448L muscles. Taken together, these results suggest that FKRP could have other unknown functions, for example, in regards to laminin α2 localization, and that factors other than glycosylation of αDG may be important for modulating disease severity.

VI. Desmin

Desminopathy is primarily (~ 80%) a dominantly inherited form of myofibrillar myopathy that leads to cardiac conduction disease, cardiomyopathy, and skeletal muscle weakness.[141,142] The initial symptoms can arise before 10 years of age but are more often reported in the 30s.[141,142] Symptoms usually begin with cardiac conduction disease and progress to cardiomyopathy with severe respiratory distress. There is currently no specific treatment for desminopathies to prevent the progressing cardiomyopathy and deterioration of the diaphragm. Desmin is the primary intermediate filament protein of muscle that connects the sarcomeres to the sarcolemma and organelles.[143] In the heart, desmin is particularly concentrated in the conduction system, at costameres, and at the desmosome in intercalated disks. Desmin-null mice develop a severe dilated cardiomyopathy from approximately 5 days of age that eventually leads to widespread cardiomyocyte degeneration, fibrosis, and calcification.[144–147] Abnormal mitochondria distribution, proliferation, and swelling are thought to be a central pathologic feature of desmin knockout cardiomyocytes and desmin-related cardiomyopathy.[148–150] The cardiomyocytes are also structurally abnormal in that the sarcomeres do not align and do not attach appropriately to the intercalated disks.[144–147] Desmin knockout mice develop a mild cardiac conduction disease probably owing to the disruption of desmosomes in the conduction system.[151] There are two transgenic mouse models that mimic dominantly inherited desminopathy in humans. The first is a 7-amino acid deletion (R173-E179) that leads to chimeric intracellular aggregates, not normally seen in desmin-null mice.[152] The second transgenic mouse is a L345P mutation that significantly alters mitochondrial morphology and function that is associated with altered Ca^{2+} handling.[153] Together, these transgenic mice suggest that different mutations in desmin can lead to strikingly different phenotypes.

VII. Laminin α2

Laminin α2-deficient mouse models have significantly expanded our understanding of congenital MD type 1A (MDC1A). Models derive from the dystrophic *dy/dy* mouse line whose pathology originates from mutations in the *LAMA2* gene.[154,155] Differences in site and type of mutation determines residual laminin α2 expression, and by extension, disease severity.[156–158] The *dy/dy* mouse possesses a moderate fibrotic and dystrophic phenotype, coincident with reduced laminin α2 expression.[156,159] The mutation responsible for the *dy/dy* phenotype is unknown; however, additional mouse models with targeted mutation (dy^{2J}/dy^{2J}) or removal (dy^{W}/dy^{W}) of the LAMA2 N-terminal domain show minor and severe dystrophic phenotypes, respectively.[156,160,161] Importantly, the severe dystrophic phenotype observed in dy^{W}/dy^{W} closely resembles that of dy^{3K}/dy^{3K} mice where laminin α2 expression is completely abolished.[156,162] While the secondary mechanisms governing MDC1A pathogenesis remain unclear, these models underscore the connection between laminin α2 expression and disease severity and will likely prove invaluable in the design of novel therapeutic strategies.

A congenital MD was found in a mixed breed dog that had an absence of the laminin isoform α2, which is predominant in, but not specific to, skeletal muscle.[163] Mutations causing laminin α2 deficiency result in central nervous system defects in humans and peripheral nerve defects in *dy* mice. It will be interesting to see if similar abnormalities are found in laminin α2-deficient dogs.

VIII. Collagen VI

Collagens are secreted triple-helical extracellular matrix proteins that maintain integrity across many tissues.[164,165] Collagen VI forms a heterotrimer composed of three alpha chains encoded by the *COL6A1*, *COL6A2*, and *COL6A3* genes. The microfibrils generated upon assembly of collagen VI molecules have been implicated in the maintenance of muscle function and provide an important link between muscle cells and the surrounding matrix through a variety of molecular interactions.[166–168] Mutations in the three collagen VI alpha chains give rise to autosomal dominant Bethlem myopathy (BM)[169–172] as well as the more severe usually autosomal recessive Ullrich congenital muscular dystrophy (UCMD).[173,174] BM is characterized by later onset, proximal muscle weakness with joint contractures, especially of the elbow, finger, and ankle, and slow progression, while UCMD typically manifests early with rapid progression of symptoms that include muscle weakness and hypotonia, joint contractures, and hyperelastic distal joints. Respiratory failure is common in later stages of UCMD and can result in early death if untreated. In both BM and UCMD, intelligence is normal and serum CK levels are normal or slightly elevated.

Investigation of collagen function led to generation of collagen VI-deficient mice through targeted disruption of the *col6a1* gene, thus abolishing assembly of the triple helix.[175] The resulting phenotype strongly resembled BM, and these mice became a useful animal model for studying the pathophysiology of both BM and UCMD.[167] Histology and Evans blue dye staining revealed early onset focal necrosis, fiber size variation, and mild central nucleation, all indicators of skeletal muscle pathology. Locomotor performance is not severely affected, though *col6a1*-deficient mice exhibit muscle weakness, and the condition progresses very slowly, similar to BM. However, the diaphragm is more severely affected than other muscle groups, similar to UCMD.[167,176,177] *Col6a1*-deficient mice also reflect the dominant character of BM, as heterozygotes show signs of myopathy, though with milder defects.[175]

The mouse model of collagen VI deficiencies has revealed important mechanistic insights into the pathophysiology of BM and UCMD, including the involvement of mitochondrial dysfunction and apoptosis,[167,176] but the relationship between these abnormalities and collagen mutations is still poorly understood. Nevertheless, *col6a1*-deficient mice have allowed testing of specific therapeutic interventions such as cyclosporine A and other pharmacological treatments to target the mitochondrial permeability transition pore.[167,178] While the severity of the deficits observed in UCMD is not captured by these mice, zebrafish models have been reported with early onset, severe myopathy that could be useful for initial screening of novel therapies, and further investigating the molecular etiology of both BM and UCMD.[179]

IX. D4Z4 Repeats and FSHD

Facioscapulohumeral muscular dystrophy (FSHD) is caused by macrosatellite DNA contraction in the D4Z4 region of telomeric 4q35.[180] The molecular events following DNA contraction include epigenetic events resulting in chromatin hypomethylation, chromatin relaxation, and expression of a highly conserved double homeobox retrogene, *DUX4*, embedded within the repeats.[181] Recent studies suggest that disease arises only when *DUX4* expression occurs in the context of an intact polyadenylation site following the last repeat on the pathogenic alleles and presents a unifying model for FSHD disease mechanism.[182,183]

Controversy surrounds the involvement of another gene proximal to 4q35, *FRG1*. Transgenic mouse studies show that overexpression of *FRG1* causes an MD similar to FSHD.[184] Disease severity increases with an increase in *FRG1* expression. To date, this is the only model that recapitulates some of the disease features, such as a degenerative muscle phenotype more pronounced in muscles similarly affected in FSHD patients, increased central nuclei, abnormal fiber morphology with increased cross-sectional area, decreased muscle size, kyphosis, and weakness.

Attempts to produce a transgenic mouse line expressing *DUX4* have not been successful, but engineering a mouse line that expresses a low level of DUX4 in adult mice to recapitulate DUX4 levels in FSHD patient muscle may prove a difficult task. DUX4 expression in muscle cells in culture results in cell death accompanied by an increased sensitivity to oxidative stress which supports the notion that DUX4 plays a role in early development but is inactivated in the adult. However, because it is generally agreed that overexpression of a gene or genes is the cause of FSHD, therapy development is now underway directed toward eliminating toxicity by targeting mRNA for destruction. Muscle structural and histological changes associated with FRG1 overexpression in the FRG1 transgenic mouse were improved with delivery of constitutively expressed RNA hairpins to direct RNA interference.[140,185] More importantly, functional rescue of muscles, as measured by endurance in treadmill running studies, was possible with AAV6 short hairpin RNA intravascular systemic delivery after the onset of disease.[140] These studies serve as a proof-of-principle for therapeutic antisense approaches for FSHD and other dominant muscle diseases.

X. DMPK and Myotonic Dystrophy

Myotonic dystrophy (DM1) is caused by a microsatellite CTG repeat expansion in the 3′UTR of the dystrophia myotonica protein kinase gene (*DMPK*).[186] Transcripts with expanded repeats accumulate in the nuclei of cells and alter the availability of proteins involved in the regulation of alternative splicing during development.[187,188] Important protein families for developmental splicing regulation that are affected by the *DMPK* mRNA repeat expansions are the muscleblind-like (MBNL) and CUGBP/Elav-like family (CELF) proteins.[187,189,190] Disruption of splicing patterns to produce embryonic transcript profiles for a host of unrelated mRNAs results in a disease phenotype with seemingly unrelated features, such as muscle stiffness (myotonia), hyperinsulinemia, gastrointestinal insufficiencies, and cognitive impairment.[186]

Approximately 20 transgenic mouse lines have been generated to dissect the molecular pathology resulting from the toxic gain-of-function mRNA expression and to produce models for testing potential therapies for DM1 (reviewed in detail[191a]). The models best suited for therapy testing are based on either expressing the CTG repeat expansion or altering the levels of MBNL1 and CUGBP1. One of the most informative and useful transgenic lines, the HSA^{LR} line, was generated from expression of CTG_{250} from the human α-skeletal actin gene placed in the *HSA* 3′UTR.[191b] The HSA^{LR} mice display profound myotonia and subsequent changes in MBNL levels that lead to altered splicing profiles. Similar to the human disease, MBNL1 can be found in nuclear accumulations of the protein with the CUG_{250} repeat *HSA* mRNA. Notably, CUGBP1 levels are not affected, and the muscle-wasting phenotype of DM1 is not reproduced in this transgenic mouse line. Additionally, expression of HSA^{LR} is limited to muscle and therefore cannot contribute to the effects in multiple organ systems associated with the human condition. Regardless of the restriction of expression of the HSA^{LR} mRNA in muscle, the HSA^{LR} mouse has been used for testing antisense therapeutic approaches and has resulted in studies showing elimination of myotonia and reversal of splicing defects with intramuscular delivery of AOs targeting the CUG repeats.[192]

Another notable transgenic line contains a large fragment of the human *DMPK* locus with an unstable expanded CTG repeat that has become larger with increasing generations to reach 1000–1800 repeats.[193] The DM500 mice display multisystemic attributes of human DM1, including splicing defects in the muscle, heart, CNS, myotonia, muscle weakness, glucose metabolism defects, and tau distribution abnormalities in the brain. The requirement for homozygosity in these mice suggests that the level of mutant *DMPK* mRNA expression is below a threshold necessary for pronounced disease and makes these mice expensive and difficult for testing potential therapies.

A third model allowed conditional tissue-specific expression of a 960 repeat expansion in the context of the last exon of *DMPK* where the expansion normally is found.[194] These mice recapitulate all muscle features of DM1, including muscle wasting, highlighting the importance of the chromosomal context for a more complete disease phenocopy. Other models focus on manipulation of the levels of the MBNL and CLEF protein family members and have limited potential for therapeutic testing related to manipulation of their expression levels. MBNL1 knockout mice recapitulate the splicing defects and myotonia observed with toxic repeat expression.[195] CUGBP1 overexpressing mice display a severe muscle-wasting phenotype not seen in repeat expansion-based models lacking the *DMPK* chromosomal context, which correlates CUGBP1 upregulation with a muscle-wasting phenotype.[196] Despite the numbers of transgenic mice made to reproduce the many features of DM1, there remain only two lines that have been used for testing potential therapies: the *HSA*^{LR} model and the DM500 model (500 repeats in large chrosomal fragment from the *DMPK* locus). This situation highlights the difficulties often associated with attempts to recapitulate human disease in the mouse with gene insertion, especially related to unstable repetitive and ultimately toxic DNA.

XI. Conclusions

The large and growing number of animal models available to study MD has proven to be an indispensible resource for understanding the pathophysiology of these disorders and for testing experimental approaches to therapy.[197] However, the genes for a number of forms of MD have not yet been identified and there do not yet exist corresponding animal models, and even those genes that have been found have not all been mutated in animals. Developing these newer models is an important goal, especially for the rare MDs where few patient samples are available. In such cases, therapies being developed for other disorders could potentially be quickly tested in animal models for safety prior to therapeutically oriented clinical trials.

Acknowledgments

Supported by grants from the Muscular Dystrophy Association (to G. L. O., G. B. B., J. R. C., and J. S. C.), by NIH Grant AR044533 (to J. S. C.), and by a grant from the Pacific Northwest Friends of FSHD (to J. R. C.).

References

1. Emery AE. The muscular dystrophies. *Lancet* 2002;**359**:687–95.
2. Bansal D, Miyake K, Vogel SS, Groh S, Chen CC, Williamson R, et al. Defective membrane repair in dysferlin-deficient muscular dystrophy. *Nature* 2003;**423**:168–72.
3. Arnett AL, Chamberlain JR, Chamberlain JS. Therapy for neuromuscular disorders. *Curr Opin Genet Dev* 2009;**19**:290–7.
4. Kunkel LM, et al. Analysis of deletions in DNA from patients with Becker and Duchenne muscular dystrophy. *Nature* 1986;**322**:73–7.
5. Monaco AP, Neve RL, Coletti-Feener C, Bertelson CJ, Kurnit DM, Kunkel LM. Isolation of candidate cDNA clones for portions of the Duchenne muscular dystrophy gene. *Nature* 1986;**323**:646–50.

6. Emery AE, Muntoni F. *Duchenne muscular dystrophy.* 3rd ed. Oxford: Oxford University Press; 2003.
7. Abmayr S, Chamberlain J. The structure and function of dystrophin. In: Winder SJ, editor. *Molecular mechanisms of muscular dystrophies.* Georgetown: Landes Biosciences; 2006. pp. 14–34.
8. Ervasti JM. Structure and function of the dystrophin-glycoprotein complex. In: Winder SJ, editor. *Molecular mechanisms of muscular dystrophies.* Georgetown: Landes Biosciences; 2006. pp. 1–13.
9. Banks GB, Chamberlain JS. The value of mammalian models for Duchenne muscular dystrophy in developing therapeutic strategies. *Curr Top Dev Biol* 2008;**84**:431–53.
10. Bulfield G, Siller WG, Wight PA, Moore KJ. X chromosome-linked muscular dystrophy (mdx) in the mouse. *Proc Natl Acad Sci USA* 1984;**81**:1189–92.
11. Sicinski P, Geng Y, Ryder-Cook AS, Barnard EA, Darlison MG, Barnard PJ. The molecular basis of muscular dystrophy in the mdx mouse: a point mutation. *Science* 1989;**244**:1578–80.
12. Im WB, Phelps SF, Copen EH, Adams EG, Slightom JL, Chamberlain JS. Differential expression of dystrophin isoforms in strains of mdx mice with different mutations. *Hum Mol Genet* 1996;**5**:1149–53.
13. Chamberlain JS, Metzger J, Reyes M, Townsend D, Faulkner JA. Dystrophin-deficient *mdx* mice display a reduced life span and are susceptible to spontaneous rhabdomyosarcoma. *FASEB J* 2007;**21**:2195–204.
14. Matsumura K, Ervasti JM, Ohlendieck K, Kahl SD, Campbell KP. Association of dystrophin-related protein with dystrophin-associated proteins in mdx mouse muscle. *Nature* 1992;**360**:588–91.
15. Takemitsu M, Ishiura S, Koga R, Kamakura K, Arahata K, Nonaka I, et al. Dystrophin-related protein in the fetal and denervated skeletal muscles of normal and mdx mice. *Biochem Biophys Res Commun* 1991;**180**:1179–86.
16. Coulton GR, Curtin NA, Morgan JE, Partridge TA. The mdx mouse skeletal muscle myopathy: II. Contractile properties. *Neuropathol Appl Neurobiol* 1988;**14**:299–314.
17. Dangain J, Vrbova G. Muscle development in mdx mutant mice. *Muscle Nerve* 1984;**7**:700–4.
18. Tanabe Y, Esaki K, Nomura T. Skeletal muscle pathology in X chromosome-linked muscular dystrophy (mdx) mouse. *Acta Neuropathol* 1986;**69**:91–5.
19. Stedman HH, Sweeney HL, Shrager JB, Maguire HC, Panettieri RA, Petrof B, et al. The mdx mouse diaphragm reproduces the degenerative changes of Duchenne muscular dystrophy. *Nature* 1991;**352**:536–9.
20. Lynch GS, Rafael JA, Hinkle RT, Cole NM, Chamberlain JS, Faulkner JA. Contractile properties of diaphragm muscle segments from old mdx and old transgenic mdx mice. *Am J Physiol* 1997;**272**:C2063–8; [Cell Physiol. 41].
21. Chapman VM, Miller DR, Armstrong D, Caskey CT. Recovery of induced mutations for X chromosome-linked muscular dystrophy in mice. *Proc Natl Acad Sci USA* 1989;**86**:1292–6.
22. Cox GA, Phelps SF, Chapman VM, Chamberlain JS. New mdx mutation disrupts expression of muscle and nonmuscle isoforms of dystrophin. *Nat Genet* 1993;**4**:87–93.
23. Danko I, Chapman V, Wolff JA. The frequency of revertants in mdx mouse genetic models for Duchenne muscular dystrophy. *Pediatr Res* 1992;**32**:128–31.
24. Rafael JA, Cox GA, Corrado K, Jung D, Campbell KP, Chamberlain JS. Forced expression of dystrophin deletion constructs reveals structure-function correlations. *J Cell Biol* 1996;**134**:93–102.
25. Gregorevic P, Allen JM, Minami E, Blankinship MJ, Haraguchi M, Meuse L, et al. rAAV6-microdystrophin preserves muscle function and extends lifespan in severely dystrophic mice. *Nat Med* 2006;**12**:787–9.

26. Odom GL, Gregorevic P, Allen JM, Finn E, Chamberlain JS. Microutrophin delivery through rAAV6 increases lifespan and improves muscle function in dystrophic dystrophin/utrophin-deficient mice. *Mol Ther* 2008;**16**:1539–45.
27. Wertz K, Fuchtbauer EM. Dmd(mdx-beta geo): a new allele for the mouse dystrophin gene. *Dev Dyn* 1998;**212**:229–41.
28. Araki E, Nakamura K, Nakao K, Kameya S, Kobayashi O, Nonaka I, et al. Targeted disruption of exon 52 in the mouse dystrophin gene induced muscle degeneration similar to that observed in Duchenne muscular dystrophy. *Biochem Biophys Res Commun* 1997;**238**:492–7.
29. Aartsma-Rus A, Fokkema I, Verschuuren J, Ginjaar I, van Deutekom J, van Ommen GJ, et al. Theoretic applicability of antisense-mediated exon skipping for Duchenne muscular dystrophy mutations. *Hum Mutat* 2009;**30**:293–9.
30. Helderman-van den Enden AT, Straathof CS, Aartsma-Rus A, den Dunnen JT, Verbist BM, Bakker E, et al. Becker muscular dystrophy patients with deletions around exon 51; a promising outlook for exon skipping therapy in Duchenne patients. *Neuromuscul Disord* 2010;**20**:251–4.
31. Kinali M, Arechavala-Gomeza V, Feng L, Cirak S, Hunt D, Adkin C, et al. Local restoration of dystrophin expression with the morpholino oligomer AVI-4658 in Duchenne muscular dystrophy: a single-blind, placebo-controlled, dose-escalation, proof-of-concept study. *Lancet Neurol* 2009;**8**:918–28.
32. Goemans NM, Tulinius M, van den Akker JT, Burm BE, Ekhart PF, Heuvelmans N, et al. Systemic administration of PRO051 in Duchenne's muscular dystrophy. *N Engl J Med* 2011;**364**:1513–22.
33. van Deutekom JC, Janson AA, Ginjaar IB, Frankhuizen WS, Aartsma-Rus A, Bremmer-Bout M, et al. Local dystrophin restoration with antisense oligonucleotide PRO051. *N Engl J Med* 2007;**357**:2677–86.
34. Chamberlain JR, Chamberlain JS. Muscling in: gene therapies for muscular dystrophy target RNA. *Nat Med* 2010;**16**:170–1.
35. Aoki Y, Nakamura A, Yokota T, Saito T, Okazawa H, Nagata T, et al. In-frame dystrophin following exon 51-skipping improves muscle pathology and function in the exon 52-deficient mdx mouse. *Mol Ther* 2010;**18**:1995–2005.
36. Kudoh H, Ikeda H, Kakitani M, Ueda A, Hayasaka M, Tomizuka K, et al. A new model mouse for Duchenne muscular dystrophy produced by 2.4 Mb deletion of dystrophin gene using Cre-loxP recombination system. *Biochem Biophys Res Commun* 2005;**328**:507–16.
37. Deconinck AE, Rafael JA, Skinner JA, Brown SC, Potter AC, Metzinger L, et al. Utrophin-dystrophin-deficient mice as a model for Duchenne muscular dystrophy. *Cell* 1997;**90**:717–27.
38. Grady RM, Teng H, Nichol MC, Cunningham JC, Wilkinson RS, Sanes JR. Skeletal and cardiac myopathies in mice lacking utrophin and dystrophin: a model for Duchenne muscular dystrophy. *Cell* 1997;**90**:729–38.
39. Tinsley J, Deconinck N, Fisher R, Kahn D, Phelps S, Gillis JM, et al. Expression of full-length utrophin prevents muscular dystrophy in mdx mice. *Nat Med* 1998;**4**:1441–4.
40. Rafael JA, Tinsley JM, Potter AC, Deconinck AE, Davies KE. Skeletal muscle-specific expression of a utrophin transgene rescues utrophin-dystrophin deficient mice. *Nat Genet* 1998;**19**:79–82.
41. Chandrasekharan K, Yoon JH, Xu Y, deVries S, Camboni M, Janssen PM, et al. A human-specific deletion in mouse Cmah increases disease severity in the mdx model of Duchenne muscular dystrophy. *Sci Transl Med* 2010;**2**:42ra54.

42. Sacco A, Mourkioti F, Tran R, Choi J, Llewellyn M, Kraft P, et al. Short telomeres and stem cell exhaustion model Duchenne muscular dystrophy in mdx/mTR mice. *Cell* 2010;**143**:1059–71.
43. Cooper BJ, Winand NJ, Stedman H, Valentine BA, Hoffman EP, Kunkel LM, et al. The homologue of the Duchenne locus is defective in X-linked muscular dystrophy of dogs. *Nature* 1988;**334**:154–6.
44. Partridge TA. Models of dystrophinopathy, pathological mechanisms and assessment of therapies. In: Brown SC, Lacy JA, editors. *Dystrophin: gene, protein and cell biology.* Cambridge: Cambridge University Press; 1997. pp. 310–31.
45. Schatzberg SJ, Olby NJ, Breen M, Anderson LV, Langford CF, Dickens HF, et al. Molecular analysis of a spontaneous dystrophin 'knockout' dog. *Neuromuscul Disord* 1999;**9**:289–95.
46. Valentine BA, Winand NJ, Pradhan D, Moise NS, de Lahunta A, Kornegay JN, et al. Canine X-linked muscular dystrophy as an animal model of Duchenne muscular dystrophy: a review. *Am J Med Genet* 1992;**42**:352–6.
47. Shimatsu Y, Katagiri K, Furuta T, Nakura M, Tanioka Y, Yuasa K, et al. Canine X-linked muscular dystrophy in Japan (CXMDJ). *Exp Anim* 2003;**52**:93–7.
48. Wang Z, Kuhr CS, Allen JM, Blankinship M, Gregorevic P, Chamberlain JS, et al. Sustained AAV-mediated dystrophin expression in a canine model of Duchenne muscular dystrophy with a brief course of immunosuppression. *Mol Ther* 2007;**15**:1160–6.
49. Meryon E. On granular and fatty degeneration of the voluntary muscles. *Med Chir Trans* 1852;**35**(73–84):1.
50. Duchenne GB. Récherches sur la paralysie musculaire pseudohypertrophique ou paralysie myo-sclerosique. *Arch Gen Med* 1868;**11**:5–25.
51. Kloepfer HW, Talley C. Autosomal recessive inheritance of Duchennetype muscular dystrophy. *Ann Hum Genet* 1958;**22**:138–43.
52. Hoffman EP, Brown Jr. RH, Kunkel LM. Dystrophin: the protein product of the Duchenne muscular dystrophy locus. *Cell* 1987;**51**:919–28.
53. Ervasti JM, Campbell KP. Membrane organization of the dystrophin-glycoprotein complex. *Cell* 1991;**66**:1121–31.
54. Roberds SL, Leturcq F, Allamand V, Piccolo F, Jeanpierre M, Anderson RD, et al. Missense mutations in the adhalin gene linked to autosomal recessive muscular dystrophy. *Cell* 1994;**78**:625–33.
55. Bonnemann CG, Modi R, Noguchi S, Mizuno Y, Yoshida M, Gussoni E, et al. Beta-sarcoglycan (A3b) mutations cause autosomal recessive muscular dystrophy with loss of the sarcoglycan complex. *Nat Genet* 1995;**11**:266–73.
56. Noguchi S, McNally EM, Ben Othmane K, Hagiwara Y, Mizuno Y, Yoshida M, et al. Mutations in the dystrophin-associated protein gamma-sarcoglycan in chromosome 13 muscular dystrophy. *Science* 1995;**270**:819–22.
57. Nigro V, de Sa Moreira E, Piluso G, Vainzof M, Belsito A, Politano L, et al. Autosomal recessive limb-girdle muscular dystrophy, LGMD2F, is caused by a mutation in the delta-sarcoglycan gene. *Nat Genet* 1996;**14**:195–8.
58. Homburger F, Baker JR, Nixon CW, Wilgram G. New hereditary disease of Syrian hamsters. Primary, generalized polymyopathy and cardiac necrosis. *Arch Intern Med* 1962;**110**:660–2.
59. Nigro V, Piluso G, Belsito A, Politano L, Puca AA, Papparella S, et al. Identification of a novel sarcoglycan gene at 5q33 encoding a sarcolemmal 35 kDa glycoprotein. *Hum Mol Genet* 1996;**5**:1179–86.
60. Nigro V, Okazaki Y, Belsito A, Piluso G, Matsuda Y, Politano L, et al. Identification of the Syrian hamster cardiomyopathy gene. *Hum Mol Genet* 1997;**6**:601–7.

61. Awad SS, Welty JD. Comparisons of hemodynamics throughout the life span of the Bio 14.6 cardiomyopathic with the F1B normal hamster. *Comp Biochem Physiol A Comp Physiol* 1990;**97**:487–91.
62. Sakamoto A, Ono K, Abe M, Jasmin G, Eki T, Murakami Y, et al. Both hypertrophic and dilated cardiomyopathies are caused by mutation of the same gene, delta-sarcoglycan, in hamster: an animal model of disrupted dystrophin-associated glycoprotein complex. *Proc Natl Acad Sci USA* 1997;**94**:13873–8.
63. Sakamoto A. Molecular pathogenesis of severe cardiomyopathy in the TO-2 hamster. *Exp Clin Cardiol* 2003;**8**:143–6.
64. Hack AA, Groh ME, McNally EM. Sarcoglycans in muscular dystrophy. *Microsc Res Tech* 2000;**48**:167–80.
65a. Coral-Vazquez R, Cohn RD, Moore SA, Hill JA, Weiss RM, Davisson RL, et al. Disruption of the sarcoglycan-sarcospan complex in vascular smooth muscle: a novel mechanism for cardiomyopathy and muscular dystrophy. *Cell* 1999;**98**:465–74.
65b. Burbach JA, Schlenker EH, Johnson JL. Morphometry, histochemistry, and contractility of dystrophic hamster diaphragm. *Am J Physiol* 1987;**253**:R275–84.
65c. Wansapura JP, Millay DP, Dunn RS, Molkentin JD, Benson DW. Magnetic resonance imaging assessment of cardiac dysfunction in delta-sarcoglycan null mice. *Neuromuscul Disord* 2011;**21**:68–73.
66. Hack AA, Ly CT, Jiang F, Clendenin CJ, Sigrist KS, Wollmann RL, et al. Gamma-sarcoglycan deficiency leads to muscle membrane defects and apoptosis independent of dystrophin. *J Cell Biol* 1998;**142**:1279–87.
67. Ben Hamida M, Ben Hamida C, Zouari M, Belal S, Hentati F. Limb-girdle muscular dystrophy 2C: clinical aspects. *Neuromuscul Disord* 1996;**6**:493–4.
68. Durbeej M, Cohn RD, Hrstka RF, Moore SA, Allamand V, Davidson BL, et al. Disruption of the beta-sarcoglycan gene reveals pathogenetic complexity of limb-girdle muscular dystrophy type 2E. *Mol Cell* 2000;**5**:141–51.
69. Araishi K, Sasaoka T, Imamura M, Noguchi S, Hama H, Wakabayashi E, et al. Loss of the sarcoglycan complex and sarcospan leads to muscular dystrophy in beta-sarcoglycan-deficient mice. *Hum Mol Genet* 1999;**8**:1589–98.
70. Cohn RD, Durbeej M, Moore SA, Coral-Vazquez R, Prouty S, Campbell KP. Prevention of cardiomyopathy in mouse models lacking the smooth muscle sarcoglycan-sarcospan complex. *J Clin Invest* 2001;**107**:R1–7.
71. Durbeej M, Campbell KP. Muscular dystrophies involving the dystrophin-glycoprotein complex: an overview of current mouse models. *Curr Opin Genet Dev* 2002;**12**:349–61.
72. Chan YM, Bonnemann CG, Lidov HG, Kunkel LM. Molecular organization of sarcoglycan complex in mouse myotubes in culture. *J Cell Biol* 1998;**143**:2033–44.
73. Duggan DJ, Manchester D, Stears KP, Mathews DJ, Hart C, Hoffman EP. Mutations in the delta-sarcoglycan gene are a rare cause of autosomal recessive limb-girdle muscular dystrophy (LGMD2). *Neurogenetics* 1997;**1**:49–58.
74. Klinge L, Dekomien G, Aboumousa A, Charlton R, Epplen JT, Barresi R, et al. Sarcoglycanopathies: can muscle immunoanalysis predict the genotype? *Neuromuscul Disord* 2008;**18**:934–41.
75. Vainzof M, Passos-Bueno MR, Canovas M, Moreira ES, Pavanello RC, Marie SK, et al. The sarcoglycan complex in the six autosomal recessive limb-girdle muscular dystrophies. *Hum Mol Genet* 1996;**5**:1963–9.
76. Duclos F, Straub V, Moore SA, Venzke DP, Hrstka RF, Crosbie RH, et al. Progressive muscular dystrophy in alpha-sarcoglycan-deficient mice. *J Cell Biol* 1998;**142**:1461–71.

77. Patel ND, Jannapureddy SR, Hwang W, Chaudhry I, Boriek AM. Altered muscle force and stiffness of skeletal muscles in alpha-sarcoglycan-deficient mice. *Am J Physiol Cell Physiol* 2003;**284**:C962–8.
78. Deitz K, Morrison JA, Kline K, Guo LT, Shelton GD. Sarcoglycan-deficient muscular dystrophy in a Boston Terrier. *J Vet Intern Med* 2008;**22**:476–80.
79. Schatzberg S. In: *Proceedings of the American College of Veterinary Internal Medicine, 22nd annual forum, Dallas, TX*; 2003.
80. Berchtold MW, Brinkmeier H, Muntener M. Calcium ion in skeletal muscle: its crucial role for muscle function, plasticity, and disease. *Physiol Rev* 2000;**80**:1215–65.
81. Turner PR, Fong PY, Denetclaw WF, Steinhardt RA. Increased calcium influx in dystrophic muscle. *J Cell Biol* 1991;**115**:1701–12.
82. Vandebrouck C, Martin D, Colson-Van SM, Debaix H, Gailly P. Involvement of TRPC in the abnormal calcium influx observed in dystrophic (mdx) mouse skeletal muscle fibers. *J Cell Biol* 2002;**158**:1089–96.
83. Culligan K, Banville N, Dowling P, Ohlendieck K. Drastic reduction of calsequestrin-like proteins and impaired calcium binding in dystrophic mdx muscle. *J Appl Physiol* 2002;**92**:435–45.
84. Dowling P, Doran P, Ohlendieck K. Drastic reduction of sarcalumenin in Dp427 (dystrophin of 427 kDa)-deficient fibres indicates that abnormal calcium handling plays a key role in muscular dystrophy. *Biochem J* 2004;**379**:479–88.
85. Divet A, Huchet-Cadiou C. Sarcoplasmic reticulum function in slow- and fast-twitch skeletal muscles from mdx mice. *Pflugers Arch* 2002;**444**:634–43.
86. Divet A, Lompre AM, Huchet-Cadiou C. Effect of cyclopiazonic acid, an inhibitor of the sarcoplasmic reticulum Ca-ATPase, on skeletal muscles from normal and mdx mice. *Acta Physiol Scand* 2005;**184**:173–86.
87. Alderton JM, Steinhardt RA. Calcium influx through calcium leak channels is responsible for the elevated levels of calcium-dependent proteolysis in dystrophic myotubes. *J Biol Chem* 2000;**275**:9452–60.
88. Mallouk N, Jacquemond V, Allard B. Elevated subsarcolemmal Ca2 + in mdx mouse skeletal muscle fibers detected with Ca2+-activated K+ channels. *Proc Natl Acad Sci USA* 2000;**97**:4950–5.
89. Turner PR, Westwood T, Regen CM, Steinhardt RA. Increased protein degradation results from elevated free calcium levels found in muscle from mdx mice. *Nature* 1988;**335**:735–8.
90. Gissel H. The role of Ca2 + in muscle cell damage. *Ann N Y Acad Sci* 2005;**1066**:166–80.
91. Iwata Y, Katanosaka Y, Arai Y, Shigekawa M, Wakabayashi S. Dominant-negative inhibition of Ca2 + influx via TRPV2 ameliorates muscular dystrophy in animal models. *Hum Mol Genet* 2009;**18**:824–34.
92. Millay DP, Goonasekera SA, Sargent MA, Maillet M, Aronow BJ, Molkentin JD. Calcium influx is sufficient to induce muscular dystrophy through a TRPC-dependent mechanism. *Proc Natl Acad Sci USA* 2009;**106**:19023–8.
93. Klinge L, Laval S, Keers S, Haldane F, Straub V, Barresi R, et al. From T-tubule to sarcolemma: damage-induced dysferlin translocation in early myogenesis. *FASEB J* 2007;**21**:1768–76.
94. Lennon NJ, Kho A, Bacskai BJ, Perlmutter SL, Hyman BT, Brown Jr. RH. Dysferlin interacts with annexins A1 and A2 and mediates sarcolemmal wound-healing. *J Biol Chem* 2003;**278**:50466–73.
95. Bashir R, Britton S, Strachan T, Keers S, Vafiadaki E, Lako M, et al. A gene related to Caenorhabditis elegans spermatogenesis factor fer-1 is mutated in limb-girdle muscular dystrophy type 2B. *Nat Genet* 1998;**20**:37–42.

96. Liu J, Aoki M, Illa I, Wu C, Fardeau M, Angelini C, et al. Dysferlin, a novel skeletal muscle gene, is mutated in Miyoshi myopathy and limb girdle muscular dystrophy. *Nat Genet* 1998;**20**:31–6.
97. Ho M, Post CM, Donahue LR, Lidov HG, Bronson RT, Goolsby H, et al. Disruption of muscle membrane and phenotype divergence in two novel mouse models of dysferlin deficiency. *Hum Mol Genet* 2004;**13**:1999–2010.
98. Klinge L, Harris J, Sewry C, Charlton R, Anderson L, Laval S, et al. Dysferlin associates with the developing T-tubule system in rodent and human skeletal muscle. *Muscle Nerve* 2010;**41**:166–73.
99. Beckmann JS, Spencer M. Calpain 3, the "gatekeeper" of proper sarcomere assembly, turnover and maintenance. *Neuromuscul Disord* 2008;**18**:913–21.
100. Duguez S, Bartoli M, Richard I. Calpain 3: a key regulator of the sarcomere? *FEBS J* 2006;**273**:3427–36.
101. Sorimachi H, Hata S, Ono Y. Calpain chronicle—an enzyme family under multidisciplinary characterization. *Proc Jpn Acad Ser B Phys Biol Sci* 2011;**87**:287–327.
102. Kramerova I, Kudryashova E, Venkatraman G, Spencer MJ. Calpain 3 participates in sarcomere remodeling by acting upstream of the ubiquitin-proteasome pathway. *Hum Mol Genet* 2005;**14**:2125–34.
103. Kramerova I, Kudryashova E, Wu B, Ottenheijm C, Granzier H, Spencer MJ. Novel role of calpain-3 in the triad-associated protein complex regulating calcium release in skeletal muscle. *Hum Mol Genet* 2008;**17**:3271–80.
104. Huang Y, de Morree A, van Remoortere A, Bushby K, Frants RR, Dunnen JT, et al. Calpain 3 is a modulator of the dysferlin protein complex in skeletal muscle. *Hum Mol Genet* 2008;**17**:1855–66.
105. Linke WA, Kruger M. The giant protein titin as an integrator of myocyte signaling pathways. *Physiology* 2010;**25**:186–98.
106. Kruger M, Linke WA. The giant protein titin: a regulatory node that integrates myocyte signaling pathways. *J Biol Chem* 2011;**286**:9905–12.
107. Gautel M. The sarcomeric cytoskeleton: who picks up the strain? *Curr Opin Cell Biol* 2011;**23**:39–46.
108. Haravuori H, Vihola A, Straub V, Auranen M, Richard I, Marchand S, et al. Secondary calpain3 deficiency in 2q-linked muscular dystrophy: titin is the candidate gene. *Neurology* 2001;**56**:869–77.
109. Richard I, Roudaut C, Marchand S, Baghdiguian S, Herasse M, Stockholm D, et al. Loss of calpain 3 proteolytic activity leads to muscular dystrophy and to apoptosis-associated IkappaBalpha/nuclear factor kappaB pathway perturbation in mice. *J Cell Biol* 2000;**151**:1583–90.
110. Kramerova I, Kudryashova E, Tidball JG, Spencer MJ. Null mutation of calpain 3 (p94) in mice causes abnormal sarcomere formation in vivo and in vitro. *Hum Mol Genet* 2004;**13**:1373–88.
111. Fanin M, Nascimbeni AC, Fulizio L, Trevisan CP, Meznaric-Petrusa M, Angelini C. Loss of calpain-3 autocatalytic activity in LGMD2A patients with normal protein expression. *Am J Pathol* 2003;**163**:1929–36.
112. Vainzof M, de Paula F, Tsanaclis AM, Zatz M. The effect of calpain 3 deficiency on the pattern of muscle degeneration in the earliest stages of LGMD2A. *J Clin Pathol* 2003;**56**:624–6.
113. Tagawa K, Taya C, Hayashi Y, Nakagawa M, Ono Y, Fukuda R, et al. Myopathy phenotype of transgenic mice expressing active site-mutated inactive p94 skeletal muscle-specific calpain, the gene product responsible for limb girdle muscular dystrophy type 2A. *Hum Mol Genet* 2000;**9**:1393–402.

114. Ojima K, Kawabata Y, Nakao H, Nakao K, Doi N, Kitamura F, et al. Dynamic distribution of muscle-specific calpain in mice has a key role in physical-stress adaptation and is impaired in muscular dystrophy. *J Clin Invest* 2010;**120**:2672–83.
115. Ojima K, Ono Y, Ottenheijm C, Hata S, Suzuki H, Granzier H, et al. Non-proteolytic functions of calpain-3 in sarcoplasmic reticulum in skeletal muscles. *J Mol Biol* 2011;**407**:439–49.
116. Ermolova N, Kudryashova E, Difranco M, Vergara J, Kramerova I, Spencer MJ. Pathogenity of some limb girdle muscular dystrophy mutations can result from reduced anchorage to myofibrils and altered stability of calpain 3. *Hum Mol Genet* 2011;**20**:3331–45.
117. Charton K, Daniele N, Vihola A, Roudaut C, Gicquel E, Monjaret F, et al. Removal of the calpain 3 protease reverses the myopathology in a mouse model for titinopathies. *Hum Mol Genet* 2010;**19**:4608–24.
118. Huebsch KA, Kudryashova E, Wooley CM, Sher RB, Seburn KL, Spencer MJ, et al. Mdm muscular dystrophy: interactions with calpain 3 and a novel functional role for titin's N2A domain. *Hum Mol Genet* 2005;**14**:2801–11.
119. Garvey SM, Rajan C, Lerner AP, Frankel WN, Cox GA. The muscular dystrophy with myositis (mdm) mouse mutation disrupts a skeletal muscle-specific domain of titin. *Genomics* 2002;**79**:146–9.
120. Hance JE, Fu SY, Watkins SC, Beggs AH, Michalak M. Alpha-actinin-2 is a new component of the dystrophin-glycoprotein complex. *Arch Biochem Biophys* 1999;**365**:216–22.
121. Otey CA, Carpen O. Alpha-actinin revisited: a fresh look at an old player. *Cell Motil Cytoskeleton* 2004;**58**:104–11.
122. Chiu C, Bagnall RD, Ingles J, Yeates L, Kennerson M, Donald JA, et al. Mutations in alpha-actinin-2 cause hypertrophic cardiomyopathy: a genome-wide analysis. *J Am Coll Cardiol* 2010;**55**:1127–35.
123. North KN, Yang N, Wattanasirichaigoon D, Mills M, Easteal S, Beggs AH. A common nonsense mutation results in alpha-actinin-3 deficiency in the general population. *Nat Genet* 1999;**21**:353–4.
124. Yang N, MacArthur DG, Gulbin JP, Hahn AG, Beggs AH, Easteal S, et al. ACTN3 genotype is associated with human elite athletic performance. *Am J Hum Genet* 2003;**73**:627–31.
125. Niemi AK, Majamaa K. Mitochondrial DNA and ACTN3 genotypes in Finnish elite endurance and sprint athletes. *Eur J Hum Genet* 2005;**13**:965–9.
126. Papadimitriou ID, Papadopoulos C, Kouvatsi A, Triantaphyllidis C. The ACTN3 gene in elite Greek track and field athletes. *Int J Sports Med* 2008;**29**:352–5.
127. Roth SM, Walsh S, Liu D, Metter EJ, Ferrucci L, Hurley BF. The ACTN3 R577X nonsense allele is under-represented in elite-level strength athletes. *Eur J Hum Genet* 2008;**16**:391–4.
128. Clarkson PM, Devaney JM, Gordish-Dressman H, Thompson PD, Hubal MJ, Urso M, et al. ACTN3 genotype is associated with increases in muscle strength in response to resistance training in women. *J Appl Physiol* 2005;**99**:154–63.
129. Moran CN, Yang N, Bailey ME, Tsiokanos A, Jamurtas A, MacArthur DG, et al. Association analysis of the ACTN3 R577X polymorphism and complex quantitative body composition and performance phenotypes in adolescent Greeks. *Eur J Hum Genet* 2007;**15**:88–93.
130. MacArthur DG, Seto JT, Chan S, Quinlan KG, Raftery JM, Turner N, et al. An Actn3 knockout mouse provides mechanistic insights into the association between alpha-actinin-3 deficiency and human athletic performance. *Hum Mol Genet* 2008;**17**:1076–86.
131. Seto JT, Lek M, Quinlan KG, Houweling PJ, Zheng XF, Garton F, et al. Deficiency of {alpha}-actinin-3 is associated with increased susceptibility to contraction-induced damage and skeletal muscle remodeling. *Hum Mol Genet* 2011;**20**:2914–27.

132. Beltran-Valero de Bernabe D, Voit T, Longman C, Steinbrecher A, Straub V, Yuva Y, et al. Mutations in the FKRP gene can cause muscle-eye-brain disease and Walker-Warburg syndrome. *J Med Genet* 2004;**41**:e61.
133. Brockington M, Blake DJ, Prandini P, Brown SC, Torelli S, Benson MA, et al. Mutations in the fukutin-related protein gene (FKRP) cause a form of congenital muscular dystrophy with secondary laminin alpha2 deficiency and abnormal glycosylation of alpha-dystroglycan. *Am J Hum Genet* 2001;**69**:1198–209.
134. Brown SC, Torelli S, Brockington M, Yuva Y, Jimenez C, Feng L, et al. Abnormalities in alpha-dystroglycan expression in MDC1C and LGMD2I muscular dystrophies. *Am J Pathol* 2004;**164**:727–37.
135. Muntoni F, Torelli S, Brockington M. Muscular dystrophies due to glycosylation defects. *Neurotherapeutics* 2008;**5**:627–32.
136. Kawahara G, Guyon JR, Nakamura Y, Kunkel LM. Zebrafish models for human FKRP muscular dystrophies. *Hum Mol Genet* 2010;**19**:623–33.
137. Wang CH, Chan YM, Tang RH, Xiao B, Lu P, Keramaris-Vrantsis E, et al. Post-natal knockdown of fukutin-related protein expression in muscle by long-term RNA interference induces dystrophic pathology [corrected]. *Am J Pathol* 2011;**178**:261–72.
138. Ackroyd MR, Skordis L, Kaluarachchi M, Godwin J, Prior S, Fidanboylu M, et al. Reduced expression of fukutin related protein in mice results in a model for fukutin related protein associated muscular dystrophies. *Brain* 2009;**132**:439–51.
139. Chan YM, Keramaris-Vrantsis E, Lidov HG, Norton JH, Zinchenko N, Gruber HE, et al. Fukutin-related protein is essential for mouse muscle, brain and eye development and mutation recapitulates the wide clinical spectrums of dystroglycanopathies. *Hum Mol Genet* 2010;**19**:3995–4006.
140. Bortolanza S, Nonis A, Sanvito F, Maciotta S, Sitia G, Wei J, et al. AAV6-mediated systemic shRNA delivery reverses disease in a mouse model of facioscapulohumeral muscular dystrophy. *Mol Ther* 2011; Aug 9. doi: 10.1038/mt.2011.153. [Epub ahead of print].
141. van Spaendonck-Zwarts K, van Hessem L, Jongbloed JD, de Walle HE, Capetanaki Y, van der Kooi AJ, et al. Desmin-related myopathy: a review and meta-analysis. *Clin Genet* 2011;**80**:354–66.
142. Goldfarb LG, Dalakas MC. Tragedy in a heartbeat: malfunctioning desmin causes skeletal and cardiac muscle disease. *J Clin Invest* 2009;**119**:1806–13.
143. Dalakas MC, Park KY, Semino-Mora C, Lee HS, Sivakumar K, Goldfarb LG. Desmin myopathy, a skeletal myopathy with cardiomyopathy caused by mutations in the desmin gene. *N Engl J Med* 2000;**342**:770–80.
144. Milner DJ, Weitzer G, Tran D, Bradley A, Capetanaki Y. Disruption of muscle architecture and myocardial degeneration in mice lacking desmin. *J Cell Biol* 1996;**134**:1255–70.
145. Li Z, Colucci-Guyon E, Pincon-Raymond M, Mericskay M, Pournin S, Paulin D, et al. Cardiovascular lesions and skeletal myopathy in mice lacking desmin. *Dev Biol* 1996;**175**:362–6.
146. Thornell L, Carlsson L, Li Z, Mericskay M, Paulin D. Null mutation in the desmin gene gives rise to a cardiomyopathy. *J Mol Cell Cardiol* 1997;**29**:2107–24.
147. Milner DJ, Taffet GE, Wang X, Pham T, Tamura T, Hartley C, et al. The absence of desmin leads to cardiomyocyte hypertrophy and cardiac dilation with compromised systolic function. *J Mol Cell Cardiol* 1999;**31**:2063–76.
148. Milner DJ, Mavroidis M, Weisleder N, Capetanaki Y. Desmin cytoskeleton linked to muscle mitochondrial distribution and respiratory function. *J Cell Biol* 2000;**150**:1283–98.
149. Weisleder N, Soumaka E, Abbasi S, Taegtmeyer H, Capetanaki Y. Cardiomyocyte-specific desmin rescue of desmin null cardiomyopathy excludes vascular involvement. *J Mol Cell Cardiol* 2004;**36**:121–8.

150. Maloyan A, Robbins J. Autophagy in desmin-related cardiomyopathy: thoughts at the halfway point. *Autophagy* 2010;**6**:665–66.
151. Schrickel JW, Stockigt F, Krzyzak W, Paulin D, Li Z, Lubkemeier I, et al. Cardiac conduction disturbances and differential effects on atrial and ventricular electrophysiological properties in desmin deficient mice. *J Interv Card Electrophysiol* 2010;**28**:71–80.
152. Wang X, Osinska H, Dorn 2nd GW, Nieman M, Lorenz JN, Gerdes AM, et al. Mouse model of desmin-related cardiomyopathy. *Circulation* 2001;**103**:2402–7.
153. Kostareva A, Gudkova A, Sjoberg G, Kiselev I, Moiseeva O, Karelkina E, et al. Desmin mutations in a St. Petersburg cohort of cardiomyopathies. *Acta Myol* 2006;**25**:109–15.
154. Xu H, Wu X-R, Wewer UM, Engvall E. Murine muscular dystrophy caused by a mutation in the laminin α2 (*Lama2*) gene. *Nat Genet* 1994;**8**:297–302.
155. Helbling-Leclerc A, Zhang X, Topaloglu H, Cruaud C, Tesson F, Weissenbach J, et al. Mutations in the laminin alpha 2-chain gene (LAMA2) cause merosin-deficient congenital muscular dystrophy. *Nat Genet* 1995;**11**:216–8.
156. Guo LT, Zhang XU, Kuang W, Xu H, Liu LA, Vilquin JT, et al. Laminin alpha2 deficiency and muscular dystrophy; genotype-phenotype correlation in mutant mice. *Neuromuscul Disord* 2003;**13**:207–15.
157. Geranmayeh F, Clement E, Feng LH, Sewry C, Pagan J, Mein R, et al. Genotype-phenotype correlation in a large population of muscular dystrophy patients with LAMA2 mutations. *Neuromuscul Disord* 2010;**20**:241–50.
158. Gawlik KI, Durbeej M. Skeletal muscle laminin and MDC1A: pathogenesis and treatment strategies. *Skelet Muscle* 2011;**1**:9.
159. Xu H, Christmas P, Wu XR, Wewer UM, Engvall E. Defective muscle basement membrane and lack of M-laminin in the dystrophic dy/dy mouse. *Proc Natl Acad Sci USA* 1994;**91**:5572–6.
160. Sunada Y, Bernier SM, Utani A, Yamada Y, Campbell KP. Identification of a novel mutant transcript of laminin α2 chain gene responsible for muscular dystrophy and dysmyelination in dy^{2J} mice. *Hum Mol Genet* 1995;**4**:1055–61.
161. Kuang W, Xu H, Vachon PH, Liu L, Loechel F, Wewer UM, et al. Merosin-deficient congenital muscular dystrophy. Partial genetic correction in two mouse models. *J Clin Invest* 1998;**102**:844–52.
162. Miyagoe Y, Hanaoka K, Nonaka I, Hayasaka M, Nabeshima Y, Arahata K, et al. Laminin α2 chain-null mutant mice by targeted disruption of the *Lama2* gene: a new model of merosin (laminin 2)-deficient congenital muscular dystrophy. *FEBS Lett* 1997;**415**:33–9.
163. Shelton GD, Liu LA, Guo LT, Smith GK, Christiansen JS, Thomas WB, et al. Muscular dystrophy in female dogs. *J Vet Intern Med* 2001;**15**:240–4.
164. Bonaldo P, Russo V, Bucciotti F, Doliana R, Colombatti A. Structural and functional features of the alpha 3 chain indicate a bridging role for chicken collagen VI in connective tissues. *Biochemistry* 1990;**29**:1245–54.
165. Kuo H-J, Maslen CL, Keene DR, Glanville RW. Type VI collagen anchors endothelial basement membranes by interacting with type IV collagen. *J Biol Chem* 1997;**272**:26522–9.
166. Schessl J, Zou Y, Bönnemann CG. Congenital muscular dystrophies and the extracellular matrix. *Semin Pediatr Neurol* 2006;**13**:80–9.
167. Bernardi P, Bonaldo P. Dysfunction of mitochondria and sarcoplasmic reticulum in the pathogenesis of collagen VI muscular dystrophies. *Ann N Y Acad Sci* 2008;**1147**:303–11.
168. Specks U, Mayer U, Nischt R, Spissinger T, Mann K, Timpl R, et al. Structure of recombinant N-terminal globule of type VI collagen alpha 3 chain and its binding to heparin and hyaluronan. *EMBO J* 1992;**11**:4281–90.
169. Bethlem J, van Wijngaarden GK. Benign myopathy, with autosomal dominant inheritance: a report on three pedigrees. *Brain* 1976;**99**:91–100.

170. Jöbsis GJ, Boers JM, Barth PG, de Visser M. Genetic localization of Bethlem myopathy. *Neurology* 1996;**46**:779–82.
171. Jöbsis GJ, Keizers H, Vreijling JP, de Visser M, Speer MC, Wolterman RA, et al. Type VI collagen mutations in Bethlem myopathy, an autosomal dominant myopathy with contractures. *Nat Genet* 1996;**14**:113–5.
172. Speer MC, Tandan R, Rao PN, Fries T, Stajich JM, Bolhuis PA, et al. Evidence for locus heterogeneity in the Bethlem myopathy and linkage to 2q37. *Hum Mol Genet* 1996;**5**:1043–6.
173. Camacho Vanegas O, Bertini E, Zhang R-Z, Petrini S, Minosse C, Sabatelli P, et al. Ullrich scleroatonic muscular dystrophy is caused by recessive mutations in collagen type VI. *Proc Natl Acad Sci USA* 2001;**98**:7516–21.
174. Demir E, Sabatelli P, Allamand V, Ferreiro A, Moghadaszadeh B, Makrelouf M, et al. Mutations in COL6A3 cause severe and mild phenotypes of Ullrich congenital muscular dystrophy. *Am J Hum Genet* 2002;**70**:1446–58.
175. Bonaldo P, Braghetta P, Zanetti M, Piccolo S, Volpin D, Bressan GM. Collagen VI deficiency induces early onset myopathy in the mouse: an animal model for Bethlem myopathy. *Hum Mol Genet* 1998;**7**:2135–40.
176. Irwin WA, Bergamin N, Sabatelli P, Reggiani C, Megighian A, Merlini L, et al. Mitochondrial dysfunction and apoptosis in myopathic mice with collagen VI deficiency. *Nat Genet* 2003;**35**:367–71.
177. Bertini E, Guglielmina P. Collagen type VI and related disorders: Bethlem myopathy and Ullrich scleroatonic muscular dystrophy. *Eur J Paediatr Neurol* 2002;**6**:193–8.
178. Tiepolo T, Angelin A, Palma E, Sabatelli P, Merlini L, Nicolosi L, et al. The cyclophilin inhibitor Debio 025 normalizes mitochondrial function, muscle apoptosis and ultrastructural defects in Col6a1−/− myopathic mice. *Br J Pharmacol* 2009;**157**:1045–52.
179. Telfer WR, Busta AS, Bönnemann CG, Feldman EL, Dowling JJ. Zebrafish models of collagen VI-related myopathies. *Hum Mol Genet* 2010;**19**:2433–44.
180. de Greef JC, Lemmers RJ, van Engelen BG, Sacconi S, Venance SL, Frants RR, et al. Common epigenetic changes of D4Z4 in contraction-dependent and contraction-independent FSHD. *Hum Mutat* 2009;**30**:1449–59.
181. Kowaljow V, Marcowycz A, Ansseau E, Conde CB, Sauvage S, Matteotti C, et al. The DUX4 gene at the FSHD1A locus encodes a pro-apoptotic protein. *Neuromuscul Disord* 2007;**17**:611–23.
182. Lemmers RJ, van der Vliet PJ, Klooster R, Sacconi S, Camano P, Dauwerse JG, et al. A unifying genetic model for facioscapulohumeral muscular dystrophy. *Science* 2010;**329**:1650–3.
183. van der Maarel SM, Tawil R, Tapscott SJ. Facioscapulohumeral muscular dystrophy and DUX4: breaking the silence. *Trends Mol Med* 2011;**17**:252–8.
184. Gabellini D, D'Antona G, Moggio M, Prelle A, Zecca C, Adami R, et al. Facioscapulohumeral muscular dystrophy in mice overexpressing FRG1. *Nature* 2006;**439**:973–7.
185. Wallace LM, Garwick-Coppens SE, Tupler R, Harper SQ. RNA interference improves myopathic phenotypes in mice over-expressing FSHD region gene 1 (FRG1). *Mol Ther* 2011 (in press).
186. Harper PS, Johnson K. Myotonic dystrophy. In: Scriver CR, Beaudet AL, Sly WS, Valle D, editors. *The metabolic and molecular bases of inherited disease*, Vol. III. New York: McGraw-Hill; 2001. pp. 5525–50.
187. Miller JW, Urbinati CR, Teng-Umnuay P, Stenberg MG, Byrne BJ, Thornton CA, et al. Recruitment of human muscleblind proteins to (CUG)(n) expansions associated with myotonic dystrophy. *EMBO J* 2000;**19**:4439–48.

188. Taneja KL, McCurrach M, Schalling M, Housman D, Singer RH. Foci of trinucleotide repeat transcripts in nuclei of myotonic dystrophy cells and tissues. *J Cell Biol* 1995;**128**:995–1002.
189. Timchenko LT, Miller JW, Timchenko NA, DeVore DR, Datar KV, Lin L, et al. Identification of a (CUG)n triplet repeat RNA-binding protein and its expression in myotonic dystrophy. *Nucleic Acids Res* 1996;**24**:4407–14.
190. Fardaei M, Rogers MT, Thorpe HM, Larkin K, Hamshere MG, Harper PS, et al. Three proteins, MBNL, MBLL and MBXL, co-localize in vivo with nuclear foci of expanded-repeat transcripts in DM1 and DM2 cells. *Hum Mol Genet* 2002;**11**:805–14.
191a. Gomes-Pereira M, Cooper TA, Gourdon G. Myotonic dystrophy mouse models: towards rational therapy development. *Trends Mol Med* 2011;**17**:506–17.
191b. Mankodi A, Logigian E, Callahan L, McClain C, White R, Henderson D, et al. Myotonic dystrophy in transgenic mice expressing an expanded CUG repeat. *Science* 2000;**289**:1769–73.
192. Wheeler TM, Sobczak K, Lueck JD, Osborne RJ, Lin X, Dirksen RT, et al. Reversal of RNA dominance by displacement of protein sequestered on triplet repeat RNA. *Science* 2009;**325**:336–9.
193. Seznec H, Agbulut O, Sergeant N, Savouret C, Ghestem A, Tabti N, et al. Mice transgenic for the human myotonic dystrophy region with expanded CTG repeats display muscular and brain abnormalities. *Hum Mol Genet* 2001;**10**:2717–26.
194. Orengo JP, Chambon P, Metzger D, Mosier DR, Snipes GJ, Cooper TA. Expanded CTG repeats within the DMPK 3′ UTR causes severe skeletal muscle wasting in an inducible mouse model for myotonic dystrophy. *Proc Natl Acad Sci USA* 2008;**105**:2646–51.
195. Kanadia RN, Johnstone KA, Mankodi A, Lungu C, Thornton CA, Esson D, et al. A muscleblind knockout model for myotonic dystrophy. *Science* 2003;**302**:1978–80.
196. Ward AJ, Rimer M, Killian JM, Dowling JJ, Cooper TA. CUGBP1 overexpression in mouse skeletal muscle reproduces features of myotonic dystrophy type 1. *Hum Mol Genet* 2010;**19**:3614–22.
197. Muir LA, Chamberlain JS. Emerging strategies for cell and gene therapy of the muscular dystrophies. *Expert Rev Mol Med* 2009;**11**:e18.

Acute Phase Proteins in Animals

Carolyn Cray

Division of Comparative Pathology, Department of Pathology, University of Miami Miller School of Medicine, Miami, Florida, USA

Acute phase proteins (APP) were first identified in the early 1900s as early reactants to infectious disease. They are now understood to be an integral part of the acute phase response (APR) which is the cornerstone of innate immunity. APP have been shown to be valuable biomarkers as increases can occur with inflammation, infection, neoplasia, stress, and trauma. All animals—from fish to mammals—have demonstrable APP, but the type of major APP differs by species. While the primary application of these proteins in a clinical setting is prognostication, studies in animals have demonstrated relevance to diagnosis and detection and monitoring for subclinical disease. APP have been well documented in laboratory, companion, and large animals. With the advent of standardized and automated

Progress in Molecular Biology
and Translational Science, Vol. 105
DOI: 10.1016/B978-0-12-394596-9.00005-6

1877-1173/12 $35.00

assays, these biomarkers are available for use in all fields of veterinary medicine as well as basic and clinical research.

The acute phase response or APR is a term given to a collective of reactions to tissue injury resulting from infection, trauma, neoplasia, inflammation, and stress. The response is formulated by a number of different acute phase proteins (APP) that vary in magnitude and type among animal species. These systemic responders act as part of the innate immune defense system with the goal of reestablishing homeostasis and promoting healing. Several publications have documented the role of APP in response to experimental and infectious stimuli in animal models.[1–11] Over the past decade, APP have been more widely applied as biomarkers of inflammation in human and veterinary medicine.

I. The Acute Phase Response

The APR is part of broad response of the innate immune or early-defense system. The production of APP is preceded by the perhaps better known local response involving the creation of proinflammatory cytokines such as IL-1, IL-6, and TNF-α. These signal hepatocytes to produce the myriad of APP which then enhance chemotaxis provide other immunomodulatory effects, impart specific bacteriostatic support, and assist in the reduction of oxidative damage (Fig. 1, Table I).

A. Innate Immunity

The innate immune system serves to prevent infection, eliminate potential pathogens, and initiate the inflammatory response.[12] It is often described as compromised of distinct defenses including anatomic barriers (i.e., skin, mucous membranes), physiologic barriers (i.e., low pH and temperature to inhibit bacterial growth), phagocytic barriers (i.e., tissue-based cells that can ingest pathogens), and inflammatory barriers. It is the latter barrier that serves as the backup in the event of the failure of the other arms of the innate immune system and also perhaps is the most recognized given the myriad of symptoms including edema, redness, and fever that are produced with the complex interactions of cytokines, chemokines, and APP.

With the increased knowledge of the inflammatory process, it is recognized that this represents a cascade of very specific and effective immune responses that also initiate several metabolic events.[13,14] The innate immune response is often characterized as nonspecific as it is contrasted to the adaptive immune

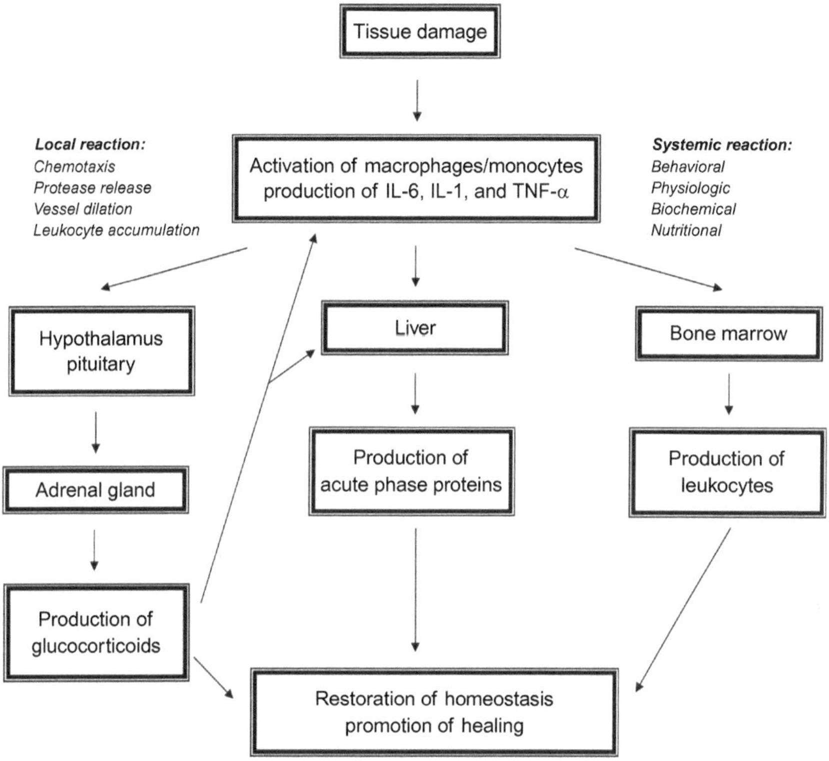

FIG. 1. The acute phase response.

response which involves antigen-specific responses via T and B cells. While the antigen receptor and its required gene recombination events are a more recent evolutionary acquisition of vertebrates, the innate immune system represents the core from which the adaptive immune system was based. Additionally, these two systems are often examined as mediating separate events. There is a true interface between the APR and adaptive immunity. Yoo and colleagues, with the use of lipopolysaccharide (LPS) treatment of mice, demonstrated that the APR mobilizes antigen presentation by hepatocytes via MHC class I.[13] Others have demonstrated that the elimination of toll-like receptors affects both innate and adaptive immune responses.[15,16]

B. Induction of the APR

The APR is initiated primarily by tissue macrophages, blood monocytes, and dendritic cells at the site of tissue damage when particular chemical structures in damaged tissue or infectious agents are presented for detection

TABLE I
LIST OF COMMON ACUTE PHASE PROTEINS AND THEIR COMMONLY USED ACRONYMS AND ACTIVITIES

Acronym	Acute phase protein	Activities
AAT	Alpha-1 antitrypsin	Protease inhibition
ACT	Alpha-1 antichymotrypsin	Protease inhibition
AGP	Alpha-1 acid glycoprotein	Bind drugs and LPS
ALB	Albumin	Transport protein
A2M	Alpha-2 macroglobulin	Protease inhibition
CP	Ceruloplasmin	Transport copper, protect from iron-mediated oxidative injury
CRP	C-reactive protein	Enhance opsonization, activate complement, induce cytokines, inhibit chemotaxis
FIB	Fibrinogen	Substrate for fibrin, tissue repair
HP	Haptoglobin	Bind hemoglobin, bacteriostatic
MAP	Pig major acute phase protein	Trypsin inhibition
SAA	Serum amyloid A	Chemotaxis, anti-inflammatory activity
SAP	Serum amyloid P	Enhance opsonization, activate complement
TN	Transferrin	Sequestration of iron

by these first-line defense cells.[17] These cells undergo a rapid activation leading to the production of the primary cytokines IL-1, TNF-α, and IL-6 which not only incite the APR but also result in the chemotactic recruitment of additional cells to the affected area to rapidly augment the response. The area of major APP production is the liver; specifically, the induction of APP expression in hepatocytes. APP have also been noted to be produced in other tissues. The production of the major APP serum amyloid A (SAA) has been demonstrated in the gastrointestinal tract, mammary gland, kidneys, and airways.[18,19]

The inflammatory stimuli may result in a differential production of cytokines, and the absence of some of the cytokines may abrogate the APR. In a study of LPS-induced inflammation of the mouse lung, the use of neutralizing antibody to IL-6 but not TNF-α resulted in the cessation of the APR.[20] In IL-6-deficient mice, the intraperitoneal injection of LPS resulted in an overproduction of TNF-α but still a significant decrease in the production of SAA.[21] Similarly, in a mouse model of bacterial pneumonia, SAA increases were completely abrogated in IL-6-deficient mice.[22] In wild-type mice, IL-6 levels were increased in both the liver and the lungs of infected mice within 24 h of bacterial instillation.

The apparent conduit between the cytokines and the expression of APP are the transcription factors STAT3 and RelA. It was shown that these factors form a complex after interaction with IL-1 and IL-6 and then induce the transcription of the SAA gene.[23] In a mouse model where the levels of STAT3 were significantly reduced, the SAA response was abrogated.[24] *In vitro* studies further demonstrated that the loss of either STAT3 or RelA prevented SAA production in a hepatocyte cell line.[22]

C. Other Effects of the APR

While specific APP will be discussed in the following sections, it should be noted that the APR is often defined by a myriad of changes not limited to the well-known biomarkers alone (Fig. 1). In addition to the immune system, both the nervous and endocrine systems are also involved and stimulated by the cytokine production resulting in behavioral, physiologic, biochemical, and nutritional changes.[25] IL-6 stimulates the corticotrophin-releasing hormone (CRH) synthesis in the hypothalamus which, in turn, modulates ACTH release from the pituitary and glucocorticoid production in the adrenal gland; this is the so called hypothalamic–pituitary–adrenal axis. In IL-6-deficient mice injected with LPS, CRH, and glucocorticoid levels were significantly lower than wild-type controls.[26]

Other host responses include fever, fatigue, hypoglycemia, anorexia, lipolysis, and muscle catabolism. In the LPS injection model, there is nearly a 10% loss in body weight by day 1.[21] Both IL-6 and neural mechanisms have been implicated in this response.[21,27] This is concomitant with a significant decrease in food intake and serum glucose. Fever is thought also to be a response to IL-6 through the production of other mediators including prostaglandins which affect the hypothalamus.[28,29] Decreases in serum iron and zinc concentrations are also often observed with their redistribution to the tissues where they assist in antioxidant activity.[25]

Leukocytosis is initially mediated through the production of cortisol by the removal of marginal pool cells from the tissues and placement in the blood in what is often called an immediate stress response. The true leukocytosis follows some days after the effects of the proinflammatory cytokine response and its positive effect on the bone marrow, and often involves enhanced levels of both lymphocytes and neutrophils.

Glucocorticoids are known to mediate upregulatory effects on the APR. Experiments have been completed with the application of exogenous dexamethasone into different models of APR; increased APP transcription, IL-6 receptor, and hepatocyte proliferation have been demonstrated.[30] Downregulation occurs as the initial stimulus (tissue damage) is addressed. Cytokines can be inhibited by other cytokines and excess cortisol.[30,31] Notably, in addition to their direct roles in addressing the inflammatory tissue and resultant tissue damages, APP also often have anti-inflammatory activity. For example, SAA was found to inhibit myeloperoxidase release, migration, and free oxygen radical production in neutrophils.[32]

II. Acute Phase Proteins

It is thought that the APR results in changes in more than 200 proteins and is defined as those in which there is a minimum 25% change in concentration.[33,34] APP are categorized as positive or negative. Positive APP increase in concentration and include most of the described proteins. Negative APP decrease in concentration and include albumin and transferrin (in nonavian species). APP are further classified on the basis of the magnitude of their increase.[3,8,35] Major APP represent those proteins that increase 10- to 1000-fold; most have negligible levels in normal animals. These are often the first responders with rapid increases of large magnitude within 24–48 h of the stimulus. Moderate APP have higher basal levels than major APP and increase 5- to 10-fold with peak levels occurring 3 or more days after the stimulus. Minor APP increase slowly with increases often less than twofold. It should be considered that these classifications are not formalized as such they are dependent on both the type of assay performed and the type of stimulus. In various reviews, there are often discordant classifications of some APP. Of note, the classification of APP by these categories differs by species; this will be further discussed in Section IV. In many cases, the biological activities of the APP are only partially known. Table I provides a list of the most commonly referenced APP.

A. C-Reactive Protein

C-reactive protein (CRP) was the first APP to be identified in the blood of humans and nonhuman primates (NHP) infected with *Streptococcus pneumoniae*. The "C" fraction (C polysaccharide) of the bacteria was found to react with CRP.[36] It is one of the most well-known APP given its use in humans after myocardial infarction.[37] As CRP has become more understood, the diagnostic utility has grown to include wellness assessments on clinically normal humans as well as patients with underlying diseases including diabetes and coronary heart disease.[14,38,39]

CRP is composed of five subunits combining to form a pentameric structure.[40] It acts as an opsonin binding polysaccharide residues on bacteria, fungi, and parasites to activate complement and phagocytosis.[41] CRP has also been described to result in the induction of cytokines and interact with Fc-gamma receptors prompting the hypothesis that CRP is also interactive in the adaptive immune response.[42] Serum amyloid P (SAP), a major APP of mice, is considered to have biological functions and structural similarities to CRP.[43]

CRP is a major APP of the canine species and a canine-specific ELISA is commercially available.[44] Recently, a human immunoturbidimetric assay for CRP has been validated for use with canine sera.[45] This appears to be manufacturer specific as many other human CRP reagents fail to provide any cross-reactivity.[44,46]

B. Serum Amyloid A

SAA is normally found complexed with lipoproteins and different isoforms have been described with numbers varying by species.[47] SAA represents one of the most conserved proteins among mammals supporting the premise that it has a basic and essential role in the innate immune system. In species with CRP, SAA often mirrors the magnitude and time course of the response.[48] Studies have demonstrated SAA activity in chemotaxis of leukocytes as well as the induction of additional proinflammatory cytokines.[49,50] In addition, SAA has also been described to induce extracellular matrix degrading enzymes which can assist in tissue repair.[47]

Assays for SAA in animals had been traditionally in the ELISA format. More recently, an immunoturbidimetric assay for human SAA was described with excellent reactivity for equine SAA.[51] Similar assays have been described for use with other species.[52]

C. Haptoglobin

Haptoglobin (HP) is composed of 2α and 2β subunits. There are different subtypes and these may vary with species.[8] HP binds free hemoglobin which may be released during various autoimmune, infectious, or inherited diseases.[53] This HP–hemoglobin complex is phagocytosed by macrophages via the CD163 receptor. In mouse models, clearance is rapid with a half-life of less than 50 min.[53] In HP-deficient mice, high levels of hemoglobin are found to accumulate in the kidney.[54] The effective binding to CD163 has been found to stimulate cytokine production, and HP has also been linked to protease activity and immune suppression.[55] Interestingly, HP is found both in mammals and in fish but not in chickens and in frogs where another hemoglobin-binding protein has been identified called PIT54.[56]

Assays for HP in animals are often via spectrophotometry using the natural affinity of HP for hemoglobin.[57] Immunoassays for human HP which show cross-reactivity with animal HP are also available but are not widely implemented.[58]

D. Other APP

Alpha 1-acid glycoprotein (AGP) is a highly glycosylated protein which has the ability to bind drugs including heparin, serotonin, steroids, and histamine.[59] This protein has also been implicated in anti-inflammatory responses and has been described to bind LPS.[59,60] Assays for AGP in animals are commonly performed via radial immunodiffusion making this particular APP unenviable for use as a rapid diagnostic test. A feline-specific immunoturbidimetric assay has been described.[61]

Fibrinogen, like HP, is present in the serum of all vertebrates and has been recognized to be part of the APR for several years. It is a large protein composed of six polypeptide chains and a glycoprotein which is converted by thrombin into fibrin and is essential in the coagulation cascade.[62] Many assays are based on heat precipitation methodology.[63]

Ceruloplasmin (CP) is a glycoprotein which binds most of the free serum copper. In the inflammatory process, it acts as a protectant against damage by free iron which promotes free radical oxidation.[64] ELISA-based assays are available for quantitation of this biomarker.

Pig major acute phase protein (Pig-MAP) is one of the main APP in swine. It is a glycoprotein that appears to be unique to this species; it functions as a trypsin inhibitor.[65] ELISA methodology is validated for quantitation of this protein.[66]

Protease inhibitors include MAP as well as alpha 1-antitrypsin, alpha 1-antichymotrypsin, and alpha 2-macroglobulin (A2M).[3] Proteases are released during damage as well as from responding neutrophils; these inhibitors are protective from immune-mediated damage.

E. Negative Acute Phase Proteins

Albumin is the major negative APP in all species. As the most abundant protein in the serum, it serves as a source of nutrients and a regulator of osmotic pressure. Decreases can be attributed to protein loss due to kidney or gastrointestinal disease or edema and also due to decreased synthesis related to liver disease or malnutrition.[33] In mammals, the iron transporter transferrin (TN) is also a negative APP, although in avian species it has been shown to increase with the APR.[67,68]

F. Quantitation of APP

Given the broad range of APP, the differences which occur at the species levels (see Section IV), and the rapid changes in technology over the past 25 years, it is understandable that many of the publications referenced in this review have used a myriad of techniques to assess the APR. At the research level, proteomic profiling has been utilized. With the maturation of immunoassays, immunodiffusion tests, ELISA, and immunoturbidimetric assays have been employed. Where possible, colorimetric assays have also been implemented. Lastly, the APR has been broadly judged by the use of serum protein electrophoresis. This method provides quantitation of albumin and then fractions of APP which are grouped as alpha 1, alpha 2, beta, and gamma globulins based on their migration characteristics.[33] Major benefits of this method are not having the need for species-specific reagents and the assessment of the complete APR—albumin and the groups of APP. Protein electrophoresis has been commonly implemented in avian species as an adjunct test to infectious disease assessments.[69,70] A negative to this assay is the limits of detection as fraction sensitivity is at the g/dL level rather than the mg/L level obtained by specific APP assays.[71]

APP assays have become more available to researchers and veterinarians in recent years. In many cases, labor intensive and costly ELISA-based assays have been replaced by automated assays that have heterologous cross-reactivity with many species. Examples of this type of assay include the Eiken SAA immunoturbidimetric assay (a human assay which reactive with horse and cat but not dog SAA) and the Randox reagents for CRP (a human assay which is reactive with dog CRP).[45,72] These assays afford a commercial source of reagents negating lengthy developmental time. In other cases, species-specific reagents are preferred. These include an immunoturbidimetric assay for feline AGP and a pig-MAP-specific ELISA.[61,66] These needs are based more on the unique APP in particular species which defies the routine use of human or other reagents.

The decision to use a heterologous assay is based on a validation process that ensures not only a low coefficient of variation but also reproducibility. In some cases, a species-specific standard curve can be developed with use with these cross-reactive reagents when such control reagents are commercially available.[5,45] Lacking that, patient-derived controls can be obtained through routine practice and store indefinitely to serve as assessors of different reagent lots. Recently, a four-step validation approach was proposed for these types of assays.[5] These steps include assessment of analytical characteristics (i.e., linearity, imprecision), comparison of a small set samples from normal and abnormal animals, evaluation of the assay in an actual clinical setting in which sensitivity and specificity can be judged, and assessment of the clinical and research impact of the test.

Adoption of these types of methods also allows for reproducible results among different laboratories which allows for the first steps toward assay standardization. Within this very review, reference intervals have not been presented due to the diverse types of assays that have been used in the various research studies. This problem was recognized more than 10 years ago at which time a proposal was made to harmonize the types of assays performed among major international laboratories as well as to distribute a reference sample for intralaboratory testing.[73] This was formalized into the development of a Concerted Action Group as well as a Colloquium of Acute Phase Proteins where assay information and research findings could be more regularly distributed.[74]

The positive implementation of APP assays is linked to the development of reference intervals. As discussed above, this is very much method specific but other physiological differences also have a role. CRP levels were examined over a several-day period in normal dogs, and they were found to vary more than 10-fold; no pattern of circadian rhythm was detected.[75] Others reported a low biological variability in CRP, HP, AGP, and FIB in dogs[76] Age has been demonstrated to influence APP production. In a rat model of LPS stimulation, significantly decreased HP and AGP were observed in aged animals.[77] In dogs injected with turpentine oil, significantly higher levels of CRP were found in dogs over 3 months versus 1 month of age.[78] Lower resting levels of AGP were observed in these very young dogs versus older dogs prompting the proposal that age should be considered when using and interpretating APP results.[79] Similar recommendations were made for newborn dairy calves that gave high levels of APP post-birth followed by a significant decrease through day 21 and eventual stabilization.[80]

It should also be recognized that APP levels may also be husbandry and strain/breed related as well as reflective of altered physiology or treatment status. Germ-free mice, injected with LPS, were observed to have a lower SAA response.[81] Similar differences were observed in dogs housed in clean versus dirty husbandry conditions.[82] In addition, SAP low responder and higher responder strains of mice were described; this was not strongly linked to the H-2 complex.[83] Differences in pig-MAP levels were also described between Large White and Meishan breeds.[84] Pregnancy was also reported to alter CRP levels by nearly 10-fold in dogs.[85] Additionally, glucocorticoid treatment and other drugs have described to result in a persistent HP increase.[2]

III. Clinical Value of APP

The origin of the APP stimulus can be infection, neoplasia, trauma, inflammation/immunologic reaction, and stress (Table II). In general, APP increases are of high magnitude and can be detected within the first day of stimulus; this rivals other traditional markers of inflammation. In veterinary medicine, applications have been documented in diagnosis, prognosis, detection of subclinical disease, monitoring levels of stress, and viewing the progression of chronic inflammatory processes.

A. Comparison to Other Markers of APR

An experimental model of inflammation using the injection of turpentine oil in dogs provides data which demonstrate the differences among the various inflammatory markers (Fig. 2).[86] Levels of IL-6 and TNF-α were increased within a few hours with maximal response at 12–24 h postinjection. While an early increase (marginal response) in total white blood cells (WBC) was observed, the higher magnitude response was not present until days 4–7. A gradual decrease in serum albumin was present through day 14. The maximal decrease was less than 50% of day 0 values. CRP levels were increased 400-fold by day 2 postinjection and normalized by day 14. Serum

TABLE II
SUMMARY OF INFLAMMATORY STIMULI THAT CAN INDUCE AN APR

Stimuli
Experimental—inflammatory agents such as turpentine oil
Infection—bacterial, viral, parasitic
Surgery and trauma
Other trauma including burns
Autoimmune diseases
Toxins
Neoplasia
Stress

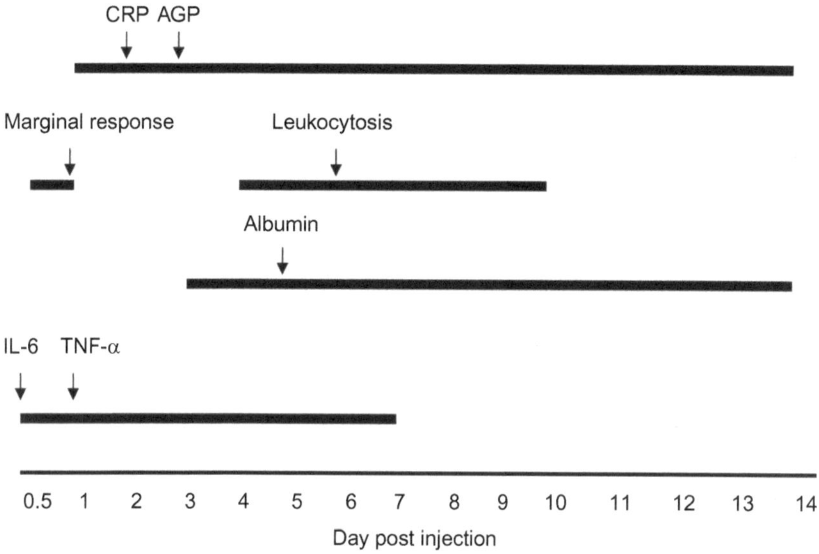

FIG. 2. Comparison of expression of acute inflammatory biomarkers after the injection of turpentine oil in a dog. Arrows indicate time of peak expression. Horizontal lines indicate period of expression.

AGP was increased approximately twofold on a similar timeline with maximal levels on day 3. Similar results can be found in a study of LPS injection in a model of arthritis in horses.[87] An early increase in WBC occurs within 24 h of the stimulus. SAA increases by 24 h with peak values at 48 h and elevated levels through day 5.

APP are commonly compared to WBC and fibrinogen as they are long-standing hallmarks of inflammatory processes. In a large study of samples from dogs with various inflammatory processes, only a weak correlation ($r = 0.44$) was observed between CRP levels and WBC counts but no correlation to the presence of band neutrophils which generally indicate bone marrow stimulation.[88] In a smaller study, a poor correlation ($r = 0.34$) was observed between the CRP and the presence of band neutrophils.[89] A positive correlation was observed between HP and FIB in dogs with various diseases.[90] In a study of horses with various conditions, SAA and FIB were found to be consistently elevated in those animals with bacterial infection, whereas low or normal total WBC and neutrophil counts were observed.[91]

In our own laboratory, we have compared inflammatory markers in both dogs and horses with various diseases (C. Cray, 2011). The mean total WBC values, while significantly higher than the control group, did not exceed normal reference intervals for this determination. Similarly, there was a mild decrease in the A/G ratio and a mild increase in the fibrinogen between the clinical groups. Changes for all these determinations were less than twofold. Notably, a 70- and 20-fold mean increase was observed for SAA and CRP in horses and dogs, respectively. It is this magnitude of difference which makes the clinical use of APP especially attractive.[5] As major APP are normally either not present or present in very low levels, increases are rapid and persistent as other minor and moderate APP are produced during later stages of inflammation.

B. Diagnosis

As APP can be increased by a myriad of different stimuli, it is difficult to state that APP, alone, can lead to a diagnosis (Table II). As discussed above, APP can provide more and earlier information regarding an ongoing inflammatory process over traditional assessments of WBC and fibrinogen. APP can also assist in providing a differential diagnosis. For example, LPS appears to produce the strongest APP response so that animals with high-magnitude APP increases can be suspect of bacterial infection. Thus, elevated APP can provide a basis for an inflammatory etiology which can aid in the development of a diagnosis and early treatment intervention.

SAA was found to differentiate the etiology of colic in horses.[92] Those animals with more serious enteritis, colitis, and peritonitis had a 50- or more fold increase in SAA versus those with obstruction, perforation, and ulcers. Similarly, a disorder in cattle called traumatic reticuloperitonitis (related to foreign body ingestion) is difficult to diagnose with just a general examination and often requires radiology and surgery.[93] The use of SAA and HP quantitation provided 100% sensitivity and 86% specificity. Lastly, a large study was undertaken to examine cattle with acute and chronic disorders.[35] SAA levels were found to be consistently fivefold higher in acutely ill animals versus those with chronic processes and were reported to provide 100% sensitivity in the discrimination between these conditions. The slower forming HP response provided a specificity of 76%. The combined use of APP has been the subject of study, and an APP value index has been proposed for diagnosis and monitoring of clinically ill humans and animals.[94]

C. Prognosis

Given the high magnitude of response, duration of response, and the short half-life of APP, it is readily understood why APP are favored in prognostication. Studies regarding prognosis have included examination of mortality as well as short-term response to treatment. A large study of the general disorder colic in horses demonstrated a nearly twofold higher proportion of nonsurvivors had marked increases in SAA versus survivors.[92] In another study, SAA levels from horses naturally infected with equine influenza virus were observed to be strongly associated with disease severity with eventual normalization of SAA levels with uncomplicated convalescence.[95] Similar changes were observed in horses with a positive response to treatment for joint disease.[96]

In canine medicine, critically ill animals presenting with trauma, sepsis, or pancreatitis demonstrated elevated CRP levels, although no differences were observed between survivors and nonsurvivors.[97] Using a single CRP assessment, similar results were observed in dogs diagnosed with sepsis or systemic inflammatory response syndrome.[98] As expected, CRP levels decreased with recovery thus predicting survival in 94% of the patients. Acute abdomen syndrome was reported to result in up to 200-fold increases in CRP.[99] Nonsurvivors were observed to have significantly higher levels of CRP than survivors.

APP levels have also been used as prognostic indicators in a variety of infections, inflammatory, and neoplastic conditions. In one study, AGP was monitored in dogs with pyometra.[100] Nearly fourfold increases in AGP were observed and animals with higher AGP levels had prolonged hospitalization. CRP, SAA, AGP, and HP levels were observed to be elevated in dogs with steroid-responsive meningitis.[101] Decreasing levels of APP were reported with response to treatment as well as identifying relapse. CRP was also reported to be a valuable management tool of inflammatory bowel disease, idiopathic arthritis, and autoimmune hemolytic anemia as well as post operative management.[102–105]

Prognostic value has also been identified in neoplasia. Dogs with multicentric lymphoma demonstrated higher levels of CRP, and 90% of the animals had normal levels with remission although there was a high variability among the patients.[106] In contrast, AGP levels did not serve as a prognostic marker in cats with lymphoma.[107] In a mouse model of colorectal cancer with a novel immunotherapy treatment, SAA and SAP levels were higher in nonresponders.[108]

We have also proposed implementing APP assays as an aid to determine the humane endpoint for different animal models of disease.[1] In most studies, the endpoint of an experimental model is often loosely judged on subjective measures and based on the desire of the investigator to obtain endpoint information which may often be clouded by the severe clinical disease in the animal. APP, as a reflection of stress and severe physiologic changes, may have sensitivity in determining an endpoint. Studies are underway to address this proposal.

D. Subclinical Disease

APP have been found to be useful in the detection of subclinical disease. In a large study conducted in cows, clinical signs of inflammatory processes occurred with elevated SAA in 26% of the assessments, whereas normal SAA levels were observed in clinically normal animals in 95% of the assessments.[109] These results are similar to that proposing APP as indicators of calf herd health.[110] The group with a higher incidence of disease and lower weight gain had significantly higher levels of HP, SAA, and FIB. Increases in HP were also observed before the onset of severe clinical disease in pigs.[111] In a similar study, pig-MAP was observed to increase before the onset of postweaning multisystemic wasting syndrome.[112]

These concepts have led to the implementation of APP quantitation in improving food safety. Toussaint and colleagues suggested the implementation of an APP index that could be used at the slaughterhouse level.[113] In another study, HP levels were compared in normal cattle versus culled animals and carcasses removed during slaughterhouse inspection due to gross abnormities.[114] Higher HP levels were found in the suspected or confirmed diseased groups. Assays for pig-MAP and HP have also been adapted for use with meat juice of pigs perhaps making this a more viable test panel for the assessment of food safety.[115]

Our laboratory has studied the application of the herd health concept to the monitoring program of laboratory mice.[71] Mice and rats in animal colonies are potentially subject to numerous viral, bacterial, and parasitic infections. Most institutions complete labor intensive and costly screening programs for each pathogen usually by specific serology and PCR detection methods. In our study, we examined mice experimentally infected with Sendai virus and mouse parvovirus and found no significant elevations in SAA, SAP, CRP, and HP. Additionally, APP levels in sentinel mice from colonies with several different endemic infections were also found to be not elevated.

E. Stress

There are several reports of the utility of APP as indicators of stress where these biomarkers rival more traditional assessments of serum cortisol and fecal corticosterone. Most of these studies have involved the use of large animals. Two breeds of cows that differed in nutrition and level of human handling were examined.[116] The semi-feral cows that were subject to more nutritional stress demonstrated more than a sixfold increase in SAA. The effects of animal density in housing were also examined in pigs.[117] In this study, significant increases in pig-MAP were observed in those animals subjected to a higher stocking rate for two 4-day periods over 26 days. These increases occurred after the second period and were concurrent with the detection of increased antioxidant activity in the serum. Similar changes were reported in pigs subjected to different transport conditions, cows subject to transportation and new housing, and newly weaned calves.[118–121] APP expression was also studied in mice after stress induced by acoustic and immobilization methods.[122] CRP levels were increased nearly threefold in animals with acute stress and more than fourfold with chronic exposure.

F. Chronic Inflammation

The goal of the APR is to facilitate tissue repair and the restoration of homeostasis. Under some conditions, however, inflammatory processes are prolonged and can result in further damage. Recent terms have been coined including "metainflammatory" and "parainflammatory" and refer to the observation that chronic elevations of APP have been associated with the onset of different syndromes including atherosclerosis, cardiovascular conditions, obesity, asthma, and diabetes.[123,124] It is suggested that the classic APR results in tissue repair and the metainflammatory APR are initiated with the same mechanisms, but self-tissues are the target rather than pathogens or other inflammatory stimuli in the latter process.

CRP has been associated with cardiovascular disease in humans, and CRP levels are often used to monitor treatment regimens.[125] Extravascular inflammation has been proposed as the basis for cardiac damage and the promotion of atherosclerosis, and this may be related to altered HDL particles which occur with increased SAA expression.[123] Interestingly, however, a study of a genetic mouse model of atherosclerosis revealed that the recurrent exposure of mice to potent inflammatory agents did not affect the development of disease.[126] In contrast, exposure of the same strain of mice to chronic unpredictable stress including temperature changes, light cycle changes, and agitation resulted in the promotion of atherosclerotic lesions as well as an elevated CRP level.[127] Obesity has also been proposed as a chronic inflammatory process in animals and humans.[123,128] CRP and HP have been found to be elevated in obese dogs concurrent with insulin resistance and the development of diabetes, and APP have been proposed to be valuable monitors in case management.[128] Similarly, HP expression was found to be increased in rats that later developed type 1 diabetes.[129] The chronic expression of SAA has been described as the cause of systemic amyloidosis, and this overstimulation has been linked to infectious and autoimmune diseases.[130] Amyloidosis has been reported in both humans and animals. The fibril protein AA is derived from the SAA and deposited in tissues as amyloid where it gradually affects tissue function. In early subclinical stages of disease, SAA levels have been reported to be elevated in NHP and genetically predisposed Abyssinian cats.[131,132]

IV. APP in Animals

Major, moderate, and minor APP differ by species. This information is presented in Table III and has been drawn from several references.[1–4,7,35,133–136] Although acute inflammation and APP are common to all species, numerous studies have shown that APP do appear to have differential sensitivity to the types of diseases and disorders which are present in each species. In addition, the unique environment and usage of some species further dictate the usage of APP quantitation. For example, some animals are viewed in a clinical sense as a

TABLE III
LIST OF APP BY ANIMAL SPECIES

Species	Major APP	Moderate and minor APP
Cat	AGP, SAA	FIB, HP
Chicken	None	AGP, CP, FIB, HP (PIT54), SAA, TN
Cow	HP, SAA	AGP, CP, CRP, FIB
Dog	CRP, SAA	AGP, CP, FIB, HP
Goat	HP, SAA	AGP, FIB
Horse	SAA	AGP, CP, FIB, HP
Human	CRP, SAA	AGP, FIB, HP
Mouse	HP, SAA, SAP	CRP, FIB
Nonhuman primate	CRP	A2M, FIB, HP, SAA
Pig	HP, MAP, SAA	AGP,CP, CRP, FIB
Rabbit	CRP, HP, SAA	AGP, CP, FIB, TN
Rat	AGP, A2M	CP, CRP, FIB, HP
Sheep	HP, SAA	AGP, CP, CRP, FIB

colony or herd (i.e., laboratory animals, cows, etc.) which create an opportunity for possible novel applications of APP assessment (see Section III.D). Many prototypic studies on animal APP have been conducted using inflammatory agents such as LPS, croton oil, and turpentine oil. These represent the gold standard of the inflammatory process and are included in the following discussion sections as well as studies of diseased animals (Table II).

A. Rodents and Rabbits

As presented in the previous sections, much of what is known about APP and the APR has been examined in rodent models and transgenic mice. APP have not only been well characterized in these animals, but they also have served as biomarkers in specific animal models of disease. There are several comprehensive reviews on rodent APP.[1,134,136]

The APR of mice has been studied with the injection of the inflammatory agent LPS.[48] SAA and CRP levels were found to significantly increase within 24 h; SAP and HP increased by 48 h. This was concomitant with a decrease in serum albumin. Decreasing positive APP levels were observed with the recovery of the animal (i.e., weight gain, clinical appearance) by day 4. Several models of infection have also been examined. Experimental infection with rodent infectious agents Sendai virus and mouse parvovirus was found to not induce an APR; this was also examined in mice naturally infected with several agents which commonly occur in laboratory animal colonies.[48] In contrast, experimental infection with malaria was observed to result in increased SAP levels.[137] SAP and HP levels were documented in an experimental model of *Trypanosoma* infection.[138] Increases were present early postinfection as well as with the onset of posttreatment encephalopathy.

Significant increases in SAA, SAP, and HP were observed in mice subjected to 20% burn injury; these levels were not completely comparable to that induced with LPS where higher SAA levels were observed.[139] Fourfold increases in HP were observed in mice subjected to lethal doses of irradiation.[140] SAP was also found to be increased in a mouse model of arthritis.[141] Mice injected with a hepatotoxin demonstrated very early transient increases in SAA and SAP, whereas repeated exposure with the same agent resulted in consistent increases at all time points.[142] CRP increases were most pronounced within 24 h of mouse exposure to scorpion venom.[143]

Experimental models of acute inflammation using Freund's adjuvant (CFA) and turpentine oil revealed the major APP of rats.[144–146] CRP and HP increases were observed with CFA injection, but only HP was observed to increase with a single dose of hepatotoxin.[144] A marked increase in A2M was observed after intramuscular injection with turpentine oil with maximal levels by day 2 and resolution by day 6.[145] With repeated turpentine oil injection, AGP levels also peaked at 2 days.[146] Injection of organophosphorus compounds resulted in a twofold increase in HP, sixfold increase in A2M and FIB, and eightfold increase in AGP within 24 h.[147] Thirty-fold increases in A2M were also observed in rats inoculated with *Staphylococcus aureus*.[148] An acute septicemia model resulted in elevated CP levels, and the provision of exogenous AGP was observed to decrease the severity of the inflammatory response.[149] Trauma related to surgery for castration or oophorohysterectomy induced similar levels.[148]

Rabbit APP studies have concentrated on experimental inflammatory stimuli. Animals subjected to turpentine oil treatment demonstrated 50-fold increases in CRP by 36 hours with resolution by 96 h.[150] CP was also observed to increase. In a similar model, CRP, SAA, HP, and CP were observed to increase with a decrease in albumin.[151] TN, a negative APP in most mammals, was also found to increase. Notably, both hepatic and extrahepatic SAA expressions were observed in rabbits injected with turpentine oil, LPS, and casein as well as multiple injections with silver nitrate.[152,153] APP expression has also been studied in other rabbit models. In an arthritis model, CRP was observed to increase immediately after intra-articular injection with normalization 10 days later.[141] In our own studies of rabbits with suspected naturally acquired infection with *Encephalitozoon cuniculi*, we observed up to 30-fold increases in CRP in approximately 50% of the animals with renal, ocular, and neurological signs (C. Cray, 2011).

B. Companion Animals

Small animals including canine and feline species predominate the companion animal category. Both species are readily used in laboratory animal medicine in model development also occupy the major portion of the U.S. veterinary market with ownership of over 70 million cats and dogs.[154] Although APP have been well documented in the literature, these assays are not widely used in the United States likely partly due to their poor availability. However, our laboratory has demonstrated a utility for APP quantitation in dogs in a large specialty veterinary practice. In a study of over 200 dogs, a mean increase of 20-fold and 7-fold was observed for CRP and HP, respectively (C. Cray, 2011). These represented animals with varied diseases and disorders including neoplasia, trauma, infection, and autoimmune disorders. Our data are supported by numerous reports of specific diseases and disorders, part of which are reviewed in the following paragraphs and have been previously reviewed by others.[2]

In a study of the intramuscular injection of turpentine oil in dogs, CRP levels were increased 400-fold by day 2 postinjection and normalization by day 14. CRP increases can certainly be observed in animals with natural infection and disease, but the magnitude of increase is not as common. In a large study of dogs ($n = 928$) with various diseases, CRP was found to be a major APP with a weak correlation with total WBC counts and a weak negative correlation with albumin.[88]

Consistently, high levels of CRP were observed in cases with pyometra, polyarthritis, pancreatitis, and panniculitis. Many metabolic diseases were observed not to have significant increases including epilepsy, hypothyroidism, rhinitis, bronchitis, and diabetes. AGP levels were examined in dogs with various diseases.[79] Increases were consistently observed in dogs with acute heartworm disease, viral infections including parvovirus and distemper, pyometra, and renal failure.

Nasal disease in dogs can arise from several different etiologies. A recent report demonstrated significant increases in CRP and HP in symptomatic animals.[155] Differences among APP were present among the groups with aspergillosis, chronic rhinitis, and neoplasia not only with CRP and HP but also in AGP and SAA levels. Versus other inflammatory stimuli, it is notable that all of these presentations resulted in less than an 80-fold increase in CRP.

Consistent APP increases are observed with infectious diseases in dogs. Two examples are ehrlichiosis and babesiosis which are syndromes related to infection with hematoprotozoans. In an experimental model of infection with *Ehrlichia canis*, the mean CRP increase was 400-fold with peak levels occurring with evidence of *Ehrlichia* replication and seroconversion.[156] Another study detailed APP levels in naturally infected dogs. Whereas 100-fold increases in CRP were found in all infected dogs, significantly, higher levels of HP, CRP, and SAA were observed in those dogs with the more advanced myelosuppressive infection.[157] In animals with naturally occurring babesiosis, CRP and CP levels were significantly increased; maximal levels were linked to disease severity.[158] Naturally infected dogs with leishmaniasis, another protozoan, were also studied.[159] CRP was increased 30- to 60-fold in infected animals with the highest levels present in symptomatic animals. Significant increases in HP and CP were also observed.

Several studies have been conducted examining APP expression with neoplastic disease. In a studies of dogs with lymphoma, 68% had abnormal levels of CRP.[106] Both CRP and HP were increased in dogs with lymphoma, myeloma, and leukemia.[160] In dogs with mammary neoplasia, significant increases in CRP, SAA, and HP were observed in only severe metastatic cases or those with ulcerated tumors.[161] HP was also reported to be increased in dogs with mammary tumors, but neither HP nor CRP levels were predictive of malignancy.[162]

Trauma, via surgery, has been documented to increase APP.[105,163] Basic surgeries including tooth extraction and removal of superficial tumors were observed to increase CRP levels two- threefold.[163] In dogs with postovariohysterectomy complications (often bacterial infections), 200-fold increases in CRP were observed.[105] Whereas animals with uncomplicated surgery demonstrated an early rise and normalization by 10 days postsurgery, elevated levels were observed in those with pyometra for more than 17 days even with appropriate antibiotic treatment. Increased serum AGP levels were also found to be correlative to pyometra.[100]

Experimental models of trauma show similar changes. In a drug-induced model of acute gastric mucosal injury, significant increases in CRP and SAA were observed on day 1 with normalization by day 4.[164] Increases in fibrinogen and HP continued through the terminal endpoint of day 7. In a model of pulmonic stenosis, approximately 30% of the animals demonstrated elevated CRP levels and were those animals that more likely had clinical signs.[165] After balloon valvuloplasty, CRP increased twofold. Similar increases were observed in dogs with naturally acquired chronic valvular disease.[166]

Inflammatory processes in dogs may have an infectious or noninfectious etiology which is the case for canine idiopathic arthritis. CRP levels were markedly increased in all of the study group dogs with values ranging from 3- to 20-fold over healthy controls.[104] CRP was also found to be increased nearly 50-fold in dogs with acute pancreatitis, a disease with a high mortality rate.[167] In dogs diagnosed with autoimmune hemolytic anemia, 100-fold increases in CRP and 5-fold increases in AGP were reported.[103] Steroid-responsive meningitis arteritis represents another autoimmune-based disease in dogs. CRP, SAA, AGP, and HP levels were significantly increased in all study patients.[101] More than 100-fold increases in CRP and 1000-fold increases in SAA were observed.

The clinical utility of feline APP has been reviewed in other publications.[2,4] The major APP in cats include SAA and AGP, and mild to moderate HP increases are also observed. In an experimental model of either LPS or turpentine oil injection, significant increases in SAA were observed by 8 h with AGP and HP increased by 24 h postinjection.[168] No changes in CRP were observed. In addition, 10-fold increases in SAA and HP and 20-fold increases in AGP were reported in cats hospitalized for various conditions. A larger study examined SAA levels in clinically ill cats and found significant increases in those animals grouped by diagnosis: infection, endocrine disease, and neoplasia.[169] The highest levels were observed in the infection group. Others reported significant increases in cats with renal failure and feline lower urinary tract disease.[170]

Increases in AGP and HP were observed in cats with feline infectious peritonitis (FIP) and feline immunodeficiency virus as well in cats with symptoms similar to FIP.[171] A moderate increase in AGP was found to discriminate FIP cats from cats with other inflammatory processes, and a marked increase was supportive of a FIP diagnosis.[172] In a study of the efficacy of different diagnostic tests for FIP, AGP was reported to have 100% sensitivity and specificity.[173] Other studies have indicated that the assessment of AGP in peritoneal fluid—where a common effusion occurs in acutely ill cats—is correlated with the definitive diagnosis of this infectious disease.[61,171] APP changes were also found in cats infected with feline coronavirus which can progress to FIP disease; the investigators postulated that APP other than SAA, AGP, and HP might have a role in monitoring the progression of disease.[174]

The APR was also examined in cats naturally infected with *Chlamydophila* which commonly results in ocular disease. Higher levels of SAA but not AGP were observed in seropositive animals that were actively shedding the organism.

A mean increase of 1.5-fold AGP levels was reported in cats with various malignant tumors.[175] No significant differences were observed by tumor type which included carcinoma, sarcoma, and round cell tumors. Increased AGP was also consistent observed in cats with lymphoma.[107] In other studies, cats were subjected to surgery to study the APR. In these animals, significant increases in SAA, AGP, and HP were all present by 24 h and persisted until day 4.[168] Significant increases in SAA were also reported in cats suffering from various injuries.[170] Inconsistent changes in other inflammatory diseases have been reported in cats.[2,4,170] However, a 160-fold increase in SAA was reported in a clinical case report of pancreatitis.[176] SAA was found to decrease with treatment and increase with disease recurrence.

C. Large Animals

APP have been demonstrated to be of value in management of individual cases of disease in large animal species as well as in herd management. These species include horses, cows, pigs, sheep, and goats.

The major APP in horses is SAA.[135,177,178] Experimental stimuli including the injection of turpentine oil resulted in over 400-fold increases in SAA with the first few days of exposure.[178] Twofold increases in CP were also reported.[179] Marked increases were also observed in animals with clinical signs including enteritis, pneumonitis, fever, diarrhea, cellulitis, and colic.[178] The etiology of the latter disease is multifactorial, and in those animals with a primary inflammatory cause, SAA levels were increased up to 500-fold.[92]

Experimental and natural bacterial infection has also demonstrated the utility of SAA assessments. In animals experimentally infected with *Streptococcus equi* in the lung, the levels were observed to increase more than 800-fold with normalization with the decrease in clinical signs.[180] Variable concentrations were observed in foals naturally infected with *Rhodococcus equi* which is a major pathogen in horses.[181] SAA was reported to increase with experimental infection with equine herpes virus more than 100-fold.[177] SAA increases of 400-fold were observed in horses naturally infected with equine influenza virus; the highest values were observed during the acute phase of infection with the appearance of significant clinical signs.[95]

APR to surgery has been extensively studied in horses.[177,178,182] SAA increases were reported to be rapid followed by increases in FIB and HP; levels were found useful in postoperative monitoring. AGP levels were reported to increase by day 2 and normalized in uncomplicated recovery by

28 days.[183] A recent study compared the levels of SAA among horses subjected to minimal, intermediate, and extensive surgical trauma.[184] SAA and FIB differed by group, whereas total WBC counts did not achieve significance.

Joint disease is a common affliction of horses, and the APR response has been studied in experimental models.[87,185] In a noninfectious arthritis model where inflammation was induced by amphotericin B, SAA increased more than 200-fold with more mild increases in HP.[185] The injection of LPS resulted in similar increases in SAA in both the serum and the synovial fluid.[87] In animals with naturally occurring joint diseases suspected to have bacterial origin, significant increases in SAA were also observed.[96] In grass sickness, a degenerative neurological disease, several APP including AGP, A2M, HP, and CP were observed to be elevated with specific increases often dependent on clinical presentation.[186] AGP and HP appeared to be linked to the onset of clinical signs.

Both SAA and HP appear to be important APP in cows.[35] After challenge with LPS, more than a 30-fold increase in HP and a 1300-fold increase in SAA were reported by days 2–3.[187] A transient decrease in albumin was also observed. Extensive studies have been conducted monitoring APP levels during infection.[35] In a model of experimental infection with bovine respiratory syncytial virus (BRSV), strong responses in both HP and SAA were reported.[188] In animals naturally infected with BRSV as well as different pathogenic bacteria, similar increases were observed with higher levels of HP when bronchoalveolar fluid was examined.[189] APP expression in other infections including bovine herpes virus, foot-and-mouth disease, bovine tropical theileriosis, and bovine diarrhea virus has also been examined.[35,190] After bacterial or viral challenge, mild increases in CP were reported.[191] Levels of SAA and HP have also been measured in the serum as well as the milk from cows with mastitis.[192] Mild increases in CRP have also been observed in the serum.[193] Physiological changes in cows were also observed to affect APP levels. Within the week following calving, significant increases in HP and SAA were observed; this was especially prevalent in cows in their first parturition.[194] Lactating cows can commonly suffer from inflammatory disorders relative to the displacement of the abomasum, and some of these animals have secondary or concurrent diseases including hepatic lipidosis.[195] Both SAA and HP were observed to be elevated with the displacement, and HP was most associated with the development of lipidosis. In another study involving the implantation of a chamber in the paralumbar fossa during an experimental protocol, mild AGP levels were detected after 4 days.[196]

In the pig, the APP have been identified as including AGP, CP, HP, CRP, SAA, TN, and pig-MAP.[35] The injection of turpentine oil was observed to induce highest early response in CRP and HP with significant increases also in AGP and CP.[197] APP levels were also assessed during naturally occurring infections including porcine reproductive and respiratory syndrome virus,

Aujeszky's disease virus, porcine circovirus type 2, and *Mycoplasma*.[198] HP levels were the most consistently elevated among the infections. Marked increases in CRP were observed with the circovirus and *Mycoplasma* infections with moderate increases in SAA and MAP. In an experimental model of bacterial infection, both hepatic and extrahepatic expressions of SAA, CRP, MAP and HP were observed with marked expression of SAA and HP in the serum.[199] Animals with trauma (ear/tail bites) as well as arthritis also demonstrated high levels of SAA, CRP, MAP, and HP.[198] Experimental surgery was reported to result in transient increases in SAA and HP.[200] Lastly, exposure to the mycotoxin deoxynivalenol resulted in significant increases in HP and SAA after 24 h.[201] This mycotoxin is produced by the *Fusarium* spp. of fungus which is a frequent contaminant of livestock foodstuffs.

HP was reported to be a sensitive indicator of infection in sheep.[202] On the bases of postmortem diagnoses allowing for the comparison of sheep with bacterial or fungal disease versus noninfectious diseases including metabolic disorders and renal disease, HP was demonstrated to have 85% sensitivity and specificity for infectious processes. In an experimental model using intrathoracic injection of yeast, HP, CP, and FIB levels increased within the first week followed by mild decreases in albumin.[203] In another study, after challenge with *Corynebacterium pseudotuberculosis*, SAA, HP, and AGP levels were significantly increased with a more persistent increase in the latter APP.[204] CRP, HP, CP, and FIB were all increased in animals experimentally infected with *Mannheimia haemolytica*.[205] Naturally occurring infection consistently resulted in increased FIB, HP, SAA, and AGP.[206] A surgical model resulting in bronchial obstruction was utilized to study the trauma-induced APR.[207] Increases were observed in sham animals, but marked increases were especially observed with HP in the obstruction group. CP and FIB increases were also present.

The APP response of goats has not been widely studied. Injection of turpentine oil was reported to result in a marked early response of SAA with significant lagging responses including HP, AGP, FIB as well as a decrease in albumin.[208]

D. Avian, Exotic, and Wildlife Species

While several research-based studies have documented the presence of APP in various avian, exotic, and wildlife species, APP assays have not be widely adapted for use in these species. This was the subject of a recent paper which demonstrated that several of the standardized assays for CRP, SAA, and HP appear to cross-react with APP from many species, yet samples from some species demonstrated no clinically significant differences which may be a function of the low sample size or the lack of reagent cross-reactivity.[209] The information from this and other publications (discussed below) is summarized in Table IV.

TABLE IV
List of APP in Avian, Exotic, and Wildlife Species

Species	APP
Avian	AGP, CP, HP (PIT54), SAA, TN
Elephant	HP, SAA
Fish	A2M, CRP/SAP, HP, SAA, TN
Frog	HP (PIT54)
Harbor seal	CRP, HP
Impala	HP, SAA
Manatee	HP, SAA
Musk ox	HP, SAA
Nonhuman primate	CRP, HP, SAA
Sitatunga	HP
Turtle	SAA

As with other species, each wildlife species appears to have its own APP phenotype, that is, major APP. Impala, musk ox, and chimpanzee were found to be reactive on the assay for SAA.[209] The chimpanzee was also found to be reactive of the assay for CRP which is consistent with previous studies of NHP.[210,211] Treatment of NHP with recombinant IL-6 was found to result in an increase in A2M.[212] HP reactivity was found in the impala, musk ox, sitatunga, and chimpanzee.[209] Our laboratory has also found reactivity on the HP colorimetric assay with serum samples from diseased elephants (C. Cray, 2011). HP has been described in clinically healthy camels.[213] CP levels have been reported as a marker of pregnancy in the giant panda.[214] Considerably, more knowledge will be gained with increased sample size of animals with various diseases and disorders.

Studies of APP in avian species have been dominated by work with chickens.[133] TN was identified as a major APP with peak levels obtained 3 days after the injection of croton oil.[215] Injection of LPS revealed peak levels of AGP at 2 days after exposure.[216,217] The APR was also studied using infectious agents. Experimental infection with *S. aureus* resulted in increased TN and SAA, although SAA changes were more consistent and higher.[218] Infection with *Salmonella enteritidis* resulted in a 2.5-fold increase in AGP.[219] *Escherichia coli* inoculation was observed to increase TN, HP, and CP.[215,220] CP, but not HP, increased after inoculation of the parasite *Eimeria tenella*.[220] Challenge with infectious bursal disease virus, a major disease problem in chickens, resulted in 2.5- to 7-fold increases in TN.[215] Others reported significant increases in SAA and CP as well as AGP.[221,222] In ongoing studies in our laboratory, we have observed increased SAA, TN, and HP in other avian species undergoing inflammatory processes including parrots and penguins (C. Cray, unpublished observations).[218]

The evolution of APP has been the focus of several research studies in reptiles and fish. An SAA homolog was recently identified in the soft-shelled turtle.[223] Turtles were infected with a Gram-negative bacterium, and SAA levels were found to increase over 1000-fold in the liver by 2 days postinfection. Smaller increases were also observed in the kidney and spleen. Decreases in albumin were also observed supporting a role for this biomarker as a negative APP.

The innate immune response of fish is a subject of an extensive review.[224] CRP and SAP like biomarkers have been reported in several species of fish, and both *in vivo* and *in vitro* studies have identified SAA. Bacterial infection in trout has been described to result in a 3000-fold increase in SAA.[225] Earlier studies documented the diverse inflammatory response to bacteria, virus, and experimental agents.[226] Both TN and HP were identified as major APP in another teleost.[227] This indicates that TN is a positive APP as observed in avian species. A2M has also been documented in fish.[224]

APP have also been reported in marine mammals. Harr and colleagues studied the cross-reactivity of SAA, HP, AGP, and CRP assays in serum of manatees.[228] CRP and AGP assays were described to not cross-react with the test sera, although, perhaps, this indicates that these are not major APP in this species. SAA was reported to have the highest diagnostic sensitivity for inflammatory disease in this species versus FIB, HP, and total WBC. We have similarly observed SAA to be a valuable prognostic marker in stressed and injured manatees. (C. Cray, 2011). In other studies, specific CRP assays were produced for use with serum samples from harbor seals.[229] Animals with pneumonia and other inflammatory processes were found to have as much as a 50-fold increase. HP in harbor seals was characterized in two recent publications.[230,231]

V. Concluding Remarks

Dramatic increases in APP occur in response to inflammation and tissue injury, and this process is essential to the development of an innate immune response. The APR is conserved among different species which is a further reflection of its critical function. Although there is marked species variation among the proteins, this has not only resulted in a wealth of publications in different animals but has also facilitated the development of several species-specific and heterologous assays which can be employed to quantitate APP. The implementation of these assays has provided results that have demonstrated clear benefits at the research level with the association of APP with diagnosis, prognosis, the detection of subclinical disease, and applications to food safety. The excitement regarding the applications of APP has been measurable for years.[6,9,232,233] At this time, the only impediments are the need for education and acceptance of veterinary practitioners and the more ready access to these methods at the reference laboratory level. APP testing which is available today represents just the start of newly emerging biomarkers for use in animals. As technology improves and research studies ensue, a further understanding of APP and the APR in animals will result in an important step in health monitoring, understanding innate immunity, and potential therapeutic interventions of acute and chronic inflammatory diseases in animals and humans.

References

1. Cray C, Zaias J, Altman NH. Acute phase response in animals: a review. *Comp Med* 2009;**59**:517–26.
2. Ceron JJ, Eckersall PD, Martynez-Subiela S. Acute phase proteins in dogs and cats: current knowledge and future perspectives. *Vet Clin Pathol* 2005;**34**:85–99.
3. Murata H, Shimada N, Yoshioka M. Current research on acute phase proteins in veterinary diagnosis: an overview. *Vet J* 2004;**168**:28–40.
4. Paltrinieri S. The feline acute phase reaction. *Vet J* 2008;**177**:26–35.
5. Kjelgaard-Hansen M, Jacobsen S. Assay validation and diagnostic applications of major acute-phase protein testing in companion animals. *Clin Lab Med* 2011;**31**:51–70.
6. Eckersall PD. Acute phase proteins: from research laboratory to clinic. *Vet Clin Pathol* 2010;**39**:1–2.
7. Eckersall PD, Bell R. Acute phase proteins: biomarkers of infection and inflammation in veterinary medicine. *Vet J* 2010;**185**:23–7.
8. Eckersall PD. Proteins, proteomics, and the dysproteinemias. In: Kaneko JJ, Harvey JW, Bruss ML, editors. *Clinical biochemistry of domestic animals*. 6th ed. San Diego, CA: Academic Press; 2008. pp. 117–55.
9. Eckersall PD. The time is right for acute phase protein assays. *Vet J* 2004;**168**:3–5.
10. Eckersall PD. Recent advances and future prospects for the use of acute phase proteins as markers of disease in animals. *Rev Med Vet* 2000;**151**:577–84.
11. Eckersall PD. Acute phase proteins as markers of inflammatory lesions. *Comp Haematol Int* 1995;**5**:93–7.
12. Janeway CA, Travers P, Walport M, Shlomchik MJ. *Immunobiology, vol. 5*. 5th ed. New York, NY: Garland Publishing; 2001 (732).
13. Yoo JY, Desiderio S. Innate and acquired immunity intersect in a global view of the acute-phase response. *Proc Natl Acad Sci USA* 2003;**100**:1157–62.
14. Gabay C, Kushner I. Acute-phase proteins and other systemic responses to inflammation. *N Engl J Med* 1999;**340**:448–54.
15. Schnare M, Barton GM, Holt AC, Takeda K, Akira S, Medzhitov R. Toll-like receptors control activation of adaptive immune responses. *Nat Immunol* 2001;**2**:947–50.
16. Kawai T, Adachi O, Ogawa T, Takeda K, Akira S. Unresponsiveness of MyD88-deficient mice to endotoxin. *Immunity* 1999;**11**:115–22.
17. Cassatella MA. The production of cytokines by polymorphonuclear neutrophils. *Immunol Today* 1995;**16**:21–6.
18. Ramadori G, Sipe JD, Colten HR. Expression and regulation of the murine serum amyloid A (SAA) gene in extrahepatic sites. *J Immunol* 1985;**135**:3645–7.

19. Vreugdenhil AC, Dentener MA, Snoek AM, Greve JW, Buurman WA. Lipopolysaccharide binding protein and serum amyloid A secretion by human intestinal epithelial cells during the acute phase response. *J Immunol* 1999;**163**:2792–8.
20. Vernooy JH, Reynaert N, Wolfs TG, Cloots RH, Haegens A, de Vries B, et al. Rapid pulmonary expression of acute-phase reactants after local lipopolysaccharide exposure in mice is followed by an interleukin-6 mediated systemic acute-phase response. *Exp Lung Res* 2005;**31**:855–71.
21. Fattori E, Cappelletti M, Costa P, Sellitto C, Cantoni L, Carelli M, et al. Defective inflammatory response in interleukin 6-deficient mice. *J Exp Med* 1994;**180**:1243–50.
22. Quinton LJ, Jones MR, Robson BE, Mizgerd JP. Mechanisms of the hepatic acute-phase response during bacterial pneumonia. *Infect Immun* 2009;**77**:2417–26.
23. Hagihara K, Nishikawa T, Sugamata Y, Song J, Isobe T, Taga T, et al. Essential role of STAT3 in cytokine-driven NF-kappaB-mediated serum amyloid A gene expression. *Genes Cells* 2005;**10**:1051–63.
24. Alonzi T, Maritano D, Gorgoni B, Rizzuto G, Libert C, Poli V. Essential role of STAT3 in the control of the acute-phase response as revealed by inducible gene inactivation [correction of activation] in the liver. *Mol Cell Biol* 2001;**21**:1621–32.
25. Borghetti P, Saleri R, Mocchegiani E, Corradi A, Martelli P. Infection, immunity and the neuroendocrine response. *Vet Immunol Immunopathol* 2009;**130**:141–62.
26. Bethin KE, Vogt SK, Muglia LJ. Interleukin-6 is an essential, corticotropin-releasing hormone-independent stimulator of the adrenal axis during immune system activation. *Proc Natl Acad Sci USA* 2000;**97**:9317–22.
27. Sarraf P, Frederich RC, Turner EM, Ma G, Jaskowiak NT, Rivet DJ, et al. Multiple cytokines and acute inflammation raise mouse leptin levels: potential role in inflammatory anorexia. *J Exp Med* 1997;**185**:171–5.
28. Akarsu ES, House RV, Coceani F. Formation of interleukin-6 in the brain of the febrile cat: relationship to interleukin-1. *Brain Res* 1998;**803**:137–43.
29. Coceani F, Akarsu ES. Prostaglandin E2 in the pathogenesis of fever. An update. *Ann N Y Acad Sci* 1998;**856**:76–82.
30. Yeager MP, Guyre PM, Munck AU. Glucocorticoid regulation of the inflammatory response to injury. *Acta Anaesthesiol Scand* 2004;**48**:799–813.
31. Loyer P, Ilyin G, Abdel Razzak Z, Banchereau J, Dezier JF, Campion JP, et al. Interleukin 4 inhibits the production of some acute-phase proteins by human hepatocytes in primary culture. *FEBS Lett* 1993;**336**:215–20.
32. Gatt ME, Urieli-Shoval S, Preciado-Patt L, Fridkin M, Calco S, Azar Y, et al. Effect of serum amyloid A on selected *in vitro* functions of isolated human neutrophils. *J Lab Clin Med* 1998;**132**:414–20.
33. Kaneko JJ. Serum proteins and the dysproteinemias. In: Kaneko JJ, Harvey JW, Bruss ML, editors. *Clinical biochemistry of domestic animals*. 5th ed. San Diego: Academic Press; 1997. pp. 117–38.
34. Kushner I, Mackiewicz A. Acute phase proteins as disease markers. *Dis Markers* 1987;**5**:1–11.
35. Petersen HH, Nielsen JP, Heegaard PM. Application of acute phase protein measurements in veterinary clinical chemistry. *Vet Res* 2004;**35**:163–87.
36. Tillett WS, Francis T. Serological reactions in pneumonia with a non-protein somatic fraction of pneumococcus. *J Exp Med* 1930;**52**:561–71.
37. Griselli M, Herbert J, Hutchinson WL, Taylor KM, Sohail M, Krausz T, et al. C-reactive protein and complement are important mediators of tissue damage in acute myocardial infarction. *J Exp Med* 1999;**190**:1733–40.
38. Pepys MB, Hirschfield GM. C-reactive protein: a critical update. *J Clin Invest* 2003;**111**:1805–12.

39. Koenig W, Sund M, Frohlich M, Fischer HG, Lowel H, Doring A, et al. C-Reactive protein, a sensitive marker of inflammation, predicts future risk of coronary heart disease in initially healthy middle-aged men: results from the MONICA (Monitoring Trends and Determinants in Cardiovascular Disease) Augsburg Cohort Study, 1984 to 1992. *Circulation* 1999;**99**:237–42.
40. Thompson D, Pepys MB, Wood SP. The physiological structure of human C-reactive protein and its complex with phosphocholine. *Structure* 1999;**7**:169–77.
41. Mold C, Rodic-Polic B, Du Clos TW. Protection from Streptococcus pneumoniae infection by C-reactive protein and natural antibody requires complement but not Fc gamma receptors. *J Immunol* 2002;**168**:6375–81.
42. Du Clos TW, Mold C. The role of C-reactive protein in the resolution of bacterial infection. *Curr Opin Infect Dis* 2001;**14**:289–93.
43. de Haas CJ. New insights into the role of serum amyloid P component, a novel lipopolysaccharide-binding protein. *FEMS Immunol Med Microbiol* 1999;**26**:197–202.
44. Fransson BA, Bergstrom A, Wardrop KJ, Hagman R. Assessment of three automated assays for C-reactive protein determination in dogs. *Am J Vet Res* 2007;**68**:1281–6.
45. Kjelgaard-Hansen M, Jensen AL, Kristensen AT. Evaluation of a commercially available human C-reactive protein (CRP) turbidometric immunoassay for determination of canine serum CRP concentration. *Vet Clin Pathol* 2003;**32**:81–7.
46. Klenner S, Bauer N, Moritz A. Evaluation of three automated human immunoturbidimetric assays for the detection of C-reactive protein in dogs. *J Vet Diagn Invest* 2010;**22**:544–52.
47. Uhlar CM, Whitehead AS. Serum amyloid A, the major vertebrate acute-phase reactant. *Eur J Biochem* 1999;**265**:501–23.
48. Cray C, Besselsen DG, Hart JL, Yoon D, Rodriguez M, Zaias J, et al. Quantitation of acute phase proteins and protein electrophoresis in monitoring the acute inflammatory process in experimentally and naturally infected mice. *Comp Med* 2010;**60**:263–71.
49. Patel H, Fellowes R, Coade S, Woo P. Human serum amyloid A has cytokine-like properties. *Scand J Immunol* 1998;**48**:410–8.
50. Badolato R, Wang JM, Murphy WJ, Lloyd AR, Michiel DF, Bausserman LL, et al. Serum amyloid A is a chemoattractant: induction of migration, adhesion, and tissue infiltration of monocytes and polymorphonuclear leukocytes. *J Exp Med* 1994;**180**:203–9.
51. Jacobsen S, Kjelgaard-Hansen M, Hagbard Petersen H, Jensen AL. Evaluation of a commercially available human serum amyloid A (SAA) turbidometric immunoassay for determination of equine SAA concentrations. *Vet J* 2006;**172**:315–9.
52. Tamamoto T, Ohno K, Ohmi A, Goto-Koshino Y, Tsujimoto H. Verification of measurement of the feline serum amyloid A (SAA) concentration by human SAA turbidimetric immunoassay and its clinical application. *J Vet Med Sci* 2008;**70**:1247–52.
53. Levy AP, Asleh R, Blum S, Levy NS, Miller-Lotan R, Kalet-Litman S, et al. Haptoglobin: basic and clinical aspects. *Antioxid Redox Signal* 2010;**12**:293–304.
54. Fagoonee S, Gburek J, Hirsch E, Marro S, Moestrup SK, Laurberg JM, et al. Plasma protein haptoglobin modulates renal iron loading. *Am J Pathol* 2005;**166**:973–83.
55. Langlois MR, Delanghe JR. Biological and clinical significance of haptoglobin polymorphism in humans. *Clin Chem* 1996;**42**:1589–600.
56. Wicher KB, Fries E. Haptoglobin, a hemoglobin-binding plasma protein, is present in bony fish and mammals but not in frog and chicken. *Proc Natl Acad Sci USA* 2006;**103**:4168–73.
57. Harvey JW. Comparison between serum haptoglobin and alpha-2-globulin concentrations in dogs. *Vet Clin Pathol* 1986;**15**:4–5.
58. Tecles F, Subiela SM, Petrucci G, Panizo CG, Ceron JJ. Validation of a commercially available human immunoturbidimetric assay for haptoglobin determination in canine serum samples. *Vet Res Commun* 2007;**31**:23–36.

59. Fournier T, Medjoubi-N N, Porquet D. Alpha-1-acid glycoprotein. *Biochim Biophys Acta* 2000;**1482**:157–71.
60. Moore DF, Rosenfeld MR, Gribbon PM, Winlove CP, Tsai CM. Alpha-1-acid (AAG, orosomucoid) glycoprotein: interaction with bacterial lipopolysaccharide and protection from sepsis. *Inflammation* 1997;**21**:69–82.
61. Bence LM, Addie DD, Eckersall PD. An immunoturbidimetric assay for rapid quantitative measurement of feline alpha-1-acid glycoprotein in serum and peritoneal fluid. *Vet Clin Pathol* 2005;**34**:335–41.
62. Mosesson MW. Fibrinogen and fibrin structure and functions. *J Thromb Haemost* 2005;**3**:1894–904.
63. Millar HR, Simpson JG, Stalker AL. An evaluation of the heat precipitation method for plasma fibrinogen estimation. *J Clin Pathol* 1971;**24**:827–30.
64. Roeser HP, Lee GR, Nacht S, Cartwright GE. The role of ceruloplasmin in iron metabolism. *J Clin Invest* 1970;**49**:2408–17.
65. Lampreave F, Gonzalez-Ramon N, Martinez-Ayensa S, Hernandez MA, Lorenzo HK, Garcia-Gil A, et al. Characterization of the acute phase serum protein response in pigs. *Electrophoresis* 1994;**15**:672–6.
66. Pineiro M, Lampreave F, Alava MA. Development and validation of an ELISA for the quantification of pig major acute phase protein (Pig-MAP). *Vet Immunol Immunopathol* 2009;**127**:228–34.
67. Gomme PT, McCann KB, Bertolini J. Transferrin: structure, function and potential therapeutic actions. *Drug Discov Today* 2005;**10**:267–73.
68. Tohjo H, Yadatsu M, Uchida E, Niiyama M, Syuto B, Moritsu Y, et al. Polyacrylamide gel electrophoretic serum protein patterns of acute inflammation induced by intramuscular injection of turpentine in young broiler chickens. *J Vet Med Sci* 1996;**58**:267–8.
69. Cray C, Wack A, Arheart KL. Invalidity of albumin measurement of specimens from clinically ill birds by bromcresol green methodology. *J Av Med Surg* 2011;**25**:14–22.
70. Cray C, Tatum L. Application of protein electrophoresis in avian diagnostic testing. *J Av Med Surg* 1998;**12**:4–10.
71. Cray C, Besselsen DG, Hart JL, Yoon D, Rodriguez MR, Zaias J, et al. Application of acute phase protein quantitation and protein electrophoresis in monitoring the acute inflammatory process in experimentally and naturally infected mice. *Comp Med* 2010;**60**(4):263–71.
72. Jacobsen S, Kjelgaard-Hansen M, Hagbard Petersen H, Jensen AL. Evaluation of a commercially available human serum amyloid A (SAA) turbidometric immunoassay for determination of equine SAA concentrations. *Vet J* 2006;**172**:315–9.
73. Eckersall PD, Duthie S, Toussaint MJ, Gruys E, Heegaard P, Alava M, et al. Standardization of diagnostic assays for animal acute phase proteins. *Adv Vet Med* 1999;**41**:643–55.
74. Skinner JG. International standardization of acute phase proteins. *Vet Clin Pathol* 2001;**30**:2–7.
75. Otabe K, Sugimoto T, Jinbo T, Honda M, Kitao S, Hayashi S, et al. Physiological levels of C-reactive protein in normal canine sera. *Vet Res Commun* 1998;**22**:77–85.
76. Kjelgaard-Hansen M, Mikkelsen L, Kristensen AT, Jensen AL. Study on biological variability of five acute-phase reactants in dogs. *Comp Clin Pathol* 2003;**12**:69–74.
77. Gomez CR, Acuna-Castillo C, Perez C, Leiva-Salcedo E, Riquelme DM, Ordenes G, et al. Diminished acute phase response and increased hepatic inflammation of aged rats in response to intraperitoneal injection of lipopolysaccharide. *J Gerontol A Biol Sci Med Sci* 2008;**63**:1299–306.
78. Hayashi S, Jinbo T, Iguchi K, Shimizu M, Shimada T, Nomura M, et al. A comparison of the concentrations of C-reactive protein and alpha1-acid glycoprotein in the serum of young and adult dogs with acute inflammation. *Vet Res Commun* 2001;**25**:117–26.

79. Yuki M, Itoh H, Takase K. Serum alpha-1-acid glycoprotein concentration in clinically healthy puppies and adult dogs and in dogs with various diseases. *Vet Clin Pathol* 2010;**39**:65–71.
80. Orro T, Jacobsen S, Lepage JP, Niewold T, Alasuutari S, Soveri T. Temporal changes in serum concentrations of acute phase proteins in newborn dairy calves. *Vet J* 2008;**176**(2):182–7.
81. Ikeda M, Hamada K, Sumitomo N, Okamoto H, Sakakibara B. Serum amyloid A, cytokines, and corticosterone responses in germfree and conventional mice after lipopolysaccharide injection. *Biosci Biotechnol Biochem* 1999;**63**:1006–10.
82. Yamamoto S, Tagata K, Nagahata H, Ishikawa Y, Morimatsu M, Naiki M. Isolation of canine C-reactive protein and characterization of its properties. *Vet Immunol Immunopathol* 1992;**30**:329–39.
83. Mortensen RF, Beisel K, Zeleznik NJ, Le PT. Acute-phase reactants of mice. II. Strain dependence of serum amyloid P-component (SAP) levels and response to inflammation. *J Immunol* 1983;**130**:885–9.
84. Clapperton M, Bishop SC, Pineiro M, Campbell FM, Glass EJ. The association between plasma levels of acute phase proteins, haptoglobin, alpha-1 acid glycoprotein (AGP), pig-MAP, transthyretin and serum amyloid A (SAA) in Large White and Meishan pigs. *Vet Immunol Immunopathol* 2007;**119**:303–9.
85. Kuribayashi T, Shimada T, Matsumoto M, Kawato K, Honjyo T, Fukuyama M, et al. Determination of serum C-reactive protein (CRP) in healthy beagle dogs of various ages and pregnant beagle dogs. *Exp Anim* 2003;**52**:387–90.
86. Yamashita K, Fujinaga T, Miyamoto T, Hagio M, Izumisawa Y, Kotani T. Canine acute phase response: relationship between serum cytokine activity and acute phase protein in dogs. *J Vet Med Sci* 1994;**56**:487–92.
87. Jacobsen S, Niewold TA, Halling-Thomsen M, Nanni S, Olsen E, Lindegaard C, et al. Serum amyloid A isoforms in serum and synovial fluid in horses with lipopolysaccharide-induced arthritis. *Vet Immunol Immunopathol* 2006;**110**:325–30.
88. Nakamura M, Takahashi M, Ohno K, Koshino A, Nakashima K, Setoguchi A, et al. C-reactive protein concentration in dogs with various diseases. *J Vet Med Sci* 2008;**70**:127–31.
89. Burton SA, Honor DJ, Mackenzie AL, Eckersall PD, Markham RJ, Horney BS. C-reactive protein concentration in dogs with inflammatory leukograms. *Am J Vet Res* 1994;**55**:613–8.
90. Solter PF, Hoffmann WE, Hungerford LL, Siegel JP, St Denis SH, Dorner JL. Haptoglobin and ceruloplasmin as determinants of inflammation in dogs. *Am J Vet Res* 1991;**52**:1738–42.
91. Hulten C, Demmers S. Serum amyloid A (SAA) as an aid in the management of infectious disease in the foal: comparison with total leucocyte count, neutrophil count and fibrinogen. *Equine Vet J* 2002;**34**:693–8.
92. Vandenplas ML, Moore JN, Barton MH, Roussel AJ, Cohen ND. Concentrations of serum amyloid A and lipopolysaccharide-binding protein in horses with colic. *Am J Vet Res* 2005;**66**:1509–16.
93. Nazifi S, Ansari-Lari M, Asadi-Fardaqi J, Rezaei M. The use of receiver operating characteristic (ROC) analysis to assess the diagnostic value of serum amyloid A, haptoglobin and fibrinogen in traumatic reticuloperitonitis in cattle. *Vet J* 2009;**182**:315–9.
94. Gruys E, Toussaint MJ, Niewold TA, Koopmans SJ, van Dijk E, Meloen RH. Monitoring health by values of acute phase proteins. *Acta Histochem* 2006;**108**:229–32.
95. Hulten C, Sandgren B, Skioldebrand E, Klingeborn B, Marhaug G, Forsberg M. The acute phase protein serum amyloid A (SAA) as an inflammatory marker in equine influenza virus infection. *Acta Vet Scand* 1999;**40**:323–33.
96. Jacobsen S, Thomsen MH, Nanni S. Concentrations of serum amyloid A in serum and synovial fluid from healthy horses and horses with joint disease. *Am J Vet Res* 2006;**67**:1738–42.
97. Chan DL, Rozanski EA, Freeman LM. Relationship among plasma amino acids, C-reactive protein, illness severity, and outcome in critically ill dogs. *J Vet Intern Med* 2009;**23**:559–63.

98. Gebhardt C, Hirschberger J, Rau S, Arndt G, Krainer K, Schweigert FJ, et al. Use of C-reactive protein to predict outcome in dogs with systemic inflammatory response syndrome or sepsis. *J Vet Emerg Crit Care* 2009;**19**:450–8.
99. Galezowski AM, Snead EC, Kidney BA, Jackson ML. C-reactive protein as a prognostic indicator in dogs with acute abdomen syndrome. *J Vet Diagn Invest* 2010;**22**:395–401.
100. Hagman R. Serum alpha-1-acid glycoprotein concentrations in 26 dogs with pyometra. *Vet Clin Pathol* 2011;**40**:52–9.
101. Lowrie M, Penderis J, Eckersall PD, McLaughlin M, Mellor D, Anderson TJ. The role of acute phase proteins in diagnosis and management of steroid-responsive meningitis arteritis in dogs. *Vet J* 2009;**182**:125–30.
102. Jergens AE, Schreiner CA, Frank DE, Niyo Y, Ahrens FE, Eckersall PD, et al. A scoring index for disease activity in canine inflammatory bowel disease. *J Vet Intern Med* 2003;**17**:291–7.
103. Mitchell KD, Kruth SA, Wood RD, Jefferson B. Serum acute phase protein concentrations in dogs with autoimmune hemolytic anemia. *J Vet Intern Med* 2009;**23**:585–91.
104. Ohno K, Yokoyama Y, Nakashima K, Setoguchi A, Fujino Y, Tsujimoto H. C-reactive protein concentration in canine idiopathic polyarthritis. *J Vet Med Sci* 2006;**68**:1275–9.
105. Dabrowski R, Kostro K, Lisiecka U, Szczubial M, Krakowski L. Usefulness of C-reactive protein, serum amyloid A component, and haptoglobin determinations in bitches with pyometra for monitoring early post-ovariohysterectomy complications. *Theriogenology* 2009;**72**:471–6.
106. Nielsen L, Toft N, Eckersall PD, Mellor DJ, Morris JS. Serum C-reactive protein concentration as an indicator of remission status in dogs with multicentric lymphoma. *J Vet Intern Med* 2007;**21**:1231–6.
107. Correa SS, Mauldin GN, Mauldin GE, Mooney SC. Serum alpha 1-acid glycoprotein concentration in cats with lymphoma. *J Am Anim Hosp Assoc* 2001;**37**:153–8.
108. Vafadar-Isfahani B, Laversin SA, Ahmad M, Ball G, Coveney C, Lemetre C, et al. Serum biomarkers which correlate with failure to respond to immunotherapy and tumor progression in a murine colorectal cancer model. *Proteomics Clin Appl* 2010;**4**:682–96.
109. Karreman HJ, Wentink GH, Wensing T. Using serum amyloid A to screen dairy cows for sub-clinical inflammation. *Vet Q* 2000;**22**:175–8.
110. Ganheim C, Alenius S, Persson Waller K. Acute phase proteins as indicators of calf herd health. *Vet J* 2007;**173**:645–51.
111. Harding JC, Baarsch MJ, Murtaugh MP. Association of tumour necrosis factor and acute phase reactant changes with post arrival disease in swine. *Zentralbl Veterinarmed B* 1997;**44**:405–13.
112. Grau-Roma L, Heegaard PM, Hjulsager CK, Sibila M, Kristensen CS, Allepuz A, et al. Pig-major acute phase protein and haptoglobin serum concentrations correlate with PCV2 viremia and the clinical course of postweaning multisystemic wasting syndrome. *Vet Microbiol* 2009;**138**:53–61.
113. Toussaint MJ, van Ederen AM, Gruys E. Implication of clinical pathology in assessment of animal health and in animal production and meat inspection. *Comp Haematol Int* 1995;**5**:149–57.
114. Saini PK, Riaz M, Webert DW, Eckersall PD, Young CR, Stanker LH, et al. Development of a simple enzyme immunoassay for blood haptoglobin concentration in cattle and its application in improving food safety. *Am J Vet Res* 1998;**59**:1101–7.
115. Pineiro M, Gymnich S, Knura S, Pineiro C, Petersen B. Meat juice: an alternative matrix for assessing animal health by measuring acute phase proteins. Correlations of pig-MAP and haptoglobin concentrations in pig meat juice and plasma. *Res Vet Sci* 2009;**87**:273–6.
116. Saco Y, Fina M, Gimenez M, Pato R, Piedrafita J, Bassols A. Evaluation of serum cortisol, metabolic parameters, acute phase proteins and faecal corticosterone as indicators of stress in cows. *Vet J* 2008;**177**:439–41.

117. Marco-Ramell A, Pato R, Pena R, Saco Y, Manteca X, Ruiz de la Torre JL, et al. Identification of serum stress biomarkers in pigs housed at different stocking densities. *Vet J* 2011; doi:10.1016/j.tvjl.2011.01.003.
118. Pineiro M, Pineiro C, Carpintero R, Morales J, Campbell FM, Eckersall PD, et al. Characterisation of the pig acute phase protein response to road transport. *Vet J* 2007;**173**:669–74.
119. Hickey MC, Drennan M, Earley B. The effect of abrupt weaning of suckler calves on the plasma concentrations of cortisol, catecholamines, leukocytes, acute-phase proteins and *in vitro* interferon-gamma production. *J Anim Sci* 2003;**81**:2847–55.
120. Arthington JD, Eichert SD, Kunkle WE, Martin FG. Effect of transportation and commingling on the acute-phase protein response, growth, and feed intake of newly weaned beef calves. *J Anim Sci* 2003;**81**:1120–5.
121. Lomborg SR, Nielsen LR, Heegaard PM, Jacobsen S. Acute phase proteins in cattle after exposure to complex stress. *Vet Res Commun* 2008;**32**:575–82.
122. Depke M, Steil L, Domanska G, Volker U, Schutt C, Kiank C. Altered hepatic mRNA expression of immune response and apoptosis-associated genes after acute and chronic psychological stress in mice. *Mol Immunol* 2009;**46**:3018–28.
123. Venteclef N, Jakobsson T, Steffensen KR, Treuter E. Metabolic nuclear receptor signaling and the inflammatory acute phase response. *Trends Endocrinol Metab* 2011;**22**(8):333–43.
124. Medzhitov R. Origin and physiological roles of inflammation. *Nature* 2008;**454**:428–35.
125. Correale M, Brunetti ND, De Gennaro L, Di Biase M. Acute phase proteins in atherosclerosis (acute coronary syndrome). *Cardiovasc Hematol Agents Med Chem* 2008;**6**:272–7.
126. Ko KW, Corry DB, Brayton CF, Paul A, Chan L. Extravascular inflammation does not increase atherosclerosis in apoE-deficient mice. *Biochem Biophys Res Commun* 2009;**384**:93–9.
127. Zhang T, Chen Y, Liu H, Zhou Z, Zhai Y, Yang J. Chronic unpredictable stress accelerates atherosclerosis through promoting inflammation in apolipoprotein E knockout mice. *Thromb Res* 2010;**126**:386–92.
128. German AJ, Ryan VH, German AC, Wood IS, Trayhurn P. Obesity, its associated disorders and the role of inflammatory adipokines in companion animals. *Vet J* 2010;**185**:4–9.
129. Kruger AJ, Yang C, Tam SW, Hinerfeld D, Evans JE, Green KM, et al. Haptoglobin as an early serum biomarker of virus-induced autoimmune type 1 diabetes in biobreeding diabetes resistant and LEW1.WR1 rats. *Exp Biol Med (Maywood)* 2010;**235**:1328–37.
130. Westermark GT, Westermark P. Serum amyloid A and protein AA: molecular mechanisms of a transmissible amyloidosis. *FEBS Lett* 2009;**583**:2685–90.
131. DiBartola SP, Reiter JA, Cornacoff JB, Kociba GJ, Benson MD. Serum amyloid A protein concentration measured by radial immunodiffusion in Abyssinian and non-Abyssinian cats. *Am J Vet Res* 1989;**50**:1414–7.
132. MacGuire JG, Christe KL, Yee JL, Kalman-Bowlus AL, Lerche NW. Serologic evaluation of clinical and subclinical secondary hepatic amyloidosis in rhesus macaques (Macaca mulatta). *Comp Med* 2009;**59**:168–73.
133. Chamanza R, van Veen L, Tivapasi MT, Toussaint MJ. Acute phase proteins in the domestic fowl. *Worlds Poult Sci J* 1999;**55**:61–71.
134. French T. Acute phase proteins. In: Loeb WF, Quimby FW, editors. *The clinical chemistry of laboratory animals*. Oxford: Pergamon Press; 1989. pp. 201–35.
135. Jacobsen S, Andersen PH. The acute phase protein serum amyloid A (SAA) as a marker of inflammation in horses. *Eq Vet Educ* 2007;**19**:38–46.
136. Schreiber G, Tsykin A, Aldred AR, Thomas T, Fung WP, Dickson PW, et al. The acute phase response in the rodent. *Ann N Y Acad Sci* 1989;**557**:61–85 [discussion 85–6].
137. Balmer P, McMonagle F, Alexander J, Phillips RS. Experimental erythrocytic malaria infection induces elevated levels of serum amyloid P production in mice. *Immunol Lett* 2000;**72**:147–52.

138. Eckersall PD, Gow JW, McComb C, Bradley B, Rodgers J, Murray M, et al. Cytokines and the acute phase response in post-treatment reactive encephalopathy of Trypanosoma brucei brucei infected mice. *Parasitol Int* 2001;**50**:15–26.
139. Duan X, Yarmush DM, Berthiaume F, Jayaraman A, Yarmush ML. A mouse serum two-dimensional gel map: application to profiling burn injury and infection. *Electrophoresis* 2004;**25**:3055–65.
140. Burger M, Knyszynski A. Haptoglobin and irradiation in mice. *Radiat Res* 1971;**46**:613–20.
141. Hunneyball IM, Spowage M, Crossley MJ, Rowe IF, Baltz ML. Acute phase protein changes in antigen-induced mono-articular arthritis in rabbits and mice. *Clin Exp Immunol* 1986;**65**:311–8.
142. Lockwood JF, Rutherford MS, Myers MJ, Schook LB. Induction of hepatic acute-phase protein transcripts: differential effects of acute and subchronic dimethylnitrosamine exposure *in vivo*. *Toxicol Appl Pharmacol* 1994;**125**:288–95.
143. Pessini AC, de Souza AM, Faccioli LH, Gregorio ZM, Arantes EC. Time course of acute-phase response induced by Tityus serrulatus venom and TsTX-I in mice. *Int Immunopharmacol* 2003;**3**:765–74.
144. Giffen PS, Turton J, Andrews CM, Barrett P, Clarke CJ, Fung KW, et al. Markers of experimental acute inflammation in the Wistar Han rat with particular reference to haptoglobin and C-reactive protein. *Arch Toxicol* 2003;**77**:392–402.
145. Jinbo T, Sakamoto T, Yamamoto S. Serum alpha2-macroglobulin and cytokine measurements in an acute inflammation model in rats. *Lab Anim* 2002;**36**:153–7.
146. Honjo T, Kuribayashi T, Matsumoto M, Yamazaki S, Yamamoto S. Kinetics of {alpha}2-macroglobulin and {alpha}1-acid glycoprotein in rats subjected to repeated acute inflammatory stimulation. *Lab Anim* 2010;**44**:150–4.
147. Ivanovic-Matic S, Poznanovic G, Grigorov I, Dinic S, Mihailovic M, Grdovic N, et al. The organophosphate-induced acute-phase response is characterized by synthesis of alpha 1-acid glycoprotein that exhibits an immunomodulatory effect. *J Appl Toxicol* 2008;**28**:63–71.
148. Jinbo T, Motoki M, Yamamoto S. Variation of serum alpha2-macroglobulin concentration in healthy rats and rats inoculated with Staphylococcus aureus or subjected to surgery. *Comp Med* 2001;**51**:332–5.
149. Osikov MV, Makarov EV, Krivokhizhina LV. Effects of alpha1-acid glycoprotein on hemostasis in experimental septic peritonitis. *Bull Exp Biol Med* 2007;**144**:178–80.
150. Giclas PC, Manthei U, Strunk RC. The acute phase response of C3, C5, ceruloplasmin, and C-reactive protein induced by turpentine pleurisy in the rabbit. *Am J Pathol* 1985;**120**:146–56.
151. Mackiewicz A, Ganapathi MK, Schultz D, Samols D, Reese J, Kushner I. Regulation of rabbit acute phase protein biosynthesis by monokines. *Biochem J* 1988;**253**:851–7.
152. Rygg M, Husby G, Marhaug G. Differential expression of rabbit serum amyloid A genes in response to various inflammatory agents. *Scand J Immunol* 1993;**38**:417–22.
153. Ray A, Ray BK. Persistent expression of serum amyloid A during experimentally induced chronic inflammatory condition in rabbit involves differential activation of SAF, NF-kappa B, and C/EBP transcription factors. *J Immunol* 1999;**163**:2143–50.
154. American Veterinary Medical Association. *U.S. Pet Ownership & Demographics Sourcebook.* 2007 ed. Schaumburg, IL: AVMA; 2007.
155. Sheahan D, Bell R, Mellanby RJ, Gow AG, Friend E, Heller J, et al. Acute phase protein concentrations in dogs with nasal disease. *Vet Rec* 2010;**167**:895–9.
156. Shimada T, Ishida Y, Shimizu M, Nomura M, Kawato K, Iguchi K, et al. Monitoring C-reactive protein in beagle dogs experimentally inoculated with Ehrlichia canis. *Vet Res Commun* 2002;**26**:171–7.
157. Mylonakis ME, Ceron JJ, Leontides L, Siarkou VI, Martinez S, Tvarijonaviciute A, et al. Serum acute phase proteins as clinical phase indicators and outcome predictors in naturally occurring canine monocytic ehrlichiosis. *J Vet Intern Med* 2011;**25**(4):811–7.

158. Ulutas B, Bayramli G, Ulutas PA, Karagenc T. Serum concentration of some acute phase proteins in naturally occurring canine babesiosis: a preliminary study. *Vet Clin Pathol* 2005;**34**:144–7.
159. Martinez-Subiela S, Tecles F, Eckersall PD, Ceron JJ. Serum concentrations of acute phase proteins in dogs with leishmaniasis. *Vet Rec* 2002;**150**:241–4.
160. Mischke R, Waterston M, Eckersall PD. Changes in C-reactive protein and haptoglobin in dogs with lymphatic neoplasia. *Vet J* 2007;**174**:188–92.
161. Tecles F, Caldin M, Zanella A, Membiela F, Tvarijonaviciute A, Subiela SM, et al. Serum acute phase protein concentrations in female dogs with mammary tumors. *J Vet Diagn Invest* 2009;**21**:214–9.
162. Planellas M, Bassols A, Siracusa C, Saco Y, Gimenez M, Pato R, et al. Evaluation of serum haptoglobin and C-reactive protein in dogs with mammary tumors. *Vet Clin Pathol* 2009;**38**:348–52.
163. Yamamoto S, Shida T, Miyaji S, Santsuka H, Fujise H, Mukawa K, et al. Changes in serum C-reactive protein levels in dogs with various disorders and surgical traumas. *Vet Res Commun* 1993;**17**:85–93.
164. Bayramli G, Ulutas B. Acute phase protein response in dogs with experimentally induced gastric mucosal injury. *Vet Clin Pathol* 2008;**37**:312–6.
165. Saunders AB, Smith BE, Fosgate GT, Suchodolski JS, Steiner JM. Cardiac troponin I and C-reactive protein concentrations in dogs with severe pulmonic stenosis before and after balloon valvuloplasty. *J Vet Cardiol* 2009;**11**:9–16.
166. Rush JE, Lee ND, Freeman LM, Brewer B. C-reactive protein concentration in dogs with chronic valvular disease. *J Vet Intern Med* 2006;**20**:635–9.
167. Mansfield CS, James FE, Robertson ID. Development of a clinical severity index for dogs with acute pancreatitis. *J Am Vet Med Assoc* 2008;**233**:936–44.
168. Kajikawa T, Furuta A, Onishi T, Tajima T, Sugii S. Changes in concentrations of serum amyloid A protein, alpha 1-acid glycoprotein, haptoglobin, and C-reactive protein in feline sera due to induced inflammation and surgery. *Vet Immunol Immunopathol* 1999;**68**:91–8.
169. Hansen AE, Schaap MK, Kjelgaard-Hansen M. Evaluation of a commercially available human serum amyloid A (SAA) turbidimetric immunoassay for determination of feline SAA concentration. *Vet Res Commun* 2006;**30**:863–72.
170. Sasaki K, Ma Z, Khatlani TS, Okuda M, Inokuma H, Onishi T. Evaluation of feline serum amyloid A (SAA) as an inflammatory marker. *J Vet Med Sci* 2003;**65**:545–8.
171. Duthie S, Eckersall PD, Addie DD, Lawrence CE, Jarrett O. Value of alpha 1-acid glycoprotein in the diagnosis of feline infectious peritonitis. *Vet Rec* 1997;**141**:299–303.
172. Paltrinieri S, Giordano A, Tranquillo V, Guazzetti S. Critical assessment of the diagnostic value of feline alpha1-acid glycoprotein for feline infectious peritonitis using the likelihood ratios approach. *J Vet Diagn Invest* 2007;**19**:266–72.
173. Giori L, Giordano A, Giudice C, Grieco V, Paltrinieri S. Performances of different diagnostic tests for feline infectious peritonitis in challenging clinical cases. *J Small Anim Pract* 2011;**52**:152–7.
174. Giordano A, Spagnolo V, Colombo A, Paltrinieri S. Changes in some acute phase protein and immunoglobulin concentrations in cats affected by feline infectious peritonitis or exposed to feline coronavirus infection. *Vet J* 2004;**167**:38–44.
175. Selting KA, Ogilvie GK, Lana SE, Fettman MJ, Mitchener KL, Hansen RA, et al. Serum alhpa 1-acid glycoprotein concentrations in healthy and tumor-bearing cats. *J Vet Intern Med* 2000;**14**:503–6.
176. Tamamoto T, Ohno K, Ohmi A, Seki I, Tsujimoto H. Time-course monitoring of serum amyloid A in a cat with pancreatitis. *Vet Clin Pathol* 2009;**38**:83–6.

177. Pepys MB, Baltz ML, Tennent GA, Kent J, Ousey J, Rossdale PD. Serum amyloid A protein (SAA) in horses: objective measurement of the acute phase response. *Equine Vet J* 1989;**21**:106–9.
178. Nunokawa Y, Fujinaga T, Taira T, Okumura M, Yamashita K, Tsunoda N, et al. Evaluation of serum amyloid A protein as an acute-phase reactive protein in horses. *J Vet Med Sci* 1993;**55**:1011–6.
179. Okumura M, Fujinaga T, Yamashita K, Tsunoda N, Mizuno S. Isolation, characterization, and quantitative analysis of ceruloplasmin from horses. *Am J Vet Res* 1991;**52**:1979–85.
180. Hobo S, Niwa H, Anzai T. Evaluation of serum amyloid A and surfactant protein D in sera for identification of the clinical condition of horses with bacterial pneumonia. *J Vet Med Sci* 2007;**69**:827–30.
181. Cohen ND, Chaffin MK, Vandenplas ML, Edwards RF, Nevill M, Moore JN, et al. Study of serum amyloid A concentrations as a means of achieving early diagnosis of Rhodococcus equi pneumonia. *Equine Vet J* 2005;**37**:212–6.
182. Jacobsen S, Jensen JC, Frei S, Jensen AL, Thoefner MB. Use of serum amyloid A and other acute phase reactants to monitor the inflammatory response after castration in horses: a field study. *Equine Vet J* 2005;**37**:552–6.
183. Taira T, Fujinaga T, Tamura K, Izumi M, Itoh H, Tsunoda N, et al. Isolation and characterization of alpha 1-acid glycoprotein from horses, and its evaluation as an acute-phase reactive protein in horses. *Am J Vet Res* 1992;**53**:961–5.
184. Jacobsen S, Nielsen JV, Kjelgaard-Hansen M, Toelboell T, Fjeldborg J, Halling-Thomsen M, et al. Acute phase response to surgery of varying intensity in horses: a preliminary study. *Vet Surg* 2009;**38**:762–9.
185. Hulten C, Gronlund U, Hirvonen J, Tulamo RM, Suominen MM, Marhaug G, et al. Dynamics in serum of the inflammatory markers serum amyloid A (SAA), haptoglobin, fibrinogen and alpha2-globulins during induced noninfectious arthritis in the horse. *Equine Vet J* 2002;**34**:699–704.
186. Milne EM, Doxey DL, Kent JE, Pemberton A. Acute phase proteins in grass sickness (equine dysautonomia). *Res Vet Sci* 1991;**50**:273–8.
187. Jacobsen S, Andersen PH, Toelboell T, Heegaard PM. Dose dependency and individual variability of the lipopolysaccharide-induced bovine acute phase protein response. *J Dairy Sci* 2004;**87**:3330–9.
188. Heegaard PM, Godson DL, Toussaint MJ, Tjornehoj K, Larsen LE, Viuff B, et al. The acute phase response of haptoglobin and serum amyloid A (SAA) in cattle undergoing experimental infection with bovine respiratory syncytial virus. *Vet Immunol Immunopathol* 2000;**77**:151–9.
189. Angen O, Thomsen J, Larsen LE, Larsen J, Kokotovic B, Heegaard PM, et al. Respiratory disease in calves: microbiological investigations on trans-tracheally aspirated bronchoalveolar fluid and acute phase protein response. *Vet Microbiol* 2009;**137**:165–71.
190. Nazifi S, Razavi SM, Esmailnejad Z, Gheisari H. Study on acute phase proteins (haptoglobin, serum amyloid A, fibrinogen, and ceruloplasmin) changes and their diagnostic values in bovine tropical theileriosis. *Parasitol Res* 2009;**105**:41–6.
191. Stabel JR, Spears JW, Brown Jr. TT. Effect of copper deficiency on tissue, blood characteristics, and immune function of calves challenged with infectious bovine rhinotracheitis virus and Pasteurella hemolytica. *J Anim Sci* 1993;**71**:1247–55.
192. Eckersall PD, Young FJ, McComb C, Hogarth CJ, Safi S, Weber A, et al. Acute phase proteins in serum and milk from dairy cows with clinical mastitis. *Vet Rec* 2001;**148**:35–41.
193. Lee WC, Hsiao HC, Wu YL, Lin JH, Lee YP, Fung HP, et al. Serum C-reactive protein in dairy herds. *Can J Vet Res* 2003;**67**:102–7.

194. Humblet MF, Guyot H, Boudry B, Mbayahi F, Hanzen C, Rollin F, et al. Relationship between haptoglobin, serum amyloid A, and clinical status in a survey of dairy herds during a 6-month period. *Vet Clin Pathol* 2006;**35**:188–93.
195. Guzelbektes H, Sen I, Ok M, Constable PD, Boydak M, Coskun A. Serum amyloid A and haptoglobin concentrations and liver fat percentage in lactating dairy cows with abomasal displacement. *J Vet Intern Med* 2010;**24**:213–9.
196. Walker JL, Clarke CR, Lessley BA, Hague CM. Effect of *Pasteurella haemolytica* infection on alpha 1-acid glycoprotein and albumin concentrations in serum and subcutaneous tissue chamber fluid of calves. *Res Vet Sci* 1994;**56**:158–63.
197. Eckersall PD, Saini PK, McComb C. The acute phase response of acid soluble glycoprotein, alpha(1)-acid glycoprotein, ceruloplasmin, haptoglobin and C-reactive protein, in the pig. *Vet Immunol Immunopathol* 1996;**51**:377–85.
198. Parra MD, Fuentes P, Tecles F, Martinez-Subiela S, Martinez JS, Munoz A, et al. Porcine acute phase protein concentrations in different diseases in field conditions. *J Vet Med B Infect Dis Vet Public Health* 2006;**53**:488–93.
199. Skovgaard K, Mortensen S, Boye M, Poulsen KT, Campbell FM, Eckersall PD, et al. Rapid and widely disseminated acute phase protein response after experimental bacterial infection of pigs. *Vet Res* 2009;**40**:23.
200. Jacobson M, Lindberg JE, Lindberg R, Segerstad CH, Wallgren P, Fellstrom C, et al. Intestinal cannulation: model for study of the midgut of the pig. *Comp Med* 2001;**51**:163–70.
201. Mikami O, Kubo M, Murata H, Muneta Y, Nakajima Y, Miyazaki S, et al. The effects of acute exposure to deoxynivalenol on some inflammatory parameters in miniature pigs. *J Vet Med Sci* 2011;**73**:665–71.
202. Skinner JG, Roberts L. Haptoglobin as an indicator of infection in sheep. *Vet Rec* 1994;**134**:33–6.
203. Pfeffer A, Rogers KM, O'Keeffe L, Osborn PJ. Acute phase protein response, food intake, liveweight change and lesions following intrathoracic injection of yeast in sheep. *Res Vet Sci* 1993;**55**:360–6.
204. Eckersall PD, Lawson FP, Bence L, Waterston MM, Lang TL, Donachie W, et al. Acute phase protein response in an experimental model of ovine caseous lymphadenitis. *BMC Vet Res* 2007;**3**:35.
205. Ulutas PA, Ozpinar A. Effect of *Mannheimia (Pasteurella) haemolytica* infection on acute-phase proteins and some mineral levels in colostrum-breast milk-fed or colostrum-breast milk-deprived sheep. *Vet Res Commun* 2006;**30**:485–95.
206. Nikunen S, Hartel H, Orro T, Neuvonen E, Tanskanen R, Kivela SL, et al. Association of bovine respiratory disease with clinical status and acute phase proteins in calves. *Comp Immunol Microbiol Infect Dis* 2007;**30**:143–51.
207. Pfeffer A, Rogers KM. Acute phase response of sheep: changes in the concentrations of ceruloplasmin, fibrinogen, haptoglobin and the major blood cell types associated with pulmonary damage. *Res Vet Sci* 1989;**46**:118–24.
208. Gonzalez FH, Tecles F, Martinez-Subiela S, Tvarijonaviciute A, Soler L, Ceron JJ. Acute phase protein response in goats. *J Vet Diagn Invest* 2008;**20**:580–4.
209. Bertelsen MF, Kjelgaard-Hansen M, Grondahl C, Heegaard PM, Jacobsen S. Identification of acute phase proteins and assays applicable in nondomesticated mammals. *J Zoo Wildl Med* 2009;**40**:199–203.
210. Jinbo T, Ami Y, Suzaki Y, Kobune F, Ro S, Naiki M, et al. Concentrations of C-reactive protein in normal monkeys (*Macaca irus*) and in monkeys inoculated with *Bordetella bronchiseptica* R-5 and measles virus. *Vet Res Commun* 1999;**23**:265–74.
211. De Pablos V, Barcia C, Martinez S, Gomez A, Ros-Bernal F, Zamarro-Parra J, et al. MPTP administration increases plasma levels of acute phase proteins in non-human primates (*Macaca fascicularis*). *Neurosci Lett* 2009;**463**:37–9.

212. Myers LA, Boyce JT, Robison RL. The tolerability and pharmacology of interleukin-6 administered in combination with GM-CSF or G-CSF in the rhesus monkey. *Toxicology* 1995;**101**:157–66.
213. Nazifi S, Saeb M, Ghasemian O, Esmailnezhad Z. Evaluation of serum haptoglobin in clinically healthy Iranian camels (*Camelus dromedarius*). *Comp Clin Pathol* 2006;**15**:195–7.
214. Willis EL, Kersey DC, Durrant BS, Kouba AJ. The acute phase protein ceruloplasmin as a non-invasive marker of pseudopregnancy, pregnancy, and pregnancy loss in the giant panda. *PLoS One* 2011;**6**:e21159.
215. Xie H, Newberry L, Clark FD, Huff WE, Huff GR, Balog JM, et al. Changes in serum ovotransferrin levels in chickens with experimentally induced inflammation and diseases. *Avian Dis* 2002;**46**:122–31.
216. Nakamura K, Mitarai Y, Yoshioka M, Koizumi N, Shibahara T, Nakajima Y. Serum levels of interleukin-6, alpha1-acid glycoprotein, and corticosterone in two-week old chickens inoculated with *Escherichia coli* lipopolysaccharide. *Poult Sci* 1998;**77**:908–11.
217. Takahashi K, Miyake N, Ohta T, Akiba Y, Tamura K. Changes in plasma alpha 1-acid glycoprotein concentration and selected immune response in broiler chickens injected with *Escherichia coli* lipopolysaccharide. *Br Poult Sci* 1998;**39**:152–5.
218. Chamanza R, Toussaint MJ, van Ederen AM, van Veen L, Hulskamp-Koch C, Fabri TH. Serum amyloid A and transferrin in chicken. A preliminary investigation of using acute-phase variables to assess diseases in chickens. *Vet Q* 1999;**21**:158–62.
219. Dunkley CS, McReynolds JL, Dunkley KD, Njongmeta LN, Berghman LR, Kubena LF, et al. Molting in *Salmonella enteritidis*-challenged laying hens fed alfalfa crumbles. IV. Immune and stress protein response. *Poult Sci* 2007;**86**:2502–8.
220. Georgieva TM, Koinarski VN, Urumova VS, Marutsov PD, Christov TT, Nikolov J, et al. Effects of *Escherichia coli* infection and *Eimeria tenella* invasion on blood concentrations of some positive acute phase proteins (haptoglobin (PIT 54), fibrinogen, and ceruloplasmin) in chickens. *Rev Med Vet* 2010;**161**:84–9.
221. Inoue M, Satoh W, Murakami H. Plasma alpha 1-acid glycoprotein in chickens infected with infectious bursal disease virus. *Avian Dis* 1997;**41**:164–70.
222. Nazifi S, Dadras H, Hoseinian SA, Ansari-Lari M, Masoudian M. Measuring acute phase proteins (haptoglobin, ceruloplasmin, serum amyloid A, and fibrinogen) in healthy and infectious bursal disease virus-infected chicks. *Comp Clin Pathol* 2010;**19**:283–6.
223. Zhou X, Wang L, Feng H, Guo Q, Dai H. Acute phase response in Chinese soft-shelled turtle (*Trionyx sinensis*) with *Aeromonas hydrophila* infection. *Dev Comp Immunol* 2011;**35**:441–51.
224. Bayne CJ, Gerwick L. The acute phase response and innate immunity of fish. *Dev Comp Immunol* 2001;**25**:725–43.
225. Raida MK, Buchmann K. Innate immune response in rainbow trout (Oncorhynchus mykiss) against primary and secondary infections with *Yersinia ruckeri* O1. *Dev Comp Immunol* 2009;**33**:35–45.
226. Gerwick L, Steinhauer R, Lapatra S, Sandell T, Ortuno J, Hajiseyedjavadi N, et al. The acute phase response of rainbow trout (*Oncorhynchus mykiss*) plasma proteins to viral, bacterial and fungal inflammatory agents. *Fish Shellfish Immunol* 2002;**12**:229–42.
227. Ahmad R, Khan KA, Hasnain A, Qayyus S. Distribution of major serum proteins in an air breathing teleost, *Channa punctata Bl.* (Channidae: Channiformes). *Biomed Res* 2007;**18**:123–8.
228. Harr K, Harvey J, Bonde R, Murphy D, Lowe M, Menchaca M, et al. Comparison of methods used to diagnose generalized inflammatory disease in manatees (*Trichechus manatus latirostris*). *J Zoo Wildl Med* 2006;**37**:151–9.
229. Funke C, King DP, Brotheridge RM, Adelung D, Stott JL. Harbor seal (*Phoca vitulina*) C-reactive protein (C-RP): purification, characterization of specific monoclonal antibodies

and development of an immuno-assay to measure serum C-RP concentrations. *Vet Immunol Immunopathol* 1997;**59**:151–62.
230. Rosenfeld H, Lassen S, Prange A. Characterization of haptoglobin in the blood plasma of harbor seals (*Phoca vitulina*). *J Proteome Res* 2009;**8**:2923–32.
231. Kakuschke A, Erbsloeh HB, Griesel S, Prange A. Acute phase protein haptoglobin in blood plasma samples of harbour seals (*Phoca vitulina*) of the Wadden Sea and of the isle Helgoland. *Comp Biochem Physiol B Biochem Mol Biol* 2010;**155**:67–71.
232. Kent J. Acute phase proteins: their use in veterinary diagnosis. *Br Vet J* 1992;**148**:279–82.
233. Ceron JJ, Martinez-Subiela S, Ohno K, Caldin M. A seven-point plan for acute phase protein interpretation in companion animals. *Vet J* 2008;**177**:6–7.

Animal Models of Hemophilia

DENISE E. SABATINO,[*]
TIMOTHY C. NICHOLS,[†]
ELIZABETH MERRICKS,[†]
DWIGHT A. BELLINGER,[†]
ROLAND W. HERZOG,[‡] AND
PAUL E. MONAHAN[§]

[*]*Division of Hematology, The Children's Hospital of Philadelphia, Philadelphia, Pennsylvania, USA*

[†]*Francis Owen Blood Research Laboratory, University of North Carolina, Chapel Hill, North Carolina, USA*

[‡]*Department of Pediatrics, University of Florida, Gainesville, Florida, USA*

[§]*Department of Pediatrics, University of North Carolina, Chapel Hill, North Carolina, USA*

Progress in Molecular Biology
and Translational Science, Vol. 105
DOI: 10.1016/B978-0-12-394596-9.00006-8

1877-1173/12 $35.00

The X-linked bleeding disorder hemophilia is caused by mutations in coagulation factor VIII (hemophilia A) or factor IX (hemophilia B). Unless prophylactic treatment is provided, patients with severe disease (less than 1% clotting activity) typically experience frequent spontaneous bleeds. Current treatment is largely based on intravenous infusion of recombinant or plasma-derived coagulation factor concentrate. More effective factor products are being developed. Moreover, gene therapies for sustained correction of hemophilia are showing much promise in preclinical studies and in clinical trials. These advances in molecular medicine heavily depend on availability of well-characterized small and large animal models of hemophilia, primarily hemophilia mice and dogs. Experiments in these animals represent important early and intermediate steps of translational research aimed at development of better and safer treatments for hemophilia, such a protein and gene therapies or immune tolerance protocols. While murine models are excellent for studies of large groups of animals using genetically defined strains, canine models are important for testing scale-up and for long-term follow-up as well as for studies that require larger blood volumes.

I. The Hemophilia A Mice

A. Hemophilia A

Hemophilia A is an X-linked bleeding disorder due to a deficiency in factor VIII (FVIII). It is the most prevalent form of hemophilia representing 80% of hemophilia cases. Genetic defects in the *F8* gene that cause hemophilia A include point mutations, deletions, insertions, and gene inversions.[1] Current treatment for hemophilia is protein replacement given either prophylactic or in response to bleeding episodes. The major complication of this treatment is the development of an immune response to the protein that occurs in 25–30% of patients.

FVIII is a critical component of the intrinsic blood coagulation pathway. The FVIII protein is composed of A1–A2–B–A3–C1–C2 domains. The secreted form of the protein is a heterodimer composed of a 90- to 200-kDa heavy chain (A1–A2–B) and an 80-kDa light chain (A3–C1–C2). The B-domain is not essential for FVIII activity. Upon cleavage by thrombin, FVIII forms a heterotrimer consisting of an A1 subunit, A2 subunit and A3–C1–C2 subunit that is the active form of the protein. Activated FVIII (FVIIIa) is a cofactor for factor IX (FIX) in the tenase complex. In the presence of FVIIIa, phospholipid and Ca^{2+}, FIXa cleaves FX to generate the activated form (FXa). FVIII is stabilized by vWF in the circulation and the concentration of FVIII is 100–200 ng/ml in the plasma.

The murine FVIII (mFVIII) is 74% identical to human FVIII (hFVIII).[2] The A and C domains are 84% and 93% identical, respectively, while the B-domain is the most divergent (55% identical). The domain structures as well as the processing and activation of FVIII are similar in mouse and humans. The ability to assess hemostasis in a mouse model is an important tool for preclinical studies for novel therapeutics for hemophilia A.

B. The Hemophilia A Mouse Model

While a colony of the naturally occurring hemophilia A dog model had been established in 1947, hemophilia mouse models were not available until the 1990s (all murine and canine as well as alternative animal models of hemophilia are summarized in Table I).[22] As novel therapeutics including gene therapy approaches were being developed, initial studies were performed in wild-type mice or the hemophilia A dog model. The wild-type mice were not able to provide information about hemostasis while the studies in the dog model were limited by the expense and the long generation time of the dog. Almost 10 years after the hFVIII cDNA sequence was described,[23] the mFVIII cDNA sequence was cloned.[2] Shortly thereafter, Dr. Haig Kazazian and colleagues generated mice with FVIII deficiency using standard gene targeting techniques to establish the first murine model of hemophilia.[3]

After isolation of a genomic clone containing exons 15–22 of the mFVIII gene, two separate neomycin cassettes were used to generate the hemophilia A mice, one targeting exon 16 and a second targeting exon 17 (both in the A3 domain of the light chain). The exon 16 knockout mice were generated by insertion of the neomycin cassette into the 3′ end of that exon causing abnormal splicing of the RNA.[3] The exon 17 knockout mice have the neomycin cassette inserted into the 5′ end of exon 17 that leads to exon skipping. Mutations in both of these exons cause hemophilia A with a severe phenotype ($<1\%$ FVIII residual activity). Analysis of liver RNA demonstrated that the exon 16 knockout mice have a truncated protein and exon 17 mice have a truncated or partially deleted protein.[24] While there is no detectable light chain in either knockout strain, these mice do have low levels of heavy chain protein in the circulation.[24,25]

Phenotypically, the exon 16 and exon 17 knockout mice are indistinguishable. The homozygous knockout female and hemizygous knockout male mice have $<1\%$ activity while the carrier females have been reported to express

TABLE I

ANIMAL MODELS AND STRAINS WITH SEVERE HEMOPHILIA A OR B

Species	Strain	Genotype	Reference
Mouse	Hemophilia A	Exon 16 knockout	3,24
	• C57BL6/129		3,24
	• BALB/c		4
	• CD4 deficient		98
Mouse	Hemophilia A	Exon 17 knockout	3,24
Mouse	Hemophilia A	Transgenic for human F.VIII with missense mutation R593C; exon 17 knockout	5
Mouse	Hemophilia A	Transgenic for human MHC II allele HLA-DRB1*1501; exon 17 knockout	6
Mouse	Hemophilia B	Targeted deletion	7
	• C57BL6		339
	• BALB/c		339
	• C3H/HeJ		339
	Hemophilia B	Knock-in for human F.IX with crm + missense mutation R333Q	8
	Hemophilia B	Knock-in for human F.IX with early stop codon at amino acid residue 29	9
Mouse	Hemophilia B (C57BL6 or C3H/HeJ)	Targeted deletion plus transgenic for human FIX:	10,11
		• crm − missense mutation G381E	
		• crm + missense mutation R180W	
		• Early stop codon at residue 29	
		• Late stop codon at residue 338	
Rat	Hemophilia A	Missense mutation L176P	12,13
Dog	Hemophilia A (Chapel Hill)	Exon 22 inversion	14,15
Dog	Hemophilia A (Queens)	Exon 22 inversion	16
Dog	Hemophilia A (Alabama)	Unknown	17,18
Dog	Hemophilia B (Chapel Hill)	CRM − missense mutation G379E	19
Dog	Hemophilia B (Alabama)	Partial deletion/early stop codon/low mRNA	20
Sheep	Hemophilia A	Early stop codon/frameshift	21

~75% of normal FVIII activity. In contrast to humans with severe hemophilia, the mice do not appear to have frequent spontaneous bleeding episodes; however, bleeding in response to trauma is often lethal. However, in a subcolony of the exon 17 mice it was reported that spontaneous bleeding that was lethal occurred in 22% of hemizygous males and 10% of homozygous females.[26] Due to their fragile phenotype, gentle handling of these animals minimizes fatal bleeding events. The use of forceps while handling mice, ear tags and surgical procedures all increase the likelihood of trauma induced internal or external bleeding. After tail transection, affected hemophilia A mice have excessive bleeding and die within several hours, while carrier females and wild-type mice survive the procedure. Homozygous females survive pregnancy and delivery of the litters with rare complications. Thus, breedings between hemizygous affected males and homozygous affected females are feasible with litter sizes of 5–8 pups.

While the original hemophilia A mice were established on a C57BL6/129sv genetic background, other lines have been established on different strains. The hemophilia A exon 16 mice have been backcrossed onto the C57BL/6, Balb/C, and nonobese diabetic/severe combined immunodeficiency (NOD–SCID) backgrounds. The NOD–SCID hemophilia A mice have impaired B and T cell lymphocyte development and provide the opportunity to evaluate novel therapeutics in the absence of an immune response to FVIII.[27]

C. Assessment of Hemostasis in Hemophilia A Mouse Models

Assessment of coagulation function in the hemophilia mice after protein infusion or expression of a FVIII transgene has been hampered by variation in the assays that are performed. The hemophilia A mice have < 1% FVIII activity which provides a control background for detection of even low levels of FVIII activity. The human, canine, porcine as well as the mFVIII protein are efficacious in the hemophilia A mouse, however, differences in the specific activity and/or stability of these proteins from different species has been observed.[28–30] Species-specific reagents may provide the most appropriate assessment of FVIII activity.

In vivo assessment of FVIII function is performed using a variety of different assays including the tail vein transection assay as well as more recently developed models of thrombosis and hemostasis.[31] The tail transection assay measures both venous and arterial bleeding.[32] Variation in the bleeding with this assay is often due to differences in the amount of the tail that is clipped, how the blood is collected and how long the blood is collected.[33] Despite this variation, there is consensous that prolonged bleeding without intervention is lethal. The cremaster injury model measures thrombus formation following laser injury to a small arteriole in the cremaster muscle by monitoring platelet accumulation at the site of injury.[34] The ferric chloride carotid artery occlusion model uses ferric chloride to induce an injury to the vessel wall and then measures blood flow and the time to occlusion using a Doppler probe.[35] The primary difference between these two *in vivo* injury models is that the cremaster model assesses the microcirculation while the ferric chloride model measures thrombus formation in the macrocirculation. The use of both *in vitro* and *in vivo* assays to determine FVIII function in the hemophilia A mouse model provides important tools for development of therapeutics.

D. Arthropathy in the Murine Model

Hemophilia patients experience arthropathy and impaired wound healing secondary to joint hemorrhage. While these are nonlethal complications of the disease, these sequela have a debilitating impact on the quality of life in these patients. Without regular protein treatments, hemophilia A patients have 30–35 joint hemorrhages per year that causes significant morbidity.[36] By age 25, 90% of hemophilia A patients have chronic degenerative changes in 1–6 joints.[37] The hemophilia A mouse has been used to develop a model of the bleeding into the joints that causes inflammation of the synovial membrane leading to structural damage within the joint termed hemophilic synovitis. In controlled studies a blunt trauma is delivered to one knee using a spring loaded device which causes joint swelling due to bleeding and inflammation. Acute morbidity is observed that is assessed by inactivity, immobility, and weight loss. Different parameters that are measured experimentally include: knee diameter, joint capsule bleeding score assessed by visual bleeding score, changes in body mass and histology using a grading scheme.[38,39] Two different injury models were developed: one that induced a mechanical trauma to the joint and then required that the animals walk on a rotating rod apparatus and a second model induced hemarthrosis by puncture with a 30-gauge needle which was thought may mimic blood-induced arthropathy.[40] Using these models, recombinant hFVIII and human factor VIIa (hFVIIa) were evaluated for their ability to prevent acute joint bleeding in the hemophilia A mouse.[41] Joint swelling and acute hemarthrosis was reduced along with synovitis; however, the treatments did not completely prevent bleeding into the joint space. Clinically, this correlates with findings that hemophilia A pediatric patients that receive regular recombinant FVIII (rFVIII) treatments are 83% less likely to develop bone and cartilage changes compared to episodic treatments for bleeding.[42] Using these mouse models to measure the impact of bleeding on the joint is challenging due to the subjective nature of the scoring of the pathology. Furthermore, the mouse model may not mimic the weight bearing impact on the joints in humans. Nonetheless, these models may provide an opportunity to study the pathogenesis of joint disease and evaluate novel therapies designed to prevent joint bleeding.

E. Immune Responses and Tolerance Induction to FVIII in the Hemophilia A Mouse

Since the major complication in hemophilia A patients is the development of neutralizing antibodies (inhibitors) to FVIII, the mouse model became a valuable tool for investigating the immune response to FVIII. The hemophilia A mouse has several advantages as a model for studying immune responses. First, the immune responses can be studied in a genetically homogeneous strain. Second, there is less influence of environmental factors in mouse studies. Third, murine-specific reagents are readily available and studies can be designed to manipulate the immune system to investigate the mechanisms involved in the immune response.

1. Characterization of the Immune Response to FVIII in Hemophilia A Mice

Early studies in the hemophilia A mouse model demonstrated that the immune response to hFVIII in mice was similar to what was observed in hemophilia A patients. It was known that the inhibitors in patients are IgG antibodies that require CD4+ T helper cells.[43,44] Initial studies in the hemophilia A mouse (C57BL/6 exon 16 and exon 17 knockout mice) demonstrated a dose-dependent IgG antibody response that could be detected after one or two intravenous infusions of hFVIII protein and increased after subsequent doses.[45] There was a linear correlation between anti-hFVIII IgG antibody titers representing both neutralizing and nonneutralizing antibodies and the neutralizing antibodies (inhibitors) measured in a functional FVIII assay (Bethesda units). Wild-type mice (C57BL/6) may develop an immune response to hFVIII with antibody titers that are significantly lower (or undetectable) than in C57BL/6 hemophilia A mice.[45] Since CD4+T cells specific to FVIII would be eliminated by clonal deletion during ontogenesis of the immune system, these antibody responses may reflect species specific differences in the proteins since the mFVIII and hFVIII protein are 84% and 93% identical in the A and C domains, respectively.[2,23] Thus, wild-type mice may have a similar immune response to some cross-reacting material positive (crm+) patients that have mutations in FVIII that result in synthesis of a nonfunctional protein.

A more detailed characterization of the immune response to hFVIII in the hemophilia A mice that were infused intermittently and followed over a long period of time similar to hemophilia A patients revealed similarities between human and murine responses. Hemophilia A mice (exon 17) that were administered hFVIII protein intravenously over the course of months developed anti-hFVIII inhibitory antibodies.[46] In contrast to the 25–30% of hemophilia A patients that develop inhibitors, all of the hemophilic mice developed antibodies to the hFVIII which can be attributed to their identical genetic background with the same mutation. However, even in this controlled study there was considerable variation among the antibody titers in the mice. IgG subclasses were homologous to inhibitors in hemophilia A patients with primarily IgG1 (equivalent to human IgG4) and also IgG2a or IgG2c (equivalent to the human IgG1). Cytokine profiles agreed with the IgG subclass analysis of the anti-hFVIII antibody response with IL-10 as the predominant Th2 cytokine.[46] CD4 + T cell epitopes were observed throughout all of the domains of the hFVIII protein. In hemophilia A mice the blockade of B7/CD28 costimulatory pathways (of T cells) prevented the synthesis of anti-FVIII antibodies, supporting the role of CD4 + T cells in the development of the immune response to FVIII.[45]

The immune response to FVIII is strain dependent. C57BL/6 exon 16 hemophilia A mice have a more robust immune response to FVIII protein than Balb/C mice with higher anti-FVIII IgG antibody titers.[47] These strain differences pose a limitation of the mouse model since observations in one strain do not always hold in a different strain. This heterogeneity in immune response is similar to observations in humans.[48]

2. Transgenic Models in the Hemophilia A Mouse for Immunology Studies

While the hemophilia A mouse provides a model to study immune responses to FVIII, it has several limitations. First, humans have different mutations in the factor VIII gene that impact the immune response to the protein. The hemophilic mice make a truncated protein (exon 16 or exon 17) or have a partial deletion (exon 17) while some patients synthesize a full-length protein that is nonfunctional and others do not make any protein. Second, these studies characterize the immune response to hFVIII peptides that are presented by mouse major histocompatibility complex (MHC) II. In an effort to develop improved mouse models for the study of immune responses to hFVIII, humanized transgenic mice were bred onto the hemophilia A mouse.

One strategy to study the immunology of FVIII has been to express mutations that are found in the hemophilia A population. One missense mutation that causes a mild form of hemophilia is an arginine to cystine mutation at position 593 (R593C). These patients are particularly interesting because a subset of these patients make inhibitors to FVIII. Hemophilia A mice that express a hFVIII-R593C transgene driven by the murine albumin enhancer/promoter were generated and bred onto the mFVIII knockout mouse.[5] While these mice have detectable hFVIII mRNA in the liver, no hFVIII protein was detected in the plasma. These mice do not develop antibodies or a T cell response to hFVIII. Thus, these mice do not mimic the inhibitor formation observed in some patients with this mutation. The differences in the mouse and human responses may be due to underlying genetic factors that contribute to the formation of inhibitors.

A mouse model was also created that humanized the HLA MHC Class II molecule in the hemophilia A mice. These mice express the human HLA-DRB1*1501 in the exon 17 knockout mice.[6] While this HLA only represents a subset of the human population, this haplotype is associated with an increased risk of inhibitor formation in hemophilia A patients.[49] These mice provide an *in vivo* model for identifying immunodominant epitopes by isolation of T and B cell clones and studies of CD4 + T cell responses.

Mice that are knockout for mFVIII (exon 17 knockout) but express the full-length hFVIII cDNA have also been described.[50] The expression of hFVIII is liver specific since it is directed by the murine albumin enhancer/promoter. While a transgenic line had detectable hFVIII mRNA but no detectable hFVIII protein in the circulation, the mice were immunologically tolerant to hFVIII upon challenge with recombinant hFVIII. However, the mice do develop an immune response to a modified form of hFVIII that was shown to be more immunogenic in the exon 17 mFVIII knockout mice. These mice may be a valuable model because they do not make antibodies to the native protein and can be used to evaluate novel FVIII molecules that may represent neoantigens in the setting of protein replacement.

3. B Cell Epitope Mapping

The study of B cell responses to FVIII in humans is difficult since the lymphoid tissues are not accessible and humans are genetically heterogenous which influences the immune response. Furthermore, functional and structural studies of B cell epitopes is complicated by the polyclonality of the inhibitory antibodies. The hemophilia A mouse model provides the opportunity to identify epitopes within FVIII and determine their functional significance. After infusion of hFVIII into hemophilia A mice, splenocytes were harvested for generation of anti-FVIII B cell hybridomas. Epitope mapping was performed to determine the domain specificity in the context of a hybrid human FVIII–porcine FVIII (pFVIII) molecule that retains the structure and function of the molecule. As observed with humans, the predominant epitopes for anti-FVIII inhibitory antibodies reside in the A2 and C2 domains.[51] Notably, 40% of the B cell clones that were analyzed were not inhibitory antibodies thus a significant portion of anti-FVIII antibodies may be nonneutralizing but may contribute to the pathogenic effect. Further functional studies demonstrated how these antibodies inhibit hemostasis by blocking the binding of FVIII and FVIIIa to phospholipids and vWF or prevent activation of FVIII by thrombin and factor Xa or blocking release of vWF from FVIII after thrombin activation.[52] These findings correlate with the human studies that observe a predominant portion of anti-FVIII antibodies are specific to the A2 and C2 domains; however, it is not known whether the mouse accurately predicts the specific human epitopes and their prevalence. Mutagenesis of specific epitopes in the A2 (or C2) domains may reduce the immunogenicity while maintaining function.[53] However, the safety of introducing FVIII variants into the hemophilia population has not been determined.

F. Safety and Efficacy of Novel Protein Therapeutics

The development of novel protein therapeutics for hemophilia raises the concern for the immunogenicity of a new treatment product. Plasma-derived purified clotting factors contain different formulations of not only clotting factor but additional proteins; in the case of FVIII, human vWF is present that can influence the immune response to the FVIII. In contrast, rFVIII products are not formulated with vWF. For hemophilia A patients, the only protein that may be foreign to the immune system is FVIII while these patients have normal vWF protein. A comparison of plasma-derived with recombinant full-length FVIII in the hemophilia A mouse (Balb/C) suggests that the prevalence of inhibitors to rFVIII appeared to be higher than the immune response to plasma-derived FVIII.[47,54,55] Notably, the hemophilia A mice develop an immune response to human vWF. The presence of human vWF may provide antigen competition thus impacting the immune response to the FVIII products; however, human vWF administered with rFVIII does not reduce the inhibitory antibody titer. These studies provide insight into the possible mechanisms involved in the differential immune response between rFVIII and plasma-derived FVIII that may involve antigen competition and the influence of the splenic cytokine environment. In fact, the presence of human FIX (hFIX) spiked into rFVIII diminishes the immune response to the rFVIII protein.[47] While the murine model provides an opportunity to characterize immune responses that are not possible in humans, these studies illustrate the challenges of using a murine model to interpret potentially clinically relevant findings. These differential immune responses observed in the mouse model remains to be defined in humans.

Historically, purified plasma-derived pFVIII was developed as an alternative product to hFVIII due to the availability of porcine plasma. Later, it was favored as a bypass agent for the treatment of patients with inhibitors due to the low cross-reactivity of hFVIII inhibitors to pFVIII that provided a treatment for 90% of bleeding episodes in patients with inhibitors.[56,57] The side effects to plasma-derived pFVIII included allergic reactions likely due to other porcine proteins in the plasma material and contamination with porcine parvovirus have led to the development of recombinant pFVIII. No differences were observed in the immunogenicity of plasma-derived (Hyate:C) and recombinant (OBI-1) pFVIII compared in hemophilia A mice with prior exposure to hFVIII.[58] Further assessment of pFVIII using a recombinant hFVIII/pFVIII by domain swapping showed that while the overall immunogenicity of pFVIII was similar to hFVIII, there were differences in the neutralizing abilities of the antibodies generated to the two proteins.[59] Recombinant hybrid pFVIII/hFVIII may provide a platform for development of a protein that is less immunogenic for use in protein therapy or exploited in gene therapy approaches.

Recent approaches to improve FVIII protein therapeutics have focused on prolonging the half-life of the FVIII protein to reduce the injection frequency required for prophylaxis in hemophilia A patients.[60] The half-life of FVIII is 10–12 h which requires three infusions per week to maintain FVIII above > 1% to prevent spontaneous bleeding. While direct PEGylation of random amine residues in the FVIII protein has not produced a functional FVIII product, site-specific PEGylation of cysteine residues that have been engineered into the surface of the FVIII molecule by site-directed mutagenesis has led to identification of a FVIII variant protein that maintains coagulation activity. In the hemophilia A mouse the FVIII with a modest twofold extension of the half-life that was significantly longer in vWF-deficient mice suggesting that the PEGylation does not extend the half-life of FVIII stabilized by vWF.[61] Another FVIII formulation termed PEGlip noncovalently binds FVIII to the outer surface of PEGylated liposomes. After treatment with PEGlip-FVIII the survival of hemophilia A mice following tail transection was significantly higher than mice treated with standard FVIII formulations.[62–64] Interestingly, the enhanced efficacy could not entirely be explained by the increased circulating half-life. The mechanism responsible for these improvements in efficacy appears to rely on the presence of platelets.[64] These studies supported several Phase I/II clinical trials for PEGylated FVIII or PEGlip-FVIII that have demonstrated that these products are not more immunogenic than current FVIII treatments and appear to provide prolonged bleed-free periods.[65] Another approach uses monomeric Fc fusion technology to stabilize the FVIII protein (rFVIIIFc). *In vivo* studies suggest that a twofold increase in the half-life of rFVIIIFc that is currently being evaluated in a PhaseII/III clinical trial.[66] The challenge of developing these approaches is that modification of the FVIII protein can interfere with the interaction of FVIII with vWF, FIX, and FX that can result in loss of coagulant activity. Ultimately, a long-acting FVIII molecule that has reduced immunogenicity would provide a new generation of protein products for hemophilia.

G. Tolerance Induction Strategies in the Hemophilia A Mouse

Since inhibitor formation is the major complication of hemophilia A, the ability to establish tolerance to FVIII would be vital to inhibitor patients. The current strategy for inducing tolerance in these patients is immune tolerance induction (ITI) that requires frequent high dose infusions of protein over the course of months or years. The success rate of this approach is 50–90%.[67] Studies in hemophilia A mice to elucidate the mechanism of tolerance induction with ITI suggest that an irreversible dose-dependent inhibition rather than stimulation of memory B cells occurs at high FVIII doses.[68] Alternative approaches that are more cost effective and have a better success rate will be important for providing treatment options for patients.

Since all hemophilia A mice make immune response to hFVIII, it is a stringent model for evaluating tolerance induction approaches. Various strategies to induce tolerance in the setting of autoimmune disease have been investigated in mouse disease models. The immune response to FVIII in hemophilia is different than the setting of autoimmune disease in which an immune response develops to self-antigens. In hemophilia tolerance to the FVIII protein was never established in some patients since the immune system had no previous exposures to the protein. Several recent studies demonstrate the possibility of using tolerance approaches for hemophilia A.

Mucosal tolerance has evolved as a mechanism to avoid immune responses to food antigens. The administration of antigens via oral or nasal routes has been used to retolerize animals in the setting of autoimmune disease.[69] The dose of the antigen influences the mechanism of tolerance induction with low doses inducing active suppression while high antigen doses resulting in T cell anergy or deletion.[70,71] The oral or nasal administration of the C2 domain of the FVIII protein that contains immunodominant epitopes induces tolerance to C2 but not to the entire FVIII protein.[72] Thus, the epitopes in other domains contribute significantly to the inhibitor development. Cytokine profiling suggested that an active suppression mechanism involving regulatory CD4+ T cells was responsible for the tolerance. This was supported by the observation that the tolerance was not maintained if the treatment was suspended. Thus, using this strategy with multiple domains or the entire full-length FVIII may prevent immune responses to FVIII; however, long-term treatment may be necessary to maintain tolerance.

A B cell tolerance approach takes advantage of using B cells as tolerogenic antigen-presenting cells. In this approach, activated syngenic B cells are transduced *ex vivo* with a retrovirus expressing an IgG fusion protein that contains the N-terminus of the IgG heavy chain and a portion of the FVIII protein. The A2 and C2 domains were selected for this approach since inhibitory antibodies react to epitopes in these domains. While the use of only one of these domains had only a modest effect, the effect of using a combination of B cells expressing both domains was tolerogenic even when challenged to the entire FVIII protein.[73] In addition, mice that were preimmunized with FVIII to develop FVIII inhibitors also showed a remarkable reduction in immune response to

FVIII after B cell transplant with a >90% reduction in inhibitor titer. The mechanism of this modulation relies on MHC Class II presentation on B cells as well as CTLA-4 signaling pathways and regulatory T cells.[74]

Cell-based approaches that utilize dendritic cells for suppression of immune responses to FVIII have been explored. Immature dendritic cells promote an environment that promotes tolerance by decreased production of proinflammatory cytokines. Dendritic cells pulsed with FVIII infused into C57BL/6 hemophilia A mice prior to exposure to hFVIII protein reduced the anti-FVIII antibody response by induction of Tregs.[75] Expression of hFVIII in autologous apoptotic fibroblasts that are phagocytosed by immature dendritic cells to create a tolerogenic environment that was able to suppress the development of inhibitors to FVIII in both naïve and preimmunized hemophilia A mice.[76] Both of these strategies rely on the immature dendritic cell ability to induce a population of Tregs (CD4+CD25+ T cells).

Pharmacological approaches utilize immune suppression regimens to induce tolerance to FVIII. An anti-CD3 antibody approach prevents anti-FVIII inhibitor development by upregulation of CD4+CD25+ regulatory T cells.[4] Immune suppression with rapamycin at the time of FVIII protein infusion suppresses the inhibitor formation and induces a population of FVIII-specific Tregs.[77]

While tolerance approaches and mechanisms have been exploited in hemophilia A mice, the ability to translate these to large animal models and patients has been limited. The key feature of a new tolerance strategy would be that it is effective not only as a preventive therapy in patients prior to the onset of anti-FVIII antibodies to prevent this serious complication but also in patients that have already developed inhibitors. Further development of these approaches will be facilitated by new large animal hemophilia A models that are inhibitor prone (see Section III.B).

H. Gene- and Cell-based Therapy for Hemophilia A

Hemophilia A is an excellent candidate for gene-based therapies for several reasons. While the primary site of FVIII synthesis is the liver (hepatocytes and liver sinusoidal endothelial cell, LSEC),[78] a variety of cell types can produce functional FVIII including hepatocytes, endothelial cells, fibroblasts, and several hematopoietic lineages. The therapeutic window for FVIII expression is wide with even a small amount of clotting factor (> 1%) in the circulation being sufficient to improve the disease phenotype. In addition, FVIII can be assayed in the plasma to frequently monitor the effects on coagulation.

The development of gene therapy for hemophilia A has been challenging and lagged behind gene therapy approaches for hemophilia B already in clinical trials. From a technical standpoint, the FVIII cDNA is larger than FIX. Most gene transfer approaches use the B-domain deleted FVIII (BDD-FVIII) cDNA

(4.3 Kb) rather than the full-length cDNA (7 Kb) as a transgene. Furthermore, the risk of inhibitor formation to protein replacement therapy in hemophilia A patients is higher (25–30%) than in hemophilia B (5%) which poses a challenge to develop a therapy that not only achieves efficacy but also induces tolerance to FVIII. Early clinical trials in the late 1990s for gene-based approaches to treat hemophilia A did not achieve long-term expression of FVIII.[79] Recent preclinical studies using several approaches show promise for the next generation of gene therapy for hemophilia A.

1. Liver Targeted Expression of FVIII

Gene-based approaches for hemophilia A initially were based on the non-integrating adenoviral vector that easily accommodated the large FVIII cDNA and efficiently targeted dividing and nondividing cells. These initial studies were performed by administration of Ad-hFVIII in wild-type mice that limited the ability to demonstrate efficacy since these mice were hemostatically normal.[17,18] With the generation of the hemophilia A mouse model, the efficacy of gene delivery approaches could be assessed by not only detection of FVIII antigen in the circulation but also FVIII activity. Early generation adenoviral vectors resulted in therapeutic expression of FVIII that was followed by the onset of antibodies to FVIII transgene and the viral vector.[25,80,81] Helper-dependent ("gutted") adenoviral vectors were developed that do not encode any viral sequences. These vectors have reduced immunogenicity; however, intravenous delivery of these vectors still elicits an innate immune response that results in acute toxicity and development of anti-FVIII immune response that precludes long-term expression from these vectors.[82–85] Recent studies using hFVIII variants in an effort to reduce the vector dose to provide therapeutic levels of FVIII resulted in development of inhibitors in all of the mice thus sustained FVIII expression was not observed.[83] As the helper-dependent adenoviral vectors are evaluated in large animal models, vector-associated toxicities and immune responses to the transgene remain a safety concern.[86]

Adeno-associated viral (AAV) vectors are derived from a nonpathogenic parvovirus. The advantages of AAV vectors are that they can transduce both nondividing and dividing cells and the DNA primarily remains as an episomal form avoiding risks of insertional mutagenesis. The major disadvantage of this vector is the limited packaging capacity of approximately 5.0 kb. Initial studies focused on using AAV serotype 2 (AAV2) because it was isolated from humans and transduced human cells efficiently. In the last decade major developments in AAV biology has led to: (1) the discovery of alternate serotypes with distinct tissue tropisms and efficacy[87]; (2) modifications in the vector genomes to overcome the conversion of single-stranded to double-stranded DNA by using self-complementary vectors[88]; (3) use of directed evolution to generate novel AAV serotypes to overcome neutralizing antibodies to AAV which are prevalent in the human population[89]; and (4) development of large scale recombinant systems to produce AAV vector.[90]

Early studies of the B-domain deleted hFVIII delivered in an AAV2 vector demonstrated that a > 4.7 kb transgene construct could be packaged into rAAV; however, the vector dose was higher than had been demonstrated effective with AAV2–hFIX despite the requirement of 50-fold less antigen to achieve therapeutic levels FVIII.[91,92] Since the large FVIII cDNA (4.3 kb) is difficult to utilize in AAV vectors with limited packaging capacity, initial studies with AAV took advantage of the normal posttranscriptional processing that FVIII undergoes to form a heterodimer composed of the heavy chain and light chain, the secreted form of the protein. *In vitro* studies had demonstrated the feasibility of expressing the heavy chain and light chain from two separate expression constructs within the same cell to secrete functional FVIII.[93] In this "two-chain" delivery approach the FVIII heavy chain is delivered in one AAV vector and the FVIII light chain is delivered in a second AAV vector. This approach relies on cotransduction of the same cells with both vectors to produce functional FVIII. The use of genetic approaches to dimerize the transgene itself resulted in low levels of FVIII expression.[94] An alternate approach relied on the self-association of the individual protein chains into the heterodimer form of the protein. The initial studies with this approach demonstrated that > 100% of normal levels of FVIII in mouse models.[95] Further refinement of this approach using liver-specific promoters[96,97] and alternate AAV serotypes (AAV8 and AAV9)[98,99] provided therapeutic levels of FVIII with a lower vector dose that was necessary to move this approach into the hemophilia A dogs.[98,99] Despite the inefficiency of this approach with 5- to 10-fold higher antigen levels than activity, therapeutic levels of > 100% activity could be detected in the mouse circulation.[97,98] These observations fueled the further development of a single chain delivery approach in which the full-length B-domain deleted FVIII is packaged into one AAV vector thus theoretically resulting in the expression of a functional FVIII molecule for each vector particle. This approach was challenging because it required the use of minimal regulatory elements to facilitate packaging of the construct in the AAV vector. Analysis of AAV-FVIII packaging by analysis of fractions of AAV-FVIII using a large 5.7 kb transgene construct demonstrated that a heterogeneous population of transgene sizes was present suggesting that a heterogenous population of FVIII transgenes was being packaged into the AAV particles.[100] Nonetheless, the BDD-FVIII can be packaged into AAV albeit not efficiently but sufficiently to express functional FVIII molecules. The development of alternative AAV serotypes that more efficiently transduced liver (AAV8 and AAV9) and identification of additional small liver-specific promoters improved the levels of FVIII expression in the circulation. These studies led to large animal studies.[98,101] Proteosome inhibitors that facilitate the intracellular trafficking of the AAV particle to the nucleus rather than the proteosome for degradation, resulted in an increase in FVIII expression.[102]

Overall, these studies demonstrated that the mouse was not predictive of the therapeutic dose of AAV in the hemophilia A dogs and that the dose required to achieve therapeutic levels of FVIII in both mice and dogs was comparable for two-chain and single chain delivery which would not be predicted since each AAV vector delivering a FVIII transgene should express a functional FVIII protein. While mouse models provide the opportunity to screen AAV serotypes for tropism and efficacy, the large differences in the efficacy of AAV serotypes observed in mouse models are more moderate in large animals. Despite efforts to minimize the FVIII transgene construct for the single chain delivery, these data eludes to the inefficient packaging as a remaining obstacle for translation of this approach to humans.

The hemophilia A mouse model has provided a valuable tool for comparing AAV serotypes and transgene constructs; however, the immunocompetent hemophilia A mouse develops an immune response to FVIII of murine, human, or canine origin. This is in contrast to hepatic delivery of AAV–FIX in the hemophilia B mouse model that appears to induce tolerance on most mouse strain backgrounds.[103] While AAV is nonpathogenic and is not associated with the pathology of adenoviral infections, the human population has been exposed to wild-type AAV infections that establish immunity to these viruses that impacts the transduction and long-term expression using these vectors.[104] These hurdles must be overcome for translation of AAV-mediated gene transfer to patients.

Retroviral vectors have also been used to target hepatocytes by *in vivo* delivery in neonatal hemophilia A models. During the neonatal period the rapidly dividing cells of the neonate would provide the cell division required for retroviral integration into the host cell genome and this early period in development may favor immune tolerance. Dose-dependent immune responses to FVIII were observed that suggests that there may be a threshold of FVIII expression required to induce tolerance.[105,106] While long-term expression is observed when high FVIII levels (> 20 ng/ml) are sustained in the absence of an immune response, it is unlikely that a retroviral approach will move forward clinically for hemophilia due to concerns of insertional mutagenesis.

Nonviral gene delivery approaches for hemophilia have focused on plasmid DNA delivery using hydrodynamic delivery or, more recently, cell targeted nanoparticles to the liver. This approach has several advantages in that (1) it circumvents issues related to the viral vector such as immune responses (2) remains episomal thus avoiding insertional mutagenesis (3) ease and low cost of production and (4) ability to deliver large genes such as the FVIII gene. As with viral vectors, optimization of the expression cassettes has improved the success of this approach in mouse models. Hydrodynamic delivery efficiently delivers plasmid DNA to the liver in mouse models; however, delivery of plasmid DNA to large animals remains challenging. Initial high levels of FVIII gene expression in immunocompetent HA mice were detected after hydrodynamic delivery of a FVIII expression cassette; however, a humoral immune response to FVIII was observed that shortens the duration of FVIII expression.[107] This anti-FVIII immune response was similar to the response observed to FVIII protein with primarily IgG1 along with some IgG2a, IgG2b. Transient immunomodulation with cyclophosphamide resulted in high levels of FVIII expression that declined after removal of the immune suppression. This response was species independent since similar findings whether a mouse, human, or canine transgene was delivered. A comparison of additional transient immune suppression regimens demonstrated that CTLA4Ig plus anti-CD40 ligand, inducible costimulator (ICOS) or anti-CD3 were effective at preventing inhibitor formation and ensuring persistent FVIII expression.[108–110] Regulatory T cells (CD4+CD25+FoxP3 +) are the key to preventing these immune responses to FVIII[111] and expansion of Tregs by IL-2 and anti-IL-2 antibodies facilitates the modulation of these responses.[112]

Plasmid DNA has been used to deliver the Sleeping Beauty transposon along with the FVIII transgene. In this approach the transposon provides a mechanism to integrate the transgene into the target cell genome. Initial studies using hydrodynamic delivery in hemophilia A mice resulted FVIII expression that was eliminated after development of an immune response to the FVIII protein.[113] Recent studies encapsulating the Sleeping Beauty/FVIII plasmid DNA in nanocapsules that specifically target receptors on hepatocytes by using the asialoorosomucoid (ASOR) ligand or LSECs using hyaluronan corrected the bleeding phenotype in the absence of an immune response to FVIII.[114] New approaches that target the liver using encapsulated DNA may improve the feasibility of these approaches in large animals.

2. Hematopoietic Stem Cell Approaches for Hemophilia A

The ability to express functional FVIII in various cell types has been exploited for hematopoietic stem cell approaches for hemophilia A. Early studies for hemophilia A used murine oncoretroviruses to transduce mouse bone marrow or fibroblasts *ex vivo* and transplant the cells into irradiated recipient mice; however, FVIII expression was typically short term or not detected.[115–118] The development of lentiviral vectors renewed enthusiasm for using retroviral based *ex vivo* bone marrow approaches for hemophilia A. These vectors promised the ability to transduce both dividing and nondividing cells and deliver long-term transgene expression.[119] These vectors combined with bioengineered FVIII transgenes that had increased secretion in immunodeficient hemophilia A mice and then immunocompetent mice with nonmyeloablative conditioning showed no evidence for an anti-FVIII immune response but FVIII levels remained low.[120,121]

Ex vivo retroviral delivery approaches of pFVIII to hematopoietic stem cells sought to overcome the low expression observed in early retroviral approaches.[122] pFVIII was found to express 10- to 14-fold higher than hFVIII when compared alongside each other in exogenous mammalian expression systems.[29] This high level pFVIII expression is not at the mRNA level but is due to more efficient posttranslational mechanisms that affect movement of the FVIII through the secretory pathway. Long-term FVIII expression and immune tolerance was observed under sublethal conditioning regimens after pFVIII delivery using ecotropic and dualtropic MSCV.[122] These studies also demonstrated sustained FVIII expression in hemophilia A mice with inhibitors to hFVIII, an important issue for evaluating the translation potential. While myeloablative conditioning regimens successfully established long-term FVIII expression without development of an immune response to FVIII, nonmyeloablative regimens did not result in sustained expression of FVIII due to the presence of anti-FVIII antibodies.[123] Conditioning regimens that utilize a combination of nonmyeloablative conditioning with transient immune suppression provide long-term expression of FVIII. With expression being driven by the retroviral LTR rather than lineage-specific promoters, FVIII expression was detected in the granulocytes and it is not clear if FVIII expression in this lineage has any pitfalls. More recently, the use of a porcine/human chimeric protein that contains key regions of pFVIII that lead to high levels of expression may minimize immune responses to the nonhuman transgene. However, it is not yet known what portion of the hemophilia A population that has not developed immune responses to recombinant hFVIII protein treatment might be susceptible to development of an immune response to the neoantigen. While retrovirus-based vectors are a promising gene delivery vehicle, the risks of insertional mutagenesis that has been observed in clinical trials and the use of nonmyeloablative conditioning raises safety concerns.

Hematopoietic lineage-specific approaches have focused on expressing FVIII in megakaryocytes. While FVIII is not normally synthesized in platelets, FVIII/VWF complexes bind to activated platelets. There would be several advantages to expressing FVIII in platelets. First, the platelets are activated at the site of an injury where the α-granules are released which would provide FVIII at the site to facilitate the ongoing clot formation. Second, the FVIII within the platelets would be protected from circulating antibodies to FVIII that would neutralize its activity. Initial investigations in transgenic mouse models demonstrated that the hFVIII transgene driven by the glycoprotein 1ba promoter[124] or the glycoprotein IIb promoter[125] corrected the bleeding phenotype in hemophilia A/transgenic mice. These studies supported lentivirus delivery of hFVIII into the bone marrow of hemophilia A mice which corrected the bleeding phenotype without any evidence of an immune response to the transgene.[126] However, the hFVIII used in these studies does not produce a stable clot thus FVIII variants with higher specific activity or stability may improve the efficacy of this approach.[126,127] These studies led to a large animal study in the hemophilia A dogs.[128]

3. FVIII Expression in Endothelial Cells

Ex vivo approaches using lentiviruses to target FVIII expression to endothelial cells have also been explored. Blood outgrowth endothelial cells can be isolated from peripheral blood for *ex vivo* transduction by lentiviral vectors followed by reintroduction of the modified cells. Since vWF is synthesized in endothelial cells, expression of FVIII may result in costorage of FVIII/VWF complexes within Weibel-Palade bodies that would be released at the time of vascular injury.[129–131] Transplantation of these cells intravenously[129] or subcutaneously[130] has been challenging due to an apparent loss of implanted cells. LSECs isolated from wild-type mice were successfully transplanted into NOD/SCID hemophilia A mice.[132,133] However, an endothelial cell toxin was required to engraft significant numbers of transplanted cells for phenotypic correction. The biggest challenge for these approaches may be long-term engraftment of transplanted cells.

4. Therapeutic Approaches in the Setting of Inhibitors

While one goal of cell- and gene-based approaches is to improve the hemophilic disease phenotype, it will also be critical to establish therapeutic approaches that also achieve hemostasis and facilitate immune tolerance induction in patients that have inhibitors to FVIII. The FVIII delivery approaches that have demonstrated efficacy in the presence of inhibitors are the platelet approaches,[127] AAV-mediated delivery approaches[134] and retroviral delivery of pFVIII into hematopoietic stem cells.[122] Alternative approaches based on bypass treatment strategies rely on achieving hemostasis by sidestepping the intrinsic pathway. One example is FVIIa that has been used clinically as a bypass agent for inhibitor patients for > 10 years. FXa mutants that function similarly to the inactive form of the protein may provide another approach to achieve hemostasis by bypassing the intrinsic pathway.[135]

FVIIa is a protease in the extrinsic pathway that initiates coagulation when complexed with tissue factor. This approach requires that the active form be used to facilitate hemostasis and continuous expression of the active form carries a risk of thrombosis. Transgenic mice were generated that express mouse FVIIa in hepatocytes using a liver-specific promoter (mouse transthyretin promoter) that was then crossed onto the hemophilia A mouse (or the hemophilia B mouse) to assess the safety of continuous expression of

FVIIa that improves hemostasis but avoids thrombosis.[136] Since reports described a differential affinity of human FVIIa and mFVIIa for murine tissue factor, murine FVIIa was used in these transgenic models—pointing to the importance of considering species-specific transgenes. These studies demonstrated the efficacy of continuous expression of FVIIa but also that expression of < 1500 ng/ml was safe. However, levels of FVIIa > 2000 ng/ml were associated with premature mortality thus establishing the therapeutic window. The advantages of developing the FVIIa approach is that patients will be tolerant to the expressed protein since the patients' do not have defects in the FVII gene and that it can be used to treat both hemophilia A and hemophilia B with or without inhibitors. Recombinant FVIIa has a very short half-life (2.7 h) that often requires multiple expensive protein treatments to achieve hemostasis thus continuous expression would overcome this issue. However, the continuous expression of FVIIa has a therapeutic window that raises concerns for the safety of this approach. The initial findings with AAV delivery of FVIIa in hemophilia A (and hemophilia B) mice demonstrated phenotypic correction in the presence or absence of inhibitors which led to large animal studies.[137,138] The concerns of this approach lie in the high vector dose to achieve efficacy and the safety concerns of thrombosis. The development of regulatable promoters that would allow the gene expression to be turned off (or turned on) in response to a drug may make this a more appealing approach. In conclusion, the hemophilia A mouse model provides a valuable model for evaluating novel therapeutic approaches prior to more challenging large animal studies.

I. Hemophilia A Rat Model

A hemophilia A rat model that arose from a spontaneous mutation in a colony of WAG/RijY rats at Yale University was recently described.[12,13,139] These animals have a missense mutation that results in an amino acid substitution, Leu176Pro, which lies in the A1 domain. The rats have a prolonged clotting time while other hemostatic parameters are normal. Due to the limited availability of rat-specific reagents, it is not yet known if these rats have circulating FVIII. It will also be important to determine if this model generates inhibitors to FVIII. Similar to the hemophilia A sheep model, spontaneous joint bleeds have been observed in the rats which has been difficult to clinically diagnose in the mouse and dog models. The rat model will be experimentally advantageous since it will allow for collection of a larger plasma volume for analysis than can be collected in mice and yet will allow for breeding capacities similar to mice. This new model will likely find a valuable niche among the hemophilia A animal models used for translational research.

II. The Hemophilia B Mouse Model

A. Hemophilia B

All hemophilia was thought to have the same pathogenesis, a bleeding tendency resulting from deficiency of "antihemophilic factor," until Pavlovsky reported in 1947 that it was possible in some cases to correct the clotting deficiency in the plasma of one hemophilia patient by mixing the plasma with that of another hemophilia patient.[140] By 1952 it was clear that the deficiency of a second clotting factor (initially called "Christmas factor" or "plasma thromboplastin component" and subsequently renamed factor IX) causes hemophilia in about 20% of cases (hemophilia B), while FVIII deficiency accounts for the majority of hemophilia (hemophilia A).[141,142]

FIX is produced in the liver. The proper activity and circulating kinetics of the secreted FIX are dependent upon several posttranslational modifications (PTMs), the most critical of which is the carboxylation of the 12 glutamic acid residues in the N-terminal portion of the mature protein (the FIX Gla domain) in a vitamin K-dependent process; the majority of the remaining PTMs are to the FIX activation peptide.[143,144] FIX circulates as an inactive zymogen protein. The activation of FIX by the cleavage of the FIX activation peptide from the protein (either by the complex of factor VIIa/Tissue Factor or by factor Xia) results in a protein with a light chain and a proteolytic heavy chain linked by disulfide bonds. On the phospholipid surface of activated platelets the active serine protease FIXa can activate factor X in a calcium-dependent process. The catalysis of the activation of factor X by FIXa is increased several logs by the interaction of FIXa with its cofactor,activated factor VIII (FVIIIa), in the so-called "tenase complex" (FIXa/FVIIIa/PL/Ca++). Factor X activated by the tenase complex can subsequently drive the generation of abundant thrombin that is essential to the amplication of adequate hemostatic clot formation. Accordingly, given the requirement for the cooperative function of FVIIIa and IXa in the tenase complex, it is not surprising that the bleeding phenotype that results from a given level of deficiency of either FVIII or FIX is very similar (e.g., $< 1\%$ activity of either FVIII or IX results in severe hemophilia with a similar bleeding risk).[145]

The mouse and human *F9* gene open reading frames share $\sim 80\%$ sequence homology.[146] As in humans, the mouse *F9* gene is present on the terminus of the q arm of the X chromosome (as are the mouse and human *F8* genes) and the expected X-linked transmission of hemophilia B seen in humans is observed in mice. The domain structure of FIX and the basic interactions of the protein in hemostasis are the same in mouse and human and the same global assays of coagulation and specific factor measurements that are universally used in clinical coagulation laboratories throughout the world can be adapted for use in mice. For these reasons, the study of hFIX and potential hFIX therapeutic approaches, using engineered mouse models that are either mFIX and hFIX deficient or engineered to express modified FIX variants, has been widely pursued and recently reviewed.[147] It needs to be stated that in addition to multiple informative parallels, multiple divergences exist between humans and mice in regards to hemostatic and thrombotic processes, as do differences between strains of mice, and these must be considered in experimental design. Comprehensive considerations of comparative mouse and human hemostasis are available[148–150] and strain-specific differences in normal ranges for coagulation assays are available in the Mouse Phenome Database (available at: http://phenome.jax.ohrg/pub-cgi/phenome/mpdcgi?rtn=meas/catlister&req=Cblood%20hematologyqqqcoagulation).

B. Hemophilia B Mouse Models: FIX Knockout Mice

Three different groups generated hemophilia B mouse models in rapid succession in 1997–1998. For decades the use of hemophilia B dogs for the evaluation of novel treatment approaches had provided translationally accurate insights.[151] Nevertheless, the expanding investigation of strategies aimed to correct the genetic defect in hemophilia B demanded a small animal model in which hemostatic correction could be modeled without being confounded by the background expression of mouse FIX. The usefulness of the hemophilia B model to pattern gene therapy approaches for ultimate application in other genetic defects of protein synthesis also accelerated the development of this model. Two groups generated large deletion mutations in the mouse *F9* locus by deleting either exon h (which encodes the catalytic domain of the serine protease) or both exons g and h.[152,153] The mice generated by these two approaches have no circulating FIX protein (no antigenic "cross-reactive material" in immunologic assays to detect FIX, "CRM(−)"). The remaining group pursued a "plug-and-socket" strategy as originally described by Oliver Smithies's lab.[154] The promoter and the first three exons of the FIX gene were deleted by the insertion of a *neo* gene plus a partially deleted hypoxanthine phosphoribosyl transferase minigene. The plug and socket design allows the subsequent insertion of other sequences into the same locus of correctly targeted embryonic stem cells, consistent with the investigators' goal of creating a reagent for studying structure–function relationships of recombinant FIX proteins *in vivo*.[7] This latter knockout model has in fact been used to generate a number of additional strains each of which expresses a single copy *F9* gene from the "socket" located in the endogenous single copy mouse *F9* gene locus. These strains include:

(1) mouse FIX expressing a mutation that alters the FIX Gla domain interaction with endothelium (K5AFIX)[155]

(2) defective hFIX having a missense mutation at a critical arginine in the catalytic domain (R333QFIX)[8,156]

(3) defective hFIX having a nonsense mutation in the Gla domain that results in an early stop mutation (R29XFIX)[9]

(4) hFIX wild type (hFIX-WT mouse)[156]

(5) hFIX carrying three mutations each of which increases the specific activity of the protein ("FIX Triple").[156]

The three knockout models ($FIX^{-/-}$) produce no hepatic FIX mRNA, have no circulating FIX protein (CRM −), and bleed excessively with hemostatic challenges such as tail transection. The original reports of these mice stated that 0.03–0.08 U/ml of FIX coagulant activity was detected in plasma in one-stage hFIX activity assays. This activity has been demonstrated to be artifactual and can be reduced to < 0.01 U/ml if standards for the clotting assay are constructed using uniform dilutions of the knockout mouse plasma.[8] $FIX^{-/-}$ mice breed normally and deliver normal size litters. Colonies can be maintained with breeding females having both X chromosome *F9* genes knocked out, with the dams only occasionally lost to hemorrhage in the peripartum period. Most pups survive normally to wean and into adulthood without bleeding. It is worth noting that this normal survival of mice having deleted procoagulants of the intrinsic pathway FIX contrasts with the intrauterine or early postpartum lethality that is observed in animals that fail to express the vitamin K γ-glutamyl carboxylase or procoagulant common pathway or extrinsic pathway proteins (prothrombin, factor V, factor X, Tissue Factor, factor VII) (as reviewed by Mackman).[157] Spontaneous bleeding in the $FIX^{-/-}$ mice is rarely observed, although musculoskeletal bleeding is seen, in particular after fighting with cagemates. $FIX^{-/-}$ mice will generally hemorrhage to death after tail clip, however.

C. Additional Hemophilia B Mouse Models

None of the $FIX^{-/-}$ strains discussed above transcribe FIX mRNA and therefore they do not have antigenically detectable FIX protein (i.e., they have a CRM(−) phenotype). As a result, there is no opportunity for potential epitopes derived from FIX to be expressed or presented during mouse immunologic (thymic) development. $FIX^{-/-}$ mice have been reported to make antibodies directed against recombinant mouse FIX following parenteral challenge, although not after exposure to mouse plasma (containing FIX).[158] It is desirable to study potential new human hemophilia B therapies in the $FIX^{-/-}$ mice; however, the hFIX protein stimulates anti-hFIX antibodies in some contexts, given that it appears to be a cross-species protein (xenoprotein). Importantly, inhibitor antibody formation clinically in humans with hemophilia B occurs in only 2–4% of patients exposed to FIX replacement, so that this clinical complication is almost 10-fold less commonly observed than in hemophilia A. To study issues related to immunogenicity of FIX therapies, or to circumvent the potential for antibody formation to confound *in vivo* evaluation of FIX therapies, hemophilia B mouse models have been generated that do not express mouse FIX but instead express hFIX cDNAs having a variety of underlying FIX mutations.

As mentioned above, homologous recombination strategies derived from the "plug-and-socket" $FIX^{-/-}$ strain have been used to generate mice to model immunologic "good risk" and "poor risk" scenarios as regards FIX inhibitor antibody formation. In the first case, a hFIX cDNA having a single nucleotide change results in a missense substitution of glycine for arginine at amino acid 333 (R333QhFIX mouse).[8] The defective R333QhFIX is physiologically expressed from the mouse FIX promoter and a single copy gene in the endogenous mouse *F9* locus. The defective protein circulates at 15–30% of normal hFIX levels (CRM + phenotype) but has < 1% activity and a severe bleeding phenotype. To contrast with this, a mouse was created by homologous recombination into the *F9* locus of a hFIX cDNA having a single nucleotide change that results in nonsense termination of transcription and a very early stop mutation (R29XhFIX mouse).[9] R29X is the most common mutation that leads to inhibitor formation in humans, as recorded in the database of hemophilia B mutations (available at: http://www.kcl.ac.uk/ip/petergreen/haemBdatabase.html. However, complete gene deletions lead to inhibitors with somewhat greater frequency than the stop mutation). The antibody responses in the R29XhFIX and R333QhFIX strains mimic many expected features of hFIX inhibitors, and the models can be used to evaluate efficacy and risk of protein and gene therapy approaches. As one example, there is very low risk of inhibitor antibody formation in the setting of CRM + missense R333QhFIX mutation, owing to a relative state of immunologic nonresponsiveness of CD4 + T lymphocytes to replacement FIX.[159] These good and poor risk mutation hFIX-expressing mice have now been crossed onto the HLA background of a mouse strain that does not express mouse Major histocompatability class II antigens, but instead expresses a human HLA-DRB1*1501 MHC class II allele and these "double humanized" hemophilia B mice are being used to characterize new therapeutic approaches (Dr. Genlin Hu, University of North Carolina at Chapel Hill, personal communication).

Similarly, a series of transgenic hemophilia B mice have been created by Sabatino and colleagues and provide a spectrum of mutations to examine the degree of tolerance to hFIX conferred by the underlying mutation.[10] In addition to transgenic mice expressing early (R29X) and late (R338X) stop mutations and missense mutations (R180W, G381) these investigators generated hemostatically normal mice that do not express mouse FIX but instead express two or more copies of a hFIX cDNA. An example of the value of these mice has been to demonstrate the preclinical therapeutic potential of an aminoglycoside translational readthrough strategy to generate circulating FIX from a nonsense mutation F9 gene sequence.[160] A human Phase I clinical trial of a translational readthrough strategy for hemophilia has been initiated.[161,162] (see http://www.clinicaltrials.gov/ct2/show/NCT00947193?term=hemophilia+B&rank=34).

D. Engineered Mice in the Study of FIX Expression and Circulating Kinetics

Clinical observation of patients with unusual phenotypic expression of hemophilia suggests that mechanistic insights may be found by studying the genotype: phenotype expression. Expressing the genotype of interest in cell culture may allow partial characterization of the effect of mutated gene sequences; however, in some cases the expression observed in cell lines does not correlate well with the expression observed from adult liver, so that study in animals is a more accurate or revealing test system.[163,164] Transcriptional regulation of FIX gene expression has been modeled extensively in engineered mouse models. Mapping of transcription factor binding and regulation of the FIX promoter has been modeled in mice that recapitulate the hemophilia B Leyden phenotype, demonstrating age-dependent and sex-specific postpubertal increases in FIX expression.[165–167] Elements in the 5′ and 3′ untranslated regions of the FIX gene that direct natural gender-independent age-associated increases in FIX expression have also been elucidated in mice with engineered FIX gene sequences.[163]

Hemophilic dogs have been used as an excellent resource for the preclinical testing of replacement clotting factors over the last several decades, demonstrating circulating kinetics for FVIII, FIX, and factor VIIa that are very similar to the kinetics in humans.[151,168] In recent years there has been a tremendous increase in the investigation of clotting proteins that are either genetically or chemically modified to achieve improved therapeutic potential.[169] The most common efforts are to increase the circulating survival of the hemostatic protein and/or to increase the specific activity. FVIII and FIX knockout mice have been increasingly used for preclinical pharmacokinetic/pharmacodynamic screening of the *in vivo* properties of candidate proteins to inform rational choice of protein to examine in the more scarce hemophilic dog resource. In these applications it is clear that the mouse pharmacokinetics are not the same as human. For instance, the circulating half-life of FIX is reproducibly shorter in mice than in humans. Nevertheless, the *relative* kinetics of candidate proteins can be compared to the commercially available replacement protein (e.g., plasma-derived or recombinant FIX) in adequately sized cohorts of mice to generate statistically meaningful data. There has been consistency in the relative kinetics that have been demonstrated in mice and those subsequently confirmed in the dog model (i.e., agreement that the novel therapeutic is biologically equivalent, inferior, or superior to approved FIX in small and large species).[170]

E. Engineered Mice in the Study of FIX Hemostasis, Thrombosis, and Response to Injury

Hemophilia B mice have also been used for preclinical testing of relative pharmacodynamics of candidate FIX or factor VIIa proteins in hemostatic challenges.[33,171] Tail transection bleeding time assays measure initial hemostasis after tail vessel wounding. Nevertheless, the tail-clip assay alone may fail to reliably distinguish factor deficiencies.[32,172–174] (Dejana E *et al.* Contributing factors to the discrepant results of tail bleeding assays have been recently reviewed by the Animal Models Scientific Subcommittee of the International Society for Thrombosis and Hemostasis).[33] These include the central role of platelets in initial hemostasis (as distinct from soluble clotting factors), as well as variations in central tail artery constriction and dilation that may obscure discrete endpoints. Delayed hemorrhage and persistent blood loss, however, are hallmarks of hemophilic bleeding and result in decreased survival. For this reason, the tail transection bleeding time may be modified to observe persistent as well as initial hemostasis.[32,155,175] Secondary bleeding time assays have been described to measure this characteristic phenotype in humans with hemophilia,[176] as well as in a monkey model of hemophilia.[177]

Another bleeding challenge has been described recently by Whinna and colleagues that involves observation of the average time to hemostasis following a standardized saphenous vein incision.[178] The assay appears to have greatly increased sensitivity when compared to the tail-clip assay; differences in hemostasis can be demonstrated in the range between 1% and 10% FIX activity level, corresponding to the ability to demonstrate the correction from a severe hemophilia phenotype to a mild hemophilia phenotype. The model has been applied to the *in vivo* study of FIX mutations that result in gain or loss of proteolytic function as well as to the comparative *in vivo* investigation of alternative recombinant activated factor VIIa therapeutics as bypassing agents for hemophilia care.[178,179]

Hoffman and Monroe and colleagues have examined hemophilia B mice to demonstrate that wound healing is defective in hemophilia. Normal wound healing entails four overlapping phases: hemostasis, inflammation, proliferation, and remodeling or resolution.[180] The healing of standardized skin punch biopsy wounds of hemophilia B mice, as compared to hemostatically normal mice, demonstrates delayed wound closure, prominent neovascularization, delayed but subsequently prolonged iron contamination and macrophage/monocyte infiltration of the wound tissue.[181] Late hemorrhage into the tissue near the wound site is seen even after the surface wound closes. Restoring hemostasis at the time of injury, without prolonged coagulation protein replacement throughout healing, does not normalize healing in hemophilia B mice.[182] Extending the observations using this model, perivascular tissue factor (TF) is shown to be downregulated following cutaneous wounding.[183] TF expression is depressed for a longer period in hemophilic mice than in wild-type mice. It appears likely that appropriate wound remodeling requires the coordination of effective hemostasis with modulation of inflammation and appropriate vasculogenesis and that insights may be gained by modeling the ways this coordination is dysregulated in hemophilia B mice.

The cutaneous wound model suggests a number of paradigms for the study of healing in the presence of impaired hemostasis; however, clinically the most common site of hemophilic bleeding and degeneration is in the joints, resulting in chronic synovial and osteochondral arthropathy. The distinctive pattern of hemophilic bleeding can be understood by recognizing that hemostatic potential is tissue specific. Extrinsic pathway tissue factor (TF), the primary initiator of blood coagulation, is expressed at high levels in the brain and heart, as well as in the testis, uterus, placenta, and kidney.[157,184] Tissue factor is expressed at low levels in healthy skeletal muscle and joints.[185,186] At baseline the potential for extrinsic pathway (TF/FVIIa) initiation of coagulation is compromised by low TF in joints and when intrinsic pathway proteins are also deficient (e.g., FVIII or IX in hemophilia A and B) joints are particularly vulnerable to local hemorrhage.

As discussed above (Arthropathy in the murine model—Hemophilia A) when a reproducible trauma is used to induce bleeding into the joint of hemophilic mice the characteristic histologic and radiologic changes of human hemarthropathy are reproduced.[38,187] The clinical, histopathologic, and radiologic changes initially described in hemophilia A mice by Valentino and colleagues are also seen in the hemophilia B mice.[187,188] Importantly, these hemophilic changes are not seen in hemostatically normal mice. The effect of potential therapies, including FIX gene therapy, to reduce or prevent the most relevant end-organ pathology of hemophilia B (i.e., the degenerative effect of bleeding into joints), can be used as the endpoint of interest in preclinical modeling. Even more intriguing is the potential to study the action of FIX to effect wound healing within the joint. For example, coagulant proteins are present in biological fluids including synovial fluid,[189] pleural fluid,[190] and lymph.[190,191] While it is conceivable that these proteins exist in these locations as merely an ultrafiltrate relative to their plasma concentrations, as merely an ultrafiltrate, the hypothesis that synovial fluid FIX could protect the hemophilic joint via an extravascular hemostatic function has been investigated in hemophilia B mice. When FVIII or IX was injected directly into the joint at the time of induced joint bleeding, the joints developed minimal synovitis. This protection was seen despite the fact that no factor activity was detectable in the circulation throughout 72 h following the intraarticular (I.A.) dose. In fact, hemophilic mice subjected to the same induced joint hemorrhage and treated intravenously with FIX developed more joint degeneration than those treated directly into the synovial space.[187] Subsequent investigations have established that I.A. FIX expressed from an AAV gene therapy vector expressing exclusively in the joint can also protect the joint from progression of bleeding-induced synovitis, whether used to treat a joint with preexisting hemophilic synovitis or a joint that is naïve to previous blood exposure.[192] Correlating findings in the mouse joint injury with the cutaneous injury may yield valuable mechanistic insights of the action of clotting factors in locations outside of the circulation, as well as general mechanisms of wound healing.

Clotting factor knockout mice have been used to study occlusive thrombus formation and pathologic clotting. Not surprisingly, an intact classical intrinsic coagulation pathway is required for stable occlusive thrombus formation, as demonstrated in ferric chloride carotid artery large vessel injury and in laser-induced mesenteric arteriolar small vessel injury (recently reviewed).[193,194]

An example that demonstrates several of these uses of mouse models of FIX expression to elucidate FIX biology is the study of the effect of substitutions in the FIX Gla domain and their effect upon *in vivo* FIX function. Darrel Stafford and colleagues described FIX single amino acid substitution mutants K5A or V10K, and the effect of these mutations to greatly decrease the binding of FIX apparently resulting from their greatly decreased binding to endothelial collagen IV, an endothelial basement membrane protein.[195,196] The substitution of an arginine (K5R) instead of an alanine (K5A) at amino acid 5 resulted in a FIX mutant with increased specific binding to collagen IV. Recombinant FIX protein stocks of each of these variants were generated, purified, and infused into hemophilia B mice for *in vivo* pharmacokinetic comparison.[197] The K5R FIX disappeared from the circulation rapidly after infusion with greatly decreased area under the FIX activity curve kinetics relative to wild-type FIX. The decreased binding mutant FIXs, delivered as recombinant K5A hFIX protein to $FIX^{-/-}$ mice[198] or as combined K5A/V10K hFIX gene therapy vectors to $FIX^{-/-}$/T cell CD4 knockout mice,[199] demonstrated twofold greater survival. These findings suggested, for example, that there might be value to incorporate the K5A, for instance, into recombinant FIX protein, if indeed the level of plasma FIX activity is a surrogate measure for hemostatic protection, as is often assumed in clinical practice. Nevertheless, the possibility of a functional purpose underlying the observed interaction of FIX with collagen IV was explored by Stafford and colleagues by generating a mouse that expressed low-binding K5AFIX instead of wild-type FIX.[155] Although the K5AFIX mouse has

20% higher circulating FIX protein *and* activity than wild type, the mouse displays a mildly hemophilic phenotype. This phenotype was demonstrated using several of the methods discussed above including the demonstration of prolonged primary and secondary bleeding in the tail transection challenge, slow time to occlusive thrombus formation in the ferric chloride-treated mesenteric arterioles, cutaneous wound closure time that was intermediate between that of FIX−/− mice and hemostatically normal mice, all suggesting that there is likely a physiologic role for the binding of FIX to collagen IV.[155] These insights would certainly have been difficult to achieve relying upon *in vitro* studies or upon studies in hemostatically normal animals.

F. Tolerance and Immunogenicity of FIX Studied in Hemophilia B Mice

As has been mentioned earlier in this chapter, the development of antibodies that neutralize the activity of replacement clotting factor is the most common serious complication that arises from hemophilia treatment. The development of "inhibitor" antibodies occurs at a much lower incidence in severe hemophilia B (~2–4% of patients) when compare to severe hemophilia A (~20–30% incidence of low and higher titer inhibitors), but can be a devastating complication, in particular because FIX inhibitors are difficult to eliminate with immune tolerance induction regimens and can be associated with life-threatening anaphylactic responses.[200] The ability to extract clues about inhibitor etiology or treatment through examination of large population epidemiologic data or registries is being vigorously pursued in regards to FVIII inhibitors, but is unrealistic for FIX inhibitors due to their rarity. The ability to mechanistically study FIX neutralizing antibodies in a mouse model that faithfully models the hFIX inhibitor phenotype could be valuable—if only enough were understood about hFIX inhibitors to judge how representative such a mouse model were. Some areas in which the hemophilic mouse inhibitor phenotype appears to mimic FIX inhibitor in humans will be reviewed in brief; more comprehensive discussions are available.[147,200,201]

The only strong determinant of FIX inhibitor risk that is unequivocally determined in hemophilia B is the patient's FIX genotype (family history is an established risk, however, clearly associated with the underlying shared FIX mutation). As discussed above, the underlying FIX mutation is a principal determinant of antibody response to FIX in hemophilia B mice: greater degrees of loss of FIX coding sequence and lack of circulating CRM(+) protein correlate with greater inhibitor risk.[8–11,159,202] Anti-FIX IgG has been reported to develop in some strains of mice after exposure to intravenous hFIX,[159,203] intravenous mouse FIX,[158] subcutaneous hFIX given with adjuvant,[204] or intramuscular (IM) AAV-FIX gene therapy.[158,159,205–207] In each of these contexts the antibodies are mostly IgG1 isotype. Mouse IgG1 is the homologue of human IgG4; hFIX inhibitors are reported to be mostly IgG4 isotype, and so the mouse inhibitor response appears to parallel the human response. Low titers of IgG2a and IgG2b sometimes have been observed in mice. Taken together, the results are consistent that FIX elicits a Th2 lymphocyte-dependent immunoglobulin response which is associated with the high affinity antibody production.[159,208,209] FIX-specific IgE has also been demonstrated by RAST reaction in some individuals (although not all) who exhibit immediate hypersensitivity to FIX exposure.[210] IgE anti-FIX has recently been described in hemophilia B mice crossed onto the particularly immunologically intolerant C3H background (see below).[211]

Human hemophilia A cohort studies suggest associations of FVIII inhibitor development with polymorphisms in cytokine immune response genes interleukin 10 (IL-10),[212] tumor necrosis factor-α,[213] and in the CTLA-4 gene[214] and weaker associations with MHC class II phenotype. The number of studied hemophilia B inhibitor patients studied is too small to establish the same associations between FIX inhibitors and MHC II or other potential genetic factors.[49,215] Nevertheless, genetic studies in inbred mouse strains that have developed antibodies after exposure to species-mismatched FIX demonstrate linkage to genes for immune response modifiers including IL-10.[203,216]

The MHC class II phenotype of a patient should determine which FIX-derived peptides are presented to the T-helper cells. Very striking differences exist between strains of hemophilic mice having different MHC backgrounds in regards to tolerance of hFIX. For example, C57Bl/6 (MHC H-2^b) $FIX^{-/-}$ mice tolerate repeated recombinant hFIX protein infusion, and following liver-directed hFIX gene therapy and they develop tolerance to the protein that is not broken by subsequent challenge with skeletal muscle-directed FIX gene therapy.[217] At the other end of the spectrum, C3H (MHC H-2^k) $FIX^{-/-}$ mice develop IgG1 inhibitors after liver-directed FIX gene therapy and may also develop IgE-associated anaphylactic reactions and death following repeated intravenous hFIX challenges.[211] CD4+ T cell-dependent B cell antibody response to FIX has been demonstrated in several strains of mice and immunodominant epitopes for T cell stimulation that are specific to each strain have been identified. Immunodominant epitopes for each strain have been mapped in the catalytic domain of the protein.[201,204,209] In this respect, as well, it appears that mice may be useful models for studying hFIX inhibitors, because epitopes targeted in humans are primarily in the catalytic domain.[218] A final suggestion that consistencies between the mouse and human epitopes support the use of the hemophilia B mice to model inhibitor development is that immunodominant FIX epitopes in very similar regions of the catalytic domain are recognized by a novel hemophilia B mouse model that is engineered to express human HLA-DRB1*1501 MHC class II allele and hFIX gene sequence. (Personal communication, Dr. Genlin Hu, University of North Carolina at Chapel Hill.)

These and other investigations in mouse models are consistent with a mechanism wherein autoreactive T cells, capable of recognizing FIX protein, are neither deleted nor anergic, but are maintained in an unresponsive condition. Regulatory CD4+CD25+FoxP3-expressing T-regulatory cells suppress antibody formation to FIX. *In vivo* depletion of CD4+CD25+T_{regs} has led to loss of tolerance evidenced by antibody formation.[103] Two recent investigations explored the potential to use mucosal tolerance to achieve immune deviation towards a regulatory T cell immunologic environment in inhibitor prone C3H $FIX^{-/-}$ mice. The first approach demonstrated that intranasal peptide administration of a hFIX-specific CD4+ T cell epitope reduced inhibitor formation.[201] The second approach fed bioencapsulated cholera toxin β-subunit-fused FIX, resulting in control of inhibitor formation and preventing anaphylactic reactions.[211] These experiments, along with others that seek to pharmacologically deviate the immune response following FIX exposure towards IL-10 and TGF-β dominated toleragenic immune responses, present intriguing challenges for translation of mouse results to care of humans.[219]

Somewhat more disappointing have been attempts to use hemophilia B mice to model an apparent CD8+ cytotoxic T-lymphocyte (CTL) response that was seen in a human clinical trial of FIX gene therapy (see below).[220–223] Mouse strain-specific class MHC I responses that direct CD8+ T cell responses to FIX epitopes on FIX can be mapped, if present.[224] Nevertheless, there has proved to be limited ability to study a human trial subject's CTL response against a viral gene therapy vector derived from a virus, the normal host of which is human, as will be discussed below.

G. Engineered Mice for the Study of FIX Gene- and Cell-Based Therapy

The most common use of hemophilia B mice has been to explore potential gene correction strategies. Preclinical studies in the hemophilia B mice have been central to the approaches used in two completed and two ongoing human clinical trials, and the ways in which hemophilia B mouse studies are driving current investigations will be the focus of this section. The reader is also referred to multiple reviews of hemophilia B gene therapy preclinical modeling and clinical translation.[104,147,225–227]

The strategies studied in hemophilia B mice include *ex vivo* approaches, in which the corrective gene is transferred to cells (autologous or allogeneic) outside the body and the cells expressing the therapeutic gene subsequently delivered to the subject. Autologous keratinocytes,[228] embryonic stem cells directed toward hepatic endodermal differentiation,[229] allogeneic hepatocytes,[230] hematopoietic stem cells,[231] hematopoietic stem cells directed toward erythroid differentiation,[232] allogeneic megakaryocytes[233] and encapsulated primary myoblasts[234–236] have all been investigated for phenotypic correction of FIX-deficient mice. In general, the degree of correction achieved by these strategies has been from <1% to 10% of normal FIX levels; stem cell approaches have in some cases achieved advantageous longevity of expression[231,232] and FIX tolerance induction.[229,231,232] In addition, the definition of efficacy of the megakaryocyte approach needs special consideration, given that the plasma FIX activity may not reflect the relevant endpoint. In the megakaryocyte approach, it is estimated that 90% of the total FIX is stored in the platelets, and is releasable upon activation of the platelet, and is therefore immediately available at the required surface for thrombin generation.

In vivo gene correction approaches studied in hemophilia B mice have delivered the FIX gene either as naked or chemically formulated DNA[175,237–240] or using a virus vector for gene delivery. Naked DNA delivery is inefficient. Achieving FIX expression at levels adequate to achieve hemostatic correction has required driving target cell entry by the therapeutic nucleic acid using hydrodynamic pressure,[175,241] electrical current,[239] or ultrasound.[242] Several groups are examining whether these enhancements to gene delivery can be scaled for efficient and safe use in large animals and humans.

Early approaches to viral vector-mediated FIX correction used oncoretroviral vectors and adenovirus vectors. Primarily due to safety concerns related to retrovirus and adenovirus vectors, lentivirus and AAV vectors have progressed for hemophilia B applications. Proficient and pantropic transduction by lentivirus vectors led to initial enthusiasm for these vectors. With time, the efficient cell entry by lentiviral vectors was recognized to have disadvantages, because efficient transduction and expression of FIX by antigen-presenting cells led to neutralizing antibody formation against the therapeutic transgene. This problem was overcome by the incorporation in the lentivirus of micro-RNA sequences to silence FIX expression in a cell-specific fashion in the APCS, avoiding inhibitor formation while allowing continued FIX expression from the liver.[243] In recent years, however, concerns regarding the occurrence of insertional mutagenesis (oncogenesis) have prompted the development of integration-defective lentiviral vectors (IDLV). Although this innovation undoubtedly increases the safety of the lentivirus, a very substantial loss of efficiency of expression is incurred, and substantial optimization will be required before the IDLV are likely to move beyond the mouse model.[244,245]

Preclinical data derived in hemophilia B mice has been central to the development of two completed human clinical trials using AAV2 vectors to deliver the FIX gene to skeletal muscle and liver. The wild-type AAV2 upon which most AAV vectors have been based commonly infects most individual in childhood but causes no human disease. On the first of these two trials, no safety concerns were raised and expression of FIX in transduced skeletal muscle tissue was demonstrated for as long as 3.7 years after AAV2 delivery in one subject, although none of 8 subjects had measurable circulating FIX activity.[246] Transient partial correction of hemophilia B with FIX activity > 10% was seen at the highest dose level in a subsequent trial of AAV2–FIX directed to the liver. Evidence of asymptomatic hepatocellular inflammation was observed, however, by 6 weeks postadministration and coincided with the loss of FIX expression.[247] Subsequent study has determined that AAV capsid epitopes presented on the surface of successfully transduced cells were most likely the target of memory T cell CTL-mediated elimination.[248]

The results of this trial have prompted the return to study of the hemophilia B mice with results that highlight both the usefulness and the limitation of the mouse model. As the investigators of the liver-directed clinical trial have modeled in subsequent studies, hemophilia mouse models do not predict the occurrence of the CTL response. Presumably the CTL response was mediated by preexisting memory CD8 + T cells: such cells will be present in humans as the natural host of wild-type AAV, but mice do not naturally host this primate virus.[248] Although artificial mouse systems have been bioengineered that confirm the proposed mechanism (CTL-mediated elimination of AAV capsid-presenting cells), these systems will not predict parameters to allow the complication to be avoided in future trials. The investigators further concluded that alternative serotypes of AAV would be unlikely to evade capsid-specific immune responses. Unfortunately, this conclusion also cannot be tested in the hemophilia B mouse.[248]

Other groups responded to the clinical trial results by returning to the hemophilia B mice to examine and optimize AAV vector design and to address the hypothesis that more highly efficient FIX vectors might direct hepatic FIX expression without requiring as large a load of AAV capsid as precipitated the CTL response in the clinical trial. In 2001, novel AAV vectors were reported that incorporated a "self-complementary" genomic form (scAAV) that bypasses the rate-limiting requirement for DNA second-strand synthesis prior to expression and could be used to deliver therapeutic transgenes.[249] Optimization of elements for incorporation in a FIX expression cassette for liver-targeting scAAV were subsequently determined *in vivo* in mice, and the potential for significant reduction of AAV particle number confirmed.[250,251] Optimization of the mammalian codon usage in the FIX cDNA further improved expression and allowed further reduction of AAV particle number. Immunohistochemical staining of the pattern of hepatic expression of endogenous hFIX from the native mouse promoter could be examined in the R333QhFIX mouse. This demonstrated that physiologic FIX expression occured in a homogeneous FIX expression across the liver parenchyma.[250] scAAV vectors also demonstrated this physiologic FIX expression pattern, whereas expression from the conventional single-strand AAV2–FIX vector occurred in an intense pattern in only a small subset of cells, in which expression was intense. In 2002, AAV8 serotype vectors were described and quickly shown to have enhanced tropism for liver, further enhancing the efficiency of FIX gene delivery.[87] Preclinical studies of a self-complementary AAV serotype 8 codon-optimized FIX vector (scAAV8FIXco) in hemophilia B mice and hemostatically normal primates led to a human clinical trial sponsored by St Jude Children's Research Hospital and conducted at the University College London.[252] In 2010, the first report of sustained partial correction of human hemophilia with a gene therapy vector was reported from this trial,[253] with least five of six patients on this trial converted from severe to moderate hemophilia B.

H. Hemophilia B Mice and Future Directions

An interesting development on the most recent scAAV8FIXco trial is that after safety was demonstrated at lower doses, AAV dose was once again escalated to the range used in the previous single-strand AAV2 trial. Once again an apparent immune-mediated transient inflammation of vector-transduced liver was observed, although in contrast to the previous trial persistent FIX expression was not eliminated.[254] Taken together, these important trials define a threshold load of AAV capsid that has consistently stimulated capsid-specific cytotoxic lymphocyte recognition and potential transaminitis in humans. The limitation of hemophilia B mice to predict risk of CTL response to the AAV vector is demonstrated. On the other hand, the tasks for the next cycle of investigation—from preclinical application in mice to human application in gene therapy, and back to mice again—may now include:

(1) Improve further the efficiency of delivery of FIX correction, so that true cure can be achieved using doses that do butt up against known toxicity. This is particularly important for the hemophilia patient population, which has an extremely high incidence of underlying hepatic disease. A bioengineered FIX mouse has recently been reported by Shu-Wha Lin's group that has a gene knock-in of a *F9* gene incorporating an Arg338Ala (previously characterized)[199,255,256] and two additional gain of function mutations.[156,257] Mice with endogenous expression of this increased specific activity "FIX Triple" have a hemostatic phenotype that is not complicated by inappropriate activation of coagulation. Incorporating similar gain of function FIX variants into AAV vectors may permit improved correction of hemophilia while further decreasing vector capsid load. Alternatively,

tyrosine capsid mutant AAV vectors have already been demonstrated to correct in hemophilia B mice while permitting dose-reduction, potentially decreasing immunogenicity, using a capsid improvement that is globally applicable and not directed to any one target organ.[258]

(2) Test therapeutic approaches for the population that is not served by current hemophilia gene therapy approaches, for example, those with neutralizing antibodies directed against either FIX or the AAV capsid. Multiple cell-based strategies would circumvent widely prevalent naturally occurring antibodies against AAV and adenovirus vectors.

(3) Explore the potential hemostatic role that FIX plays at sites of action within tissues. Although our laboratory assays are standardized to measure plasma factor activity as a marker of hemostatic potential, experimental results in hemophilia mice suggest that understanding the potential for FIX function at cell surfaces has therapeutic implications, whether it be the surface of the FIX gene-expressing platelet[233] or the surface of the FIX-expressing fibroblast-like synoviocyte.[187]

(4) Further develop the recent approach of targeted integration for safe and persistent gene therapy.[259]

In summary, engineered FIX mice have become an essential reagent to the study of the intersection of coagulation, inflammation, and immunity in hemophilia and hemostasis.

III. The Hemophilia A and B Dogs

A. Background to the Canine Models

The primary benefits, value, and significance of dogs with hemophilia A and B is that they have provided the investigational basis to study the pathophysiology of blood coagulation, hemostasis, and thrombosis, to safely and successfully translate many experimental therapies into clinical practice,[260–265] and to continue preclinical evaluation of promising new strategies and therapies.[207,266,99,86,267–276] Several therapeutic agents, developed by performing research in the dogs with hemophilia, have been successfully introduced into human clinical practice and have a multi-year track record of safety and efficacy worldwide.[260–265] This means that these bleeder dogs are recognized as being a valid animal model of human-like hemophilia. It also means that when novel investigative treatments are shown to be safe and efficacious in the dogs with hemophilia, the data will reliably have a strong positive preclinical predictive value for subsequent safe and successful translation to human medicine. Because of this history of successful translational research, many investigators and advisory boards regard these dogs as an essential for preclinical testing of new treatments for hemophilia A and hemophilia B (MASAC Recommendations #137 and #160, http://www.hemophilia.org/research/masac/masac_all.htm).

B. Hemophilia A Dogs

The Chapel Hill hemophilia A colony, established in 1947, appears to be the longest maintained strain of a serious genetic disease in dogs.[14] The prototype animals were Irish Setters, but outbreeding was required to attain hybrid vigor. Both homozygous females and hemizygous males are maintained for breeding.[22] The causative molecular defect appears to be due to an aberrant transcript in exon 22 that is reminiscent of the inversion defect that accounts for nearly half of human hemophilia A.[15] A separate strain of hemophilia A dogs maintained at Queens University, Kingston ON also has an intron 22 inversion type defect.[16] The plasma level of FVIII:Ag in both strains is < 0.005 U/ml. Much of the basic understanding of hemophilia as a disease was first made in these animals. As in humans, severe disease is associated with levels of FVIII less than 1% of normal coagulant activity and the clinical severity is inversely proportional to the circulating FVIII level in plasma. Likewise, spontaneous bleeding typically occurs in joints and soft tissues but can occur at any anatomic site.[277–279] Their availability played a key role in developing an understanding of the hemophilic clotting defect, which led to the establishment of modern methods for diagnosis and treatment including the partial thromboplastin time test, the one-stage factor assay system for FVIII, and the "home" treatment management system consisting of an early alert at the first sign of hemorrhage with immediate intensive plasma or concentrate treatment.[280–284] These advances for detection, characterization, treatment, and prevention of hemophilic disorders over the past six decades significantly improved the quality and duration of life for patients with hemophilia from likely death at an age of < 20 years with concurrent painful and crippling arthritis to a nearly normal life span with relative sparring of joints.[285,42]

Despite the improved life expectancy and quality of life currently enjoyed by humans and animals with bleeding disorders, significant limitations in treatment persist. There are several interrelated areas of ongoing translational research in hemophilia A dogs that address these limitations. The first exploits strains of hemophilia A dogs that develop inhibitory antibodies to FVIII, the most common and most feared complication of replacement therapy.[286,287] These studies focus on mechanisms of inhibitor development and strategies for treatment and prevention. The second is a recent major advance that allows for the production of large amounts of recombinant canine FVIII (rcFVIII).[30] This rcFVIII has provided a much needed new treatment for bleeding as well as a consistent antigen stimulus for the inhibitor dogs to test the hypothesis that polymorphisms in FVIII protein induce inhibitors in susceptible patients.[288] In a third area of research, liver (and other organ) transplantation has provided a phenotypic cure for canine and human hemophilia.[289–295] Fourth, novel approaches that modulate FVIII gene expression are in preclinical and clinical testing.[296,297]

C. Gene Therapy in Hemophilia A Dogs

Gene transfer remains an important and active area of research and clinical investigation. All three initial human trials for gene therapy hemophilia A were stopped due to low-level expression or vector toxicity.[298–300] Two of these three trials were tested in the Chapel Hill hemophilia A dogs and their outcomes were presaged by the results of these preclinical trials.[301,302] Fortunately, no inhibitor to FVIII was detectable after gene transfer in these patients. Several gene therapy strategies have been tried using retroviral, adenoviral,[303,304,82] lentiviral,[305] and AAV vectors.[306,307] Currently, novel serotypes of AAV are showing continued progress with expression up to 20% of normal, levels that would prevent spontaneous bleeding in hemophilia A patients.[98] Importantly, a strategy that coadministered the proteasome inhibitor bortezomib with AAV vectors expressing the relatively oversized FVIII transgene have documented multi-year expression.[102] This breakthrough has the potential to expand the applicability of AAV vectors for gene therapy in general but especially for large cDNAs such as FVIII. Most importantly, all animals undergoing gene transfer continue to be monitored for the long-term (i.e., years) safety and efficacy of all of these new methods. An unexpected but a highly desirable outcome in these gene transfer studies has been transgene expression over 10 years albeit at low levels but without detectible toxicity.[168,308] These observations are particularly important when considering the occurrence of hepatocellular carcinoma and angiosarcomas in mice with the lysosomal storage disease mucopolysaccharidosis type VII (MPSVII) after treatment with rAAV vectors.[309,310] Insertional mutagenesis is probably the operative mechanism but other possibilities include overexpression of a human transgene in rodents and colony contamination by oncogenic viruses. To address this issue we are screening all AAV-treated Chapel Hill hemophilia A (or hemophilia B described below) dogs for evidence of tumor formation or other pathology at the sites of gene transfer. The availability of relevant animal models that survive for 10 years after gene therapy with a species-specific transgene has proven invaluable to assess toxicity over time.

D. Hemophilia B Dogs

The Chapel Hill strain of hemophilia B strain of dogs, maintained since 1966, was derived from the Guelph strain. The prototype animals were Keagles, a mixed beagle breed. As in hemophilia A, both the affected male and female genotypes are maintained for breeding. These animals have the severest bleeder state in the colony, with no detectable FIX:Ag in their plasma.[311] This characteristic makes them an outstanding model for gene therapy studies. The genetic defect is a missense mutation in a highly conserved residue of the catalytic domain of FIX; it results in a complete lack of detectable protein in the circulation of the affected animals. This mutation (G∧A at nucleotide 1477) results in the substitution of glutamic acid for glycine-379 (G379E), a mutation that occurs at a residue that has remained invariant in serine proteases from bacteria to humans. The profound effect of the mutation on FIX levels suggest that the glycine at position 379 (position 211 in the chymotrypsinogen numbering system) plays a critical role in the integrity of the trypsin-like serine proteases in general.[19] A second strain of hemophilia B dogs, originally described in Lhasa Apso dogs at Auburn, has a deletion mutation and is prone to inhibitor formation.[20,312] The well-described phenotype and genotype of these hemophilia B dogs make them very desirable for developing widely-used basic assays for FIX,[313] studying the pathophysiology of hemophilia B, and testing replacement therapies and gene therapy strategies. Several broad current areas of research in the hemophilia B dogs include muscle- and liver-based strategies for gene transfer and the associated immune response, novel delivery strategies, and new FIX molecules with prolonged half-life.[314,315]

E. Gene Therapy in Hemophilia B Dogs and in Humans

The first two Phase I clinical trials in humans with severe hemophilia B, utilizing muscle- or liver-directed gene therapy with AAV vectors, were safe and successfully completed after having originally been approved based on the safety and efficacy demonstrated in several animal models including the FOBRL hemophilia B dogs.[316,317,247] The intramuscular (IM) trial tested for safety, which was documented at the expense of dosing for efficacy.[317] The ultimate doses of rAAV used in the human trials were chosen after consideration of data obtained in the hemophilia B dogs.[312,318,319] Also based on studies in hemophilic dogs, the initial muscle trial was limited to subjects whose disease was due to a missense mutation (i.e., the trial excluded those with nonsense mutations or gene deletions). As observed in the dogs, none of the eight patients who received IM gene therapy has shown evidence of an inhibitor to date or any other significant side effects.[316,317] Muscle biopsies of injection sites performed 2–42 months after vector administration to patients confirmed gene transfer (as evidenced by Southern blot) and transgene expression (as evidenced by immunohistochemical staining).[246] Preexisting high-titer antibodies to AAV did not prevent gene transfer or expression. Plasma levels of FIX were mostly less than 1–2% of normal. The vector was administered by separate needle injections and higher dosing seemed impractical; in addition, studies in the dogs suggested that higher doses might induce inhibitor formation.[318] The preclinical information on vector dosing and patient selection provided by the hemophilia dogs were essential data for safely translating gene therapy to the clinic and achieving the goal of safety in these Phase 1 studies.

The liver-based trial of AAV delivery of FIX was initially promising and exciting as it resulted in FIX levels of 10–12%.[247] Unfortunately, there was a gradual decline in FIX accompanied by a transient asymptomatic elevation of liver transaminases that resolved without treatment, both of which were associated with expansion of AAV capsid-specific CD8 + T cells in the circulation. The increased presence of these cells suggested that the transduced hepatocytes were being destroyed by cell-mediated immunity-targeting antigens of the AAV capsid, an event that was not predicted by the preclinical studies in any other species including rodents, dogs, and nonhuman primates.

While these studies have been safe, the major barrier remains the immune responses to vector capsid. Further studies are in progress in the hemophilic dogs to develop a model of this immune response.[320] The development of reagents that rigorously characterize the canine immune response have enabled the identification of operative mechanisms in the hemophilia B dogs as well as the development of methods for preventing inhibitor formation.[207,311,318,320] Future liver-based studies in humans may require optimization of the expression cassette[250] or immunomodulation to achieve long-term expression. The US Food and Drug Administration (FDA) has recently approved an immunomodulation protocol. As with hemophilia A, there is a significant need for additional preclinical and clinical studies to achieve successful gene therapy of hemophilia B in humans.

F. Development of Alternative Treatment Modalities in Hemophilia B Dogs

Related to gene therapy, novel delivery strategies for muscle-based gene transfer have been studied in the hemophilia B dogs.[270,268,321] These methods have achieved muscle transduction in entire limbs and reasonable expression levels of at least 10%. This novel method also holds great promise for treating many genetic disorders including muscle disorders such as muscular dystrophy.

Furthermore, novel recombinant FIX preparations are being tested that have significantly prolonged half lives.[314,322] If shown to be safe and efficacious in human trials, the immediate practical outcome is that hemophilia B patients may be able to go from twice a week intravenous injections to weekly or monthly injections, possibly even subcutaneously.[323,324] For a patient or a parent of a young child with hemophilia B, the improvement in quality of life associated with less frequent needle punctures would be profound.

Finally, as with hemophilia A, organ transplantation produces a phenotypic cure.[325–327] While curative, this treatment option is limited by the number of donors and the need for immunosuppression to prevent rejection of the transplanted organ.

G. Normal Donor Dogs

Normal dogs are maintained as a "walking blood bank" with compatible blood types for preparation of replacement products for treatment of bleeding. The affected dogs of all strains are severe bleeders with an average of approximately 5 spontaneous bleeds per year.[168,279,324] Clinical bleeding requires immediate replacement therapy with pooled normal canine plasma for the hemophilic animals. Without transfusion therapy hemophilia A and B are fatal at an early age. The new capacity to produce rcFVIII has been an important addition to the treatment options for the hemophilia A dogs.[30] At present, recombinant canine FIX is not available in sufficient quantities for treatment. The maintenance and breeding programs require constant close attention by specially trained personnel for the detection of early hemorrhages and their management.

H. Summary of the Benefits of Research with Large Animal Models of Hemophilia

Work at the FOBRL over the past six decades and in other laboratories includes studies in dogs and humans with hemophilia that have contributed to considerable progress in the basic understanding of these diseases and the development of safe and successful treatments. But their development has not been without setbacks, and limitations persist. For hemophilia patients, antihemophilia products are expensive (typically $100,000 per year or more for an adult) and not readily available in less developed countries.[328] For most patients worldwide, replacement therapy, if available at all, is usually used to treat acute bleeding on demand rather than to provide the benefits of continuous prophylactic coverage.[42,329] Furthermore, the products' half-life in the circulation is relatively short and quite variable among patients (e.g., 8–23 h for FVIII), mandating repeated venipunctures or port implantation for prophylactic administration.[263,330] Last, for the plasma-derived material, despite the introduction of viral inactivation techniques that have eliminated the risk of HIV transmission, there are ongoing concerns about other transmissible agents (e.g., prions) that may not be eliminated.[331–336] While there has been considerable progress in gene therapy for hemophilia, several barriers remain that are being addressed with the dogs with hemophilia A and hemophilia B.[337] The success with correction of defects due to single genetic mutations such as hemophilia dogs may presage great expansion of the role of gene therapy in clinical medicine, both for genetic and acquired diseases. Additional large animals models (such as canine and ovine) are available and are being further characterized to aid in this task.[21,338] For example, the recently reestablished line of hemophilia A sheep shows a severe bleeding phenotype with no circulating FVIII antigen due to a premature stop codon and frameshift in exon 14.[21] Sheep models have been particularly useful for *in utero* gene and gene therapy studies, thereby adding to the range of potential studies in hemophilic animals for development of novel approaches to therapy.

References

1. Antonarakis SE, Kazazian HH, Gitschier J, Hutter P, de Moerloose P, Morris MA. Molecular etiology of factor VIII deficiency in hemophilia A. *Adv Exp Med Biol* 1995;**386**:19–34.
2. Elder B, Lakich D, Gitschier J. Sequence of the murine factor VIII cDNA. *Genomics* 1993;**16**:374–9.
3. Bi L, Lawler AM, Antonarakis SE, High KA, Gearhart JD, Kazazian Jr. HH. Targeted disruption of the mouse factor VIII gene produces a model of haemophilia A. *Nat Genet* 1995;**10**:119–21.
4. Waters B, Qadura M, Burnett E, Chegeni R, Labelle A, Thompson P, et al. Anti-CD3 prevents factor VIII inhibitor development in hemophilia A mice by a regulatory CD4+CD25+-dependent mechanism and by shifting cytokine production to favor a Th1 response. *Blood* 2009;**113**:193–203.
5. Bril WS, van Helden PM, Hausl C, Zuurveld MG, Ahmad RU, Hollestelle MJ, et al. Tolerance to factor VIII in a transgenic mouse expressing human factor VIII cDNA carrying an Arg(593) to Cys substitution. *Thromb Haemost* 2006;**95**:341–7.
6. Reipert BM, Steinitz KN, van Helden PM, Unterthurner S, Schuster M, Ahmad RU, et al. Opportunities and limitations of mouse models humanized for HLA class II antigens. *J Thromb Haemost* 2009;**7**(Suppl. 1):92–7.
7. Lin HF, Maeda N, Smithies O, Straight DL, Stafford DW. A coagulation factor IX-deficient mouse model for human hemophilia B. *Blood* 1997;**90**:3962–6.
8. Jin DY, Zhang TP, Gui T, Stafford DW, Monahan PE. Creation of a mouse expressing defective human factor IX. *Blood* 2004;**104**:1733–9.
9. Hu G, Monahan PE, Gui T, Velander WH, Stafford DW. Examination of differential T regulatory lymphocyte responses following human fix gene therapy in a novel mouse model expressing defective human factor IX compared to factor IX knockout mice. *Haemophilia* 2010;**16**:93.
10. Sabatino DE, Armstrong E, Edmonson S, Liu YL, Pleimes M, Schuettrumpf J, et al. Novel hemophilia B mouse models exhibiting a range of mutations in the Factor IX gene. *Blood* 2004;**104**:2767–74.

11. Cao O, Hoffman BE, Moghimi B, Nayak S, Cooper M, Zhou S, et al. Impact of the underlying mutation and the route of vector administration on immune responses to factor IX in gene therapy for hemophilia B. *Mol Ther* 2009;**17**:1733–42.
12. Booth CJ, Brooks MB, Rockwell S. Spontaneous coagulopathy in inbred WAG/RijYcb rats. *Comp Med* 2010;**60**:25–30.
13. Booth CJ, Brooks MB, Rockwell S, Murphy JW, Rinder HM, Zelterman D, et al. WAG-F8 (m1Ycb) rats harboring a factor VIII gene mutation provide a new animal model for hemophilia A. *J Thromb Haemost* 2010;**8**:2472–7.
14. Graham JB, Buckwalter JA, Hartley LJ, Brinkhous KM. Canine hemophilia: observations on the course, the clotting anomaly, and the effects of blood transfusion. *J Exp Med* 1949;**90**:97–111.
15. Lozier JN, Dutra A, Pak E, Zhou N, Zheng Z, Nichols TC, et al. The Chapel Hill hemophilia A dog colony exhibits a factor VIII gene inversion. *Proc Natl Acad Sci USA* 2002;**99**:12991–6.
16. Hough C, Kamisue S, Cameron C, Notley C, Tinlin S, Giles A, et al. Aberrant splicing and premature termination of transcription of the FVIII gene as a cause of severe canine hemophilia A: similarities with the intron 22 inversion mutation in human hemophilia. *Thromb Haemost* 2002;**87**:659–65.
17. Connelly S, Gardner JM, Lyons RM, McClelland A, Kaleko M. Sustained expression of therapeutic levels of human factor VIII in mice. *Blood* 1996;**87**:4671–7.
18. Connelly S, Gardner JM, McClelland A, Kaleko M. High-level tissue-specific expression of functional human factor VIII in mice. *Hum Gene Ther* 1996;**7**:183–95.
19. Evans JP, Brinkhous KM, Brayer GD, Reisner HM, High KA. Canine Hemophilia B resulting from a point mutation with unusual consequences. *Proc Natl Acad Sci USA* 1989;**86**:10095–9.
20. Mauser AE, Whitlark J, Whitney KM, Lothrop Jr. CD. A deletion mutation causes hemophilia B in Lhasa Apso dogs. *Blood* 1996;**88**:3451–5.
21. Porada CD, Sanada C, Long CR, Wood JA, Desai J, Frederick N, et al. Clinical and molecular characterization of a re-established line of sheep exhibiting hemophilia A. *J Thromb Haemost* 2009;**8**:276–85.
22. Brinkhous KM, Graham JB. Hemophilia in the female dog. *Science* 1950;**111**:723–4.
23. Gitschier J, Wood WI, Goralka TM, Wion KL, Chen EY, Eaton DH, et al. Characterization of the human factor VIII gene. *Nature* 1984;**312**:326–30.
24. Bi L, Sarkar R, Naas T, Lawler AM, Pain J, Shumaker SL, et al. Further characterization of factor VIII-deficient mice created by gene targeting: RNA and protein studies. *Blood* 1996;**88**:3446–50.
25. Sarkar R, Gao GP, Chirmule N, Tazelaar J, Kazazian Jr. HH. Partial correction of murine hemophilia A with neo-antigenic murine factor VIII. *Hum Gene Ther* 2000;**11**:881–94.
26. Muchitsch EM, Turecek PL, Zimmermann K, Pichler L, Auer W, Richter G, et al. Phenotypic expression of murine hemophilia. *Thromb Haemost* 1999;**82**:1371–3.
27. Gilles JG, Vanzieleghem B, Saint-Remy JM. Animal models to explore mechanisms of tolerance induction to FVIII: SCID mice and SCID-FVIII-KO mice. *Haematologica* 2000;**85**:103–6 (Discussion: 106–107).
28. Doering C, Parker ET, Healey JF, Craddock HN, Barrow RT, Lollar P. Expression and characterization of recombinant murine factor VIII. *Thromb Haemost* 2002;**88**:450–8.
29. Doering CB, Healey JF, Parker ET, Barrow RT, Lollar P. High level expression of recombinant porcine coagulation factor VIII. *J Biol Chem* 2002;**277**:38345–9.
30. Sabatino DE, Freguia CF, Toso R, Santos A, Merricks EP, Kazazian Jr. HH, et al. Recombinant canine B-domain-deleted FVIII exhibits high specific activity and is safe in the canine hemophilia A model. *Blood* 2009;**114**:4562–5.
31. Furie B, Furie BC. In vivo thrombus formation. *J Thromb Haemost* 2007;**5**(Suppl. 1):12–7.

32. Broze Jr. GJ, Yin ZF, Lasky N. A tail vein bleeding time model and delayed bleeding in hemophiliac mice. *Thromb Haemost* 2001;**85**:747–8.
33. Greene TK, Schiviz A, Hoellriegl W, Poncz M, Muchitsch EM. Towards a standardization of the murine tail bleeding model. *J Thromb Haemost* 2010;**8**:2820–2.
34. Falati S, Gross P, Merrill-Skoloff G, Furie BC, Furie B. Real-time in vivo imaging of platelets, tissue factor and fibrin during arterial thrombus formation in the mouse. *Nat Med* 2002;**8**:1175–81.
35. Farrehi PM, Ozaki CK, Carmeliet P, Fay WP. Regulation of arterial thrombolysis by plasminogen activator inhibitor-1 in mice. *Circulation* 1998;**97**:1002–8.
36. Manco-Johnson MJ, Nuss R, Geraghty S, Funk S, Kilcoyne R. Results of secondary prophylaxis in children with severe hemophilia. *Am J Hematol* 1994;**47**:113–7.
37. Aledort LM, Haschmeyer RH, Pettersson H. A longitudinal study of orthopaedic outcomes for severe factor-VIII-deficient haemophiliacs. The Orthopaedic Outcome Study Group. *J Intern Med* 1994;**236**:391–9.
38. Valentino LA, Hakobyan N. Histological changes in murine haemophilic synovitis: a quantitative grading system to assess blood-induced synovitis. *Haemophilia* 2006;**12**:654–62.
39. Valentino LA, Hakobyan N, Kazarian T, Jabbar KJ, Jabbar AA. Experimental haemophilic synovitis: rationale and development of a murine model of human factor VIII deficiency. *Haemophilia* 2004;**10**:280–7.
40. Mejia-Carvajal C, Hakobyan N, Enockson C, Valentino LA. The impact of joint bleeding and synovitis on physical ability and joint function in a murine model of haemophilic synovitis. *Haemophilia* 2008;**14**:119–26.
41. Hakobyan N, Enockson C, Cole AA, Sumner DR, Valentino LA. Experimental haemophilic arthropathy in a mouse model of a massive haemarthrosis: gross, radiological and histological changes. *Haemophilia* 2008;**14**:804–9.
42. Manco-Johnson MJ, Abshire TC, Shapiro AD, Riske B, Hacker MR, Kilcoyne R, et al. Prophylaxis versus episodic treatment to prevent joint disease in boys with severe hemophilia. *N Engl J Med* 2007;**357**:535–44.
43. Reding MT, Wu H, Krampf M, Okita DK, Diethelm-Okita BM, Key NS, et al. CD4 + T cell response to factor VIII in hemophilia A, acquired hemophilia, and healthy subjects. *Thromb Haemost* 1999;**82**:509–15.
44. Singer ST, Addiego Jr. JE, Reason DC, Lucas AH. T lymphocyte proliferative responses induced by recombinant factor VIII in hemophilia A patients with inhibitors. *Thromb Haemost* 1996;**76**:17–22.
45. Qian J, Borovok M, Bi L, Kazazian Jr. HH, Hoyer LW. Inhibitor antibody development and T cell response to human factor VIII in murine hemophilia A. *Thromb Haemost* 1999;**81**:240–4.
46. Wu H, Reding M, Qian J, Okita DK, Parker E, Lollar P, et al. Mechanism of the immune response to human factor VIII in murine hemophilia A. *Thromb Haemost* 2001;**85**:125–33.
47. Qadura M, Waters B, Burnett E, Chegeni R, Hough C, Othman M, et al. Immunoglobulin isotypes and functional anti-FVIII antibodies in response to FVIII treatment in Balb/c and C57BL/6 haemophilia A mice. *Haemophilia* 2010;**17**:288–95.
48. Santos A, Annichino-Bizzacchi JM, Ozelo MC. Inhibitors of factor VIII in hemophilia. *N Engl J Med* 2009;**361**:309–10 (Author reply 310).
49. Hay CR, Ollier W, Pepper L, Cumming A, Keeney S, Goodeve AC, et al. HLA class II profile: a weak determinant of factor VIII inhibitor development in severe haemophilia A. UKHCDO Inhibitor Working Party. *Thromb Haemost* 1997;**77**:234–7.
50. van Helden PM, Unterthurner S, Hermann C, Schuster M, Ahmad RU, Schiviz AN, et al. Maintenance and break of immunological tolerance against human factor VIII in a new transgenic hemophilic mouse model. *Blood* 2011;**118**:3698–707.

51. Healey JF, Parker ET, Barrow RT, Langley TJ, Church WR, Lollar P. The humoral response to human factor VIII in hemophilia A mice. *J Thromb Haemost* 2007;**5**:512–9.
52. Lollar P. Molecular characterization of the immune response to factor VIII. *Vox Sang* 2002;**83** (Suppl. 1):403–8.
53. Parker ET, Healey JF, Barrow RT, Craddock HN, Lollar P. Reduction of the inhibitory antibody response to human factor VIII in hemophilia A mice by mutagenesis of the A2 domain B-cell epitope. *Blood* 2004;**104**:704–10.
54. Behrmann M, Pasi J, Saint-Remy JM, Kotitschke R, Kloft M. Von Willebrand factor modulates factor VIII immunogenicity: comparative study of different factor VIII concentrates in a haemophilia A mouse model. *Thromb Haemost* 2002;**88**:221–9.
55. Delignat S, Dasgupta S, Andre S, Navarrete AM, Kaveri SV, Bayry J, et al. Comparison of the immunogenicity of different therapeutic preparations of human factor VIII in the murine model of hemophilia A. *Haematologica* 2007;**92**:1423–6.
56. Barrow RT, Lollar P. Neutralization of antifactor VIII inhibitors by recombinant porcine factor VIII. *J Thromb Haemost* 2006;**4**:2223–9.
57. Hay CR. Porcine factor VIII: current status and future developments. *Haemophilia* 2002;**8** (Suppl. 1):24–7 (Discussion: 28–32).
58. Parker ET, Craddock HN, Barrow RT, Lollar P. Comparative immunogenicity of recombinant B domain-deleted porcine factor VIII and Hyate:C in hemophilia A mice presensitized to human factor VIII. *J Thromb Haemost* 2004;**2**:605–11.
59. Healey JF, Parker ET, Barrow RT, Langley TJ, Church WR, Lollar P. The comparative immunogenicity of human and porcine factor VIII in haemophilia A mice. *Thromb Haemost* 2009;**102**:35–41.
60. Yatuv R, Robinson M, Dayan-Tarshish I, Baru M. The use of PEGylated liposomes in the development of drug delivery applications for the treatment of hemophilia. *Int J Nanomedicine* 2010;**5**:581–91.
61. Mei B, Pan C, Jiang H, Tjandra H, Strauss J, Chen Y, et al. Rational design of a fully active, long-acting PEGylated factor VIII for hemophilia A treatment. *Blood* 2010;**116**:270–9.
62. Baru M, Carmel-Goren L, Barenholz Y, Dayan I, Ostropolets S, Slepoy I, et al. Factor VIII efficient and specific non-covalent binding to PEGylated liposomes enables prolongation of its circulation time and haemostatic efficacy. *Thromb Haemost* 2005;**93**:1061–8.
63. Dayan I, Robinson M, Baru M. Enhancement of haemostatic efficacy of plasma-derived FVIII by formulation with PEGylated liposomes. *Haemophilia* 2009;**15**:1006–13.
64. Pan J, Liu T, Kim JY, Zhu D, Patel C, Cui ZH, et al. Enhanced efficacy of recombinant FVIII in noncovalent complex with PEGylated liposome in hemophilia A mice. *Blood* 2009;**114**:2802–11.
65. Spira J, Plyushch OP, Andreeva TA, Andreev Y. Prolonged bleeding-free period following prophylactic infusion of recombinant factor VIII reconstituted with pegylated liposomes. *Blood* 2006;**108**:3668–73.
66. Dumont JA, Kamphaus GD, Fraley C, Ashworth T, Franck HG, Merricks EP, et al. Factor VIII-Fc fusion protein shows extended half-life and hemostatic activity in hemophilia a dogs. *Blood* 2009;**114**(22):228 (In: American Society of Hematology Annual Meeting).
67. Brackmann HH, Oldenburg J, Schwaab R. Immune tolerance for the treatment of factor VIII inhibitors—twenty years' 'bonn protocol'. *Vox Sang* 1996;**70**(Suppl. 1):30–5.
68. Hausl C, Ahmad RU, Sasgary M, Doering CB, Lollar P, Richter G, et al. High-dose factor VIII inhibits factor VIII-specific memory B cells in hemophilia A with factor VIII inhibitors. *Blood* 2005;**106**:3415–22.
69. Weiner HL. Oral tolerance for the treatment of autoimmune diseases. *Annu Rev Med* 1997;**48**:341–51.

70. Chen Y, Inobe J, Marks R, Gonnella P, Kuchroo VK, Weiner HL. Peripheral deletion of antigen-reactive T cells in oral tolerance. *Nature* 1995;**376**:177–80.
71. Chen Y, Kuchroo VK, Inobe J, Hafler DA, Weiner HL. Regulatory T cell clones induced by oral tolerance: suppression of autoimmune encephalomyelitis. *Science* 1994;**265**:1237–40.
72. Rawle FE, Pratt KP, Labelle A, Weiner HL, Hough C, Lillicrap D. Induction of partial immune tolerance to factor VIII through prior mucosal exposure to the factor VIII C2 domain. *J Thromb Haemost* 2006;**4**:2172–9.
73. Lei TC, Scott DW. Induction of tolerance to factor VIII inhibitors by gene therapy with immunodominant A2 and C2 domains presented by B cells as Ig fusion proteins. *Blood* 2005;**105**:4865–70.
74. Scott DW. Gene therapy for immunological tolerance: using 'transgenic' B cells to treat inhibitor formation. *Haemophilia* 2010;**16**:89–94.
75. Ragni MV, Wu W, Liang X, Hsieh CC, Cortese-Hassett A, Lu L. Factor VIII-pulsed dendritic cells reduce anti-factor VIII antibody formation in the hemophilia A mouse model. *Exp Hematol* 2009;**37**:744–54.
76. Su RJ, Epp A, Latchman Y, Bolgiano D, Pipe SW, Josephson NC. Suppression of FVIII inhibitor formation in hemophilic mice by delivery of transgene modified apoptotic fibroblasts. *Mol Ther* 2010;**18**:214–22.
77. Moghimi B, Sack BK, Nayak S, Markusic DM, Mah CS, Herzog RW. Tolerance induction to factor VIII by transient co-administration with rapamycin. *J Thromb Haemost* 2011;**9**:1524–33.
78. Hollestelle MJ, Thinnes T, Crain K, Stiko A, Kruijt JK, van Berkel TJ, et al. Tissue distribution of factor VIII gene expression in vivo—a closer look. *Thromb Haemost* 2001;**86**:855–61.
79. Chuah MK, Collen D, VandenDriessche T. Clinical gene transfer studies for hemophilia A. *Semin Thromb Hemost* 2004;**30**:249–56.
80. Connelly S, Andrews JL, Gallo AM, Kayda DB, Qian J, Hoyer L, et al. Sustained phenotypic correction of murine hemophilia A by in vivo gene therapy. *Blood* 1998;**91**:3273–81.
81. Gallo-Penn AM, Shirley PS, Andrews JL, Kayda DB, Pinkstaff AM, Kaloss M, et al. In vivo evaluation of an adenoviral vector encoding canine factor VIII: high-level, sustained expression in hemophiliac mice. *Hum Gene Ther* 1999;**10**:1791–802.
82. Brown BD, Shi CX, Powell S, Hurlbut D, Graham FL, Lillicrap D. Helper-dependent adenoviral vectors mediate therapeutic factor VIII expression for several months with minimal accompanying toxicity in a canine model of severe hemophilia A. *Blood* 2004;**103**:804–10.
83. Cerullo V, Seiler MP, Mane V, Cela R, Clarke C, Kaufman RJ, et al. Correction of murine hemophilia A and immunological differences of factor VIII variants delivered by helper-dependent adenoviral vectors. *Mol Ther* 2007;**15**:2080–7.
84. Rawle FE, Shi CX, Brown B, McKinven A, Tinlin S, Graham FL, et al. Heterogeneity of the immune response to adenovirus-mediated factor VIII gene therapy in different inbred hemophilic mouse strains. *J Gene Med* 2004;**6**:1358–68.
85. Reddy PS, Sakhuja K, Ganesh S, Yang L, Kayda D, Brann T, et al. Sustained human factor VIII expression in hemophilia A mice following systemic delivery of a gutless adenoviral vector. *Mol Ther* 2002;**5**:63–73.
86. McCormack Jr. WM, Seiler MP, Bertin TK, Ubhayakar K, Palmer DJ, Ng P, et al. Helper-dependent adenoviral gene therapy mediates long-term correction of the clotting defect in the canine hemophilia A model. *J Thromb Haemost* 2006;**4**:1218–25.
87. Gao GP, Alvira MR, Wang L, Calcedo R, Johnston J, Wilson JM. Novel adeno-associated viruses from rhesus monkeys as vectors for human gene therapy. *Proc Natl Acad Sci USA* 2002;**99**:11854–9.

88. Ward NJ, Buckley SM, Waddington SN, Vandendriessche T, Chuah MK, Nathwani AC, et al. Codon optimization of human factor VIII cDNAs leads to high-level expression. *Blood* 2011;**117**:798–807.
89. Kwon I, Schaffer DV. Designer gene delivery vectors: molecular engineering and evolution of adeno-associated viral vectors for enhanced gene transfer. *Pharm Res* 2008;**25**:489–99.
90. Wright JF. New adeno-associated virus strategies to support momentum in the clinic. *Hum Gene Ther* 2011;**22**:519–21.
91. Gnatenko DV, Saenko EL, Jesty J, Cao LX, Hearing P, Bahou WF. Human factor VIII can be packaged and functionally expressed in an adeno-associated virus background: applicability to haemophilia A gene therapy. *Br J Haematol* 1999;**104**:27–36.
92. Herzog RW, High KA. Adeno-associated virus-mediated gene transfer of factor IX for treatment of hemophilia B by gene therapy. *Thromb Haemost* 1999;**82**:540–6.
93. Yonemura H, Sugawara K, Nakashima K, Nakahara Y, Hamamoto T, Mimaki I, et al. Efficient production of recombinant human factor VIII by co-expression of the heavy and light chains. *Protein Eng* 1993;**6**:669–74.
94. Chao H, Sun L, Bruce A, Xiao X, Walsh CE. Expression of human factor VIII by splicing between dimerized AAV vectors. *Mol Ther* 2002;**5**:716–22.
95. Burton M, Nakai H, Colosi P, Cunningham J, Mitchell R, Couto L. Coexpression of factor VIII heavy and light chain adeno-associated viral vectors produces biologically active protein. *Proc Natl Acad Sci USA* 1999;**96**:12725–30.
96. Sarkar R, Tetreault R, Gao G, Wang L, Bell P, Chandler R, et al. Total correction of hemophilia A mice with canine FVIII using an AAV 8 serotype. *Blood* 2004;**103**:1253–60.
97. Scallan CD, Liu T, Parker AE, Patarroyo-White SL, Chen H, Jiang H, et al. Phenotypic correction of a mouse model of hemophilia A using AAV2 vectors encoding the heavy and light chains of FVIII. *Blood* 2003;**102**:3919–26.
98. Sabatino DE, Lange AM, Altynova ES, Sarkar R, Zhou S, Merricks EP, et al. Efficacy and safety of long-term prophylaxis in severe hemophilia A dogs following liver gene therapy using AAV vectors. *Mol Ther* 2011;**19**:442–9.
99. Sarkar R, Mucci M, Addya S, Tetreault R, Bellinger DA, Nichols TC, et al. Long-term efficacy of adeno-associated virus serotypes 8 and 9 in hemophilia A dogs and mice. *Hum Gene Ther* 2006;**17**:427–39.
100. Lu H, Chen L, Wang J, Huack B, Sarkar R, Zhou S, et al. Complete correction of hemophilia A with adeno-associated viral vectors containing a full-size expression cassette. *Hum Gene Ther* 2008;**19**:648–54.
101. Jiang H, Lillicrap D, Patarroyo-White S, Liu T, Qian X, Scallan CD, et al. Multi-year therapeutic benefit of AAV serotypes 2, 6 and 8 delivering factor VIII to hemophilia A mice and dogs. *Blood* 2006;**108**:107–15.
102. Monahan PE, Lothrop CD, Sun J, Hirsch ML, Kafri T, Kantor B, et al. Proteasome inhibitors enhance gene delivery by AAV virus vectors expressing large genomes in hemophilia mouse and dog models: a strategy for broad clinical application. *Mol Ther* 2010;**18**:1907–16.
103. Cao O, Dobrzynski E, Wang L, Nayak S, Mingle B, Terhorst C, et al. Induction and role of regulatory CD4+CD25+ T cells in tolerance to the transgene product following hepatic in vivo gene transfer. *Blood* 2007;**110**:1132–40.
104. Mingozzi F, High KA. Therapeutic in vivo gene transfer for genetic disease using AAV: progress and challenges. *Nat Rev Genet* 2011;**12**:341–55.
105. VandenDriessche T, Vanslembrouck V, Goovaerts I, Zwinnen H, Vanderhaeghen ML, Collen D, et al. Long-term expression of human coagulation factor VIII and correction of hemophilia A after in vivo retroviral gene transfer in factor VIII-deficient mice. *Proc Natl Acad Sci USA* 1999;**96**:10379–84.

106. Xu L, Mei M, Ma X, Ponder KP. High expression reduces an antibody response after neonatal gene therapy with B domain-deleted human factor VIII in mice. *J Thromb Haemost* 2007;**5**:1805–12.
107. Ye P, Thompson AR, Sarkar R, Shen Z, Lillicrap DP, Kaufman RJ, et al. Naked DNA transfer of Factor VIII induced transgene-specific, species-independent immune response in hemophilia A mice. *Mol Ther* 2004;**10**:117–26.
108. Miao CH, Ye P, Thompson AR, Rawlings DJ, Ochs HD. Immunomodulation of transgene responses following naked DNA transfer of human factor VIII into hemophilia A mice. *Blood* 2006;**108**:19–27.
109. Peng B, Ye P, Blazar BR, Freeman GJ, Rawlings DJ, Ochs HD, et al. Transient blockade of the inducible costimulator pathway generates long-term tolerance to factor VIII after nonviral gene transfer into hemophilia A mice. *Blood* 2008;**112**:1662–72.
110. Peng B, Ye P, Rawlings DJ, Ochs HD, Miao CH. Anti-CD3 antibodies modulate anti-factor VIII immune responses in hemophilia A mice after factor VIII plasmid-mediated gene therapy. *Blood* 2009;**114**:4373–82.
111. Miao CH, Harmeling BR, Ziegler SF, Yen BC, Torgerson T, Chen L, et al. CD4+FOXP3+ regulatory T cells confer long-term regulation of factor VIII-specific immune responses in plasmid-mediated gene therapy-treated hemophilia mice. *Blood* 2009;**114**:4034–44.
112. Liu CL, Ye P, Yen BC, Miao CH. In vivo expansion of regulatory T cells with IL-2/IL-2 mAb complexes prevents anti-factor VIII immune responses in hemophilia A mice treated with factor VIII plasmid-mediated gene therapy. *Mol Ther* 2011;**19**:1511–20.
113. Ohlfest JR, Frandsen JL, Fritz S, Lobitz PD, Perkinson SG, Clark KJ, et al. Phenotypic correction and long-term expression of factor VIII in hemophilic mice by immunotolerization and nonviral gene transfer using the Sleeping Beauty transposon system. *Blood* 2005;**105**:2691–8.
114. Kren BT, Unger GM, Sjeklocha L, Trossen AA, Korman V, Diethelm-Okita BM, et al. Nanocapsule-delivered Sleeping Beauty mediates therapeutic Factor VIII expression in liver sinusoidal endothelial cells of hemophilia A mice. *J Clin Invest* 2009;**119**:2086–99.
115. Chuah MK, Brems H, Vanslembrouck V, Collen D, Vandendriessche T. Bone marrow stromal cells as targets for gene therapy of hemophilia A. *Hum Gene Ther* 1998;**9**:353–65.
116. Dwarki VJ, Belloni P, Nijjar T, Smith J, Couto L, Rabier M, et al. Gene therapy for hemophilia A: production of therapeutic levels of human factor VIII in vivo in mice. *Proc Natl Acad Sci USA* 1995;**92**:1023–7.
117. Evans GL, Morgan RA. Genetic induction of immune tolerance to human clotting factor VIII in a mouse model for hemophilia A. *Proc Natl Acad Sci USA* 1998;**95**:5734–9.
118. Hoeben RC, Einerhand MP, Briet E, van Ormondt H, Valerio D, van der Eb AJ. Toward gene therapy in haemophilia A: retrovirus-mediated transfer of a factor VIII gene into murine haematopoietic progenitor cells. *Thromb Haemost* 1992;**67**:341–5.
119. Kootstra NA, Matsumura R, Verma IM. Efficient production of human FVIII in hemophilic mice using lentiviral vectors. *Mol Ther* 2003;**7**:623–31.
120. Moayeri M, Hawley TS, Hawley RG. Correction of murine hemophilia A by hematopoietic stem cell gene therapy. *Mol Ther* 2005;**12**:1034–42.
121. Moayeri M, Ramezani A, Morgan RA, Hawley TS, Hawley RG. Sustained phenotypic correction of hemophilia a mice following oncoretroviral-mediated expression of a bioengineered human factor VIII gene in long-term hematopoietic repopulating cells. *Mol Ther* 2004;**10**:892–902.
122. Gangadharan B, Parker ET, Ide LM, Spencer HT, Doering CB. High-level expression of porcine factor VIII from genetically modified bone marrow-derived stem cells. *Blood* 2006;**107**:3859–64.

123. Ide LM, Gangadharan B, Chiang KY, Doering CB, Spencer HT. Hematopoietic stem-cell gene therapy of hemophilia A incorporating a porcine factor VIII transgene and nonmyeloablative conditioning regimens. *Blood* 2007;**110**:2855–63.
124. Yarovoi HV, Kufrin D, Eslin DE, Thornton MA, Haberichter SL, Shi Q, et al. Factor VIII ectopically expressed in platelets: efficacy in hemophilia A treatment. *Blood* 2003;**102**:4006–13.
125. Shi Q, Wilcox DA, Fahs SA, Weiler H, Wells CW, Cooley BC, et al. Factor VIII ectopically targeted to platelets is therapeutic in hemophilia A with high-titer inhibitory antibodies. *J Clin Invest* 2006;**116**:1974–82.
126. Shi Q, Wilcox DA, Fahs SA, Fang J, Johnson BD, Du LM, et al. Lentivirus-mediated platelet-derived factor VIII gene therapy in murine haemophilia A. *J Thromb Haemost* 2007;**5**:352–61.
127. Shi Q, Fahs SA, Wilcox DA, Kuether EL, Morateck PA, Mareno N, et al. Syngeneic transplantation of hematopoietic stem cells that are genetically modified to express factor VIII in platelets restores hemostasis to hemophilia A mice with preexisting FVIII immunity. *Blood* 2008;**112**:2713–21.
128. Du LM, Franck HG, Merricks EP, Nurden P, Jensen ES, Haberichter SL, et al. De novo synthesis and storage of human factor VIII in platelets reduces bleeding in canine hemophilia A. *Blood* 2010;**116**:906 (In: American Society of Hematology Annual Meeting).
129. Lin Y, Chang L, Solovey A, Healey JF, Lollar P, Hebbel RP. Use of blood outgrowth endothelial cells for gene therapy for hemophilia A. *Blood* 2002;**99**:457–62.
130. Matsui H, Shibata M, Brown B, Labelle A, Hegadorn C, Andrews C, et al. Ex vivo gene therapy for hemophilia A that enhances safe delivery and sustained in vivo factor VIII expression from lentivirally engineered endothelial progenitors. *Stem Cells* 2007;**25**:2660–9.
131. van den Biggelaar M, Bouwens EA, Kootstra NA, Hebbel RP, Voorberg J, Mertens K. Storage and regulated secretion of factor VIII in blood outgrowth endothelial cells. *Haematologica* 2009;**94**:670–8.
132. Follenzi A, Benten D, Novikoff P, Faulkner L, Raut S, Gupta S. Transplanted endothelial cells repopulate the liver endothelium and correct the phenotype of hemophilia A mice. *J Clin Invest* 2008;**118**:935–45.
133. Kumaran V, Benten D, Follenzi A, Joseph B, Sarkar R, Gupta S. Transplantation of endothelial cells corrects the phenotype in hemophilia A mice. *J Thromb Haemost* 2005;**3**:2022–31.
134. Finn JD, Ozelo MC, Sabatino DE, Franck HW, Merricks EP, Crudele JM, et al. Eradication of neutralizing antibodies to factor VIII in canine hemophilia A after liver gene therapy. *Blood* 2010;**116**:5842–8.
135. Bunce MW, Toso R, Camire RM. Zymogen-like factor Xa variants restore thrombin generation and effectively bypass the intrinsic pathway in vitro. *Blood* 2011;**117**:290–8.
136. Aljamali MN, Margaritis P, Schlachterman A, Tai SJ, Roy E, Bunte R, et al. Long-term expression of murine activated factor VII is safe, but elevated levels cause premature mortality. *J Clin Invest* 2008;**118**:1825–34.
137. Margaritis P, Arruda VR, Aljamali M, Camire RM, Schlachterman A, High KA. Novel therapeutic approach for hemophilia using gene delivery of an engineered secreted activated Factor VII. *J Clin Invest* 2004;**113**:1025–31.
138. Margaritis P, Roy E, Aljamali MN, Downey HD, Giger U, Zhou S, et al. Successful treatment of canine hemophilia by continuous expression of canine FVIIa. *Blood* 2009;**113**:3682–9.
139. Monahan PE. The expanding menagerie: animal models of hemophilia A. *J Thromb Haemost* 2010;**8**:2469–71.
140. Pavlovsky A. Contribution to the pathogenesis of hemophilia. *Blood* 1947;**2**:185–91.
141. Biggs R, Douglas AS, Macfarlane RG, Dacie JV, Pitney WR, Merskey . Christmas disease: a condition previously mistaken for haemophilia. *Br Med J* 1952;**2**:1378–82.

142. Aggeler PM, White SG, Glendening MB, Page EW, Leake TB, Bates G. Plasma thromboplastin component (PTC) deficiency; a new disease resembling hemophilia. *Proc Soc Exp Biol Med* 1952;**79**:692–4.
143. Monahan PE, Di Paola J. Recombinant factor IX for clinical and research use. *Semin Thromb Hemost* 2010;**36**:498–509.
144. Kaufman RJ. Post-translational modifications required for coagulation factor secretion and function. *Thromb Haemost* 1998;**79**:1068–79.
145. Makris M. Is VIII worse than IX? *Blood* 2009;**114**:750–1.
146. Sarkar G, Koeberl DD, Sommer SS. Direct sequencing of the activation peptide and the catalytic domain of the factor IX gene in six species. *Genomics* 1990;**6**:133–43.
147. Monahan PE. Factor IX: insights from knock-out and genetically engineered mice. *Thromb Haemost* 2008;**100**:563–75.
148. Emeis JJ, Jirouskova M, Muchitsch EM, Shet AS, Smyth SS, Johnson GJ. A guide to murine coagulation factor structure, function, assays, and genetic alterations. *J Thromb Haemost* 2007;**5**:670–9.
149. Matsuo O, Lijnen HR, Ueshima S, Kojima S, Smyth SS. A guide to murine fibrinolytic factor structure, function, assays, and genetic alterations. *J Thromb Haemost* 2007;**5**:680–9.
150. Jirouskova M, Shet AS, Johnson GJ. A guide to murine platelet structure, function, assays, and genetic alterations. *J Thromb Haemost* 2007;**5**:661–9.
151. Brinkhous KM, Sigman JL, Read MS, Stewart PF, McCarthy KP, Timony GA, et al. Recombinant human factor IX: replacement therapy, prophylaxis, and pharmacokinetics in canine hemophilia B. *Blood* 1996;**88**:2603–10.
152. Wang L, Zoppe M, Hackeng TM, Griffin JH, Lee KF, Verma IM. A factor IX-deficient mouse model for hemophilia B gene therapy. *Proc Natl Acad Sci USA* 1997;**94**:11563–6.
153. Kundu RK, Sangiorgi F, Wu LY, Kurachi K, Anderson WF, Maxson R, et al. Targeted inactivation of the coagulation factor IX gene causes hemophilia B in mice. *Blood* 1998;**92**:168–74.
154. Detloff PJ, Lewis J, John SW, Shehee WR, Langenbach R, Maeda N, et al. Deletion and replacement of the mouse adult beta-globin genes by a "plug and socket" repeated targeting strategy. *Mol Cell Biol* 1994;**14**:6936–43.
155. Gui T, Reheman A, Ni H, Gross PL, Yin F, Monroe D, et al. Abnormal hemostasis in a knock-in mouse carrying a variant of factor IX with impaired binding to collagen type IV. *J Thromb Haemost* 2009;**7**:1843–51.
156. Kao CY, Lin CN, Yu IS, Tao MH, Wu HL, Shi GY, et al. FIX-Triple, a gain-of-function factor IX variant, improves haemostasis in mouse models without increased risk of thrombosis. *Thromb Haemost* 2010;**104**:355–65.
157. Mackman N. Tissue-specific hemostasis in mice. *Arterioscler Thromb Vasc Biol* 2005;**25**:2273–81.
158. Fields PA, Arruda VR, Armstrong E, Chu K, Mingozzi F, Hagstrom JN, et al. Risk and prevention of anti-factor IX formation in AAV-mediated gene transfer in the context of a large deletion of F9. *Mol Ther* 2001;**4**:201–10.
159. Zhang TP, Jin DY, Wardrop 3rd RM, Gui T, Maile R, Frelinger JA, et al. Transgene expression levels and kinetics determine risk of humoral immune response modeled in factor IX knockout and missense mutant mice. *Gene Ther* 2007;**14**:429–40.
160. Yang C, Feng J, Song W, Wang J, Tsai B, Zhang Y, et al. A mouse model for nonsense mutation bypass therapy shows a dramatic multiday response to geneticin. *Proc Natl Acad Sci USA* 2007;**104**:15394–9.
161. Welch EM, Barton ER, Zhuo J, Tomizawa Y, Friesen WJ, Trifillis P, et al. PTC124 targets genetic disorders caused by nonsense mutations. *Nature* 2007;**447**:87–91.

162. Hirawat S, Welch EM, Elfring GL, Northcutt VJ, Paushkin S, Hwang S, et al. Safety, tolerability, and pharmacokinetics of PTC124, a nonaminoglycoside nonsense mutation suppressor, following single- and multiple-dose administration to healthy male and female adult volunteers. *J Clin Pharmacol* 2007;**47**:430–44.
163. Kurachi S, Deyashiki Y, Takeshita J, Kurachi K. Genetic mechanisms of age regulation of human blood coagulation factor IX. *Science* 1999;**285**:739–43.
164. Inoue Y, Peters LL, Yim SH, Inoue J, Gonzalez FJ. Role of hepatocyte nuclear factor 4alpha in control of blood coagulation factor gene expression. *J Mol Med* 2006;**84**:334–44.
165. Sweeney JD, Hoernig LA. Age-dependent effect on the level of factor IX. *Am J Clin Pathol* 1993;**99**:687–8.
166. Boland EJ, Liu YC, Walter CA, Herbert DC, Weaker FJ, Odom MW, et al. Age-specific regulation of clotting factor IX gene expression in normal and transgenic mice. *Blood* 1995;**86**:2198–205.
167. Brady JN, Notley C, Cameron C, Lillicrap D. Androgen effects on factor IX expression: in-vitro and in-vivo studies in mice. *Br J Haematol* 1998;**101**:273–9.
168. Nichols TC, Dillow AM, Franck HW, Merricks EP, Raymer RA, Bellinger DA, et al. Protein replacement therapy and gene transfer in canine models of hemophilia A, hemophilia B, von willebrand disease, and factor VII deficiency. *ILAR J* 2009;**50**:144–67.
169. Pipe SW. Hemophilia: new protein therapeutics. *Hematology Am Soc Hematol Educ Program* 2010;**2010**:203–9.
170. Ostergaard H, Bjelke JR, Hansen L, Petersen LC, Pedersen AA, Elm T, et al. Prolonged half-life and preserved enzymatic properties of factor IX selectively PEGylated on native N-glycans in the activation peptide. *Blood* 2011;**118**:2333–41.
171. Hoffman M. Animal models of bleeding and tissue repair. *Haemophilia* 2008;**14**(Suppl. 3):62–7.
172. Dejana E, Callioni A, Quintana A, de Gaetano G. Bleeding time in laboratory animals. II—a comparison of different assay conditions in rats. *Thromb Res* 1979;**15**:191–7.
173. Gailani D, Lasky NM, Broze Jr. GJ. A murine model of factor XI deficiency. *Blood Coagul Fibrinolysis* 1997;**8**:134–44.
174. Tsakiris DA, Scudder L, Hodivala-Dilke K, Hynes RO, Coller BS. Hemostasis in the mouse (Mus musculus): a review. *Thromb Haemost* 1999;**81**:177–88.
175. Keravala A, Chavez CL, Hu G, Woodard LE, Monahan PE, Calos MP. Long-term phenotypic correction in factor IX knockout mice by using PhiC31 integrase-mediated gene therapy. *Gene Ther* 2011;**18**:842–8.
176. Borchgrevink CF, Waaler BA. The secondary bleeding time. A new method for the differentiation of hemorrhagic diseases. *Acta Med Scand* 1958;**162**:361–74.
177. Tomokiyo K, Nakatomi Y, Araki T, Teshima K, Nakano H, Nakagaki T, et al. A novel therapeutic approach combining human plasma-derived Factors VIIa and X for haemophiliacs with inhibitors: evidence of a higher thrombin generation rate in vitro and more sustained haemostatic activity in vivo than obtained with Factor VIIa alone. *Vox Sang* 2003;**85**:290–9.
178. Buyue Y, Whinna HC, Sheehan JP. The heparin-binding exosite of factor IXa is a critical regulator of plasma thrombin generation and venous thrombosis. *Blood* 2008;**112**:3234–41.
179. Whinna HC, Monroe III DM. A novel murine hemostasis model: studies on recombinant factor VIIa and NN1731. *J Thromb Haemost* 2009;**7**:PP-MO-392.
180. Hoffman M, Monroe DM. Wound healing in haemophilia—breaking the vicious cycle. *Haemophilia* 2010;**16**(Suppl. 3):13–8.
181. Hoffman M, Harger A, Lenkowski A, Hedner U, Roberts HR, Monroe DM. Cutaneous wound healing is impaired in hemophilia B. *Blood* 2006;**108**:3053–60.

182. McDonald A, Hoffman M, Hedner U, Roberts HR, Monroe DM. Restoring hemostatic thrombin generation at the time of cutaneous wounding does not normalize healing in hemophilia B. *J Thromb Haemost* 2007;**5**:1577–83.
183. McDonald AG, Yang K, Roberts HR, Monroe DM, Hoffman M. Perivascular tissue factor is down-regulated following cutaneous wounding: implications for bleeding in hemophilia. *Blood* 2008;**111**:2046–8.
184. Pawlinski R, Fernandes A, Kehrle B, Pedersen B, Parry G, Erlich J, et al. Tissue factor deficiency causes cardiac fibrosis and left ventricular dysfunction. *Proc Natl Acad Sci USA* 2002;**99**:15333–8.
185. Drake TA, Morrissey JH, Edgington TS. Selective cellular expression of tissue factor in human tissues. Implications for disorders of hemostasis and thrombosis. *Am J Pathol* 1989;**134**:1087–97.
186. Fleck RA, Rao LV, Rapaport SI, Varki N. Localization of human tissue factor antigen by immunostaining with monospecific, polyclonal anti-human tissue factor antibody. *Thromb Res* 1990;**59**:421–37.
187. Sun J, Hakobyan N, Valentino LA, Feldman BL, Samulski RJ, Monahan PE. Intraarticular factor IX protein or gene replacement protects against development of hemophilic synovitis in the absence of circulating factor IX. *Blood* 2008;**112**:4532–41.
188. Hakobyan N, Enockson C, Cole AA, Rick Sumner D, Valentino LA. Experimental haemophilic arthropathy in a mouse model of a massive haemarthrosis: gross, radiological and histological changes. *Haemophilia* 2008;**14**:804–9.
189. Chang P, Aronson DL, Borenstein DG, Kessler CM. Coagulant proteins and thrombin generation in synovial fluid: a model for extravascular coagulation. *Am J Hematol* 1995;**50**:79–83.
190. Fareed J, Callas DD, Hoppensteadt D, Bermes Jr. EW. Tissue factor antigen levels in various biological fluids. *Blood Coagul Fibrinolysis* 1995;**6**(Suppl. 1):S32–6.
191. Miller GJ, Howarth DJ, Attfield JC, Cooke CJ, Nanjee MN, Olszewski WL, et al. Haemostatic factors in human peripheral afferent lymph. *Thromb Haemost* 2000;**83**:427–32.
192. Monahan PE, Sun J. Hemophilic synovitis: factor VII and the potential role of extravascular factor VIIa. *Thromb Res* 2010;**125**(Suppl. 1):S63–6.
193. Gailani D, Renne T. Intrinsic pathway of coagulation and arterial thrombosis. *Arterioscler Thromb Vasc Biol* 2007;**27**:2507–13.
194. Gui T, Reheman A, Funkhouser WK, Bellinger DA, Hagaman JR, Stafford DW, et al. In vivo response to vascular injury in the absence of factor IX: examination in factor IX knockout mice. *Thromb Res* 2007;**121**:225–34.
195. Cheung WF, Hamaguchi N, Smith KJ, Stafford DW. The binding of human factor IX to endothelial cells is mediated by residues 3–11. *J Biol Chem* 1992;**267**:20529–31.
196. Wolberg AS, Stafford DW, Erie DA. Human factor IX binds to specific sites on the collagenous domain of collagen IV. *J Biol Chem* 1997;**272**:16717–20.
197. Gui T, Lin HF, Hoffman M, David LS, Stafford DW. In vivo behavior of factor IX and certain gla domain mutants evaluated in coagulation factor IX-deficient mice. *Blood* 2000;**96**:35a.
198. Begbie ME, Mamdani A, Gataiance S, Eltringham-Smith LJ, Bhakta V, Hortelano G, et al. An important role for the activation peptide domain in controlling factor IX levels in the blood of haemophilia B mice. *Thromb Haemost* 2005;**94**:1138–47.
199. Schuettrumpf J, Herzog RW, Schlachterman A, Kaufhold A, Stafford DW, Arruda VR. Factor IX variants improve gene therapy efficacy for hemophilia B. *Blood* 2005;**105**:2316–23.
200. DiMichele D. Inhibitor development in haemophilia B: an orphan disease in need of attention. *Br J Haematol* 2007;**138**:305–15.

201. Cao O, Armstrong E, Schlachterman A, Wang L, Okita DK, Conti-Fine B, et al. Immune deviation by mucosal antigen administration suppresses gene-transfer-induced inhibitor formation to factor IX. *Blood* 2006;**108**:480–6.
202. Fields PA, Armstrong E, Hagstrom JN, Arruda VR, Murphy ML, Farrell JP, et al. Intravenous administration of an E1/E3-deleted adenoviral vector induces tolerance to factor IX in C57BL/6 mice. *Gene Ther* 2001;**8**:354–61.
203. Lozier JN, Tayebi N, Zhang P. Mapping of genes that control the antibody response to human factor IX in mice. *Blood* 2005;**105**:1029–35.
204. Greenwood R, Wang B, Midkiff K, White 2nd GC, Lin HF, Frelinger JA. Identification of T-cell epitopes in clotting factor IX and lack of tolerance in inbred mice. *J Thromb Haemost* 2003;**1**:95–102.
205. Chao H, Monahan PE, Liu Y, Samulski RJ, Walsh CE. Sustained and complete phenotype correction of hemophilia B mice following intramuscular injection of AAV1 serotype vectors. *Mol Ther* 2001;**4**:217–22.
206. Cohn EF, Zhuo J, Kelly ME, Chao HJ. Efficient induction of immune tolerance to coagulation factor IX following direct intramuscular gene transfer. *J Thromb Haemost* 2007;**5**:1227–36.
207. Arruda VR, Schuettrumpf J, Herzog RW, Nichols TC, Robinson N, Lotfi Y, et al. Safety and efficacy of factor IX gene transfer to skeletal muscle in murine and canine hemophilia B models by adeno-associated viral vector serotype 1. *Blood* 2004;**103**:85–92.
208. Fields PA, Kowalczyk DW, Arruda VR, Armstrong E, McCleland ML, Hagstrom JN, et al. Role of vector in activation of T cell subsets in immune responses against the secreted transgene product factor IX. *Mol Ther* 2000;**1**:225–35.
209. Chen J, Wu Q, Yang P, Hsu HC, Mountz JD. Determination of specific CD4 and CD8 T cell epitopes after AAV2- and AAV8-hF.IX gene therapy. *Mol Ther* 2006;**13**:260–9.
210. Dioun AF, Ewenstein BM, Geha RS, Schneider LC. IgE-mediated allergy and desensitization to factor IX in hemophilia B. *J Allergy Clin Immunol* 1998;**102**:113–7.
211. Verma D, Moghimi B, LoDuca PA, Singh HD, Hoffman BE, Herzog RW, et al. Oral delivery of bioencapsulated coagulation factor IX prevents inhibitor formation and fatal anaphylaxis in hemophilia B mice. *Proc Natl Acad Sci USA* 2010;**107**:7101–6.
212. Astermark J, Oldenburg J, Pavlova A, Berntorp E, Lefvert AK. Polymorphisms in the IL10 but not in the IL1beta and IL4 genes are associated with inhibitor development in patients with hemophilia A. *Blood* 2006;**107**:3167–72.
213. Astermark J, Oldenburg J, Carlson J, Pavlova A, Kavakli K, Berntorp E, et al. Polymorphisms in the TNFA gene and the risk of inhibitor development in patients with hemophilia A. *Blood* 2006;**108**:3739–45.
214. Astermark J, Wang X, Oldenburg J, Berntorp E, Lefvert AK. Polymorphisms in the CTLA-4 gene and inhibitor development in patients with severe hemophilia A. *J Thromb Haemost* 2007;**5**:263–5.
215. Oldenburg J, Picard JK, Schwaab R, Brackmann HH, Tuddenham EG, Simpson E. HLA genotype of patients with severe haemophilia A due to intron 22 inversion with and without inhibitors of factor VIII. *Thromb Haemost* 1997;**77**:238–42.
216. Zhang HG, High KA, Wu Q, Yang P, Schlachterman A, Yu S, et al. Genetic analysis of the antibody response to AAV2 and factor IX. *Mol Ther* 2005;**11**:866–74.
217. Hoffman BE, Dobrzynski E, Wang L, Hirao L, Mingozzi F, Cao O, et al. Muscle as a target for supplementary factor IX gene transfer. *Hum Gene Ther* 2007;**18**:603–13.
218. Christophe OD, Lenting PJ, Cherel G, Boon-Spijker M, Lavergne JM, Boertjes R, et al. Functional mapping of anti-factor IX inhibitors developed in patients with severe hemophilia B. *Blood* 2001;**98**:1416–23.

219. Nayak S, Cao O, Hoffman BE, Cooper M, Zhou S, Atkinson MA, et al. Prophylactic immune tolerance induced by changing the ratio of antigen-specific effector to regulatory T cells. *J Thromb Haemost* 2009;**7**:1523–32.
220. Li C, Hirsch M, Asokan A, Zeithaml B, Ma H, Kafri T, et al. Adeno-associated virus type 2 (AAV2) capsid-specific cytotoxic T lymphocytes eliminate only vector-transduced cells coexpressing the AAV2 capsid in vivo. *J Virol* 2007;**81**:7540–7.
221. Li C, Hirsch M, DiPrimio N, Asokan A, Goudy K, Tisch R, et al. Cytotoxic-T-lymphocyte-mediated elimination of target cells transduced with engineered adeno-associated virus type 2 vector in vivo. *J Virol* 2009;**83**:6817–24.
222. Wang L, Figueredo J, Calcedo R, Lin J, Wilson JM. Cross-presentation of adeno-associated virus serotype 2 capsids activates cytotoxic T cells but does not render hepatocytes effective cytolytic targets. *Hum Gene Ther* 2007;**18**:185–94.
223. Li H, Murphy SL, Giles-Davis W, Edmonson S, Xiang Z, Li Y, et al. Pre-existing AAV capsid specific CD8+ T cells are unable to eliminate AAV-transduced hepatocytes. *Mol Ther* 2007;**15**:792–800.
224. Li C, Goudy K, Hirsch M, Asokan A, Fan Y, Alexander J, et al. Cellular immune response to cryptic epitopes during therapeutic gene transfer. *Proc Natl Acad Sci USA* 2009;**106**:10770–4.
225. Matrai J, Chuah MK, VandenDriessche T. Preclinical and clinical progress in hemophilia gene therapy. *Curr Opin Hematol* 2010;**17**:387–92.
226. Montgomery RR, Monahan PE, Ozelo MC. Unique strategies for therapeutic gene transfer in haemophilia A and haemophilia BWFH State-of-the-Art Session on Therapeutic Gene Transfer Buenos Aires, Argentina. *Haemophilia* 2010;**16**(Suppl. 5):29–34.
227. Viiala NO, Larsen SR, Rasko JE. Gene therapy for hemophilia: clinical trials and technical tribulations. *Semin Thromb Hemost* 2009;**35**:81–92.
228. White SJ, Page SM, Margaritis P, Brownlee GG. Long-term expression of human clotting factor IX from retrovirally transduced primary human keratinocytes in vivo. *Hum Gene Ther* 1998;**9**:1187–95.
229. Fair JH, Cairns BA, Lapaglia MA, Caballero M, Pleasant WA, Hatada S, et al. Correction of factor IX deficiency in mice by embryonic stem cells differentiated in vitro. *Proc Natl Acad Sci USA* 2005;**102**:2958–63.
230. Tatsumi K, Ohashi K, Kataoka M, Tateno C, Shibata M, Naka H, et al. Successful in vivo propagation of factor IX-producing hepatocytes in mice: potential for cell-based therapy in haemophilia B. *Thromb Haemost* 2008;**99**:883–91.
231. Bigger BW, Siapati EK, Mistry A, Waddington SN, Nivsarkar MS, Jacobs L, et al. Permanent partial phenotypic correction and tolerance in a mouse model of hemophilia B by stem cell gene delivery of human factor IX. *Gene Ther* 2006;**13**:117–26.
232. Chang AH, Stephan MT, Sadelain M. Stem cell-derived erythroid cells mediate long-term systemic protein delivery. *Nat Biotechnol* 2006;**24**:1017–21.
233. Zhang G, Shi Q, Fahs SA, Kuether EL, Walsh CE, Montgomery RR. Factor IX ectopically expressed in platelets can be stored in alpha-granules and corrects the phenotype of hemophilia B mice. *Blood* 2010;**116**:1235–43.
234. Van Raamsdonk JM, Ross CJ, Potter MA, Kurachi S, Kurachi K, Stafford DW, et al. Treatment of hemophilia B in mice with nonautologous somatic gene therapeutics. *J Lab Clin Med* 2002;**139**:35–42.
235. Wen J, Vargas AG, Ofosu FA, Hortelano G. Sustained and therapeutic levels of human factor IX in hemophilia B mice implanted with microcapsules: key role of encapsulated cells. *J Gene Med* 2006;**8**:362–9.
236. Wen J, Xu N, Li A, Bourgeois J, Ofosu FA, Hortelano G. Encapsulated human primary myoblasts deliver functional hFIX in hemophilic mice. *J Gene Med* 2007;**9**:1002–10.

237. Miao CH. A novel gene expression system: non-viral gene transfer for hemophilia as model systems. *Adv Genet* 2005;**54**:143–77.
238. Ye X, Loeb KR, Stafford DW, Thompson AR, Miao CH. Complete and sustained phenotypic correction of hemophilia B in mice following hepatic gene transfer of a high-expressing human factor IX plasmid. *J Thromb Haemost* 2003;**1**:103–11.
239. Liu F, Sag D, Wang J, Shollenberger LM, Niu F, Yuan X, et al. Sine-wave current for efficient and safe in vivo gene transfer. *Mol Ther* 2007;**15**:1842–7.
240. Jacobs F, Snoeys J, Feng Y, Van Craeyveld E, Lievens J, Armentano D, et al. Direct comparison of hepatocyte-specific expression cassettes following adenoviral and nonviral hydrodynamic gene transfer. *Gene Ther* 2008;**15**:594–603.
241. Schuttrumpf J, Milanov P, Roth S, Seifried E, Tonn T. Non-viral gene transfer results in therapeutic factor IX levels in haemophilia B mice. *Hamostaseologie* 2008;**28**(Suppl. 1): S92–5.
242. Miao CH, Brayman AA, Loeb KR, Ye P, Zhou L, Mourad P, et al. Ultrasound enhances gene delivery of human factor IX plasmid. *Hum Gene Ther* 2005;**16**:893–905.
243. Brown BD, Cantore A, Annoni A, Sergi LS, Lombardo A, Della Valle P, et al. A microRNA-regulated lentiviral vector mediates stable correction of hemophilia B mice. *Blood* 2007;**110**:4144–52.
244. Matrai J, Cantore A, Bartholomae CC, Annoni A, Wang W, Acosta-Sanchez A, et al. Hepatocyte-targeted expression by integrase-defective lentiviral vectors induces antigen-specific tolerance in mice with low genotoxic risk. *Hepatology* 2011;**53**:1696–707.
245. Matsui H, Hegadorn C, Ozelo M, Burnett E, Tuttle A, Labelle A, et al. A microRNA-regulated and GP64-pseudotyped lentiviral vector mediates stable expression of FVIII in a murine model of Hemophilia A. *Mol Ther* 2011;**19**:723–30.
246. Jiang H, Pierce GF, Ozelo MC, de Paula EV, Vargas JA, Smith P, et al. Evidence of multiyear factor IX expression by AAV-mediated gene transfer to skeletal muscle in an individual with severe hemophilia B. *Mol Ther* 2006;**14**:452–5.
247. Manno CS, Pierce GF, Arruda VR, Glader B, Ragni M, Rasko JJ, et al. Successful transduction of liver in hemophilia by AAV-Factor IX and limitations imposed by the host immune response. *Nat Med* 2006;**12**:342–7.
248. Mingozzi F, Maus MV, Hui DJ, Sabatino DE, Murphy SL, Rasko JE, et al. CD8(+) T-cell responses to adeno-associated virus capsid in humans. *Nat Med* 2007;**13**:419–22.
249. McCarty DM, Monahan PE, Samulski RJ. Self-complementary recombinant adeno-associated virus (scAAV) vectors promote efficient transduction independently of DNA synthesis. *Gene Ther* 2001;**8**:1248–54.
250. Wu Z, Sun J, Zhang T, Yin C, Yin F, Van Dyke T, et al. Optimization of self-complementary AAV vectors for liver-directed expression results in sustained correction of hemophilia B at low vector dose. *Mol Ther* 2008;**16**:280–9.
251. Nathwani AC, Gray JT, Ng CY, Zhou J, Spence Y, Waddington SN, et al. Self-complementary adeno-associated virus vectors containing a novel liver-specific human factor IX expression cassette enable highly efficient transduction of murine and nonhuman primate liver. *Blood* 2006;**107**:2653–61.
252. Nathwani AC, Rosales C, McIntosh J, Rastegarlari G, Nathwani D, Raj D, et al. Long-term safety and efficacy following systemic administration of a self-complementary AAV vector encoding human FIX pseudotyped with serotype 5 and 8 capsid proteins. *Mol Ther* 2011;**19**:876–85.
253. Nathwani AC, Tuddenham E, Rosales C, McIntosh J, Riddell A, Rustagi PK, et al. Early clinical results following administration of a low dose of a novel self complementary adneno-associated viral vector encoding human factor IX in two subjects with severe hemophilia B. *Blood* 2010;**116**:114 (Abstract 248).

254. High KA. Gene therapy for haemophilia: a long and winding road. *J Thromb Haemost* 2011;**9** (Suppl. 1):2–11.
255. Chang J, Jin J, Lollar P, Bode W, Brandstetter H, Hamaguchi N, et al. Changing residue 338 in human factor IX from arginine to alanine causes an increase in catalytic activity. *J Biol Chem* 1998;**273**:12089–94.
256. Brunetti-Pierri N, Grove N, Zuo Y, Edwards R, Palmer DJ, Cerullo V, et al. Bioengineered factor IX molecules with increased catalytic activity improve the safety and efficiacy of helper-dependent adenoviral vectors (HDAd) for hemophilia B gene therapy. *Mol Ther* 2008;**16**: S152.
257. Lin CN, Kao CY, Miao CH, Hamaguchi N, Wu HL, Shi GY, et al. Generation of a novel factor IX with augmented clotting activities in vitro and in vivo. *J Thromb Haemost* 2010;**8**:1773–83.
258. Markusic DM, Herzog RW, Aslanidi GV, Hoffman BE, Li B, Li M, et al. High-efficiency transduction and correction of murine hemophilia B using AAV2 vectors devoid of multiple surface-exposed tyrosines. *Mol Ther* 2010;**18**:2048–56.
259. Li H, Haurigot V, Doyon Y, Li T, Wong SY, Bhagwat AS, et al. In vivo genome editing restores haemostasis in a mouse model of haemophilia. *Nature* 2011;**475**:217–21.
260. Brinkhous KM, Shanbrom E, Roberts HR, Webster WP, Fekete L, Wagner RH. A new high-potency glycine-precipitated anti-hemophilic factor (AHF) concentrate: treatment of classical hemophilia and hemophilia with inhibitors. *JAMA* 1968;**205**:613–7.
261. Brinkhous KM, Sandberg H, Garris JB, Mattsson C, Palm M, Griggs TR, et al. Purified human factor VIII procoagulant protein: comparative hemostatic response after infusions into hemophilic and von Willebrand disease dogs. *Proc Natl Acad Sci USA* 1985;**82**:8752–6.
262. Brinkhous KM, Hedner U, Garris JB, Diness V, Read MS. Effect of recombinant factor VIIa on the hemostatic defect in dogs with hemophilia A, Hemophilia B, and von Willebrand disease. *Proc Natl Acad Sci USA* 1989;**86**:1382–6.
263. Brinkhous KM, Sigman JL, Read MS, Stewart PF, McCarthy KP, Timony GA, et al. Recombinant human factor IX: replacement therapy, prophylaxis, pharmacokinetics, and immunogenicity in canine hemophilia B. *Blood* 1996;**88**:2603–10.
264. White II GC, Beebe A, Nielson B. Recombinant Factor IX. *Thromb Haemost* 1997;**78**:261–5.
265. Brinkhous KM, Sandberg H, Widlund L, Read MS, Nichols TC, Sigman J, et al. Preclinical pharmacology of albumin-free B-domain deleted recombinant factor VIII. *Semin Thromb Hemost* 2002;**28**:269–72.
266. Xu L, Mei M, Haskins ME, Nichols TC, O'Donnell P, Cullen K, et al. Immune response after neonatal transfer of a human factor IX-expressing retroviral vector in dogs, cats, and mice. *Thromb Res* 2007;**120**:269–80.
267. Wang L, Calcedo R, Nichols TC, Bellinger DA, Dillow A, Verma IM, et al. Sustained correction of disease in naive and AAV2-pretreated hemophilia B dogs: AAV2/8-mediated, liver-directed gene therapy. *Blood* 2005;**105**:3079–86.
268. Arruda VR, Stedman HH, Nichols TC, Haskins ME, Nicholson M, Herzog RW, et al. Regional intravascular delivery of AAV-2-F.IX to skeletal muscle achieves long-term correction of hemophilia B in a large animal model. *Blood* 2005;**105**:3458–64.
269. Harding TC, Koprivnikar KE, Tu GH, Zayek N, Lew S, Subramanian A, et al. Intravenous administration of an AAV-2 vector for the expression of factor IX in mice and a dog model of hemophilia B. *Gene Ther* 2004;**11**:204–13.
270. Arruda VR, Stedman HH, Schuettrumph J, Jiang H, Pierce G, Nichols TC, et al. A novel method of regional intravenous delivery of AAV vector to skeletal muscle results in correction of canine hemophilia B phenotype. *Blood* 2004;**104**: (Abstract #3179).
271. Arruda VR, Scallan C, Jiang H, Couto L, Herzog RW, Nichols TC, et al. Comparison of the efficacy on gene transfer by AAV vectors delivered by distinct routes of administration to the liver of hemophilia B dogs. *Mol Ther* 2004;**9**:S103, Abstract supplement.

272. Xu L, Gao C, Sands MS, Cai SR, Nichols TC, Bellinger DA, et al. Neonatal or hepatocyte growth factor-potentiated adult gene therapy with a retroviral vector results in therapeutic levels of canine factor IX for hemophilia B. *Blood* 2003;**101**:3924–32.
273. Ehrhardt A, Xu H, Dillow AM, Bellinger DA, Nichols TC, Kay MA. A gene-deleted adenoviral vector results in phenotypic correction of canine hemophilia B without liver toxicity or thrombocytopenia. *Blood* 2003;**102**:2403–11.
274. Mount JD, Herzog RW, Tillson DM, Goodman SA, Robinson N, McCleland ML, et al. Sustained phenotypic correction of hemophilia B dogs with a factor IX null mutation by liver-directed gene therapy. *Blood* 2002;**99**:2670–6.
275. Jankowitz RG, Chapman H, Merricks E, Kloos M, Dillow A, Nichols TC, et al. *Recombinant IL-11 Increases VWF and VWF MRNA in Type 1 VWD.* 2007 (Abstract N° O-T-062, XXI ISTH, July).
276. Rawle FE, Lillicrap D. Preclinical animal models for hemophilia gene therapy: predictive value and limitations. *Semin Thromb Hemost* 2004;**30**:205–13.
277. Swanton MC. Hemophilic arthropathy in dogs. *Lab Invest* 1959;**8**:1269–77.
278. Hough AJ, Banfield WG, Sokoloff L. Cartilage in hemophilic arthropathy. Ultrastructural and microanalytical studies. *Arch Pathol Lab Med* 1976;**100**:91–6.
279. Nichols TC, Raymer RA, Franck HW, Merricks EP, Bellinger DA, DeFriess N, et al. Prevention of spontaneous bleeding in dogs with haemophilia A and haemophilia B. *Haemophilia* 2010;**16**(Suppl. 3):19–23.
280. Langdell RD, Wagner RH, Brinkhous KM. Effect of antihemophilic factor on one-stage clotting tests. *J Lab Clin Med* 1953;**41**:637–47.
281. Brinkhous KM. Plasma antihemophilic factor biological and clinical aspects. *Sang* 1954;**25**:738–41.
282. Roberts HR, Penick GD, Brinkhous KM. Intensive plasma therapy in the hemophilias. *JAMA* 1964;**190**:546–8.
283. Wagner RH, Roberts HR, Webster WP, Shanbrom E, Brinkhous KM. Glycine-precipitated antihemophilic factor concentrates and their clinical use. *Thromb Diath Haemorrh Suppl* 1968;**35**:41–8.
284. Brinkhous KM, Shanbrom E, Roberts HR, Webster WP, Fekete L, Wagner RH. A new high-potency glycine-precipitated antihemophilic factor (AHF) concentrate. Treatment of classical hemophilia and hemophilia with inhibitors. *JAMA* 1968;**205**:613–7.
285. Bolton-Maggs PH, Pasi KJ. Haemophilias A and B. *Lancet* 2003;**361**:1801–9.
286. Giles AR, Tinlin S, Hoogendoorn H, Greenwood P, Greenwood R. Development of factor VIII:C antibodies in dogs with hemophilia A (factor VIII:C deficiency). *Blood* 1984;**63**:451–6.
287. Finn JD, Ozelo MC, Sabatino DE, Franck HW, Merricks EP, Crudele JM, et al. Eradication of neutralizing antibodies to factor VIII in canine hemophilia A following liver gene therapy. *Blood* 2010;**116**:5842–8.
288. Viel KR, Ameri A, Abshire TC, Iyer RV, Watts RG, Lutcher C, et al. Inhibitors of factor VIII in black patients with hemophilia. *N Engl J Med* 2009;**360**:1618–27.
289. Webster WP, Reedick RL, Roberts HR, Penick GD. Release of factor 8 (antihaemophilic factor) from perfused organs and tissues. *Nature* 1967;**213**:1146–7.
290. Webster WP, Penick GD, Peacock EE, Brinkhous KM. Allotransplantation of spleen in hemophilia. *N C Med J* 1967;**28**:505–7.
291. Marchioro TL, Hougie C, Ragde H, Epstein RB, Thomas ED. Organ homografts for hemophilia. *Transplant Proc* 1969;**1**:316–20.
292. Marchioro TL, Hougie C, Ragde H, Epstein RB, Thomas ED. Hemophilia: role of organ homografts. *Science* 1969;**163**:188–90.

293. Webster WP, Mandel SR, Strike LE, Penick GD, Griggs TR, Brinkhous KM. Factor VIII synthesis: hepatic and renal allografts in swine with von Willebrand's disease. *Am J Physiol* 1976;**230**:1342–8.
294. Webster WP, Zukoski CF, Hutchin P, Reddick RL, Mandel SR, Penick GD. Plasma factor VIII synthesis and control as revealed by canine organ transplantation. *Am J Physiol* 1971;**220**:1147–54.
295. Bontempo FA, Lewis JH, Gorenc TJ, Spero JA, Ragni MV, Scott JP, et al. Liver transplantation in hemophilia A. *Blood* 1987;**69**:1721–4.
296. Prasad S, Lillicrap D, Labelle A, Knappe S, Keller T, Burnett E, et al. Efficacy and safety of a new-class hemostatic drug candidate, AV513, in dogs with hemophilia A. *Blood* 2008;**111**:672–9.
297. Ragni MV, Jankowitz RC, Chapman HL, Merricks EP, Kloos MT, Dillow AM, et al. A phase II prospective open-label escalating dose trial of recombinant interleukin-11 in mild von Willebrand disease. *Haemophilia* 2008;**14**:968–77.
298. Roth DA, Tawa Jr. NE, O'Brien JM, Treco DA, Selden RF. Nonviral transfer of the gene encoding coagulation factor VIII in patients with severe hemophilia A. *N Engl J Med* 2001;**344**:1735–42.
299. Powell JS, Ragni MV, White 2nd GC, Lusher JM, Hillman-Wiseman C, Moon TE, et al. Phase 1 trial of FVIII gene transfer for severe hemophilia A using a retroviral construct administered by peripheral intravenous infusion. *Blood* 2003;**102**:2038–45.
300. White GC. *Gutted Adenoviral Vector Delivering Factor VIII: Clinical Results. National Hemophilia Foundation Fifth Workshop on Gene Therapies for Hemophilia. Philadelphia PA*; 2002.
301. Fang X, Zhang WW, Sobol RE, Gomperts E, Thompson AR, Wong WY, et al. Studies in non human primate and hemophilic dogmodels of a gutless: adenovirus vector for treatment of hemophilia A. *Blood* 2000;**96**:428a.
302. Greengard JS, Bodner M, McCormack J, Edwards WR, Sensintaffar JL, Sheridan PL, et al. Sustained expression of human factor VIII from peripheral retroviral gene delivery in rabbits and dogs. *Blood* 1997;**90**:240a.
303. Gallo-Penn AM, Shirley PS, Andrews JL, Tinlin S, Webster S, Cameron C, et al. Systemic delivery of an adenoviral vector encoding canine factor VIII results in short-term phenotypic correction, inhibitor development, and biphasic liver toxicity in hemophilia A dogs. *Blood* 2001;**97**:107–13.
304. Chuah MK, Schiedner G, Thorrez L, Brown B, Johnston M, Gillijns V, et al. Therapeutic factor VIII levels and negligible toxicity in mouse and dog models of hemophilia A following gene therapy with high-capacity adenoviral vectors. *Blood* 2003;**101**:1734–43.
305. Matsui H, Shibata M, Brown B, Labelle A, Hegadorn C, Andrews C, et al. Ex Vivo Gene Therapy for Hemophilia A That Enhances Safe Delivery and Sustained In Vivo FVIII Expression From Lentivirally-engineered Endothelial Progenitors. *Stem Cells* 2007;**25**:2660–9.
306. Scallan CD, Lillicrap D, Jiang H, Qian X, Patarroyo-White SL, Parker AE, et al. Sustained phenotypic correction of canine hemophilia A using an adeno-associated viral vector. *Blood* 2003;**102**:2031–7.
307. Jiang H, Lillicrap D, Patarroyo-White S, Liu T, Qian X, Scallan CD, et al. Multiyear therapeutic benefit of AAV serotypes 2, 6, and 8 delivering factor VIII to hemophilia A mice and dogs. *Blood* 2006;**108**:107–15.
308. Niemeyer GP, Herzog RW, Mount J, Arruda VR, Tillson DM, Hathcock J, et al. Long-term correction of inhibitor-prone hemophilia B dogs treated with liver-directed AAV2-mediated factor IX gene therapy. *Blood* 2009;**113**:797–806.
309. Donsante A, Vogler C, Muzyczka N, Crawford JM, Barker J, Flotte T, et al. Observed incidence of tumorigenesis in long-term rodent studies of rAAV vectors. *Gene Ther* 2001;**8**:1343–6.

310. Donsante A, Miller DG, Li Y, Vogler C, Brunt EM, Russell DW, et al. AAV vector integration sites in mouse hepatocellular carcinoma. *Science* 2007;**317**:477.
311. Herzog R, Arruda VR, Fischer TH, Read MS, Nichols TC, High KA. Absence of circulating factor IX antigen in hemophilia B dogs of the UNC-Chapel Hill Colony. *Thromb Haemost* 2000;**84**:352–4.
312. Herzog RW, Mount JD, Arruda VR, High KA, Lothrop Jr. CD. Muscle-directed gene transfer and transient immune suppression result in sustained partial correction of canine hemophilia B caused by a null mutation. *Mol Ther* 2001;**4**:192–200.
313. Goldsmith JC, Chung KS, Roberts HR. A simple assay for human factor IX: use of canine hemophilia B plasma as substrate. *Thromb Res* 1978;**12**:497–502.
314. Ezban M, Hermith MTB, Holm PK, Hansen L, Østergaard H, Tranholm M, et al. Pharmacokinetic (PK) and pharmacodynamic (PD) properties of a new recombinant long acting factor IX (40KPEG-RFIX) product after intravenous (IV) administration to hemophilia b dogs. *J Thromb Haemost* 2009;**7**: (Abstract N° PP-TH-579 - XXII ISTH July 2009).
315. Peters RT, Low SC, Kamphaus GD, Dumont JA, Amari JV, Lu Q, et al. Prolonged activity of factor IX as a monomeric Fc fusion protein. *Blood* 2010;**115**:2057–64.
316. Kay MA, Manno CS, Ragni MV, Larson PJ, Couto LB, McClelland A, et al. Evidence for gene transfer and expression of factor IX in haemophilia B patients treated with an AAV vector [see comments]. *Nat Genet* 2000;**24**:257–61.
317. Manno CS, Chew AJ, Hutchison S, Larson PJ, Herzog RW, Arruda VR, et al. AAV-mediated factor IX gene transfer to skeletal muscle in patients with severe hemophilia B. *Blood* 2003;**101**:2963–72.
318. Herzog RW, Fields PA, Arruda VR, Brewbaker J, Armstrong E, McClintock D, et al. Influence of vector dose on factor IX-specific T and B cell responses in muscle-directed gene therapy. *Hum Gene Ther* 2002;**13**:1281–91.
319. Herzog RW, Yang EY, Couto LB, Hagstrom JN, Elwell D, Fields PA, et al. Long-term correction of canine hemophilia B by gene transfer of blood coagulation factor IX mediated by adeno-assocated viral vector. *Nat Med* 1999;**5**:56–63.
320. Buchlis GJ, Hui DJ, Finn J, Pien GC, Edmonson S, Dillow A, et al. Identification of transgene-specific CD4+FOXP3+ T cells secreting high levels of IL-10 following intravascular AAV-cF.IX gene transfer to muscle in hemophilia B dogs. *Mol Ther* 2009;**17**: (12th Annual Meeting May 2009, Abstract 643).
321. Arruda VR, Stedman H, Nichols TC, Haskins M, Schuettrumph J, Herzog RW, et al. Sustained correction of hemophilia B phenotype following intravascular delivery of AAV vector to skeletal muscle. *Mol Ther* 2002;**5**:S157 (Abstract Supplement).
322. Peters RT, Bitonti AJ. Enhanced pharmacokinetics of Factor IX as a monomeric FC-fusion. *J Thromb Haemost* 2007;**5**: (Abstract O-M-016).
323. Liles D, Landen CN, Monroe DM, Lindley CM, Read MS, Roberts HR, et al. Extravascular administration of factor IX: potential for replacement therapy of canine and human hemophilia B. *Thromb Haemost* 1997;**77**:944–8 (PMID 9184407).
324. Russell KE, Olsen EHN, Raymer RA, Merricks EP, Bellinger DA, Read MS, et al. Reduced bleeding events with subcutaneous administration of recombinant human factor IX in immune tolerant hemophilia B dogs. *Blood* 2003;**102**:4393–8.
325. Webster WP, Mandel SR, Reddick RL, Wagner JL, Penick GD. Orthotopic liver transplantation in canine hemophilia B. *Am J Physiol* 1974;**226**:496–500.
326. Merion RM, Delius RE, Campbell Jr. DA, Turcotteauthor JG. Orthotopic liver transplantation totally corrects factor IX deficiency in hemophilia B. *Surgery* 1988;**104**:929–31.
327. Delorme MA, Adams PC, Grant D, Ghent CN, Walker IR, Wallauthor WJ. Orthotopic liver transplantation in a patient with combined hemophilia A and B. *Am J Hematol* 1990;**33**:136–8.

328. Van Cott KE, Monahan PE, Nichols TC, Velander WH. Haemophilic factors produced by transgenic livestock: abundance that can enable alternative therapies worldwide. *Haemophilia* 2004;**10**(Suppl. 4):70–6.
329. Roosendaal G, Lafeber F. Prophylactic treatment for prevention of joint disease in hemophilia—cost versus benefit. *N Engl J Med* 2007;**357**:603–5.
330. van den Berg HM, De Groot PH, Fischer K. Phenotypic heterogeneity in severe hemophilia. *J Thromb Haemost* 2007;**5**(Suppl. 1):151–6.
331. Carcao M, St Louis J, Poon MC, Grunebaum E, Lacroix S, Stain AM, et al. Rituximab for congenital haemophiliacs with inhibitors: a Canadian experience. *Haemophilia* 2006;**12**:7–18.
332. Bidlingmaier C, Bergmann F, Kurnik K. Haemophilia A in two premature infants. *Eur J Pediatr* 2005;**164**:70–2.
333. Oldenburg J, Ananyeva NM, Saenko EL. Molecular basis of haemophilia A. *Haemophilia* 2004;**10**(Suppl. 4):133–9.
334. Pipe SW, Saint-Remy JM, Walsh CE. New high-technology products for the treatment of haemophilia. *Haemophilia* 2004;**10**(Suppl. 4):55–63.
335. Quintana-Molina M, Martinez-Bahamonde F, Gonzalez-Garcia E, Romero-Garrido J, Villar-Camacho A, Jimenez-Yuste V, et al. Surgery in haemophilic patients with inhibitor: 20 years of experience. *Haemophilia* 2004;**10**(Suppl. 2):30–40.
336. Tarantino MD, Collins PW, Hay CR, Shapiro AD, Gruppo RA, Berntorp E, et al. Clinical evaluation of an advanced category antihaemophilic factor prepared using a plasma/albumin-free method: pharmacokinetics, efficacy, and safety in previously treated patients with haemophilia A. *Haemophilia* 2004;**10**:428–37.
337. High KA. Gene transfer for hemophilia: can therapeutic efficacy in large animals be safely translated to patients? *J Thromb Haemost* 2005;**3**:1682–91.
338. Connelly S, Mount J, Mauser A, Gardner JM, Kaleko M, McClelland A, et al. Complete short-term correction of canine hemophilia A by in vivo gene therapy. *Blood* 1996;**88**:3846–53.
339. Mingozzi F, Liu YL, Dobrzynski E, Kaufhold A, Liu JH, Wang YQ, Arruda VR, High KA, Herzog RW. Induction of immune tolerance to coagulation factor IX antigen by in vivo hepatic gene transfer. *J Clin Invest* 2003;**111**:1347–56.

Animal Models of Lung Cancer: Characterization and Use for Chemoprevention Research

YIAN WANG,* LUCINA ROUGGLY,* MING YOU,† AND RONALD LUBET‡

*Department of Surgery, Washington University School of Medicine, St. Louis, Missouri, USA

†Cancer Center, Medical College of Wisconsin, Milwaukee, Wisconsin, USA

‡Chemoprevention Branch, National Cancer Institute, Bethesda, Maryland, USA

Of the potential sites of cancer development, cancer of the lung accounts for the highest number of cancer deaths each year in the United States (Jemal *et al.*, 2010[1]). Based on its histopathological features, lung cancer is grouped into small cell lung cancer (SCLC; ~20%) and non-SCLC (NSCLC; ~80%), which is further divided into three subtypes: squamous cell carcinoma (~30%), adenocarcinoma (~50%), and large cell lung carcinoma. Every subtype of lung cancer has a relatively low 5-year survival rate that is attributed, in part, to the fact that they are routinely diagnosed at later histologic stages. Due to this alarming statistic, it is necessary to develop not only new and effective means of treatment but also of prevention. One of the promising approaches is chemoprevention which is the use of synthetic or natural agents to inhibit the initial development of or further progression of early lung lesions (Hong and Sporn, 1997[2]). Many compounds have been identified as potentially effective chemopreventive agents using animal models. Most chemopreventive studies have been performed using mouse models which were developed to study lung adenomas or adenocarcinomas. More recently, models of squamous cell lung cancer and small cell lung cancer have also been developed. This review seeks to highlight mouse models which we helped to develop and presents the results of recent chemopreventive studies that we have performed in models of lung adenocarcinoma, squamous cell carcinoma, and small cell lung cancer.

Progress in Molecular Biology
and Translational Science, Vol. 105
DOI: 10.1016/B978-0-12-394596-9.00007-X

I. Lung Adenomas and Adenocarcinomas

Following the report by Mori in 1962[3] on the experimental induction of pulmonary tumors in mice, there have been many studies conducted to induce lung tumors using chemical carcinogens in mice such as lung adenomas induced by oral administration of dibenz(a,h)anthracene in DBA/2 mice,[4] isonicotinic acid hydrazide,[5] urethane,[6] nitrosamines,[7] and chronic inhalation of bis(chloromethyl)ether.[8] In addition, two components of tobacco smoke, benzo(a)pyrene (B[a]P)[9] and 4-(methylnitrosamino)-1-(3-pyridyl)-1-butanone (NNK, a nitrosamine),[10] have proven to be effective in inducing lung tumors in mice.

Although spontaneously occurring lung adenomas have been reported in aging mice, ~18–24 months of age, it is time-saving and cost-effective to use chemical carcinogens to induce lung adenomas because it shortens tumor latency as reviewed by Shimkin and Stoner[11] and Malkinson.[12] Wattenberg and Leong first reported their study on "inhibition of the carcinogenic action of B[a]P by flavones" in 1970.[13] More recently, numerous studies have examined the inhibition of chemical carcinogen-induced lung adenomas in mice generally, and specifically tobacco smoke-related carcinogens NNK and B[a]P. Particular emphasis in many of these studies has been on agents that might block the initiation of relevant carcinogen-induced lung adenomas in A/J mice. These models have proven useful in looking for potential chemopreventive agents.[14,15]

As shown in Fig. 1, lung adenomas are benign tumors with unique characteristics. Grossly, these tumors appear as tiny (1 mm in diameter), round, white nodules on the peripheral surface of the mouse lung (Fig. 1A). Microscopically, these tumors have relatively well-defined borders (as compared to adenocarcinomas and lung squamous cell carcinoma) with monomorphic growth in a glandular pattern and are generally comprised of well-differentiated cells which distort the alveolar septae. Occasionally, nuclear pleomorphisms can be seen in larger adenomas. Generally, the mitotic index is low in adenomas (Fig. 1C). Although lung adenomas are not commonly observed in clinical practice, this may reflect the fact that lung adenomas cause few symptoms and lung cancer patients typically present symptoms of advanced malignancy. The use of advanced imaging techniques is more likely to identify such lesions and reduces mortality from this disease.[16] Nevertheless, lung adenomas have the potential to develop into malignant tumors in the form of adenocarcinomas.[17] Thus, it would be useful to prevent the progression from benign to malignant if tumor detection occurs early.

In contrast to lung adenoma, lung adenocarcinomas in mice are malignant. These tumors appear similar to adenomas as white nodules somewhat round in shape on the peripheral surface of the lung. However, they are generally larger

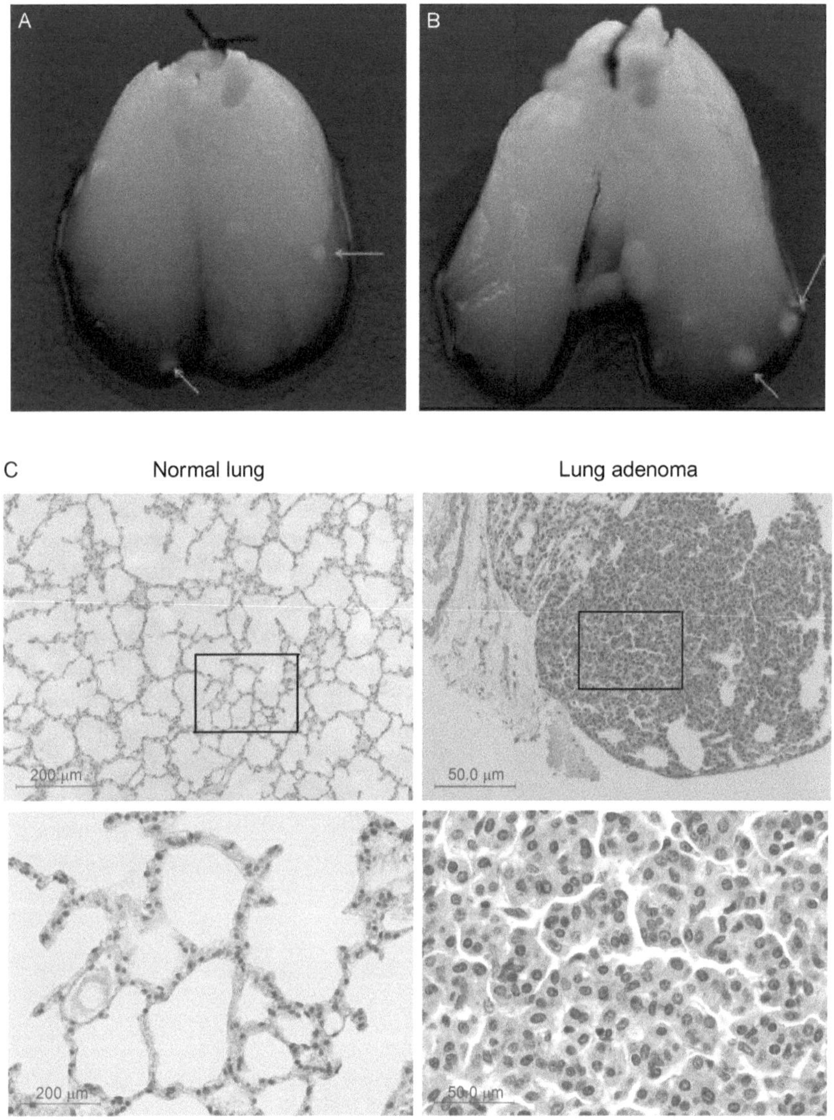

FIG. 1. Lung adenoma induced by one dose of B[a]P, 100 mg/kg BW i.p., 20 weeks, A/J female mouse with normal lung as control: (A) gross view of lung with adenoma; (B) gross view of lung with adenocarcinoma induced by vinyl carbamate , 0.32 mg per mouse, i.p. 20 weeks; (C) histopathology of lung adenomas. (See Color Insert.)

in size than lung adenomas (Fig. 1B). Histologically, lung adenocarcinomas (Fig. 2) have a malignant epithelial glandular growth pattern with acinar, papillary, or tubular patterns, increased nucleus to cytoplasm ratio, increased nuclear pleomorphisms including enlarged nuclei and prominent nucleoli, and increased mitotic indices. Although there is no distance metastasis of these tumors, they can and do have the ability to invade adjacent normal lung tissue (Fig. 2).

Lung adenocarcinomas can be induced by lung-specific chemical carcinogens in wild-type mice with varying latency periods depending on the specific strain of mice used, specific carcinogen used, etc. So far, vinyl carbamate (VC) is the most commonly used chemical carcinogen in the induction of lung adenocarcinoma in A/J mice. Lung adenocarcinomas were also induced by carcinogens in several lines of transgenic or knockout mice such as mice with a dominant-negative *p53* mutation,[18] *K-ras* knockout mice,[17,19,20] and/or mice with a heterozygous deletion of the *Ink4a/Arf* tumor suppressor genes (TSGs),[21] K-ras knockout and p53 mutated bitransgenic mice,[22] and many other strains of transgenic mice.[23–29] Although lung adenocarcinomas occurred spontaneously in certain strains of these transgenic mice, the tumor incidence was typically relatively low, and the latency was usually long. Further, it was confounded by the development of other types of malignancies. Thus, it may be difficult to use many of these transgenic mice in lung cancer chemoprevention studies due to tumors (other than lung adenocarcinoma) developing earlier and occasionally killing the host before the lung tumor is able to develop. Thus, by

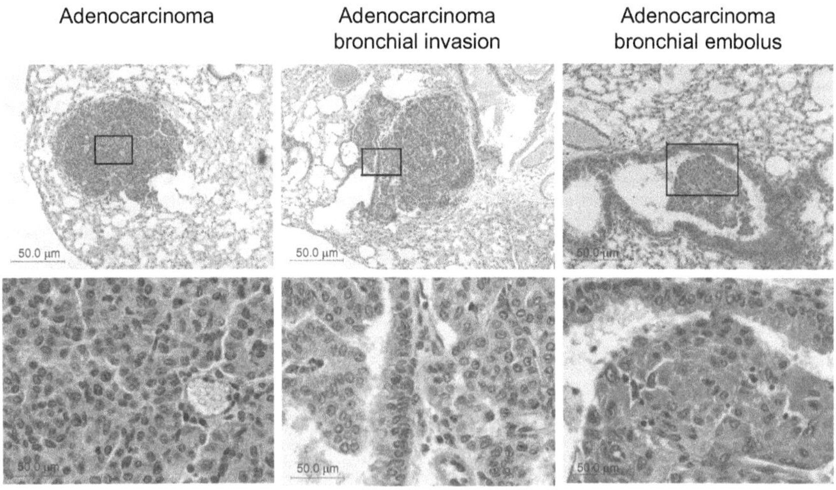

Fig. 2. Histopathology of adenocarcinomas. Histopathology of lung adenocarcinomas induced by vinyl chloride, female A/J mouse. (See Color Insert.)

using specific carcinogens, such as B[a]P, VC, urethane, or NNK, we can increase lung tumor incidence and, simultaneously, decrease latency of lung tumors, such as in mice carrying an Ala-to-Val dominant-negative p53 mutation at codon 135.[21,30–32] Alterations/mutations in *P53* or *P16 ARF* appear particularly relevant since they are often altered in human lung cancer.

Nearly 100% of mouse lung tumors contain a mutant Ki-ras oncogene (Table I). While point mutations occurred randomly in both Codon 12 and Codon 61 in lung tumors of vehicle-control mice, point mutations occurred predominately in Codon 12 in lung tumors induced by B[a]P and in Codon 61 in lung tumors induced by VC.

Different mouse strains have different levels of susceptibilities to lung tumor development when induced either spontaneously or through the use of a carcinogen. For example, C3H/J mice have low lung tumor susceptibility whereas A/J mice have high lung tumor susceptibility. When we reciprocally crossed C3H/J (females) to A/J (males) to produce C3A F1 mice and crossed A/J (females) to C3H/J (males) to produce AC3 F1 mice, we detected point mutations in the Kirsten rat sarcoma viral oncogene homologue (*K-ras*) in both spontaneous and chemically induced lung tumors in these F1 mice. Based on a 37-base-pair deletion in the second intron of the A/J *K-ras* allele, we found that over 95% of mutant *K-ras* genes were found to arise from the allele coded by the A/J parents (the susceptible parent) of both C3A and AC3 F1 mice indicating a direct linkage between the *K-ras* gene and mouse lung tumor susceptibility. This finding agrees with a major role of *K-ras* in the susceptibility of lung cancer development in humans since *K-ras* mutations are detected in 30% of lung adenocarcinomas in humans and is by far the most common mutation other than P53.[40] This further indicates that the use of carcinogens to induce lung tumors not only can reproduce the morphological phenotypes but also can reproduce the *K-ras* genotypes of corresponding human lung tumors.

II. Lung Squamous Cell Carcinoma

Most squamous cell carcinomas (SCCs) arise centrally from either the main, lobar, or segmental bronchi and ulcerate through the mucosa into the surrounding lung parenchyma.[41] Histological and cytological studies have revealed a series of changes that occur over many years and represent a morphological progression to bronchogenic carcinoma. Early changes include a basal cell hyperplasia followed by a squamous metaplasia, dysplasia, carcinoma *in situ*, and invasive SCC.[42–44] The shape and size of the neoplastic cells of lung SCC are polygonal to round with reduced amounts of cytoplasm. The nuclei are irregularly shaped with coarsely clumped chromatin.

TABLE I

K-ras Point Mutations of K-ras Gene Identified in Mouse Lung Tumors Induced by Benzo[a]pyrene (B[a]P) or Vinyl Carbamate (VC)

K-ras mutation in mouse lung tumors												
				K-ras gene point mutations								
				Codon 12 (GGT; glycine)				Codon 61 (CAA; glutamine)				
Treatment	Mouse	Average number of lung tumors per mouse	Mutated K-ras	GTT valine	GAT aspartic acid	TGT cysteine	CGT arginine	CGA arginine	CTA leucine	CAC histidine	CAT histidine	References
Vehicle	A/J		19	4	4		1	7			2	33
Vehicle	A/J		10[a]	2	2		1	2			1	34
Vehicle	C3H	7	3	1	1						1	35
Vehicle	C57BL/6J x C3H F1	23	17 (74%)	3	3	1	1	8			1	36
B[a]P	A/J		14[a]	1	4	8						34
B[a]P	A/J			25%	19%	56%						37
B[a]P	C3H x A/J F1	17	15 (88%)	4	2	9						38
VC	C57BL/6J	20	11 (55%)	2				3	4	1	2	36
VC	A/J x C57BL/6J F1	12	11 (92%)					5	6			39
VC	C57BL/6J x A/J F1	14	13 (93%)					6	7			39
VC	C57BL/6J x C3H F1	37	34 (92%)	3	8			16	7	1		36

[a]Rest of them contain unidentified mutation(s) by the method used.

The presence of uniform gaps between cells is clearly observed, although the intercellular bridges are somewhat less variable in SCCs; both confirm the squamous identity of the tumor. Bronchiolar hyperplasia/metaplasia and premalignant lesions of lung SCC were also observed. Bronchial hyperplasia is diagnosed when there are more layers of bronchiolar epithelial cells in a bronchiole. Bronchial metaplasia is diagnosed when bronchiolar epithelial cells show abnormalities in cytoplasmic and nuclear morphology. Lung SCC *in situ*, epithelial cells restricted within the bronchiole show the malignant features. As shown in Fig. 4, *N*-nitroso-tris-chloroethylurea (NTCU)-induced lung SCC in female NIH Swiss mice closely resembles human lung SCC histopathologically. Unlike mouse lung adenomas and adenocarcinomas (Figs. 1C and 2), mouse lung SCC does not form round-shaped circumscribed white nodules. At approximately 32 weeks after the first dose of NTCU, lungs became dense and less transparent as compared to normal lungs (Fig. 3A). A small percentage of lungs have shown white patches without a clear boundary on the surface ~40 to ~45 weeks after the first dose of NTCU (Fig. 3B).

Due to the prevalence of human lung SCC, the development of a relevant model is particularly important since the standard Phase II chemoprevention trials in the lung have examined modulation of squamous cell lesions of the bronchus following fluorescent imaging.[45–48] Since this approach has generated the greatest amount of clinical data on the lung, the existence of a relevant animal model is of particular importance. Previous research showed that delivery of B[a]P directly into the trachea was successful in producing mouse SCC; however, subsequent studies found these results difficult to reproduce.[49] Thus, other methods were examined. One of these methods utilized skin painting of NTCU and of *N*-nitroso-methyl-bis-chloroethylurea in order to induce mouse SCC.[50] Unlike previous methods, further attempts to reproduce the results of this study were successful.[51] Female mice of eight different strains, which included A/J, SWR/J, 129/svJ, FVB/J, AKR/J, C57BL/6J, BALB/cJ, and NIH Swiss, were acquired and were separated into two groups at 7 weeks of age. One of these groups received NTCU treatment while the other received acetone which was the solvent for NTCU. Mice were sacrificed by CO_2 asphyxiation, and their lungs were collected and fixed in solution. While three strains of mice did not develop lung SCC, five of the strains did. Interestingly, strain sensitivity to induction of squamous cell cancer by NTCU did not correspond to strain susceptibility to induction of adenomas by various carcinogens. Histopathological characteristics of the tumors present demonstrated that not only did the mice develop lung SCC, but that in many ways mouse lung SCC was similar to that of human lung SCC. Further similarities were seen when it was assessed that there were no mutations within existing tumors in the *ras* gene family

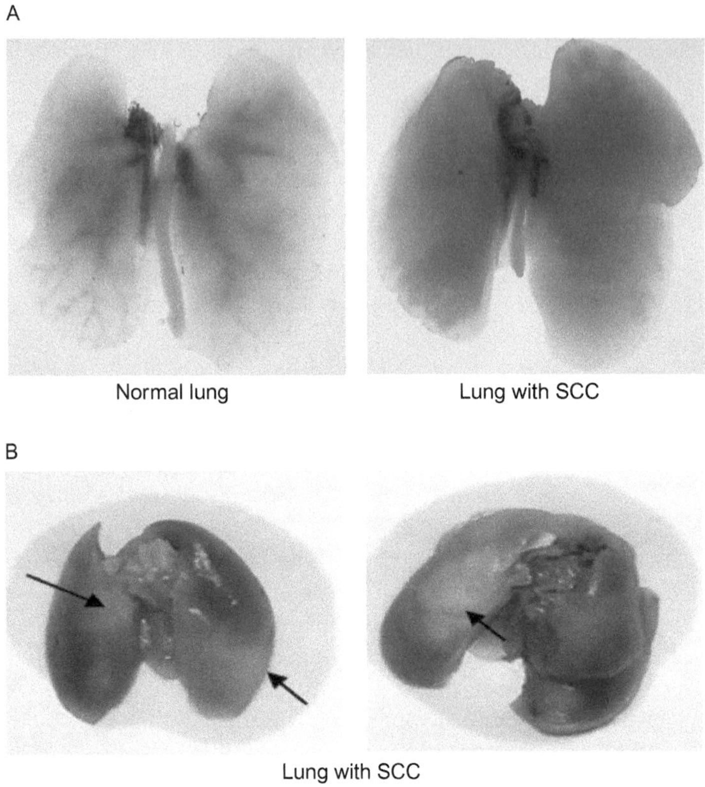

FIG. 3. Gross pathology of squamous cell carcinomas induced by NTCU in A/J mice. Adapted from Ref. 61. (See Color Insert.)

which is also consistent with that of human lung SCC. It was determined from this study that NTCU skin painting was sufficient in producing a comparable model of human lung SCC (Fig. 4) in mice.[32,51] One limitation of the model is that it requires repeated topical applications of the carcinogen NTCU. This treatment is additionally associated with some toxicity including localized skin irritation and a reduction in the rate of body weight gain. The degree of toxicity is dose dependent and strain dependent. NIH Swiss mice appear to exhibit less toxicity than A/J mice (unpublished data). We have incorporated a dominant-negative *P53* mutation into the model as well which results in a greater incidence of advanced histopathologic lesions.

Recently, lung SCC has been reported in a genetically modified mouse model. LKB1 or serine/threonine kinase 11 (also called *STK11*) is a tumor suppressor gene with inactivating somatic mutations which have been detected

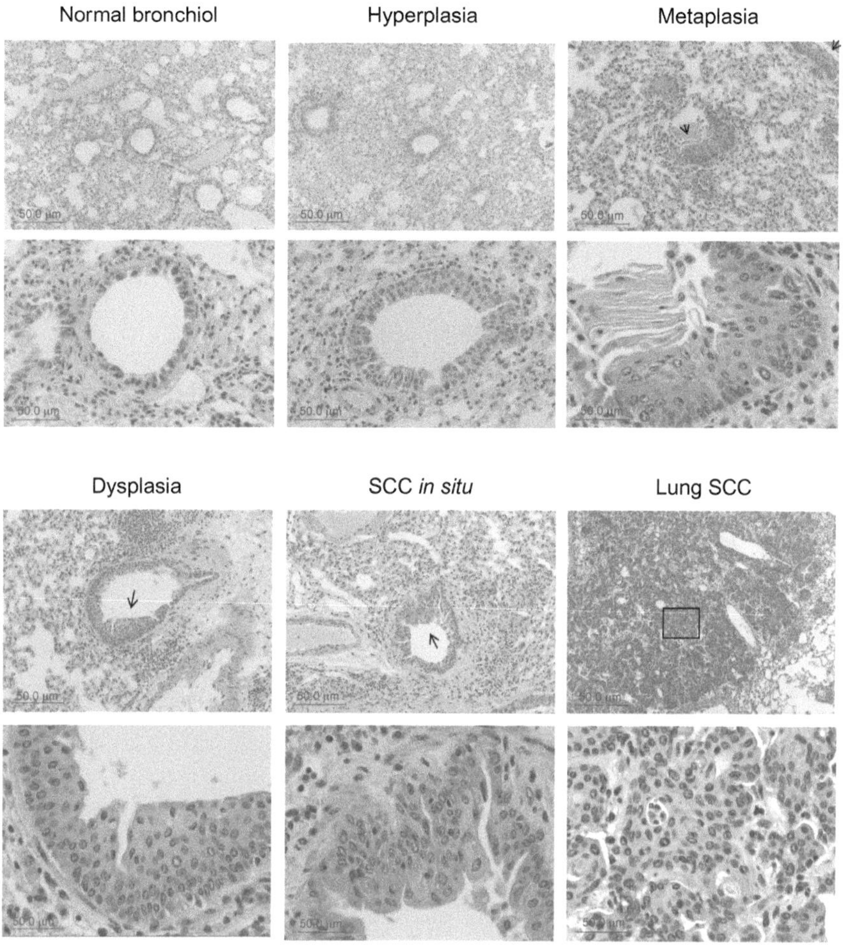

FIG. 4. Histopathology of squamous cell lesions induced by NTCU. (See Color Insert.)

in primary human lung adenocarcinomas and derivative cell lines. Ji *et al.*[52] intercrossed a conditionally activatable *Lox-Stop-Lox Kras*^{G12D} (*Ki-ras*) allele and conditional (L/L) or germline null (−/−) alleles of *Lkb1*$^{L/L}$, *p53*$^{L/L}$, *Ink4a/Arf*$^{-/-}$, or *p16*$^{Ink4a-/-}$. After transducing pulmonary cells by adeno-Cre virus through tracheal inhalation, they found that inactivation of *Lkb1* alone was not sufficient for inducing pulmonary neoplasia. They found that the strongest genetic interaction with homozygous *Lkb1* inactivation was mutated *K-ras*. Inactivation of LKB1 facilitated lung tumor growth and metastasis of all three human NSCLC (non-small cell lung cancer) subtypes: 15 out of 27 *Ki-ras Lkb1*$^{L/L}$ or *Ki-ras Lkb1*$^{L/-}$ mice harbored SCC or adenosquamous carcinoma whereas 2 out of 27 lungs showed large cell lung carcinomas. Although *K-RAS* mutation have been reported in all subtypes of human NSCLCs, the mutation rate of K-*RAS* in lung SCC is only 6% (Table S2 in Ref. 52) which make this model most directly relevant to a small subfraction of SCCs. The model may, nevertheless, be somewhat useful for preclinical chemopreventive studies.

III. Lung Small Cell Carcinoma

Despite being less prevalent than lung adenocarcinoma or SCC, small cell lung cancer (SCLC) (which represents 15–20% of all lung cancers) is a histologically distinct type of lung cancer with a generally poor prognosis.[53] This particular type of lung cancer is composed of cells which are small in size with highly pleomorphic involuted nuclei and a high nuclear/cytoplasmic ratio (Fig. 5), a high mitotic index, and is characterized by a high prevalence of mutations in Rb1 and p53 genes.[54] SCLC is believed to originate from cells residing in the epithelial lining of the bronchi, which have a neuroendocrine phenotype. They express markers of neuroendocrine differentiation such as chromogranin A, neuron-specific enolase, synaptophysin, or neural cell adhesion molecule.[55] Consequently, it was found that by inactivating these two genes using an adenovirus-mediated somatic gene transfer of Cre recombinase, small cell lung carcinoma could be produced. The resulting tumors had neuroendocrine biomarkers similar to the human disease. Further, these tumors have histologic characteristics that parallel those observed in humans (Fig. 5).

IV. Chemopreventive Applications in Preclinical Lung Cancer Models

Extensive studies have been done with these preclinical lung cancer models in searching for potential chemopreventive agents. Many of the early studies as mentioned above examined agents that might directly alter the metabolism of carcinogenic agents; for example, flavones or PEITC which can inhibit the activation of B(a)P and NNK, respectively. More recently, the focus has been on agents that might work later in tumor progression which might be more directly relevant to ex-smokers or even long-term smokers who continue to smoke. Among the compounds which we have found that routinely exhibit efficacy are certain tea polyphenols and the RXR agonist bexarotene. For example, there are many reported studies conducted on polyphenon

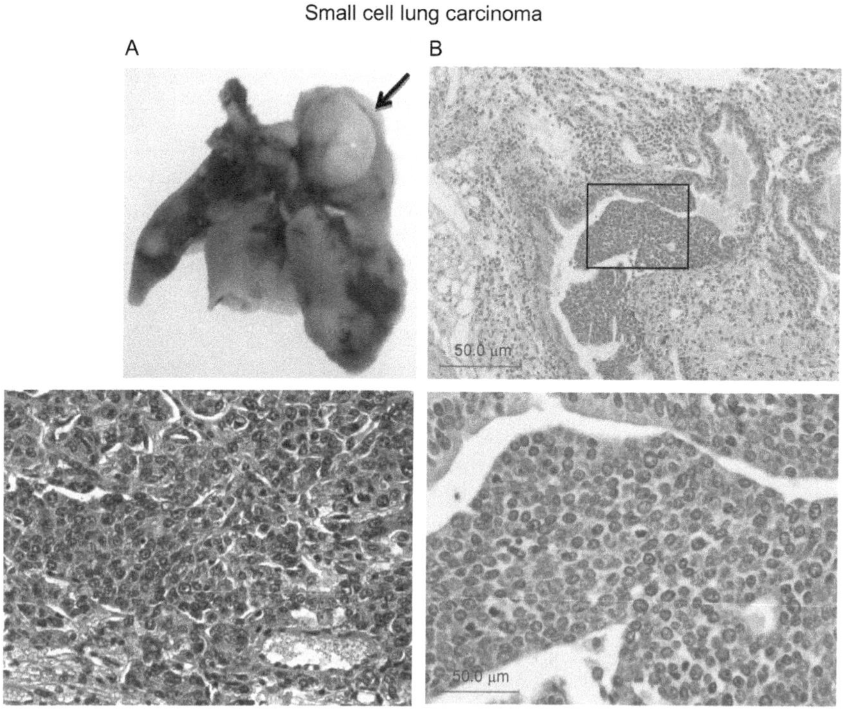

FIG. 5. Small cell carcinoma induced in bitransgenic mice with floxed Rb and p53 tumor suppressor genes. (See Color Insert.)

E (Poly E), a mixture of green tea polyphenol which have illustrated the strong chemopreventive effects of this agent as reviewed by Yang.[56] Epigallocatechin-3-gallate (EGCG) constitutes about 65% of Poly E and is also the most ample catechin in green tea. Based on this knowledge and previous research, it was inferred that EGCG given with food would have chemopreventive effects comparable to Poly E. We have evaluated Poly E in A/J mice treated with the tobacco-specific carcinogen B[a]P and found that treatment with Poly E initiated soon after expose to the carcinogen induced a significant reduction in both tumor multiplicity and tumor load (sum of tumor volume of all tumors per lung). Poly E (2%, w/w) in the diet reduced tumor multiplicity by 46% and tumor load by 94%.[57] This inhibitory effect of Poly E was also shown in B(a)P-treated mice through aerosol delivery.[58] In our more recent study, we have shown that Poly E can prevent lung SCC development in NTCU-induced lung SCC model. When compared to the NTCU control mice, it was found that, upon Poly E treatment, the percentage of more advanced pathologies CIS and

SCC were strongly decreased ($p = 0.0038$) while the percentages of less advanced pathologies (metaplasia, hyperplasia, or normal bronchial histology) increased (unpublished data).

Another agent examined in some detail was Bexarotene, a synthetic RXR agonist. We have studied Bexarotene in the chemoprevention of adenocarcinomas in wild-type mice or mice with a *K-ras* heterozygous knockout or dominant-negative *p53* mutation on an A/J background.[22] Adenocarcinomas were induced using one injection of VC in 6-week-old mice. Mice were then separated into groups according to genotype and treatment. At 16 weeks of age, mice in the treatment groups were administered Bexarotene in corn oil by oral gavage. Mice were sacrificed after 12 weeks of the treatment. We found Bexarotene treatment caused a significant reduction in tumor numbers by 28% ($p < 0.05$), 37% ($p < 0.05$), and 38% ($p < 0.05$) when compared to ($p < 0.05$) paired vehicle controls in wild-type, K-Ras$^{ko/wt}$, or p53$^{wt/Ala135Val}$ mice, respectively. Similarly, the final tumor volumes in the mice treated with Bexarotene were reduced by 26%, 48% ($p < 0.05$), and 45% ($p = 0.07$) of their respective control groups. Through microarray analysis, it was found that Bexarotene alters expression of a number of genes involved in cell cycle regulation. The efficacy of the RXR agonist has been confirmed with other RXR agonists by our laboratories, as well as others (Sporn and our unpublished data with UAB-30) and most importantly it has been shown to be active in human NSCLC.[59] With its potent inhibitory effects, Bexarotene was determined to be a very promising chemopreventive agent. However, bexarotene does increase triglyceride levels. We have recently administered bexarotene aerosol delivery and eliminated this systemic toxicity while retaining preventive efficacy.[60]

As mentioned before, lung cancer is composed of multiple histologic subtypes. There is no easy way to predict the type of lung cancer a patient may develop in their lifetime. In our next studies, we examined the effect of Bexarotene for its chemopreventive efficacy on SCLC and NSCLC. To test Bexarotene in the SCLC model, mice with double-conditional knockout of the tumor suppressor genes *p53* and *Rb* ($Trp53^{F2-10/F2-10}$; $Rb1^{F19/F19}$) were treated with Bexarotene by oral gavage 2 weeks before administration of adeno-Cre virus. Our study showed that Bexarotene decreased tumor multiplicity by 86% ($p = 0.010$) and decreased tumor load by 92% ($p = 0.039$) as compared to the control mice.[61] Similar effects were seen in lung SCC by oral administration of Bexarotene (our unpublished data). Bexarotene decreased the percentage of slides exhibiting a clear SCC histopathology by $\sim 37\%$ ($p = 0.0029$) while increasing the incidence of slides showing normal, hyperplastic, or metaplastic histologies. Immunohistochemical staining indicated that Bexarotene decreased proliferation and increased apoptosis of tumor cells. Our studies indicated that Bexarotene has a broad utility in lung cancer chemoprevention across the diverse lung pathologies as it prevents SCLC and NSCLC (adenoma, adenocarcinoma, and squamous cell carcinoma) in preclinical settings.

V. Conclusion

These animal models offer promising results for the application to human studies. The well-established measurement of lung tumor development provides us with numerous opportunities to use these preclinical tumor models to search for effective agents to prevent lung cancer development, as well as further our understanding of the molecular mechanisms through which these agents work. This will ultimately aid in developing a broader understanding of tumorigenesis inhibition and prevention and will hopefully lead to not only further studies but also clinical advancements as well.

References

1. Jemal A, Siegel R, Xu J, Ward E. Cancer statistics, 2010. *CA Cancer J Clin* 2010;**60**:277–300.
2. Hong W, Sporn M. Recent advances in chemoprevention of cancer. *Science* 1997;**278**:1073–7.
3. Mori K. Experimental induction of pulmonary tumors in mice. *Showa Igakkai Zasshi* 1962;**22**:51–2.
4. Snell KC, Stewart HL. Induction of pulmonary adenomatosis in DBA/2 mice by the oral administration of dibenz(a, h)antrhracene. *Acta Unio Int Contra Cancrum* 1963;**19**:692–4.
5. Toth B, Shubik P. Mammary tumor inhibition and lung adenoma induction by isonicotinicacid hydrazide. *Science* 1966;**152**:1376–7.
6. Kaye AM, Trainin N. Urethan carcinogenesis and nucleic acid metabolism: factors influencing lung adenoma induction. *Cancer Res* 1966;**26**:2206–12.
7. Greenblatt M, Mirvish S, So BT. Nitrosamine studies: induction of lung adenomas byconcurrent administration of sodium nitrite and secondary amines in Swiss mice. *J Natl Cancer Inst* 1971;**46**:1029–34.
8. Leong BK, Macfarland HN, Reese Jr. WH. Induction of lung adenomas by chronic inhalation of bis (chloromethyl) ether. *Arch Environ Health* 1971;**22**:663–6.
9. Rigdon RH, Neal J. Gastric carcinomas and pulmonary adenomas in mice fed benzo(a)pyrene. *Tex Rep Biol Med* 1966;**24**:195–207.
10. Hecht SS, Morse MA, Amin S, Stoner GD, Jordan KG, Choi CI, et al. Rapid single-dose model for lung tumor induction in A/J mice by 4-(methylnitrosamino)-1-(3-pyridyl)-1-butanone and the effect of diet. *Carcinogenesis* 1989;**10**:1901–4.
11. Shimkin MB, Stoner GD. Lung tumors in mice: application to carcinogenesis bioassay. *Adv Cancer Res* 1975;**21**:1–58; Review.
12. Malkinson AM. The genetic basis of susceptibility to lung tumors in mice. *Toxicology* 1989;**54**:241–71; Review. Erratum in: *Toxicology* 1989;**55**:349.
13. Wattenberg LW, Leong JL. Inhibition of the carcinogenic action of benzo(a)pyrene by flavones. *Cancer Res* 1970;**30**:1922–5.
14. Castonguay A, Pepin P, Stoner GD. Lung tumorigenicity of NNK given orally to A/J mice: its application to chemopreventive efficacy studies. *Exp Lung Res* 1991;**17**:485–99.

15. Hecht SS, Isaacs S, Trushin N. Lung tumor induction in A/J mice by the tobacco smoke carcinogens 4-(methylnitrosamino)-1-(3-pyridyl)-1-butanone and benzo[a]pyrene: a potentially useful model for evaluation of chemopreventive agents. *Carcinogenesis* 1994;**15**:2721–5.
16. The National Lung Screening Trial Research Team. Reduced lung-cancer mortality with low-dose computed tomographic screening. *N Engl J Med* 2011;**365**(5):395–409; PMID: 21714641.
17. Jackson EL, Willis N, Mercer K, Bronson RT, Crowley D, Montoya R, et al. Analysis of lung tumor initiation and progression using conditional expression of oncogenic K-ras. *Genes Dev* 2001;**15**:3243–8.
18. Lavigueur A, Maltby V, Mock D, Rossant J, Pawson T, Bernstein A. High incidence of lung, bone, and lymphoid tumors in transgenic mice overexpressing mutant alleles of the p53 oncogene. *Mol Cell Biol* 1989;**9**:3982–91.
19. Fisher GH, Wellen SL, Klimstra D, Lenczowski JM, Tichelaar JW, Lizak MJ, et al. Induction and apoptotic regression of lung adenocarcinomas by regulation of a K-Ras transgene in the presence and absence of tumor suppressor genes. *Genes Dev* 2001;**15**:3249–62.
20. Zhang Z, Wang Y, Vikis HG, Johnson L, Liu G, Li J, et al. Wildtype Kras2 can inhibit lung carcinogenesis in mice. *Nat Genet* 2001;**29**:25–33.
21. Zhang Z, Wang Y, Lantry LE, Kastens E, Liu G, Hamilton AD, et al. Farnesyltransferase inhibitors are potent lung cancer chemopreventive agents in A/J mice with a dominant-negative p53 and/or heterozygous deletion of Ink4a/Arf. *Oncogene* 2003;**22**:6257–65.
22. Wang Y, Zhang Z, Yao R, Jia D, Wang D, Lubet RA, et al. Prevention of lung cancer progression by bexarotene in mouse models. *Oncogene* 2006;**25**:1320–9.
23. Mallakin A, Sugiyama T, Taneja P, Matise LA, Frazier DP, Choudhary M, et al. Mutually exclusive inactivation of DMP1 and ARF/p53 in lung cancer. *Cancer Cell* 2007;**12**:381–94.
24. Yanagi S, Kishimoto H, Kawahara K, Sasaki T, Sasaki M, Nishio M, et al. Pten controls lung morphogenesis, bronchioalveolar stem cells, and onset of lung adenocarcinomas in mice. *J Clin Invest* 2007;**117**:2929–40.
25. Xiao W, Zhang Q, Habermacher G, Yang X, Zhang AY, Cai X, et al. U19/Eaf2 knockout causes lung adenocarcinoma, B-cell lymphoma, hepatocellular carcinoma and prostatic intraepithelial neoplasia. *Oncogene* 2008;**27**:1536–44.
26. Tanaka-Okamoto M, Hori K, Ishizaki H, Hosoi A, Itoh Y, Wei M, et al. Increased susceptibility to spontaneous lung cancer in mice lacking LIM-domain only 7. *Cancer Sci* 2009;**100**:608–16.
27. Yoneda M, Suzuki T, Nakamura T, Ajima R, Yoshida Y, Kakuta S, et al. Deficiency of antiproliferative family protein Ana correlates with development of lung adenocarcinoma. *Cancer Sci* 2009;**100**:225–32.
28. Berger AH, Niki M, Morotti A, Taylor BS, Socci ND, Viale A, et al. Identification of DOK genes as lung tumor suppressors. *Nat Genet* 2010;**42**:216–23.
29. Deng J, Fujimoto J, Ye XF, Men TY, Van Pelt CS, Chen YL, et al. Knockout of the tumor suppressor gene Gprc5a in mice leads to NF-kappaB activation in airway epithelium and promotes lung inflammation and tumorigenesis. *Cancer Prev Res (Phila)* 2010;**3**:424–37.
30. Zhang Z, Liu Q, Lantry LE, Wang Y, Kelloff GJ, Anderson MW, et al. A germ-line p53 mutation accelerates pulmonary tumorigenesis: p53-independent efficacy of chemopreventive agents green tea or dexamethasone/myo-inositol and chemotherapeutic agents taxol or adriamycin. *Cancer Res* 2000;**60**:901–7.
31. Wang Y, Zhang Z, Kastens E, Lubet RA, You M. Mice with alterations in both p53 and Ink4a/Arf display a striking increase in lung tumor multiplicity and progression: differential chemopreventive effect of budesonide in wild-type and mutant A/J mice. *Cancer Res* 2003;**63**:4389–95.
32. Wang Y, James M, Wen W, Lu Y, Szabo E, Lubet RA, et al. Chemopreventive effects of pioglitazone on chemically induced lung carcinogenesis in mice. *Mol Cancer Ther* 2010;**9**:3074–82.

33. Chen B, Liu L, Castonguay A, Maronpot RR, Anderson MW, You M. Dose-dependent ras mutation spectra in N-nitrosodiethylamine induced mouse liver tumors and 4-(methylnitrosamino)-1-(3-pyridyl)-1-butanone induced mouse lung tumors. *Carcinogenesis* 1993;**14**:1603–8.
34. You M, Candrian U, Maronpot RR, Stoner GD, Anderson MW. Activation of the Ki-ras protooncogene in spontaneously occurring and chemically induced lung tumors of the strain A mouse. *Proc Natl Acad Sci USA* 1989;**86**:3070–4.
35. Devereux TR, Anderson MW, Belinsky SA. Role of ras protooncogene activation in the formation of spontaneous and nitrosamine-induced lung tumors in the resistant C3H mouse. *Carcinogenesis* 1991;**12**:299–303.
36. Massey TE, Devereux TR, Maronpot RR, Foley JF, Anderson MW. High frequency of K-ras mutations in spontaneous and vinyl carbamate-induced lung tumors of relatively resistant B6CF1 (C57BL/6J x BALB/cJ) mice. *Carcinogenesis* 1995;**16**:1065–9.
37. Mass MJ, Jeffers AJ, Ross JA, Nelson G, Galati AJ, Stoner GD, et al. Ki-ras oncogene mutations in tumors and DNA adducts formed by benz[j]aceanthrylene and benzo[a]pyrene in the lungs of strain A/J mice. *Mol Carcinog* 1993;**8**:186–92.
38. Chen B, You L, Wang Y, Stoner GD, You M. Allele-specific activation and expression of the K-ras gene in hybrid mouse lung tumors induced by chemical carcinogens. *Carcinogenesis* 1994;**15**:2031–5.
39. You M, Wang Y, Lineen AM, Gunning WT, Stoner GD, Anderson MW. Mutagenesis of the K-ras protooncogene in mouse lung tumors induced by N-ethyl-N-nitrosourea or N-nitrosodiethylamine. *Carcinogenesis* 1992;**13**:1583–6.
40. You M, Wang Y, Stoner G, You L, Maronpot R, Reynolds SH, et al. Parental bias of Ki-ras oncogenes detected in lung tumors from mouse hybrids. *Proc Natl Acad Sci USA* 1992;**89**:5804–8.
41. Travis WD, Travis LB, Devesa SS. Lung cancer. *Cancer* 1995;**75**:191–202.
42. Auerbach O, Forman JB, Gere JB, Kassouny DY, Muehsam GE, Petrick TG, et al. Changes in the bronchial epithelium in relation to smoking and cancer of the lung; a report of progress. *N Engl J Med* 1957;**256**:97–104.
43. Auerbach O, Stout AP, Hammond EC, Garfinkel L. Changes in bronchial epithelium in relation to cigarette smoking and in relation to lung cancer. *N Engl J Med* 1961;**265**:253–67.
44. Saccomanno G, Archer VE, Auerbach O, Saunders RP, Brennan LM. Development of carcinoma of the lung as reflected in exfoliated cells. *Cancer (Phila)* 1974;**33**:256–70.
45. Lam S, MacAulay C, Le Riche JC, Dyachkova Y, Coldman A, Guillaud M, et al. A randomized phase IIb trial of anethole dithiolethione in smokers with bronchial dysplasia. *J Natl Cancer Inst* 2002;**94**:1001–9.
46. Lam S, leRiche JC, McWilliams A, Macaulay C, Dyachkova Y, Szabo E, et al. A randomized phase IIb trial of pulmicort turbuhaler (budesonide) in people with dysplasia of the bronchial epithelium. *Clin Cancer Res* 2004;**10**:6502–11.
47. McWilliams A, Lam S. New approaches to lung cancer prevention. *Curr Oncol Rep* 2002;**4**:487–94 Review. PMID: 12354360.
48. Edell E, Lam S, Pass H, Miller YE, Sutedja T, Kennedy T, et al. Detection and localization of intraepithelial neoplasia and invasive carcinoma using fluorescence-reflectance bronchoscopy: an international, multicenter clinical trial. *J Thorac Oncol* 2009;**4**:49–54.
49. Yoshimoto T, Hirao F, Sakatani M, Nishikawa H, Ogura T. Induction of squamous cell carcinoma in the lung of C57BL/6 mice by intratracheal instillation of benzo[a]pyrene with charcoal powder. *Gann* 1977;**68**:343–52.
50. Lijinsky W, Reuber M. Neoplasms of the skin and other organs observed in Swiss mice treated with nitrosoalkylureas. *J Cancer Res Clin Oncol* 1988;**114**:245–9.
51. Wang Y, Zhang Z, Yan Y, Lemon WJ, LaRegina M, Morrison C, et al. A chemically induced model for squamous cell carcinoma of the lung in mice: histopathology and strain susceptibility. *Cancer Res* 2004;**64**:1647–54.

52. Ji H, Ramsey MR, Hayes DN, Fan C, McNamara L, Kozlowski P, et al. LKB1 modulates lung cancer differentiation and metastasis. *Nature* 2007;**448**:807–10.
53. Brownson R, Chang J, Davis J. Gender and histologic type variations in smoking-related risk of lung cancer. *Epidemiology* 1992;**3**:61–4.
54. Gouyer V, Gazzeri S, Bolon I, Drevet C, Brambilla C, Brambilla E. Mechanism of retinoblastoma gene inactivation in the spectrum of neuroendocrine lung tumors. *Am J Respir Cell Mol Biol* 1998;**18**:188–96.
55. Meuwissen R, Linn SC, Linnoila RI, Zevenhoven J, Mooi WJ, Berns A. Induction of small cell lung cancer by somatic inactivation of both Trp53 and Rb1 in a conditional mouse model. *Cancer Cell* 2003;**4**:181–9.
56. Yang CS, Wang X. Green tea and cancer prevention. *Nutr Cancer* 2010;**62**:931–7; Review.
57. Yan Y, Wang Y, Tan Q, Hara Y, Yun TK, Lubet RA, et al. Efficacy of polyphenon E, red ginseng, and rapamycin on benzo(a)pyrene-induced lung tumorigenesis in A/J mice. *Neoplasia* 2006;**8**:52–8.
58. Yan Y, Cook J, McQuillan J, Zhang G, Hitzman CJ, Wang Y, et al. Chemopreventive effect of aerosolized polyphenon E on lung tumorigenesis in A/J mice. *Neoplasia* 2007;**9**:401–5.
59. Dragnev KH, Ma T, Cyrus J, Galimberti F, Memoli V, Busch AM, et al. Bexarotene plus erlotinib suppress lung carcinogenesis independent of KRAS mutations in two clinical trials and transgenic models. *Cancer Prev Res (Phila)* 2011;**4**:818–28.
60. Zhang Q, Pan J, Zhang J, Liu P, Chen R, Chen DR, et al. Aerosolized bexarotene inhibits lung tumorigenesis without increasing plasma triglyceride and cholesterol levels in mice. *Cancer Prev Res (Phila)* 2011;**4**:270–6.
61. Wang Y, Wen W, Yi Y, Zhang Z, Lubet RA, You M. Preventive effects of bexarotene and budesonide in a genetically engineered mouse model of small cell lung cancer. *Cancer Prev Res (Phila)* 2009;**2**:1059–64.

Cardiovascular Magnetic Resonance Imaging in Small Animals

RENÉ M. BOTNAR* AND MARCUS R. MAKOWSKI*,†

*Division of Imaging Sciences, King's College London, London, United Kingdom

†Department of Radiology, Charite, Berlin, Germany

Noninvasive imaging studies involving small animals are becoming increasingly important in preclinical pharmacological, genetic, and biomedical cardiovascular research. Especially small animal magnetic resonance imaging (MRI) using high field and clinical MRI systems has gained significant importance in recent years. Compared to other imaging modalities, like computer tomography, MRI can provide an excellent soft tissue contrast,

Progress in Molecular Biology
and Translational Science, Vol. 105
DOI: 10.1016/B978-0-12-394596-9.00008-1

which enables the characterization of different kinds of tissues without the use of contrast agents. In addition, imaging can be performed with high spatial and temporal resolution.

Small animal MRI cannot only provide anatomical information about the beating murine heart; it can also provide functional and molecular information, which makes it a unique imaging modality. Compared to clinical MRI examinations in humans, small animal MRI is associated with additional challenges. These included a smaller size of all cardiovascular structures and a up to ten times higher heart rate. Dedicated small animal monitoring devices make a reliable cardiac triggering and respiratory gating feasible.

MRI in combination with molecular probes enables the noninvasive imaging of biological processes at a molecular level. Different kinds of iron oxide or gadolinium-based contrast agents can be used for this purpose. Compared to other molecular imaging modalities, like single photon emission computed tomography (SPECT) and positron emission tomography (PET), MRI can also provide imaging with high spatial resolution, which is of high importance for the assessment of the cardiovascular system. The sensitivity for detection of MRI contrast agents is however lower compared to sensitivity of radiation associated techniques like PET and SPECT.

This chapter is divided into the following sections: (1) "Introduction," (2) "Principals of Magnetic Resonance Imaging," (3) "MRI Systems for Preclinical Imaging and Experimental Setup," and (4) "Cardiovascular Magnetic Resonance Imaging."

I. Introduction

Cardiovascular diseases and its consequences remain the main cause of mortality and morbidity in industrialized and developing nations. Animal models of myocardial infarction/ischemia and atherosclerosis have significantly contributed to our understanding of underlying biological and molecular process of these diseases. Beyond the characterization of diseases, these models allow the evaluation of novel medical therapies and contrast agents *in vivo*.

Noninvasive methods, that allow for the *in vivo* imaging, quantification and monitoring of cardiovascular diseases and response to therapy are becoming more important in biomedical research. Different types of preclinical imaging systems are available. Small animal computer tomography (CT) systems offer high spatial resolution in combination with short imaging times and are the preferential systems for the investigation of calcifications and calcium dense structures. Without the injection of contrast agents, CT does not provide a high soft tissue contrast. Small animal ultrasound systems offer a high soft tissue contrast in combination with high spatial and temporal resolution. Ultrasound systems suffer from the same limitations in small animals as in humans, which include the high operator and acoustic window dependence.

In the context of molecular imaging, small animal single photon emission computed tomography (SPECT) and positron emission tomography (PET) systems offer a high sensitivity for detection all contrast agents. These systems have a limited spatial and temporal resolution, which is of high importance for imaging of the cardiovascular system.

Magnetic resonance imaging (MRI) allows the characterization of different cardiovascular tissues, for example, myocardial tissue and the vascular wall, with high soft tissue contrast without the use of contrast agents. Despite the high heart and breathing rates in small animals, reliable cardiac triggering and respiratory gating is feasible. Different sequences for the evaluation of ventricular function and myocardial mass were validated in small animal and clinical MRI systems. Functional changes in myocardial and vessel wall perfusion can also be imaged with high temporal resolution. The application of untargeted contrast agents allows for the characterization of, for example, the extent of myocardial infarction with high spatial resolution. Targeted contrast agents enable the evaluation of specific cellular and subcellular markers on a molecular level.

II. Principals of MRI

A. Introduction

MRI is a nonionizing tomographic imaging modality, which can be used to generate anatomical, functional, and molecular images with excellent soft tissue contrast of the human body. The nuclear magnetic resonance phenomenon was discovered by Bloch and Purcell[1,2] in 1946. Twenty-five years later, its potential application as an imaging modality was first reported by Lauterbur and Mansfield.[3,4] This section is an introduction to the basic principles of nuclear magnetic resonance. A focus is on topics relevant to the following sections.

B. Principals of Nuclear Magnetic Resonance

The generation of a measurable signal in MRI is based on the absorption and subsequent emission of radiofrequency (RF) waves. Due to the high abundance of hydrogen ^{1}H in humans, it is used for imaging in a medical context.

The hydrogen nucleus consists of a single electron and proton. Atomic nuclei having odd atomic numbers and/or atomic weight, (e.g., hydrogen atoms) possess an angular moment, known as spin. Nuclei with both even mass and charge numbers do not exhibit spin angular momentum. This property can be displayed as a spinning motion of the nucleus about its own axis (Fig. 1).

FIG. 1. Representation of a proton spinning about its own axis and its magnetic moment vector. Adapted from Ref. 5.

Associated with the spin angular momentum is the magnetic property of the atom. This microscopic magnetic field can be represented by a vector, called the magnetic moment. The spin angular moment and the magnetic moment are related to each other by

$$\vec{\mu} = \gamma \cdot \vec{J},$$

where γ is a physical constant known as the gyromagnetic ratio, a unique constant for each nucleus possessing a spin. For ^1H it has a value of 42.58 MHz/T.

Due to thermal random motion, the direction of the magnetic field of each ^1H is random in the absence of an external magnetic field. Therefore, no measurable spontaneous magnetization exists around a macroscopic object (Fig. 2).

If an external static magnetic field (B_0) is applied along the z-axis, the nuclei acquire energy E due to the interaction with the field B_0 and the magnetic moment (Fig. 3).

The orientations of the magnetic moments align at a specific angle opposed to or along with the applied external field, known as Zeemann splitting.

The two opposing orientations display slightly different energies, with the difference being proportional to the magnitude of the applied field. Applying an additional magnetic field B_1, transitions between these states can be induced. This external field is applied perpendicular to the direction of B_0. Since an alignment along rather than opposed to the B_0 field provides a lower energy state, a higher number of nuclei are going to align along the B_0 field, which will yield a net magnetization along the z-axis, the longitudinal magnetization. The magnetic moments will align at a specific angle ($\pm 54°$) along B_0, resulting in torque that leads to a precessional movement of the magnetic moments about the B_0 field (Fig. 4). The Larmor constant characterizes this angular frequency (ω_0). To sum up the macroscopic magnetization of a spin system, a vector $\vec{M}$ is introduced. However, magnetization along the z-axis cannot be detected, therefore it is necessary to tilt the magnetization from the z-axis towards the xy-plane. This can be achieved by a RF field B_1, applied in the xy-plane.

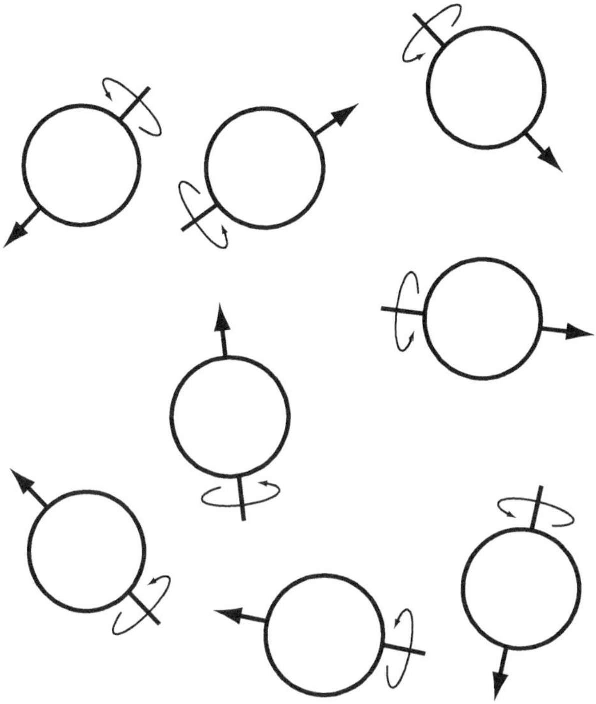

FIG. 2. In the absence of an applied magnetic field, nuclear spins in a material in free space are randomly oriented. The sum of their magnetic moments, also known as the net magnetization, is zero. Adapted from Ref. 5.

For a better understanding and visualization of the effects of B_1, the concept of a rotating frame of reference, spinning about the z-axis at frequency$\omega_0$is introduced. In this reference system the nuclear magnetization can be described as stationary. RF pulses of B_1 will shift the net magnetization away from the z-axis into the xy-plane of the rotating frame. Hence, changes in the net nuclear magnetization can be caused by applying an oscillating B_1 field with an identical Larmor frequency of the nuclei.

C. Relaxation Phenomena

The motion of $\overrightarrow{M}$ in a strong external magnetic field B_0 and the magnetic component B_1 can be described by a set of differential equations.[1]

$$\omega_0 = \gamma B_0.$$

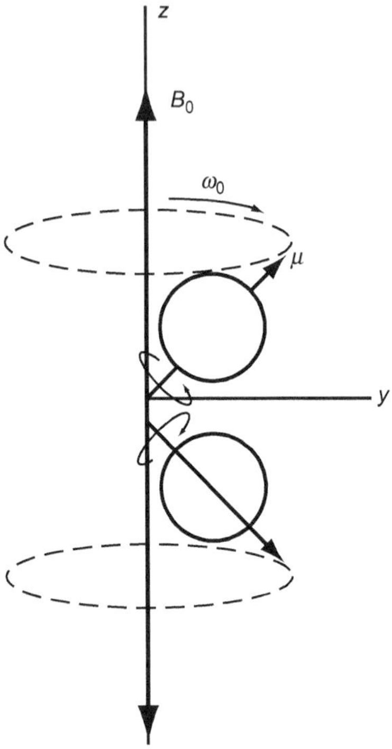

FIG. 3. If an externally magnetic field B_0 is applied, the orientations of the protons (magnetic moments) align at a specific angle along with or opposed to the field. The high- and low-energy states are referred to as "spin down" and "spin up." Adapted from Ref. 5.

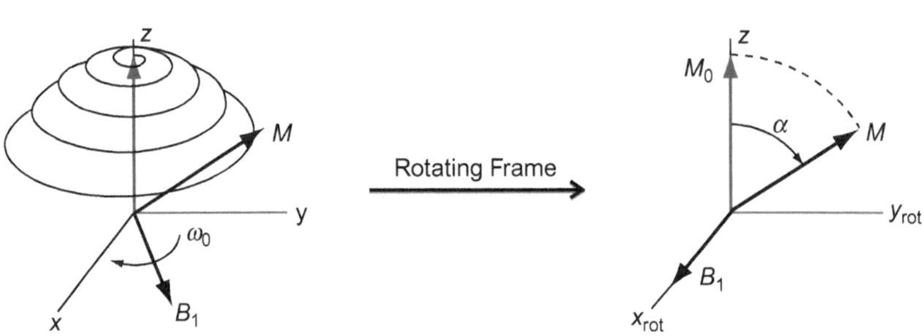

FIG. 4. Illustration of the effect of an RF pulse on the magnetization. (Left) Viewed from the stationary frame of reference, the magnetization M precesses about both the static magnetic field B_0 and the time-varying magnetic field B_1. (Right) Viewed from the reference of a rotating frame. The magnetization M precesses only about B_1. The angle between the z-axis and the magnetization, determined after the RF pulse, is termed the flip angle. Adapted from Ref. 5.

This equation describes that magnetization will precess around a magnetic field at the rate

$$\omega = \gamma|\vec{B}|.$$

This rate is correlated with the strength of the field. To describe the motion of M_0 more precisely, the different ways in which nuclear relaxation occurs must be described, namely spin–lattice (longitudinal) and spin–spin (transverse) relaxation. In the process of longitudinal relaxation M_0 reverts to its original state along the z-axis, in the process of transverse relaxation the M_{xy} component in the xy-plane decays.

These relaxation phenomena can be quantified by the characteristic time constants of decay (T_1 for longitudinal and T_2 for transverse relaxation, respectively).

$$\frac{\mathrm{d}\vec{M}}{\mathrm{d}t} = \left(\vec{M} \times \gamma\vec{B}\right) + \frac{M_0 - M_z}{T_1} - \frac{M_{xy}}{T_2},$$

$\vec{M}$ is the time-varying magnetization vector, M_{xy} is the proportion of the magnetization in the transverse plane, M_0 is the equilibrium magnetization vector and M_z is the components of the magnetization along the direction of the magnetic field $\vec{B}$. The relaxation times, T_1 and T_2, depend on the molecular environment of the spins. T_1 is always longer than T_2. For biological tissue, T_1 and T_2 can vary from nanoseconds to several seconds.

D. The Longitudinal Relaxation Time T_1

The result of spin–lattice (longitudinal) is to bring the M_0 back to its original equilibrium state along the z-axis, if it is perturbed.[5] and is described by a time constant T_1. It is termed spin–lattice relaxation, as exchange of energy between the nuclear spins and their molecular framework takes place in the process. It is described in the following differential equation.

$$M_z(t) = M_0 + (M_z - M_0)\mathrm{e}^{-t/T_1}.$$

For the case of T_1 relaxation after an inversion RF pulse, M_{z0} equals $-M_0$ and the equation for relaxation becomes

$$M_z = M_0\left(1 - 2\mathrm{e}^{-t/T_1}\right).$$

To explain the phenomena that are responsible for relaxation, it has to be taken into consideration that nuclei within a lattice structure, which are in constant motion, create a complex magnetic field. Nuclei in a higher state of energy can transfer their energy to nuclei in a lower energy state if molecular motion contributes at the resonance frequency. This phenomenon is called

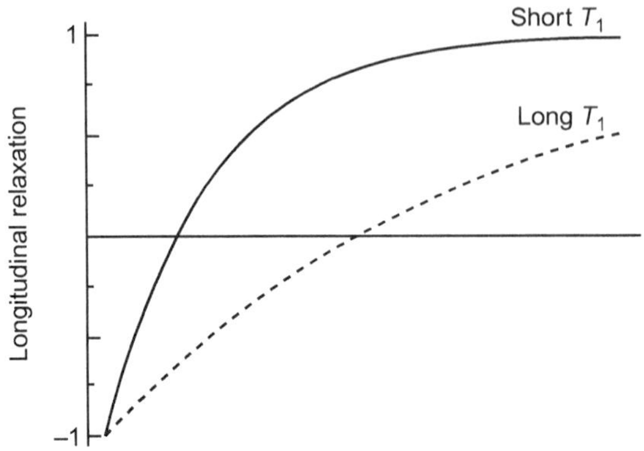

FIG. 5. Time course of the longitudinal relaxation toward its thermal equilibrium value after a 180° inversion RF pulse was applied. At the null point for the tissue, the longitudinal magnetization passes through zero. Adapted from Ref. 5.

dipole–dipole relaxation.[5] This relaxation mechanism is most efficient when the frequency of the molecular motion is equal to the resonance frequency $\tau_c = 1/\omega_0$ ($\tau_c$is the correlation time) (Fig. 5).

E. The Transverse Relaxation Time T_2

Transverse relaxation results from "spin–spin" interactions and is described by the time constant T_2 (transverse relaxation time) and leads to the return of the transverse magnetization M_{xy} to its equilibrium. Interactions between adjacent nuclear spins with identical precessional frequencies but different magnetic quantum states, allow for the exchange quantum states and result in transverse relaxation. An excited nucleus at a higher energy state can excite a nucleus in lower energy level, thereby relaxing to a lower energy state.

$$M_{xy}(t) = M_{xy}e^{-t/T_2}.$$

This equation represents the decay of the transverse magnetization to zero with the tissue-specific relaxation time T_2. This process can also be described as decoherence of the transverse nuclear spin magnetization. If an excitation RF pulse is applied magnetization is shifted into the xy-plane and the different nuclei precess at slightly different frequencies. The greater the frequency spread between the different nuclei, the faster M_{xy} decays (Fig. 6).

$$\frac{1}{T_2*} = \frac{1}{T_2} + (\text{inhomogeneity effects}).$$

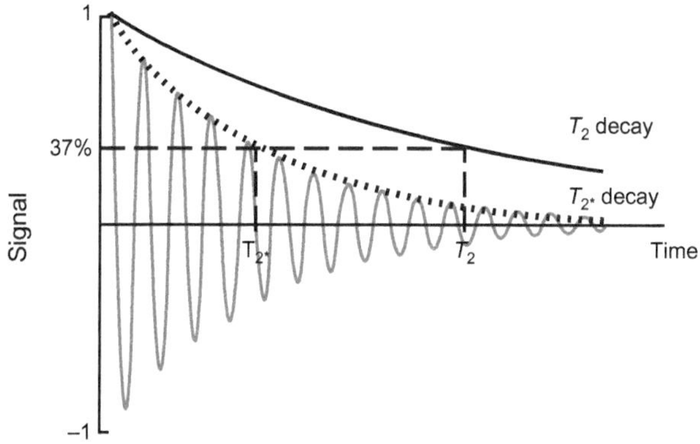

FIG. 6. In the presence of field inhomogeneities, the FID generated by an excitation RF pulse decays with the time constant T_2^* (dotted line). The solid line represents the T_2 decay. Adapted from Ref. 5.

If both T_2 relaxation and the effect of field inhomogeneities are considered, the FID generated by an RF pulse decays at a faster rate compared to what is expected by the T_2 relaxation. This decay is characterized by a time constant called the T_2^* relaxation time, which describes the tissue T_2 value and the contribution from field inhomogeneities. Unlike the T_1 and T_2 relaxation times, the T_2^* time is not an intrinsic property of a tissue. In most case, the additional inhomogeneity, and thus the value of T_2^*, are dependent on the distribution and magnitude of field inhomogeneities and on the size and shape of the voxel. As long as the field inhomogeneities do not change over time, the dephasing, and the associated portion of the MR signal decay, is reversible.

In summary, transverse and longitudinal relaxation processes are caused by fluctuating magnetic fields. Slow molecular motions contributes solely to spin–spin relaxation, components of molecular motion at the resonance frequency contribute to both transverse and longitudinal relaxation.[5]

F. The Acquisition of Signal in MRI

Spatial encoding is used to locate signal emissions (energy in the RF range) spatially in three dimensions to calculate an image. To discriminate magnetization originating from different locations, magnetic field gradients are used. These result in a variation (linear) of the magnetic field strength in space, subsequently allowing to calculate back the origin of the signal. To allow for three-dimensional (3D) encoding the image formation is separated into three steps: slice selection, phase encoding and frequency encoding.

Slice selection is an approach to image a single two-dimensional (2D) plane of the object, by exciting specifically the spins within that plane. To achieve this, a RF pulse is applied in concert with a linear gradient field perpendicular to the image plane. Only those spins with a Larmor frequency close to the frequency of the RF pulse can be excited and subsequently produce a signal.

The slice thickness is determined by the steepness of the gradient as well as the bandwidth of the RF pulse. To image a transverse plane the gradient is applied along the z-axis.

$$\omega(z) = \gamma(B_0 + zG_z).$$

The shape of the slice excited is dependent on the time-varying shape of the RF pulse used. As the use of a gradient in, for example, G_z direction results to dephasing of spins, a gradient in the opposite direction must be used to rephase the spins once the RF pulse is complete.

Subsequently to the slice selection and the application of an RF pulse to, for example, flip the longitudinal magnetization within the selected slice into the xy-plane to generate a measurable transverse proportion, the signal has to be encoded spatially in the other two dimensions.

To encode in the x-axis direction a method called frequency encoding usually is used. To achieve a spatial distinction of the different point in the imaging plane a constant magnetic field gradient G_x, is applied. The frequency of the different points along the x-axis is subsequently determined by

$$\omega(x) = \gamma(B_0 + xG_x).$$

The oscillation frequency of each point is therefore linearly related to the spatial location.

The applied gradient is altering the precession frequency along the x-axis in space resulting in a linear relation between the resonance frequency and the spatial location. A Fourier transform converts the acquired signal to the position of the contributing magnetization along the frequency-encoding direction.

To encode for the y-direction the y-gradients are used. This direction is spatially encoded by generating a spatially varying phase shift of precessing transverse magnetization. It is applied in the time interval between the RF pulse and the application of frequency encoding. The resulting spins rotate at different frequencies and after the gradient is switched off, exhibit different phases depending on their position along the phase encoding gradient. Signals originating from different locations along the y-axis can be identified subsequently by their unique initial phase.

Slice selection, frequency and phase encoding, as explained before, are the encoding techniques that allow for mapping of the 3D spatial information of the spins within the body.

A further step is needed to calculate back the received information into an image.

A dedicated receiver coil in the scanner bore detects the signal during the frequency-encoding process and saves the data in *k*-space. *k*-Space is corresponding to the spatial frequency information of the scanned object and represents its spatial position information. The spatial and frequency domain are related by the Fourier Transform and are the inverse of each other.

During image acquisition by frequency and phase encoding, the recorded signal is filling the lines of *k*-space. One line along the frequency axis is filled with each TR during the application of a phase encoding gradient. The highest signal amplitude is stored in the central area of *k*-space. The linear filling pattern of *k*-space described as "Cartesian filling pattern" and is the most used in clinical MRI.

The different lines of *k*-space can be distinguished by the phase-encoding gradient, however the frequency-encoding gradient stays the same. The more phase-encoding steps are applied the more time is needed.

III. MRI Systems for Preclinical Imaging and Experimental Setup

A. Small Animal High Field MRI Scanners

The introduction to the physical principles of MRI helps to better understand the challenges of small animal MRI. One of the most important parameters in MRI is the signal-to-noise (SNR) ratio. If the voxel size is decreased the measurable signal also decreases. To achieve the same level of detail in a mouse compared to a human, the voxel size must be approximately 3500 times smaller. To achieve this, different setups, like ultrahigh field MRI systems or dedicated microscope coils, can be used.

The increase in magnetic field strengths (B_0) leads to an increase in the signal frequency and therefore the signal that can be measured. Theoretically high gains in SNR can be achieved using ultrahigh field systems. *In vivo*, these gains are however not always fully achieved, as the relationship between field strength and SNR is very complex and additionally dependents on the sequence design that is used. Ultrahigh field MRI scanners with magnetic field strengths from 4.7 to 17.6 T are the most frequently used scanner systems for murine cardiovascular imaging.[6] Those systems are usually equipped with phased array, single loop or solenoid receiver coils.[7] Different designs of ultrahigh field MRI systems, including horizontal and vertical bore scanners, are available. In the context of cardiac imaging, vertical bore systems were shown to allow for the accurate assessment of cardiac function in healthy and diseased animals, even though the animals are in an upright and therefore not physiological position.[8]

To achieve the high spatial resolution, that is necessary for a detailed investigation of cardiac function long scan times of more than 45 min are typically required. To minimize scan time, phased-array receiver coils and parallel acquisition/acceleration techniques have been specifically designed and adjusted for small animal imaging, which can minimized imaging times to < 3 min for the functional assessment of the murine heart.[9]

B. Clinical MRI Scanners (1.5–3 T)

The murine cardiovascular MRI can also be imaged and characterized on clinical MR scanners at clinically used magnetic field strengths (1.5 or 3 T). And potential advantage of clinical MRI systems is the availability of up-to-date imaging sequences and reconstruction algorithms, which are developed for the investigation of the human cardiovascular system.[10–12]

For morphologic and spectroscopic imaging applications, ultrahigh field MRI systems can have some advantages compared to clinical MRI scanners, as a higher signal is generated. For the detection of contrast agents, a lower magnetic field strength can be beneficial as the longitudinal relaxivity (r_1) of imaging probes can decrease with an increase in field strength.

C. Receiver and Gradient Coils

As already mentioned, the major challenges for murine MRI are the small anatomic dimensions and the high heart rates. Lumen diameters and vessel wall thicknesses are 1–3 mm and 40–120 μm for the thoracic aorta in mice. The long axis dimension of a mouse heart is approximately 6–8 mm and heart rates usually vary between 400 and 600 beats/min. Therefore, cardiac imaging has to be performed using a high spatial and temporal resolution to allow for quantitative investigation of cardiac volumes and function. Imaging with high spatial/temporal resolution is associated with a loss in SNR and contrast to noise (CNR). To compensate for this loss in signal, specifically designed small animal RF coils or clinically used coils such as carotid, wrist, eye, or microscopy coils are used. These dedicated small animal RF coils are also available from specialized MRI equipment companies and some MRI vendors.

Temporal resolution is another very important parameter for the assessment of the murine cardiovascular system and especially for the assessment of functional cardiac parameters. MRI systems must therefore be equipped with fast switching and strong gradient coils. Small animal MRI systems are usually equipped with ultra-fast gradient coils (gradient strength: $\sim$400–500 mT/m, slew rate: $\sim$3000–4000 mT/m/ms) and therefore providing a temporal resolution in the order of 3–5 ms with a spatial resolution of $\sim$150–200 μm. Clinical systems equipped with gradients (gradient strength: 40–50 mT m^{-1}, slew rate: 150–250 mT m^{-1} ms^{-1}) typically allow from imaging with a temporal resolution of 9–15 ms at a spatial resolution of $\sim$150–250 μm. Clinical MRI systems

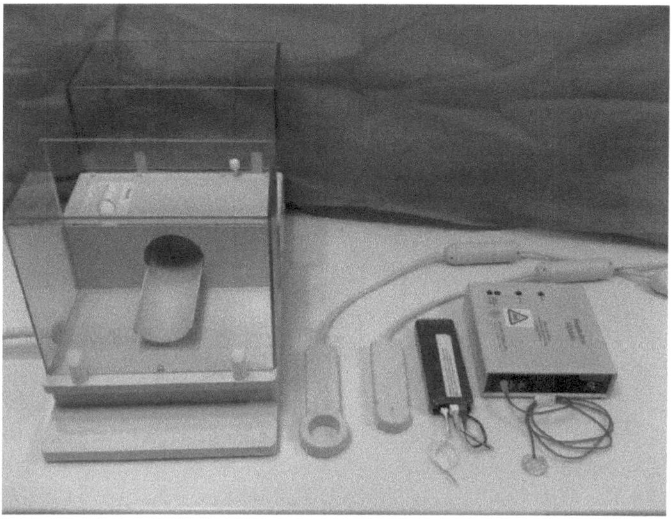

FIG. 7. Figure shows the different kinds of small animal receiver coils and animal monitoring modules that are available. On the left, a solenoid coil and two microscopy coils, are shown. The type of coils that are displayed are available from Phillips Medical Systems. On the right, an ECG/ temperature monitoring module (black box) and a respiratory monitoring module (gray box), are shown. These modules are available from SA Instruments. (See Color Insert.)

using dedicated gradient coil systems can allow to image at a temporal and spatial resolution comparable to dedicated ultrahigh field small animal systems (Fig. 7).[11]

D. Experimental Setup and Animal Preparation

Different issues have to be considered before small animal imaging experiments can be performed. First, it is important to evaluate basic study design requirements, such as the genetic background and the distress caused by the imaging experiment.

The way animals are prepared for imaging experiments and the kind of anesthesia that is chosen can alter the results of the imaging experiments significantly. This is particular important, if physiologic parameters like end-systolic volume (ESV)/end-diastolic volume (EDV) and ejection fractions (EFs) are assessed. It also has to be kept in mind, that the imaging procedure itself can influence the results of the measurements. For example, the intravenous injection of MRI contrast agents can increase the total blood volume and therefore can have an influence on the assessment of functional parameters such as the EF and cardiac volumes. If longitudinal imaging studies are performed, these effects have to be kept in mind. To ensure reproducible experimental conditions all parameters of the experiment should be kept consistent, including the lengths of imaging sessions, the dose and type of anesthesia that is given and the number of imaging sessions per animal. The severity of harm, which is caused for the animals during the imaging experiments, can vary. The injection or inhalation of anesthetics usually only causes minimal harm. The injection of contrast agents already causes moderate harm and the performance of surgical procedures causes extensive harm. If comparative studies with genetically modified animals are performed, it is of high importance that the correct control groups, for example, animals with the same genetic background, are chosen.

Various studies have indicated that slight variations in the background strain of genetically modified animals (e.g., knock-in or knock-out mice) can have significant effects on the development of diseases (e.g., severity of atherosclerosis) and on the way animals cope with imaging conditions, for example, the kind of anesthesia that is used.[13–16] The gender of the animals used also has to be taken into consideration, as it can have an effect on metabolism, pharmacokinetics, and disease development.[17–19] This variability can be explained with differences in hormone levels such as testosterone, estrogen, and hepatic enzymes.[20]

E. Animal Monitoring

During imaging sessions monitoring of physiologic parameters, such as heart rate and temperature, should always be performed. This helps to ensure that imaging conditions are as reproducible as possible. If functional cardiac parameter have to be assessed, it is important to synchronize image acquisition with the cardiac or respiratory cycle to avoid artifacts. Besides monitoring physiological functions, devices for the assessment of more invasive parameters such as arterial pressures and blood/tissue oxygen levels are also available.

If contrast agents are injected, for example, in the context of molecular imaging, it is of high importance to consider the specific properties of the injected fluids and substances, such as the type of solvent that is used, the pH level and the rate of absorption.[21]

The maximum volume of i.v. injections should not go beyond 4–6% of the blood volume of the investigated animal (an app. injection volume of < 150–200 μl for mice and 800–1000 μl for rats) and the maximum volume of intraperitoneal injection should not exceed 9–10 ml/kg body mass.[21,22]

F. Anesthesia

Murine MRI imaging studies have been performed using various kinds of anesthetics, which include inhalable and injectable anesthetics. The most commonly used inhalable anesthetic, in the context of cardiovascular imaging, is isoflurane. It was shown, that isoflurane does not significantly influence cardiovascular function and therefore provide reproducible and reliable imaging conditions.[23] The careful selection of the kind and dose of anesthetics used is of high importance, as some kinds of anesthetics have a greater influence on blood flow and cardiac function compared to others.[24–26]

Additionally, it has to be kept in mind, that the effect of anesthetics also depend on the gender, strain, and potential genetic modification of the animal. C57/BL6 is one of the most commonly investigated background strain used for preclinical cardiovascular MRI studies. For this specific strain, but also for other commonly used strains, it was shown that relatively low levels of anesthesia (e.g., 1.0–1.25% inhaled isoflurane) lead to a stable anesthesia and have minor influence on cardiac function and myocardial perfusion.[24–27] Using 1.0–1.25% inhaled isoflurane at a core temperature of 37 °C, the heart rate for C57Bl/6 mice typically varies between 450 and 550 beats/min with an left ventricular EF of 60–70%. These parameters are close to physiological conditions.

Injectable anesthetics offer some advantages over inhaled anesthetics. Injectable anesthetics are usually easier to handle, as they do not involve the technical setup inhalable anesthetics require. The time span the animal remains anesthetized is however harder to predict and control. Therefore, if this kind of anesthetic is used, animals have to be monitored very closely, for example, with a video camera in the magnetic bore, to avoid complications. It has been reported, that low doses of injectable anesthetics, like phenobarbital, do not significantly alter left ventricular structural and functional data, compared to conscious animals.[28,29]

G. Positioning of the Animal in the Bore

After the successful induction of either the inhalable or injectable anesthesia, animals can be positioned in the MRI bore on specially designed holding devices. It is not mandatory to use these kinds of devices; it however helps to place the animals in a reproducible position. This can be of high importance, if frequent follow-up scans are planned or if the animals are investigated sequentially with different imaging modalities (e.g., first the MRI scan and then the PET scan).

These positioning devices should be built to accommodate the complete animal and allow for the placement of ECG electrodes, if cardiac triggering is required and a pneumatic pillow, if the respiratory frequency is assessed. These positioning devices should keep the animal in a natural position (prone), as all additional physical and mental stress will increase the amount of anesthetics, which is needed to keep the animal anesthetized. Especially longitudinal studies, which require the animals to be scanned several times, can be very stressful. Many different factors have to be considered, which include the lengths of transport from the holding cages to the scanner, repeated injections of contrast agents or anesthetics, the experimental conditions (e.g., coronary ligation to induce a myocardial infarction) and hypothermia. Acute stress situations for the animals before the scan, like physical restraint for the cannulation of the tail vein, can induce hyperthermia and the increase in heart rate and myocardial blood flow (MBF), and therefore alter the experimental conditions.[30–32]

H. Body Temperature

With most anesthetics used, animals will lose their ability to actively regulate their core body temperature. As the temperature in the bore is usually around or slightly above room temperature, this will lead to decline of the animal's core temperature. It is therefore necessary to compensate for the loss of heat and to maintain the animals core temperature around 37 °C. To achieve this, the animal's core temperature has to be assessed, usually using a MRI compatible rectal temperature probe.[33] This information is usually fed back to a computer, which regulates the output temperature of the heating device. These heating systems can either consists of a hot air blower or a heated blanket. Different MRI compatible temperature monitoring probes with a feedback heating system are available in Europe and the United States from different vendors.

I. Cardiac Motion

For small animal cardiovascular imaging, respiratory and cardiac of motion are responsible for most imaging artifacts. Without compensation for cardiac motion, the accurate assessment functional cardiac parameters is very limited. To reduce cardiac motion artifacts, imaging sequences are in most cases synchronized to the *R*-wave of the ECG. For the accurate assessment of functional cardiac parameters, like the ESV, the EDV and EF imaging has to be performed with high temporal resolution. For an accurate volumetric analysis, temporal resolution of the data acquisition should be in the range of 5–10 ms. To reach sufficient signal, data acquisition has to be repeated over several heart cycles.

For the synchronization of cardiac MRI sequences with the cardiac cycle, a reliable detection of the *R*-wave in the ECG is of high importance. Therefore, an artifact free, strong, and continuous ECG signal has to be derived. Different specially designed small animal MRI compatible ECG gating systems are available. Most of the system use subcutaneous ECG electrodes, which are well suited for the assessment of the ECG signal with reduced artifacts from MR gradients.

Fast gradient switching and strong RF pulses, which are required for image acquisition, can however lead to an interference with and the degradation of the measured ECG signal.[34,35] Dedicated shielding for the ECG cables can lead to a reduction of artifacts. While heart rates in most large animal models, for example, swine models, are approximately the same as in humans, this does not hold true for smaller animals, especially mouse models. This does not present a problem, if dedicated small animal high field MRI scanners are used. As the heart rate of small animal models goes beyond of what can be measured in humans, clinical MRI scanners are usually not able to will detect and process these high heart rates. Software modifications can allow to overcome this limitation. Another workaround is to trigger MRI sequences to every other or third heart beat.

More advanced techniques to synchronize MRI sequences with cardiac motion have also been developed, these include acoustic methods using a fiber-optic stethoscope and self-gating approaches.[36,37] Such methods can be very useful in cases only weak ECG signals can be derived or strong gradient interferences, which can lead to an unreliable *R*-wave detection.

J. Respiratory Motion

To compensate for respiratory motion, breath hold techniques or image navigators to directly assess the motion of the diaphragm can be used in human cardiac MRI. These techniques are usually not applicable in small animal imaging. The most reliable way to compensate for respiratory motion, are dedicated small animal pneumatic pillows, comparable to breathing belts used in human imaging. These pneumatic pillows are commercially available from several different vendors.

Other more advanced techniques, like the intubation of the animal in the scanner and therefore artificial respiration can also be used to reduce artifacts arising from respiratory motion. A technique also referred to as intermittent isopressure breathhold, allows to freeze the animals respiration at a certain position. In a clinical context such a technique is used to image intubated patients, in a preclinical such a technique has only been used in the context of lung imaging applications using CT.[38]

IV. Cardiovascular MRI

A. Cardiac Functional Parameters and Myocardial Mass

All functional cardiac parameters such as the ESV, EDV, EF, stroke volume (SV), cardiac output (CO), and left ventricular mass can be derived from cardiac cine MR sequences. The assessment of left and right ventricular volumes can be performed on either a stack of short axis or long axis views, like the two chamber and four chamber view. The assessment of ventricular volumes derived from short axis views was shown to be the most reproducible, as the whole ventricle is covered from the mitral valve to the apex of the heart.

To derive quantitative volumes from cine MRI images endocardial and pericardial have to be drawn either manually or semiautomatically.[39] The end-diastole is the frame directly after the *R*-wave at which the left ventricle has the highest volume. The end-systole is the temporal frame at which the left ventricle has the smallest volume. After contours are drawn, the corresponding values are calculated by adding the areas of all endocardial segments at each cardiac phase and multiplying by each slice thickness. If the long axis views were imaged, volumes can be calculated by using the biplane ellipsoid formula.[39] Calculating the difference between left ventricular end-diastolic and left ventricular ESVs divided by the left ventricular EDV allows the computation of the left ventricular EF.

The overall mass of the myocardium can be computed by adding the differences of epicardial and endocardial areas in diastole, then multiplying the values by the slice thickness to calculate myocardial tissue volume, and finally by multiplying by the density of myocardial tissue (1.05 g/cm^3). Most of the major clinical and preclinical MRI vendors offer postprocessing packages in addition to their cardiac imaging packages, which allow to directly calculate cardiac functional values. Open source software, for example, Osirix, is also available and can be used for certain applications.

For the quantification of ventricular functional parameters in small animals, several ECG-gated spin–echo and multiphase gradient echo (cine MRI) pulse sequences have been developed.[40] These sequences allow the assessment of all relevant functional cardiac parameters, like the ESV, EDF, EF, SV, CO, and the left ventricular myocardial mass. In many preclinical studies, it is essential to assess the mass of the left ventricle, for example, in models myocardial hypertrophy. Preclinical cardiac MRI has been shown to allow for accurate noninvasive assessment of myocardial mass, compared to the ex vivo measured weight of the left ventricle.[41,42]

Cine MRI is also regarded the gold standard for the noninvasive assessment of ventricular function and myocardial wall thickening and thickness.[43] For male C57BL/6 mice ($\sim$ 10 weeks of age) typical *in vivo* parameters include an ESV of $\sim$ 15 μl, EDV of $\sim$ 40 μl, EF of $\sim$ 55–65% CO of $\sim$ 12.5 μl/min with an average heart rate of around 500 beats/min. Most preclinical cardiac MRI studies have used bright-blood cine MRI sequences for the assessment of cardiovascular function. Other types of sequences can however also be used, these include double inversion black-blood cine MRI sequences which have been shown to allow for a precise assessment of left ventricular function and mass.[42]

The assessment right ventricular function can in some cases be more challenging than the assessment of the left ventricular function. This is due to the smaller volume of the right ventricle, the thinner myocardial wall and the more pronounced myocardial trabeculation. Using high resolution (in-plane $\sim 0.1 \times 0.1$ mm) cardiac cine MRI sequences, it has been shown that the right ventricular function can be accurately quantified.[44] These type of high-resolution sequences can also be used to investigate juvenile and neonatal mouse hearts.[45]

In preclinical studies, a frequently used invasive method for the assessment of functional cardiac parameters are pressure–volume loops measured by conductance catheter. These catheters offer a higher temporal resolution compared to cardiac MRI. A direct comparison of both methods revealed a strong correlation between functional parameters in infarcted as well as normal hearts. It was however reported that the EF and the left ventricular volumes were lower using pressure–volume loops by conductance catheter.[46] Nowadays, functional cardiac MRI measurements on high field MRI systems are considered to be the most reliable readout for the assessment of cardiac function in small animals, especially for longitudinal studies.

B. Myocardial Tagging and Strain Imaging

The quantitative assessment of myocardial wall motion can be an important parameter to evaluate, especially in studies involving models of myocardial ischemia. Many techniques, which are used in preclinical MRI studies, were initiated developed for clinical applications. Some of these techniques have been scaled down and optimized for small animal imaging. These imaging sequences include myocardial tagging, velocity-encoded phase-contrast (PC) imaging, displacement-encoded imaging with stimulated echoes (DENSE) and 2D harmonic phase (HARP) analysis.

Myocardial tagging sequences allow to track the motion of specific tissue points within the myocardium and therefore allows the precise evaluation and quantification of myocardial tissue movement. This technique is based on the saturating of parallel sections within the myocardial tissue early in the myocardial cycle. The best time point to perform myocardial tagging is immediately after the detection of the R-wave in the cardiac cycle. During the relaxation and contraction of the heart, tag lines deform along the movement of the myocardium. In most cases, the saturation tags are applied making use of spatial modulation of magnetization (SPAMM) techniques. As the saturation is directly associated with the tissue magnetization, cine MRI sequences can be applied after the testing preparation pulses to image the movement and displacement of the tag lines. Displaying the tagged image series in a cinematic modus is an informative way to analyze the information and directly visualize/quantify myocardial areas with regional wall motion abnormalities. Most major vendors offer some variation of myocardial tagging pulse sequences with their cardiac packages. Grid tagging is another method for the assessment of myocardial motion. Gird tags can be generated by applying SPAMM tags in perpendicular directions.[47] This technique is however mostly used for human applications due to longer preparation times of the pulse sequence. Due to these limitations, line tagging is the preferred method in small animals.

It has to be noted, that the very high heart rate in small animals and especially in mice can lead to an increase in tagging preparation times. This can result in temporal blurring, because the time the tag preparation takes can extend beyond the duration of the end-diastole. The total imaging time required for myocardial tagging is usually comparable to the time cine imaging takes.

More than a decade ago, 2D myocardial tagging was first used in a mouse model. The measured left ventricular torsion was comparable to the torsion measured in human studies.[48] During these studies images were only acquired during the first 80% of the systole. With further developments in MRI sequence design these limitations were overcome using 2D SPAMM. This technique enabled the acquisition of tag lines during the full cardiac cycle.[42,49] Due to a much shorter cardiac cycle in small animals, SPAMM tagging only leads to minor tag fading. Myocardial tagging techniques were successfully applied in a mouse model of myocardial infarction to study the regional wall strain (Fig. 8).[51]

A variation of the SPAMM technique is complementary SPAMM (CSPAMM), which is based on a subtraction technique, which allows tracking of the tagged slice and is not limited by tag line fading in the diastole. Different variations of the implementation of CSPAMM has been investigated.[52]

Myocardial strain imaging was introduced as an extension of myocardial tagging. Myocardial wall motion analysis based on cine MRI images does represent the contraction of the myocardium in the radial direction. It does not take into account the longitudinal and circumferential direction of myocardial contraction. Therefore, strain analysis based on cine MRI images does not always represent the true extent of myocardial contractility.[50] The use of myocardial strain imaging is thought to give a more accurate evaluation of myocardial contraction in physiological and pathological conditions.[53] Myocardial strain imaging is based on the assessment of local tissue deformation which represents myocardial contractile function.[54] A strain can be described as a tensor, which incorporates the direction, length, and magnitude of change in myocardial contraction. During the cardiac systole, the direction of circumferential shortening in the short imaging plane is parallel to the epicardium. An indicator of the radial contraction is a myocardial wall thickening. The base to apex shortens in a direction parallel to the long axis of the left ventricle.

In mice, 3D strain analysis was applied by incorporating tagging information from long and short axis planes.[50] This technique combines tagging information derived from the short and long axis views. Using a spatial resolution of 0.2 mm × 0.2 mm × 1 mm, 14–16 cardiac phases could be acquired in each slice. It has to be kept in mind that imaging was performed at heart rates of 450–550 beats/min.[6] It has also to be mentioned, that one limitation of

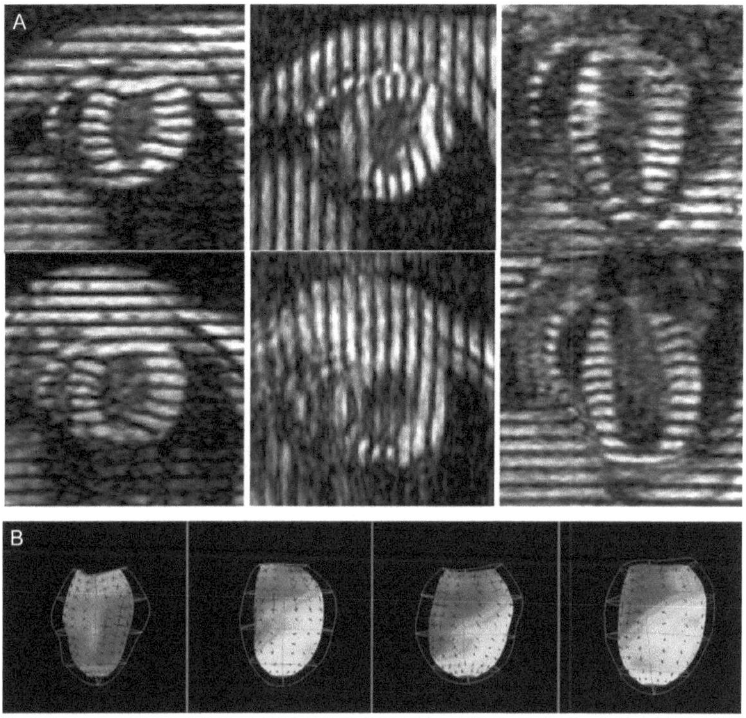

FIG. 8. (A) High resolution spatial modulation of magnetization (SPAMM) tagged magnetic resonance *in vivo* imaging at baseline and 24 h after myocardial infarction. The SPAMM stripes are 700 μm apart from each other. (B) 3D element model at end-diastole at baseline, day 1, 7, 28 after myocardial infarction. Lines represent model element boundaries. Crosses represent 3D principal strains and directions. Adapted from Ref. 50. (See Color Insert.)

myocardial tagging is the time intensive analysis of the resulting tagged MRI images. Novel algorithms, like HARP (which we introduce in the following paragraph), allow to shorten this postprocessing time significantly.

Another parameter that can be measured using MRI tagging is the left ventricular systolic torsion (the systolic twist and diastolic untwist of the myocardium).[54,55] The myocardial torsion is defined as the motion in-between the short axis planes, that occurs at the same time as the differential rotation of the myocardium around the long axis. As already indicated above, one of limitations of strain imaging and myocardial tagging is the time-consuming analysis, which has to be performed manually on the resulting tagged MRI images. HARP analysis is a technique that potentially allows to overcome this limitation.[56] The use of HARP requires significantly less intervention from the user and has a top of processing time. Analyzing the Fourier domain, HARP allows to track phase changes of off-center spectral peaks. The change in phase is associated with the in-plane motion of the respective myocardial tags. The principle of HARP has been incorporated into different MR sequences, which allow to display and analyze the myocardial strains in small animals, including mice.[57]

One example, which demonstrates the significance of myocardial strain imaging, is acute coronary ischemia. For this entity it has been shown that changes in regional strain can be measured earlier compared to nontagged cine MRI.[58] Different variations of the described techniques were used in animal models of myocardial dysfunction.[6,50,57,59] Different 2D and 3D displacement-encoded phase contrast (DENSE) sequences have been designed to analyze and quantify strain and systolic myocardial displacement at very high spatial resolutions and with short postprocessing times.[60,61] This technique uses a comparable approach as used in velocity-encoded PC imaging. DENSE techniques assess the displacement of the myocardium and therefore also allow the calculation of myocardial strain. These techniques have been applied to assess and quantify 2D strain, torsion, and twist as well as 3D myocardial displacement in mice after myocardial infarction.[60,61]

As already introduced, a different approach to assess myocardial wall motion is velocity-encoded PC imaging. This technique was initially developed and tested in mice using a 2D imaging sequence at 7 T. It was however developed further into a 3D technique, which can be applied at 17.6 T.[6,62] This technique was shown to allow for a pixel by pixel analysis of myocardial velocities. It can be run at higher spatial resolution compact to myocardial tagging (Fig. 9).

For relative measurements, like the EF, it has been demonstrated that the cardiac function in healthy control mice is comparable to values assessed in healthy humans (EF: 55–65% vs. 60–70%), (midventricular circumferential shortening: -0.2 vs. -0.15) and (normal radial thickening: 0.4 vs. -0.3).[39,42,49,54,55,61,63]

C. Imaging of Myocardial Perfusion

As already indicated genetically modified animals are increasingly used to study pathophysiology of cardiovascular diseases.[42,64–66] MBF is one of the most important readouts in cardiovascular diseases. Different disease processes, including atherosclerosis, diabetes mellitus and systemic hypertension, significantly affect myocardial perfusion.[67–69] Therefore, it is of high importance that tools are available, which allow the accurate and noninvasive assessment of MBF in small animal models.

Arterial spin labeling (ASL) is a MRI technique, which has been successfully used to measure and quantify myocardial perfusion.[27,70] ASL is based on determining the water content of blood as an endogenous tracer. The effect of

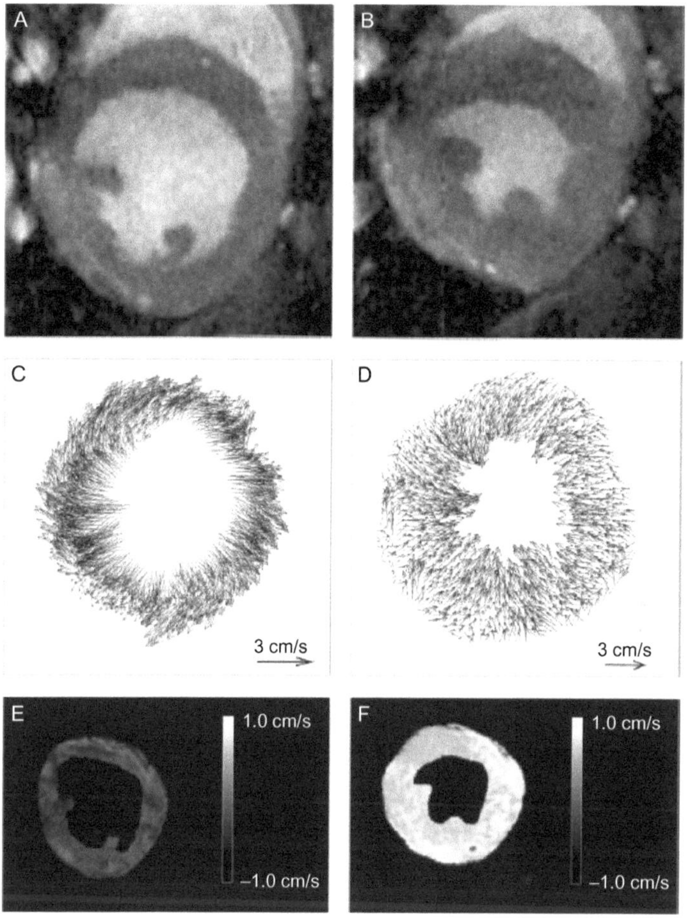

FIG. 9. *In vivo* quantitative 3D motion mapping of the murine myocardium using phase contrast magnetic resonance image at a field strength of 17.6 T. Adapted from Ref. 6.

water on the myocardial tissue T_1 relaxation time is assessed to measure myocardial perfusion. ASL has already been successfully used in mouse models of myocardial infarction. This technique also allowed to assess the effects of different kinds of anesthesia on myocardial perfusion.[27,70] These studies could quantify myocardial perfusion and measured value of 5–7 ml/g/min. These values are in the same range as values assessed invasively in mice using microspheres.[71] Even though microspheres are considered the gold standard for the assessment of myocardial perfusion, it has to be kept in mind that it is an invasive technique with some limitations, especially if small sample sizes are investigated.[72,73] Both techniques, ASL and microspheres, revealed an approximately fivefold increase in myocardial tissue perfusion in mice compared to human myocardial perfusion.

Even though spin labeling MR methods have been successfully used to characterize myocardial perfusion in small animals, imaging times of more than 20 min for a single slice can be a limitation.[27] In humans the preferred method to assess myocardial perfusion is a dynamic first-pass contrast-enhanced acquisition technique.[69] First-pass myocardial perfusion imaging using gadolinium-based contrast agents is also an attractive method for small animal imaging, as it offers a high tissue contrast and high spatial resolutions. For successful *in vivo* characterization of myocardial perfusion in mice, a spatial resolution around 0.2 mm voxel width in-plane should be reached. This can however be challenging, as first-pass perfusion imaging requires the acquisition of complete image at every single or every second heartbeat. As a magnetization preparation pulse and a short acquisition window are required, conventional MRI perfusion sequences cannot be used at the high heart rates in small animals. In the past years, novel MRI methods that allow the acceleration of data acquisition up to a factor of 8 has been introduced in clinical studies.[74] These methods are based on time domain and *k*-space undersampling (*k-t* sensitivity encoding or *k-t* SENSE). A recent study introduced first-pass contrast-enhanced myocardial perfusion imaging in mice using these methods. Mice with myocardial infarction and control mice were investigated in the study.[75] The measured and quantified values for MBF were within the expected range. This new method, represents a novel tool for the fast *in vivo* assessment of myocardial perfusion and may potentially allow image acquisition and pharmacological stress (Fig. 10).[75]

D. *In Vivo* Myocardial Tissue Characterization

In clinical practice, late gadolinium enhancement allows a differentiation between irreversibly and reversibly damaged myocardial tissue after a myocardial infarction. This kind of differentiation can be made, because contrast agents, comparable to Gd-DTPA, diffuse into the interstitial myocardial space of both viable myocardial tissues and infarcted myocardial tissues. Areas of accurate and chronic myocardial necrosis represents an area of increased distribution volume for Gd-DTPA. The decreased wash out of Gd-DTPA from these areas results in a shortening of the T_1 relaxation times of necrotic tissues compared to healthy myocardial tissue. This shortening of T_1 can be measured using inversion recovery MRI sequences. Images are usually acquired 15–30 min after the injection of the contrast agent in small animals.[76,77] Most of the contrast agents used for myocardial late enhancement imaging are injected into the tail vein.[78] It has however also been shown in the contrast agent can be directly injected into the peritoneum.[77] The most common MRI sequence used for imaging myocardial infarction is are inversion

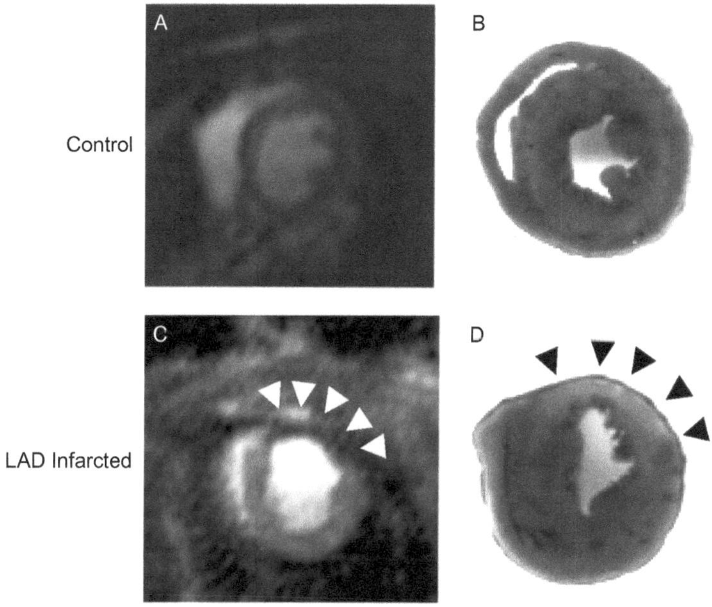

FIG. 10. First-pass myocardial perfusion images in mice. The images represent matched Evans blue sections (2% w/v, injection into the tail vein) that indicate the area at risk. Note the colocalization of large anterior perfusion defect in first-pass perfusion MRI with the decreased staining in the slice matched Evans blue section (C, D). Adapted from Ref. 75. (See Color Insert.)

recovery gradient echo pulse sequences, which are usually ECG triggered.[79] Data are in most cases acquired at end-diastole, and can however also be acquired during end-systole. To measure a high signal from Gd-DTPA, signal from the normal myocardium has to be suppressed. Therefore, the specific inversion delay is used to generate a high contrast between the contrast agent and the surrounding tissue. It is important to determine the exact inversion time to derive a strong and reliable signal. The accumulation of Gd-DTPA in necrotic myocardium also depends on the injected concentration of the contrast agent, the specific wash out of the contrast agent and the field strength used. The inversion time usually varies between 200 and 400 ms. If Gd-DTPA is used as a contrast agent, the recommended dose varies between 0.1 and 0.2 mmol/kg. Different preclinical studies have shown that the area of delayed enhancement correlates well with the area of infarcted myocardium.[77,80] The assessment of late gadolinium enhancement is not also only a useful tool for the evaluation of myocardial infarction but can also be useful for the assessment and characterization of myocarditis and myocardial fibrosis.[81,82]

E. Imaging of the Vascular Lumen

To image the vascular lumen different angiographic MRI methods exist. These include ASL, time-of-flight (TOF) angiographies and contrast-enhanced MR imaging sequences. TOF-based MR angiographies are the most widely used angiographic techniques used in small animals. The techniques are based on a slice selective RF pulse applied in such a short period, that spins in stationary or nonmoving/flowing tissues do not have sufficient time to regain longitudinal magnetization. This leads to a suppression of signal from all nonmoving tissues. Spins that are not within the saturated slice produce a positive or bright signal, when they enter the imaging plane and can therefore be imaged with a high signal. TOF angiographies have been successfully tested and used to image the carotid and cerebral vessels in different animal models.[83–88]

F. Cardiovascular Molecular Imaging

For molecular MR imaging various different imaging probes are available. These include iron oxide and gadolinium-based contrast agents, which are the most frequently used contrast agents in preclinical cardiovascular research. MR contrast agents can be targeted against various different proteins and cells. In the context of atherosclerosis different compartments and cells of the atherosclerotic plaque have been investigated using molecular MR contrast agents. One focus was put on the characterization of extracellular matrix components. During the progression of the atherosclerotic plaque, smooth muscle cells and macrophages express extracellular matrix proteins. These proteins represents one of the largest components of the atherosclerotic plaque.[89]

Recently, a new elastin-specific small molecular weight MR contrast agent was introduced that allowed the noninvasive assessment of progression and regression of plaque burden in a mouse model of accelerated atherosclerosis.[90] In addition, changes in the content of elastin, a key component of the extracellular matrix, by signal intensity and T_1 measurements, could be quantified.[90] These signal intensity measurements may allow for further quantitative characterization of plaques on a molecular level with regard to its elastin content.[90] All experiments were performed in an apoE$^{-/-}$ mouse model of atherosclerosis. This contrast agent was subsequently used in a large animal model to image coronary artery remodeling after vascular injury (Fig. 11).[91]

A different MR contrast agent that has been used to image extracellular plaque components is Gadofluorine. This gadolinium-containing macrocyclic contrast agent was shown to colocalize with lipid-rich regions in atherosclerotic plaque in an animal model of atherosclerosis.[92] It was also demonstrated that this contrast agent colocalizes with neovessel-rich regions of the plaque and with areas rich in collagenous (fibrous) material.[92,93]

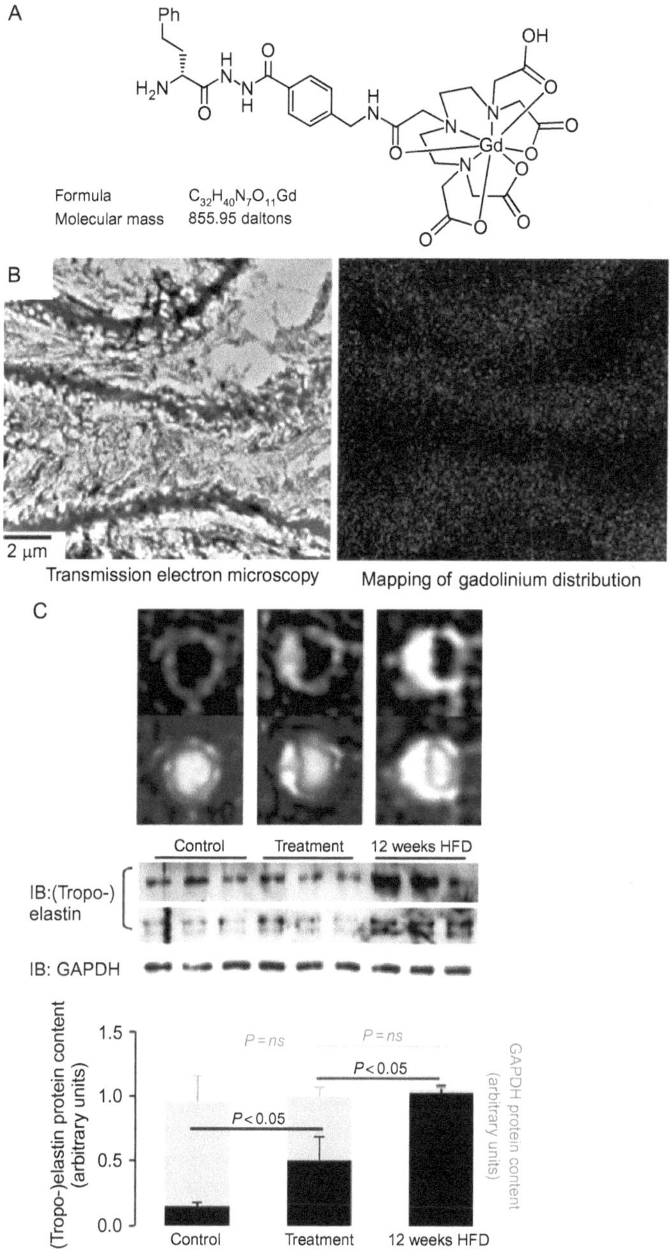
A
Ph
OH
H
O
H2N
N
N
H
O
H
N
O
N
N
O
Gd
O
O
N
O
O
O
O
O
Formula $C_{32}H_{40}N_7O_{11}Gd$
Molecular mass 855.95 daltons
B
2 μm
Transmission electron microscopy
Mapping of gadolinium distribution
C
Control
Treatment
12 weeks HFD
IB:(Tropo-) elastin
IB: GAPDH
(Tropo-)elastin protein content (arbitrary units)
1.5
1.0
0.5
0.0
P = ns
P = ns
P < 0.05
P < 0.05
GAPDH protein content (arbitrary units)
Control
Treatment
12 weeks HFD

Fibrin represents another highly important extracellular protein involved in the development of atherosclerotic plaques and in the formation of potentially embolic thrombi after rupture of the atherosclerotic plaque has occurred.[94] A low molecular weight fibrin-specific contrast agent has been successfully applied to image thrombus *in vivo* in different large animals models.[95–98] This was also the first fibrin-specific molecular MR contrast agent that has been translated into clinical studies. In these studies it was successfully used to noninvasively image fibrin-rich thrombi *in vivo*.[99] Other fibrin-specific contrast agents have also been successfully used in a preclinical setting to image fibrin in thrombus.[95]

Cellular surface molecules have also been in the center of attention. Endothelial cell adhesion molecules are potential biomarkers of atherosclerosis, that are presented on the extracellular surface very early in disease development. These markers include, intercellular adhesion molecule-1 (ICAM-1), E-selectin, and vascular cell adhesion molecule-1 (VCAM-1). A molecular MR contrast agent could be successfully targeted against VCAM-1, and allowed its sensitive detection *in vivo*.[100]

Another important process in the progression of atherosclerotic plaque is the development of novel small blood vessels in the atherosclerotic vessel wall. These neovessels also seemed to play an important role in the destabilization of atherosclerotic plaque. Different techniques to detect these neovessels can be used. The first approach is based on directly targeting unique surface proteins, like $\alpha v\beta 3$, which only expressed on the endothelial surface of these blood vessels. A MR contrast agent based on gadolinium-containing liposomes targeting $\alpha v\beta 3$ integrin has been successfully used to detect angioneogenesis in a rabbit model of atherosclerosis.[101]

A different approach, which can also be used to detect angioneogenesis is to directly assess the increased blood flow caused by these newly formed vessels. Dynamic contrast-enhanced (DCE) MRI techniques after the application of a contrast agent, like Gd-DTPA, can be used to measure these changes in the perfusion of the vascular wall.

FIG. 11. Example of a novel MRI elastin-targeting contrast agent. The usefulness of this contrast agent was investigated in mouse model of atherosclerosis. (A) Chemical structure of the elastin-targeting contrast agent. (B) The electron microscopy images show the colocalization of the gadolinium from the contrast agent with dense elastic fibers. (C) Example of delayed enhancement images (upper row) overlaid on TOF images (bottom row) of cross-sectional views of the brachiocephalic artery. On the left a control mouse is shown, in the middle in treated animal (with statin) and on the right animal that has been on the high-fat diet for 3 months. A significant increase in percentage atheroma/media volume could be measured after 3 months of HFD. The treatment with a statin resulted in a significant decrease in the percentage atheroma/media volume. Adapted from Ref. 90. (See Color Insert.)

Another focus in molecular MRI is the imaging of macrophages. Macrophages play an important role in the initiation and progression of atherosclerotic disease. The infiltration of macrophages into the fibrous cap of the atherosclerotic plaque is regarded to be one of the features of vulnerable atherosclerotic plaques, that have an increased risk of rupture. Different contrast agents can be used to image and quantify macrophages *in vivo*. In molecular MRI the most investigated contrast agents to image macrophages are based on different kinds of iron oxide particles with various coatings.[102] The first experiments that have shown the potential of molecular MRI have already been performed more than a decade ago.[46,103] Recent studies have indicated that macrophage burden and response to therapy can also be visualized and quantified *in vivo* by molecular MRI.[104] Different techniques have been developed to aid in the detection of iron oxide particles in *in vivo*.[105–107] Besides iron oxide particles it was also shown that paramagnetic Gd-based micelles can also be used to detect macrophages *in vivo*.[108]

V. Conclusion

Technical developments in preclinical cardiovascular MRI and the design of novel molecular imaging probes has made significant progress over the past years. Preclinical cardiovascular MRI allows the assessment of all important functional parameters in small animal models of cardiovascular diseases and therefore has become a valuable tool in pharmacological, genetic, and biomedical cardiovascular research. Additionally, cardiovascular MRI allows the characterization of different cardiovascular tissues without the use of contrast agents. The use of unspecific contrast agents allows a precise characterization of for example the extent of a myocardial infarction with high spatial resolution. Specifically targeted molecular contrast agents enable the characterization of pathological processes in the cardiovascular system on a molecular level.

All these techniques contributed to our better understanding of underlying molecular and biological changes in cardiovascular diseases.

References

1. Bloch F. Nuclear induction. *Phys Rev* 1946;**70**:460–74.
2. Purcell EM, Torrey HC, Pound RV. Resonance absorption by nuclear magnetic moments in a solid. *Phys Rev* 1946;**69**:37–8.
3. Lauterbur P. Image formation by induced local interactions—examples employing nuclear magnetic resonance. *Nature* 1973;**242**:190–1.
4. Mansfield P, Grannell PK. NMR 'diffraction' in solids. *J Phys C* 1973;**6**:L422–6.

5. Gadian DG. *Nuclear magnetic resonance and its applications to living systems*. New York: Oxford University Press; 1995.
6. Herold V, Morchel P, Faber C, Rommel E, Haase A, Jakob PM. In vivo quantitative three-dimensional motion mapping of the murine myocardium with PC-MRI at 17.6 t. *Magn Reson Med* 2006;**55**:1058–64.
7. Wiesmann F, Ruff J, Engelhardt S, Hein L, Dienesch C, Leupold A, et al. Dobutamine-stress magnetic resonance microimaging in mice: acute changes of cardiac geometry and function in normal and failing murine hearts. *Circ Res* 2001;**88**:563–9.
8. Schneider JE, Cassidy P, Lygate C, Tyler D, Wiesmann F, Grieve S, et al. Fast, high-resolution in vivo cine magnetic resonance imaging in normal and failing mouse hearts on a vertical 11.7 t system. *J Magn Reson Imaging* 2003;**18**:691–701.
9. Schneider JE, Lanz T, Barnes H, Medway D, Stork LA, Lygate CA, et al. Ultra-fast and accurate assessment of cardiac function in rats using accelerated MRI at 9.4 tesla. *Magn Reson Med* 2008;**59**:636–41.
10. Brockmann MA, Kemmling A, Groden C. Current issues and perspectives in small rodent magnetic resonance imaging using clinical MRI scanners. *Methods* 2007;**43**:79–87.
11. Gilson WD, Kraitchman DL. Cardiac magnetic resonance imaging in small rodents using clinical 1.5 t and 3.0 t scanners. *Methods* 2007;**43**:35–45.
12. Beuf O, Jaillon F, Saint-Jalmes H. Small-animal MRI: signal-to-noise ratio comparison at 7 and 1.5 t with multiple-animal acquisition strategies. *MAGMA* 2006;**19**:202–8.
13. Lovell DP. Variation in barbiturate sleeping time in mice. 3. Strain x environment interactions. *Lab Anim* 1986;**20**:307–12.
14. van Bogaert MJV, Groenink L, Oosting RS, Westphal KGC, van der Gugten J, Olivier B. Mouse strain differences in autonomic responses to stress. *Genes Brain Behav* 2006;**5**:139–49.
15. Shi W, Brown MD, Wang X, Wong J, Kallmes DF, Matsumoto AH, et al. Genetic backgrounds but not sizes of atherosclerotic lesions determine medial destruction in the aortic root of apolipoprotein e-deficient mice. *Arterioscler Thromb Vasc Biol* 2003;**23**:1901–6.
16. Thaker PH, Han LY, Kamat AA, Arevalo JM, Takahashi R, Lu C, et al. Chronic stress promotes tumor growth and angiogenesis in a mouse model of ovarian carcinoma. *Nat Med* 2006;**12**:939–44.
17. Daugherty A, Rateri DL. Development of experimental designs for atherosclerosis studies in mice. *Methods* 2005;**36**:129–38.
18. Curry BB. Animal models used in identifying gender-related differences. *Int J Toxicol* 2001;**20**:153–60.
19. Czerniak R. Gender-based differences in pharmacokinetics in laboratory animal models. *Int J Toxicol* 2001;**20**:161–3.
20. Kato R, Yamazoe Y. Sex-specific cytochrome p450 as a cause of sex- and species-related differences in drug toxicity. *Toxicol Lett* 1992;64–5 (Spec No:661–667).
21. Morton DB, Jennings M, Buckwell A, Ewbank R, Godfrey C, Holgate B, et al. Refining procedures for the administration of substances. Report of the bvaawf/frame/rspca/ufaw joint working group on refinement. British veterinary association animal welfare foundation/fund for the replacement of animals in medical experiments/royal society for the prevention of cruelty to animals/universities federation for animal welfare. *Lab Anim* 2001;**35**:1–41.
22. Kohn D, Wixson S, White W, Benson G, editors. *Anesthesia and analgesia in laboratory animals*. New York: Academic Press; 1997.
23. Roth DM, Swaney JS, Dalton ND, Gilpin EA, Ross J. Impact of anesthesia on cardiac function during echocardiography in mice. *Am J Physiol Heart Circ Physiol* 2002;**282**: H2134–40.
24. Janssen BJ, De Celle T, Debets JJ, Brouns AE, Callahan MF, Smith TL. Effects of anesthetics on systemic hemodynamics in mice. *Am J Physiol Heart Circ Physiol* 2004;**287**:H1618–24.

25. Zuurbier CJ, Emons VM, Ince C. Hemodynamics of anesthetized ventilated mouse models: aspects of anesthetics, fluid support, and strain. *Am J Physiol Heart Circ Physiol* 2002;**282**: H2099–105.
26. Kober F, Iltis I, Cozzone PJ, Bernard M. Cine-MRI assessment of cardiac function in mice anesthetized with ketamine/xylazine and isoflurane. *MAGMA* 2004;**17**:157–61.
27. Kober F, Iltis I, Cozzone PJ, Bernard M. Myocardial blood flow mapping in mice using high-resolution spin labeling magnetic resonance imaging: influence of ketamine/xylazine and isoflurane anesthesia. *Magn Reson Med* 2005;**53**:601–6.
28. Stein AB, Tiwari S, Thomas P, Hunt G, Levent C, Stoddard MF, et al. Effects of anesthesia on echocardiographic assessment of left ventricular structure and function in rats. *Basic Res Cardiol* 2007;**102**:28–41.
29. Berry CJ, Thedens DR, Light-McGroary K, Miller JD, Kutschke W, Zimmerman KA, et al. Effects of deep sedation or general anesthesia on cardiac function in mice undergoing cardiovascular magnetic resonance. *J Cardiovasc Magn Reson* 2009;**11**:16.
30. Brutkiewicz S, Mendonca M, Stantz K, Comerford K, Bigsby R, Hutchins G, et al. The expression level of luciferase within tumour cells can alter tumour growth upon in vivo bioluminescence imaging. *Luminescence* 2007;**22**:221–8.
31. Meijer MK, Spruijt BM, van Zutphen LFM, Baumans V. Effect of restraint and injection methods on heart rate and body temperature in mice. *Lab Anim* 2006;**40**:382–91.
32. Meijer MK, Lemmens AG, Van Zutphen BFM, Baumans V. Urinary corticosterone levels in mice in response to intraperitoneal injections with saline. *J Appl Anim Welf Sci* 2005;**8**:279–83.
33. Cassidy PJ, Schneider JE, Grieve SM, Lygate C, Neubauer S, Clarke K. Assessment of motion gating strategies for mouse magnetic resonance at high magnetic fields. *J Magn Reson Imaging* 2004;**19**:229–37.
34. Polson MJ, Barker AT, Gardiner S. The effect of rapid rise-time magnetic fields on the ECG of the rat. *Clin Phys Physiol Meas* 1982;**3**:231–4.
35. Shetty AN. Suppression of radiofrequency interference in cardiac gated MRI: a simple design. *Magn Reson Med* 1988;**8**:84–8.
36. Hiba B, Richard N, Janier M, Croisille P. Cardiac and respiratory double self-gated cine MRI in the mouse at 7 t. *Magn Reson Med* 2006;**55**:506–13.
37. Brau AC, Wheeler CT, Hedlund LW, Johnson GA. Fiber-optic stethoscope: a cardiac monitoring and gating system for magnetic resonance microscopy. *Magn Reson Med* 2002;**47**:314–21.
38. Namati E, Chon D, Thiesse J, Hoffman EA, de Ryk J, Ross A, et al. In vivo micro-ct lung imaging via a computer-controlled intermittent iso-pressure breath hold (iibh) technique. *Phys Med Biol* 2006;**51**:6061–75.
39. Dulce MC, Mostbeck GH, Friese KK, Caputo GR, Higgins CB. Quantification of the left ventricular volumes and function with cine MR imaging: comparison of geometric models with three-dimensional data. *Radiology* 1993;**188**:371–6.
40. Ruff J, Wiesmann F, Hiller KH, Voll S, von Kienlin M, Bauer WR, et al. Magnetic resonance microimaging for noninvasive quantification of myocardial function and mass in the mouse. *Magn Reson Med* 1998;**40**:43–8.
41. Slawson SE, Roman BB, Williams DS, Koretsky AP. Cardiac MRI of the normal and hypertrophied mouse heart. *Magn Reson Med* 1998;**39**:980–7.
42. Berr SS, Roy RJ, French BA, Yang Z, Gilson W, Kramer CM, et al. Black blood gradient echo cine magnetic resonance imaging of the mouse heart. *Magn Reson Med* 2005;**53**:1074–9.
43. Franco F, Thomas GD, Giroir B, Bryant D, Bullock MC, Chwialkowski MC, et al. Magnetic resonance imaging and invasive evaluation of development of heart failure in transgenic mice with myocardial expression of tumor necrosis factor-alpha. *Circulation* 1999;**99**:448–54.

44. Wiesmann F, Frydrychowicz A, Rautenberg J, Illinger R, Rommel E, Haase A, et al. Analysis of right ventricular function in healthy mice and a murine model of heart failure by in vivo MRI. *Am J Physiol Heart Circ Physiol* 2002;**283**:H1065–71.
45. Wiesmann F, Ruff J, Hiller KH, Rommel E, Haase A, Neubauer S. Developmental changes of cardiac function and mass assessed with MRI in neonatal, juvenile, and adult mice. *Am J Physiol Heart Circ Physiol* 2000;**278**:H652–7.
46. Schmitz SA, Coupland SE, Gust R, Winterhalter S, Wagner S, Kresse M, et al. Superparamagnetic iron oxide-enhanced MRI of atherosclerotic plaques in watanabe hereditable hyperlipidemic rabbits. *Invest Radiol* 2000;**35**:460–71.
47. Axel L, Dougherty L. MR imaging of motion with spatial modulation of magnetization. *Radiology* 1989;**171**:841–5.
48. Henson RE, Song SK, Pastorek JS, Ackerman JJ, Lorenz CH. Left ventricular torsion is equal in mice and humans. *Am J Physiol Heart Circ Physiol* 2000;**278**:H1117–23.
49. Zhou R, Pickup S, Glickson JD, Scott CH, Ferrari VA. Assessment of global and regional myocardial function in the mouse using cine and tagged MRI. *Magn Reson Med* 2003;**49**:760–4.
50. Young AA, French BA, Yang Z, Cowan BR, Gilson W, Berr SS, et al. Reperfused myocardial infarction in mice: 3D mapping of late gadolinium enhancement and strain. *J Cardiovasc Magn Reson* 2006;**8**:685–92.
51. Epstein FH, Yang Z, Gilson WD, Berr SS, Kramer CM, French BA. MR tagging early after myocardial infarction in mice demonstrates contractile dysfunction in adjacent and remote regions. *Magn Reson Med* 2002;**48**:399–403.
52. Stuber M, Spiegel MA, Fischer SE, Scheidegger MB, Danias PG, Pedersen EM, et al. Single breath-hold slice-following cspamm myocardial tagging. *MAGMA* 1999;**9**:85–91.
53. Spotnitz HM. Macro design, structure, and mechanics of the left ventricle. *J Thorac Cardiovasc Surg* 2000;**119**:1053–77.
54. Moore CC, McVeigh ER, Zerhouni EA. Quantitative tagged magnetic resonance imaging of the normal human left ventricle. *Top Magn Reson Imaging* 2000;**11**:359–71.
55. Moore CC, Lugo-Olivieri CH, McVeigh ER, Zerhouni EA. Three-dimensional systolic strain patterns in the normal human left ventricle: characterization with tagged MR imaging. *Radiology* 2000;**214**:453–66.
56. Osman NF, Kerwin WS, McVeigh ER, Prince JL. Cardiac motion tracking using cine harmonic phase (harp) magnetic resonance imaging. *Magn Reson Med* 1999;**42**:1048–60.
57. Heijman E, Strijkers GJ, Habets J, Janssen B, Nicolay K. Magnetic resonance imaging of regional cardiac function in the mouse. *MAGMA* 2004;**17**:170–8.
58. Kraitchman DL, Sampath S, Castillo E, Derbyshire JA, Boston RC, Bluemke DA, et al. Quantitative ischemia detection during cardiac magnetic resonance stress testing by use of fastHARP. *Circulation* 2003;**107**:2025–30.
59. Thomas D, Ferrari V, Janik M, Kim D, Pickup S, Glickson J, et al. Quantitative assessment of regional myocardial function in a rat model of myocardial infarction using tagged MRI. *MAGMA* 2004;**17**:179–87.
60. Gilson WD, Yang Z, French BA, Epstein FH. Complementary displacement-encoded MRI for contrast-enhanced infarct detection and quantification of myocardial function in mice. *Magn Reson Med* 2004;**51**:744–52.
61. Gilson WD, Yang Z, French BA, Epstein FH. Measurement of myocardial mechanics in mice before and after infarction using multislice displacement-encoded MRI with 3D motion encoding. *Am J Physiol Heart Circ Physiol* 2005;**288**:H1491–7.
62. Streif JU, Herold V, Szimtenings M, Lanz TE, Nahrendorf M, Wiesmann F, et al. In vivo time-resolved quantitative motion mapping of the murine myocardium with phase contrast MRI. *Magn Reson Med* 2003;**49**:315–21.

63. Benjelloun H, Cranney GB, Kirk KA, Blackwell GG, Lotan CS, Pohost GM. Interstudy reproducibility of biplane cine nuclear magnetic resonance measurements of left ventricular function. *Am J Cardiol* 1991;**67**:1413–20.
64. Arber S, Hunter JJ, Ross Jr. J, Hongo M, Sansig G, Borg J, et al. Mlp-deficient mice exhibit a disruption of cardiac cytoarchitectural organization, dilated cardiomyopathy, and heart failure. *Cell* 1997;**88**:393–403.
65. Koch WJ, Rockman HA, Samama P, Hamilton RA, Bond RA, Milano CA, et al. Cardiac function in mice overexpressing the beta-adrenergic receptor kinase or a beta ark inhibitor. *Science* 1995;**268**:1350–3.
66. Geisterfer-Lowrance AA, Christe M, Conner DA, Ingwall JS, Schoen FJ, Seidman CE, et al. A mouse model of familial hypertrophic cardiomyopathy. *Science* 1996;**272**:731–4.
67. Strauer BE, Motz W, Vogt M, Schwartzkopff B. Impaired coronary flow reserve in niddm: a possible role for diabetic cardiopathy in humans. *Diabetes* 1997;**46**(Suppl 2): S119–24.
68. Laine H, Raitakari OT, Niinikoski H, Pitkanen OP, Iida H, Viikari J, et al. Early impairment of coronary flow reserve in young men with borderline hypertension. *J Am Coll Cardiol* 1998;**32**:147–53.
69. Schwitter J, Nanz D, Kneifel S, Bertschinger K, Buchi M, Knusel PR, et al. Assessment of myocardial perfusion in coronary artery disease by magnetic resonance: a comparison with positron emission tomography and coronary angiography. *Circulation* 2001;**103**:2230–5.
70. Streif JU, Nahrendorf M, Hiller KH, Waller C, Wiesmann F, Rommel E, et al. In vivo assessment of absolute perfusion and intracapillary blood volume in the murine myocardium by spin labeling magnetic resonance imaging. *Magn Reson Med* 2005;**53**:584–92.
71. Trabold F, Pons S, Hagege AA, Bloch-Faure M, Alhenc-Gelas F, Giudicelli JF, et al. Cardiovascular phenotypes of kinin b2 receptor- and tissue kallikrein-deficient mice. *Hypertension* 2002;**40**:90–5.
72. Decking UK, Pai VM, Bennett E, Taylor JL, Fingas CD, Zanger K, et al. High-resolution imaging reveals a limit in spatial resolution of blood flow measurements by microspheres. *Am J Physiol Heart Circ Physiol* 2004;**287**:H1132–40.
73. Buckberg GD, Luck JC, Payne DB, Hoffman JI, Archie JP, Fixler DE. Some sources of error in measuring regional blood flow with radioactive microspheres. *J Appl Physiol* 1971;**31**:598–604.
74. Plein S, Ryf S, Schwitter J, Radjenovic A, Boesiger P, Kozerke S. Dynamic contrast-enhanced myocardial perfusion MRI accelerated with *k-t* SENSE. *Magn Reson Med* 2007;**58**:777–85.
75. Makowski MR, Jansen C, Webb I, Chiribiri A, Nagel E, Botnar R, et al. First-pass contrast-enhanced myocardial perfusion MRI in mice on a 3-t clinical MR scanner. *Magn Reson Med* 2010;**64**:1592–8.
76. Oshinski JN, Yang Z, Jones JR, Mata JF, French BA. Imaging time after Gd-DTPA injection is critical in using delayed enhancement to determine infarct size accurately with magnetic resonance imaging. *Circulation* 2001;**104**:2838–42.
77. Yang Z, Berr SS, Gilson W, Toufektsian MC, French BA. Simultaneous evaluation of infarct size and cardiac function in intact mice by contrast-enhanced cardiac magnetic resonance imaging reveals contractile dysfunction in noninfarcted regions early after myocardial infarction. *Circulation* 2004;**109**:1161–7.
78. Flacke S, Allen JS, Chia JM, Wible JH, Periasamy MP, Adams MD, et al. Characterization of viable and nonviable myocardium at MR imaging: comparison of gadolinium-based extracellular and blood pool contrast materials versus manganese-based contrast materials in a rat myocardial infarction model. *Radiology* 2003;**226**:731–8.
79. Koenig SH, Spiller M, Brown RD, Wolf GL. Relaxation of water protons in the intra- and extracellular regions of blood containing Gd(DTPA). *Magn Reson Med* 1986;**3**:791–5.

80. Kim RJ, Fieno DS, Parrish TB, Harris K, Chen EL, Simonetti O, et al. Relationship of MRI delayed contrast enhancement to irreversible injury, infarct age, and contractile function. *Circulation* 1999;**100**:1992–2002.
81. Dill T, Ekinci O, Hansel J, Kluge A, Breidenbach C, Hamm CW. Delayed contrast-enhanced magnetic resonance imaging for the detection of autoimmune myocarditis and long-term follow-up. *J Cardiovasc Magn Reson* 2005;**7**:521–3.
82. Tandri H, Saranathan M, Rodriguez ER, Martinez C, Bomma C, Nasir K, et al. Noninvasive detection of myocardial fibrosis in arrhythmogenic right ventricular cardiomyopathy using delayed-enhancement magnetic resonance imaging. *J Am Coll Cardiol* 2005;**45**:98–103.
83. Yang YM, Feng X, Yao ZW, Tang WJ, Liu HQ, Zhang L. Magnetic resonance angiography of carotid and cerebral arterial occlusion in rats using a clinical scanner. *J Neurosci Methods* 2008;**167**:176–83.
84. Reese T, Bochelen D, Sauter A, Beckmann N, Rudin M. Magnetic resonance angiography of the rat cerebrovascular system without the use of contrast agents. *NMR Biomed* 1999;**12**:189–96.
85. Gerriets T, Li F, Silva MD, Meng X, Brevard M, Sotak CH, et al. The macrosphere model: evaluation of a new stroke model for permanent middle cerebral artery occlusion in rats. *J Neurosci Methods* 2003;**122**:201–11.
86. Fehm NP, Vatankhah B, Dittmar MS, Retzl G, Schuierer G, Horn M, et al. Application of clinical scanners in rats: experimental carotid imaging using magnetic resonance imaging, spiral computed tomography, and color duplex ultrasound. *J Neuroimaging* 2005;**15**:319–25.
87. Choy M, Ganesan V, Thomas DL, Thornton JS, Proctor E, King MD, et al. The chronic vascular and haemodynamic response after permanent bilateral common carotid occlusion in newborn and adult rats. *J Cereb Blood Flow Metab* 2006;**26**:1066–75.
88. Besselmann M, Liu M, Diedenhofen M, Franke C, Hoehn M. MR angiographic investigation of transient focal cerebral ischemia in rat. *NMR Biomed* 2001;**14**:289–96.
89. Krettek A, Sukhova GK, Libby P. Elastogenesis in human arterial disease: a role for macrophages in disordered elastin synthesis. *Arterioscler Thromb Vasc Biol* 2003;**23**:582–7.
90. Makowski MR, Wiethoff AJ, Blume U, Cuello F, Warley A, Jansen CH, et al. Assessment of atherosclerotic plaque burden with an elastin-specific magnetic resonance contrast agent. *Nat Med* 2011;**17**:383–8.
91. von Bary C, Makowski M, Preissel A, Keithahn A, Warley A, Spuentrup E, et al. Mri of coronary wall remodeling in a swine model of coronary injury using an elastin-binding contrast agent. *Circ Cardiovasc Imaging* 2011;**4**:147–55.
92. Sirol M, Itskovich VV, Mani V, Aguinaldo JG, Fallon JT, Misselwitz B, et al. Lipid-rich atherosclerotic plaques detected by gadofluorine-enhanced in vivo magnetic resonance imaging. *Circulation* 2004;**109**:2890–6.
93. Ronald JA, Chen Y, Belisle AJ, Hamilton AM, Rogers KA, Hegele RA, et al. Comparison of gadofluorine-M and Gd-DTPA for noninvasive staging of atherosclerotic plaque stability using MRI. *Circ Cardiovasc Imaging* 2009;**2**:226–34.
94. Tavora F, Cresswell N, Li L, Ripple M, Burke A. Immunolocalisation of fibrin in coronary atherosclerosis: implications for necrotic core development. *Pathology* 2010;**42**:15–22.
95. Flacke S, Fischer S, Scott MJ, Fuhrhop RJ, Allen JS, McLean M, et al. Novel MRI contrast agent for molecular imaging of fibrin: implications for detecting vulnerable plaques. *Circulation* 2001;**104**:1280–5.
96. Botnar RM, Perez AS, Witte S, Wiethoff AJ, Laredo J, Hamilton J, et al. In vivo molecular imaging of acute and subacute thrombosis using a fibrin-binding magnetic resonance imaging contrast agent. *Circulation* 2004;**109**:2023–9.
97. Botnar RM, Buecker A, Wiethoff AJ, Parsons Jr. EC, Katoh M, Katsimaglis G, et al. In vivo magnetic resonance imaging of coronary thrombosis using a fibrin-binding molecular magnetic resonance contrast agent. *Circulation* 2004;**110**:1463–6.

98. Spuentrup E, Buecker A, Katoh M, Wiethoff AJ, Parsons Jr. EC, Botnar RM, et al. Molecular magnetic resonance imaging of coronary thrombosis and pulmonary emboli with a novel fibrin-targeted contrast agent. *Circulation* 2005;**111**:1377–82.
99. Spuentrup E, Botnar RM, Wiethoff A, Ibrahim T, Sebastin K, Katho M, et al. MR imaging of thrombi using EP-2104R, a fibrin specific contrast agent: Initial results in patients. *Eur Radiol* 2008;**18**:1995–2005 (1911).
100. Nahrendorf M, Jaffer FA, Kelly KA, Sosnovik DE, Aikawa E, Libby P, et al. Noninvasive vascular cell adhesion molecule-1 imaging identifies inflammatory activation of cells in atherosclerosis. *Circulation* 2006;**114**:1504–11.
101. Winter PM, Morawski AM, Caruthers SD, Fuhrhop RW, Zhang H, Williams TA, et al. Molecular imaging of angiogenesis in early-stage atherosclerosis with alpha(v)beta3-integrin-targeted nanoparticles. *Circulation* 2003;**108**:2270–4.
102. Kooi ME, Cappendijk VC, Cleutjens KB, Kessels AG, Kitslaar PJ, Borgers M, et al. Accumulation of ultrasmall superparamagnetic particles of iron oxide in human atherosclerotic plaques can be detected by in vivo magnetic resonance imaging. *Circulation* 2003;**107**:2453–8.
103. Ruehm SG, Corot C, Vogt P, Kolb S, Debatin JF. Magnetic resonance imaging of atherosclerotic plaque with ultrasmall superparamagnetic particles of iron oxide in hyperlipidemic rabbits. *Circulation* 2001;**103**:415–22.
104. Morishige K, Kacher DF, Libby P, Josephson L, Ganz P, Weissleder R, et al. High-resolution magnetic resonance imaging enhanced with superparamagnetic nanoparticles measures macrophage burden in atherosclerosis. *Circulation* 2010;**122**:1707–15.
105. Makowski MR, Varma G, Wiethoff A, Smith A, Mattock K, Jansen C, et al. Non-invasive assessment of atherosclerotic plaque progression in apoE−/− mice using susceptibility gradient mapping. *Circ Cardiovasc Imaging* 2011;**4**:295–303.
106. Korosoglou G, Weiss RG, Kedziorek DA, Walczak P, Gilson WD, Schar M, et al. Noninvasive detection of macrophage-rich atherosclerotic plaque in hyperlipidemic rabbits using "positive contrast" magnetic resonance imaging. *J Am Coll Cardiol* 2008;**52**:483–91.
107. Mani V, Briley-Saebo KC, Itskovich VV, Samber DD, Fayad ZA. Gradient echo acquisition for superparamagnetic particles with positive contrast (grasp): sequence characterization in membrane and glass superparamagnetic iron oxide phantoms at 1.5t and 3t. *Magn Reson Med* 2006;**55**:126–35.
108. Amirbekian V, Lipinski MJ, Briley-Saebo KC, Amirbekian S, Aguinaldo JG, Weinreb DB, et al. Detecting and assessing macrophages in vivo to evaluate atherosclerosis noninvasively using molecular MRI. *Proc Natl Acad Sci USA* 2007;**104**:961–6.

Animal Models of Inflammatory Bowel Disease

Atsushi Mizoguchi

Molecular Pathology Unit, Department of Pathology, Center for the Study of Inflammatory Bowel Disease, Massachusetts General Hospital, Harvard Medical School, Boston, Massachusetts, USA

Progress in Molecular Biology
and Translational Science, Vol. 105
DOI: 10.1016/B978-0-12-394596-9.00009-3

Inflammatory bowel disease (IBD) is a chronic intestinal inflammatory condition that is medicated by genetic, immune, and environmental factors. At least 66 different kinds of animal models have been established to study IBD, which are classified primarily into chemically induced, cell-transfer, congenial mutant, and genetically engineered models. These IBD models have provided significant contributions to not only dissect the mechanism but also develop novel therapeutic strategies for IBD. In addition, recent advances on genetically engineered techniques such as cell-specific and inducible knockout as well as knockin mouse systems have brought novel concepts on IBD pathogenesis to the fore. Further, mouse models, which lack some IBD susceptibility genes, have suggested more complicated mechanism of IBD than previously predicted. This chapter summarizes the distinct feature of each murine IBD model and discusses the previous and current lessons from the IBD models.

Inflammatory bowel disease (IBD) is a chronic intestinal inflammatory condition that affects approximately 1.4 million individuals in the United States.[1,2] IBD is characterized by two major forms, Crohn's disease (CD) and ulcerative colitis (UC), which are mediated by both common and different mechanisms and exhibit distinct clinical features.[1,2]

In order to dissect the mechanism underlying IBD, Dr. Kirsner and colleagues first introduced an experimental colitis model in 1957, which was induced in rabbits by sensitization to crystalline egg albumin together with small rectal instillation of dilute formalin.[3] Different types of chemically induced colitis models, primarily in rats, have since been developed.[4] In 1981, cotton-top tamarin, a marmoset native to a small geographic region of Colombia, was identified to develop colitis spontaneously when kept in the United States.[5]

This model has brought a new concept of targeting cell-homing for IBD therapy.[6] In 1990, Dr. Powrie and colleagues developed an adoptive T-cell transfer system to induce colitis in immune-deficient mice.[7] This model system has provided significant contributions to develop the novel concept on "regulatory T cells."[8] In the same year, genetically engineered rats carrying human *HLAB27* gene were identified to develop colitis.[9] A major turning point on IBD research was then made in 1993 by the discovery of spontaneous colitis in three different kinds of knockout (KO) mice; interleukin (IL)-2 KO,[10] IL-10 KO,[11] and T-cell receptor (TCR) α KO mice.[12] Indeed, there have since been well over 40 different kinds of genetically engineered KO mouse strains and congenital gene mutant mouse strains found to develop colitis and/or ileitis spontaneously (Fig. 1). These genetically engineered IBD models have provided significant contributions not only for understanding the fundamental mechanism of IBD but also for developing important interventions of IBD therapy.[13–15]

Spontaneous development of colitis/ileitis in many different kinds of genetically engineered mouse strains clearly suggests that IBD is mediated by extremely complicated mechanisms.[15] This concept is well supported by recent human genome-wide association studies that have already identified over 71 susceptibility genes in CD and 47 susceptibility genes in UC.[16–20] Genetically engineered mice lacking (KO) or overexpressing (transgenic (Tg)) some IBD susceptibility or candidate genes have been confirmed to develop intestinal inflammation spontaneously. They include IL-10 KO, STAT3 KO, XBP1 KO, IL-2Rα KO, TNFSF15 Tg, and IL-7 Tg mice. In addition, functions of some other IBD susceptibility genes have been explored using murine IBD models. This chapter summarizes the currently available mouse models of IBD, including IBD-associated genes KO mice.

I. Classification of Murine IBD Models

As shown in Fig. 1, IBD models are classified into five major groups: chemically induced model, cell-transfer model, spontaneous model, congenital (spontaneous gene mutation) model, and genetically engineered model. In addition, recent advances on genetic engineering technology have prompted us to further classify the genetically engineered model into six subgroups (Fig. 1). Conventional Tg or KO model groups are mice that are genetically engineered to continuously overexpress or lack the gene of interest in all cell types, respectively. Conditional Tg models are genetically engineered using cell-specific promoter to overexpress the gene of interest in a specific cell type. Conditional KO models are genetically engineered to lack a gene of interest in a specific cell type or/and to make the gene deletion inducible at any time when gene deletion is desired during the adult life. Knockin (KI) models are mice

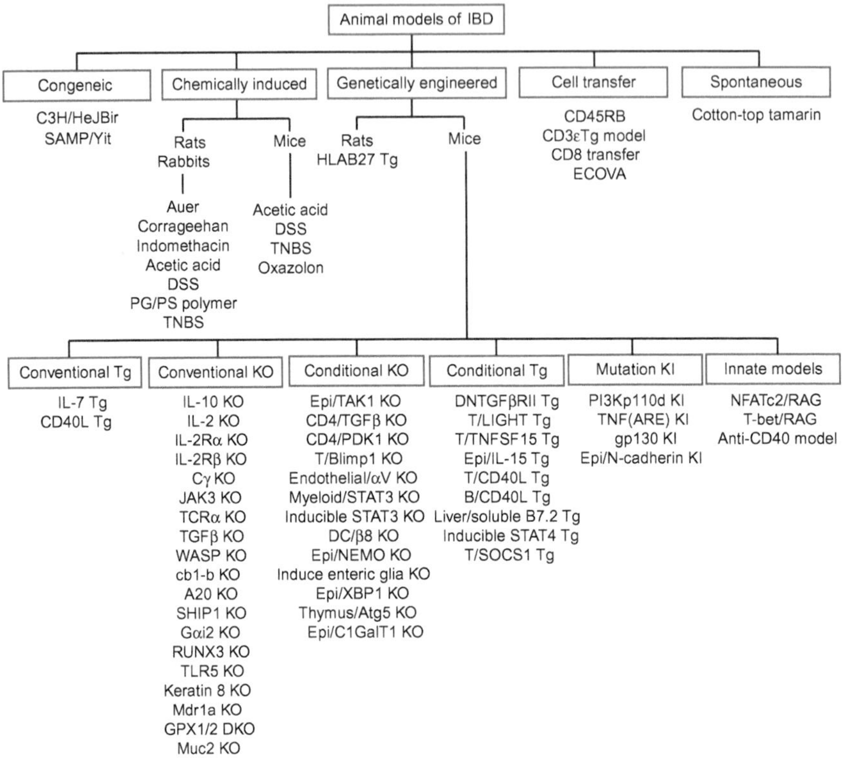

FIG. 1. Classification of animal models of IBD.

that are genetically engineered to carry a mutation in the gene of interest. Innate models are immune-deficient mouse-based system such as recombinant activation gene (RAG) KO mice that lack both T and B cells. One of the examples is T-bet-deficient RAG2 double KO mice. Mice carrying single deletion of T-bet or RAG2 do not develop intestinal inflammation spontaneously, but double deficiency of both T-bet and RAG2 causes spontaneous development of colitis in mice.[21]

II. IL-10 KO Mice

IL-10 is a well-known regulatory cytokine and represents a key IBD (both UC and CD) susceptibility gene.[16–20] IL-10 KO mice, which are genetically engineered to lack *IL-10* gene, spontaneously develop colitis after 3 months of age.[11] C3H and Balb/c backgrounds are more susceptible to this colitis as compared to C57BL/6 background. Enteric microbiota plays the major role in this colitis. Indeed, many studies in the microbiology field such as probiotics have been done using IL-10 KO mice (Table I). Development of colitis in IL-10 KO mice is abolished when they are maintained under germ-free condition and may be dependent on *Helicobacter* species.[22] Alternatively, different roles of aerobic versus anaerobic bacteria have recently been proposed in the pathogenesis of this colitis.[23] Ciprofloxacin is effective for the treatment of typhilitis (inflammation in the cecum) but not colitis in IL-10 KO mice, whereas metronidazole is beneficial for the treatment of colitis but not typhilitis in these mice.[23] The complicated participation of different bacterial species tends to make it difficult to understand the role of pattern recognition receptors in this colitis. Indeed, both protective[22,24] and deleterious[25,26] roles of TLR4 have been proposed in the colitis of IL-10 KO mice.

Since incidence of colitis in IL-10 KO mice is not so high in the absence of *Helicobater* species, nonsteroidal anti-inflammatory drugs (NSAID) such as piroxicam have been used to rapidly and reproducibly induce the development of colitis in these mice. Two-week administration of NSAID induces rapid development of colitis in IL-10 KO mice within 4–6 weeks.[27] The proxicam-induced promotion of colitis development may be mediated by an IL-13 decoy receptor IL-13Rα2.[28] Proxicam induces the expression of IL-13Rα2, which subsequently enhances Th17 responses by suppressing IL-13 bioactivity.

The colitis of IL-10 KO mice is mediated by $CD4^+$ T cells but not by B cells.[29] CTLA-4-dependent regulatory T cells (Treg) also exist in this model to suppress the progression of colitis.[30] The colitis of IL-10 KO mice has somehow been frequently referred to as "Th1-mediated" colitis. However, it would be important to mention that there has been no direct evidence on the role of IFN-γ to sustain the colitis of IL-10 KO mice.[31] In addition, IFN-γ-deficient IL-10 double KO mice develop more exacerbated form of colitis as compared to IL-10 KO mice, suggesting the rather protective role of Th1 (IFN-γ) in this colitis.[32]

IL-12p40 is associated with IL-12p35 to form IL-12p70, and with IL-12p19 to form IL-23. Protective effect of anti-IL-12p40 mAbs has been well documented in the colitis of IL-10 KO mice.[31] Alternatively, IL-12p70 (p35/p40) and IL-23 (p19/p40) may play different roles in this colitis. IL-12p35-deficient IL-10 double KO mice develop colitis at earlier age (7 weeks of age) as compared to IL-10 KO mice that develop colitis at 3 months of age.[33] In contrast, IL-12p19 (IL-23)-deficient IL-10 double KO mice are still colitis-free at 12 months of age. IL-23 promotes the production of IL-6 and IL-17 in IL-10 KO mice. Anti-IL-17 mAb together with anti-IL-6 mAb significantly improves this colitis. These findings suggest the dependence of this colitis on IL-23, IL-17, and IL-6 and the possible protective role of IL-12p70 and IFN-γ.

TABLE I
Pathogenic Versus Protective Factors in the Colitis of IL-10 KO Mice

Pathogenic	Protective
Cell types	
Memory $CD4^+$ T cells	$CD4^+Foxp3^+$ Treg, bacteria-triggering Treg
BM-retaining memory $CD4^+$ T cells	NK cells, mast cells
Endogenous factors	
Activin, adenosine H2B, CD44v7, CD52, CXCL2, CXCR4, hypoglycosylated mucin 1, IL-13Rα2, IL-23, IL-27, IL-27R, IP-10, MyD88, TLR4, TNF-α, NEMO, IKKβ in myeloid cells, histone deacetylase	CXCL10, glucagon-like peptide 2, IL-1β, IFN-γR1, IRAK-M (IL-1 receptor-associated kinase M), kinase suppressor of Ras1 (KSR1), mucin 2, MyD88, NEMO, NHE3 (Na^+H^+ exchanger), substance P, SERT (serotonin reuptake transporter), TGFβ1, thioredoxin-1, TLR4, p110δ
Exogenous factors	
Microorganisms:	*Microorganisms:*
Bacteroides vulgatus	Germ-free environment
Bifidobacterium animalis	Antibiotics mixture (amoxicillin, omeprazole, metronidazole, and clarithromycin)
Campylobacter jejuni	*Bifidobacterium infantis*
Enterococcus faecalis	*Bifidobacterium animal* subsp., CpG
Escherichia coli	*Heligmosomoides polygyrus*
E. coli with OH7:H7:K1 serotype	Lipoteichoic acid-deficient *Lactobacillus acidophilus*
Helicobacter bilis	*Lactobacillus gasseri*
Helicobacter cinaedi	*Lactobacillus plantarum*
Helicobacter hepaticus	*Lactobacillus reuteri*
Helicobacter mastomyrinus	
Helicobacter redentium	
Helicobacter rodentium	
Helicobacter trogontum	
Helicobacter typhlonius	
Herpes virus	
Others:	*Others:*
Cytolethal distending toxin	AT-1001 (zonulin peptide inhibitor)
NSAID, potato skin	ATN-161 (anti-angiogenic compound)
	Bone marrow transplant, bronelain
	Carbon monoxide (smoking), curcumin
	Dietary histidine, ethyl pyruvate
	Exogenously delivered IL-10
	ITF 2357 (histone deacetylase inhibitor), normal breast milk, NEMO inhibitor
	Shingosin 1-phosphate receptor agonist
	Tacrolimus, triptolide, vitamin D

III. IL-2 KO Mice

IL-2 is a critical cytokine that is necessary for T-cell homeostasis. IL-2 suppresses excessive T-cell responses by inducing activation-induced cell death (AICD) and elicits the function of $CD4^+$ $Foxp3^+$ Treg. Importantly, IL-2RA (IL-2 receptor α) is a potential CD susceptibility gene and *IL-2* gene is located within an IBD susceptibility locus.[16–20]

IL-2 KO mice spontaneously develop a systemic autoimmune disease characterized by colitis, gastritis, hepatitis, pneumonia, pancreatitis, nephritis, and hemolytic anemia.[10] Approximately half of IL-2 KO mice die at early age and the survivors develop colitis characterized by massive infiltration of T cells between 6 and 15 weeks of age.[10] IL-2 signaling is induced through the binding to a heterotrimer receptor composed of IL-2Rα, IL-2Rβ, and common γ chains. IL-2Rα KO, IL-2Rβ KO, and common γ chain KO mice all develop autoimmune phenotype as seen in IL-2 KO mice.[34–36] In addition, deficiency of Janus family tyrosine kinase (JAK) 3, which is an essential transducer of common γ chain, also causes spontaneous development of colitis.[35]

A distinct feature with IL-2 KO model as compared to other IBD models is the development of colitis in germ-free condition,[37] suggesting that the colitis is fully mediated by an autoimmune mechanism (Table II). Indeed, absence of MyD88 does not alter the colitis of IL-2 KO mice.[25] In addition, colitis of IL-2 KO mice fully depends on $CD4^+$ $TCR\alpha\beta^+$ T cells but not $CD8^+$ $TCR\alpha\beta^+$ T cells, $TCR\gamma\delta^+$ T cells or B cells, as indicated by the absence of colitis in IL-2-deficient RAG2 double KO mice lacking both T and B cells[38] and IL-2-deficient TCRβ double KO mice lacking $TCR\alpha\beta^+$ T cells[39] and by the presence of colitis in IL-2-deficient MHC class I double KO mice lacking $CD8^+$ T cells.[40]

DC-mediated activation of $CD4^+$ T cells through RANKL/RANK interaction is necessary for the development of colitis in IL-2 KO mice.[41] Immediately after recognition of antigen through TCR, $CD4^+$ T cells form immunological synapse (IS) that initiates the activation of major signaling machinery for promoting their activation. Protein kinase C (PKC)θ represents the key proximal molecule involved in the IS-mediated signaling. Absence of PKCθ in $CD4^+$ T cells improves the colitis of IL-2 KO mice by suppressing the

TABLE II
COLITIS DEVELOP PATTERN OF IL-2, IL-10 VERSUS TCRα KO MICE IN DIFFERENT FACILITIES

	GF	*Helicobacter*-free	SPF	Conventional
IL-2 KO	(+)	(+)	(+)	(+)
IL-10 KO	(−)	(−)	(+)	(+)
TCRα KO	(−)	(+)	(+)	(−)

proliferation of effector memory CD4$^+$ T cells.[39] In addition, CD4 T-cell-specific deletion of STAT3 (under control of CD4 promoter) improves the colitis in IL-2 KO mice. This protection is mediated by the absence of IL-17-producing Th17 T cells and in a Treg-independent manner.[42]

In addition to the important function for T effector cells, IL-2 is essential for the development and function of Treg cells. IL-2 is critically required for the development of naturally occurring Treg in the thymus. Indeed, IL-2 KO, IL-2Rα KO, and IL-2Rβ KO mice all harbor immature Treg cells that are characterized by Foxp3^{low} CD25$^-$ and by the absence of immune regulatory ability.[43] IL-2 is also necessary not only for generation of inducible Foxp3$^+$ Treg[44] but also for eliciting the function of Treg.[45] Indeed, adoptive transfer of wild-type (WT)-derived memory CD4$^+$ T cells-containing Treg can induce colitis in recipient IL-2-deficient TCRβ double KO mice but fails to do it in recipient RAG1 KO mice.[39]

IV. TCRα KO Mice

TCR, which is composed of TCRα and TCRβ chains, is required for the recognition of antigens for consequent activation of T cell-mediated adaptive immune responses. Interestingly, deletion of TCRα chains induces the spontaneous development of colitis.[12] Approximately 60% of TCRα KO mice develop Th2-mediated colitis at 6 months of age, and the inflammation is restricted primarily to the mucosa. C57BL/6 background is more susceptible to this colitis as compared to Balb/c and C3H/HeJ backgrounds.[12]

Deficiency of TCRα chains impairs normal thymic selection, resulting in the lack of Treg and in the development of unique T cells (TCRα^- β^+) that express TCRβ chains without TCRα chains.[46] CD4$^+$ TCRα^- β^+ T cells, which may primarily recognize superantigens, are immunologically functional as indicated by the TCR repertoire pattern and the cytokine profile.[47,48] The CD4$^+$ TCRα^- β^+ T cells cause the development of colitis through production of IL-4.[49–51] Unique TCR motif, which is characterized by a utilization of negatively charged amino acids in the complementary determinant region (CDR3) of TCR Vβ chains particularly Vβ8.2, may be responsible for the recognition of colitogenic antigens.[47] A major concern with TCRα KO model has been the physiological relevance because TCRα^- β^+ T cells were generally believed to specifically exist only in TCRα KO mice. However, recent advance on the single-cell analysis system has led to the identification of TCRα^- β^+ T cells in WT mice.[52] In addition, TCRα^- β^+ T cells have recently been found in humans with a specific autoimmune condition.[53]

Like observed in the majority of IBD models, TCRα KO mice fail to develop colitis when maintained in germ-free facility.[47,54] Of note, TCRα KO, but not TCRβ KO, mice develop colitis in *Helicobacter*-free environment, suggesting the independence of this spontaneous colitis on *Helicobacter* species.[55,56] Alternatively, if TCRα KO mice are infected with *Helicobacter* species, the original cytokine profile (Th2 responses) shifts to dominant Th1 responses.[56] In addition, epithelial cell permeability (as judged by mannitol transmural flux), which tends to increase in the context of infection, is not altered in TCRα KO mice.[57] Most notably, development of colitis in TCRα KO mice is abolished when maintained under conventional condition[58,59] (Table II). Colitis becomes undetectable in TCRα KO mice when they are maintained over several generations in conventional facility. In contrast, when SPF TCRα KO mice are simply transferred into conventional facility, they develop severe colitis.[59] Therefore, it is possible that hygiene hypothesis (a negative association of inflammation with enhanced exposure to microorganisms during childhood) may be applied to this model. Mechanistically, conventional condition seems to stimulate the expansion of B1 B-cell lineage cells involved in the maintenance of appropriate host/microbial interactions.[59]

As shown in Table III, several cell types and factors have been identified to play protective versus pathogenic roles in the colitis of TCRα KO mice. One of distinct features with this model as compared to other IBD models is the involvement of etiological factors associated with human UC. A negative association of UC with a prior history of appendectomy before 20 years of age has been well documented.[60] Like human UC, resection of cacal patches (mimic of human appendectomy) before 3 weeks of age suppresses the development of colitis in TCRα KO mice.[61] In addition, active smoking is negatively associated with the development of UC in humans.[60] Similarly, carbon monoxide (mimic of smoking) contributes to the suppression of colitis in TCRα KO mice.[62]

Another interesting aspect with TCRα KO mice is the dominant role of regulatory "B" cells termed Breg in the colitis.[63,64] Human UC is well characterized by the increased production of autoantibodies such as antineutrophil cytoplasmic antigen (ANCA). Similarly, development and severity of colitis in TCRα KO mice is closely correlated to the production of autoantibodies such as ANCA and antitropomyosin Ab, prompting us to raise a hypothesis of B cell/autoantibodies-mediated colitis in this model.[49] However, subsequent studies have demonstrated that B cells play a rather protective role in this colitis as indicated by the exacerbation of colitis in B-cell-deficient TCRα double KO mice as compared to TCRα KO mice.[63,65] B cells were generally considered not to be a typical source for cytokines. However, B cells, which produce large amounts of IL-10 or IL-12p70, appear in the context of colitis in TCRα KO mice and they contributed to the improvement of the disease.[65,66] These findings suggest the existence of not only pathogenic but also regulatory B cell populations in UC. Indeed, a randomized placebo-controlled trial reveals no effect of B-cell-depletion therapy (humanized anti-CD20 mAbs) on inducing the remission of moderately active UC.[67] In addition, a case report shows that B-cell depletion through administration of anti-CD20 mAb exacerbates

TABLE III
PATHOGENIC VERSUS PROTECTIVE FACTORS IN THE COLITIS OF TCRα KO MODEL

Pathogenic	Protective
Cell types	
$CD4^{+}TCR\alpha^{-}\beta^{+}$ T cells, $CD4^{+}$ T cells with negatively charged CDR3 in TCR Vβ chain, $V\beta8.2^{+}$ T cells, intestinal $IL\text{-}7R^{high}$ $CD4^{+}$ T cells, $TCR\gamma\delta^{+}$ T cells, $CD11c^{+}F4/80^{+}$ cells	IL-10-producing B cells (Breg), IL-12p70-producing B cells, B1 B cells, macrophages
Endogenous factors	
IL-4, IL-6, IL-7, IL-23, IFN-γ, TNF-α, TNFR2, MHC class II, PKCθ in $CD4^{+}$ T cell, galectin-4	Clearance of apoptotic cells, SOCS1 (suppressor of cytokine signaling), thymus leukemia (TL) antigen, AIM (apoptosis inhibitor expressed by macrophages), CD1d, CD40 on B cells, CD80 on B cells, IL-10, IL-22, IL-12p70, mucin 1
Exogenous factors	
B-cell depletion (mimic of anti-CD20 therapy in humans)	Resection of cecal patches (mimic of appendectomy in humans)
NSAID	Carbon monoxide (mimic of smoking in humans)
Epithelial self-antigens	Dexamethasone
Bacteroides vulgatus	Enhanced exposure to microorganisms during childhood
Cryptosporidum parvum	Germ-free environment
Helicbacter hepaticus	

UC in association with a decrease in IL-10 production,[68] and three case reports suggest the rapid development of UC in patients with other autoimmune diseases after B-cell-depletion therapy.[69,70]

Recent genome-wide association studies have highlighted the importance of epithelial defense pathway in IBD.[16–20] An IL-10 cytokine family member IL-22 contributes to the improvement of colitis in TCRα KO mice by stimulating STAT3-dependent production of mucin 1 in epithelial cells.[71] Since IL-22, its receptor (IL-10R2), STAT3, and mucin 1 are all UC-specific or IBD (both UC and CD) association genes,[16–20] TCRα KO mice may represent a useful model to study IBD association gene-related epithelial defense machinery.

V. TGFβ KO Mice

Transforming growth factor (TGF)β1 is a multifunctional growth factor involved in developmental, physiological, and immune regulatory processes. TGFβ1 KO mice develop necrotic inflammation in multiple organs including intestine and they are only able to survive only until 3–4 weeks of age.[72]

The early death hampered investigators to closely examine the inflammatory mechanism of TGFβ1 KO mice. To overcome this problem, conditional TGFβ1 KO mouse strains have been developed. One of the conditional systems employs a dominant negative system of TGFβ receptor type II. Overexpressions of dominant negative TGFβ receptor type II block TGFβ signaling when expressed at a sufficiently high level by interfering with the assembly of a functional signaling complex consisting of TGFβ and type I and type II TGFβ receptors. $CD4^+$ T-cell-specific overexpressions (under control of CD4 promoter) of dominant negative TGFβ receptor type II induce spontaneous development of multiorgan inflammations at 3–4 months of age, including the colon, liver, stomach, duodenum, pancreas, and kidney.[73] The inflammation is associated with an enhancement of both Th1 and Th2 responses. B cells play a protective role in this colitis as indicated by the exacerbation of colitis after depletion of B cells by anti-CD20 administration.[74] In contrast to $CD4^+$ T-cell-specific system, epithelial cell-specific overexpressions of dominant negative TGFβ receptor type II do not lead to spontaneous development of colitis although susceptibility to dextran sulfate sodium (DSS)-induced colitis is increased.[75]

TGFβ is widely expressed by leukocytes and stromal cells. Interestingly, CD4 T-cell-specific deletion of TGFβ1 leads to the spontaneous development of multiorgan inflammations (colitis, hepatitis, pneumonia, and nephritis) by 4 months of age,[76] suggesting $CD4^+$ T cells to be a major source of TGFβ1 in IBD. Although accumulating data from *in vitro* studies have suggested the requirement of TGFβ1 for differentiation of $Foxp3^+$ Treg, no abnormality of Treg is seen in the CD4-specific TGFβ1 KO mice. Of note, the colitis of CD4-specific TGFβ1 KO mice is characterized by enhancement of both Th1 and Th2 responses and by suppression of Th17 response.

VI. TAK1 KO Mice

TGFβ-activated kinase 1 (TAK1), which is activated by TLRs, IL-1, and NOD2, is an essential intermediate for the consequent activation of AP1 and NFκB1 in innate immune signaling pathways. Epithelial cell-specific deletion of TAK1 (under control of villin promoter) causes very severe epithelial damage characterized by increased apoptosis in the small intestine by 3 days of age.[77] In addition, inducible epithelial-specific TAK1 KO mice exhibit the intestinal epithelial damage within 3 days after deletion of TAK1 through administration of tomoxifen. Notably, this epithelial damage is rescued by the absence of TNFR1 but development of ileitis and colitis is in turn elicited in the TNFR1-deficient epithelial-specific TAK1 double KO mice. These findings propose that minor, but not severe, apoptotic damage in epithelial cells is responsible for the induction of ileitis and colitis.

VII. WASP KO Mice

Wiskott–Aldrich syndrome protein (WASP) is a key organizer involved in the remodeling of actin cytoskeleton to regulate cell movement, cell signaling, and cell division. Mice deficient in WASP develop colitis from 4 months of age, and 100% penetration of colitis is observed at 6 months of age.[78] This colitis is characterized by elevated expressions of IL-4 and IL-13 and decreased expression of IL-6. Adoptive cell transfer experiments indicate the requirement of IL-4-expressing $CD4^+$ T cells in the development of this colitis.[79] In addition, the regulatory function of $Foxp3^+$ Treg is impaired in WASP KO mice.[80]

VIII. P110δ Mutant Mice

Phosphatidyl inositol 3-kinase (PI3K) is a family of p85/p110 heterodimeric lipid kinases that generate second messenger signals downstream of tyrosine kinases activated by antigen receptors. A PI3K p100 subunit, p110δ is expressed predominantly in leukocytes. Mice, which are genetically engineered to carry a point mutation in the *P110δ* gene locus at position 910 (D > A), spontaneously develop mild, focal inflammation restricted to the rectum and cecum.[81] The development of colitis may be mediated by impaired activation of $Foxp3^+$ Treg.[82] In addition, the onset of colitis is exacerbated by the absence of IL-10.[83]

IX. PDK1 KO Mice

Phosphoinositide-dependent kinase 1 (PDK1), a pleckstrin-homology domain containing protein kinase, is a key downstream effector of PI3K pathway. T-cell-specific deletion (under a control of CD4 promoter) of PDK1 induces spontaneous development of colitis at 8 weeks of age.[84] This colitis is associated with the upregulation of colonic expressions of IL-17A, IL-23p19, and TNF-α. Interestingly, $CD8\alpha^+$ TCRγδ T cells, which produce IL-17A, are dramatically increased in the intraepithelial compartment but not in the lamina propria. The pathogenic role of TCRγδ T cells in this colitis is confirmed by the improvement of colitis in TCRδ-deficient PDK1 double KO mice. The dysregulated colonic TCRγδ T-cell homeostasis results from the impaired function of $Foxp3^+$ Treg by PDK1 deficiency.

X. Cbl-b KO Mice

Cbl-b belongs to a family of Casitas B-lineage lymphoma (Cbl) proteins consisting of c-Cbl, Cbl-b, and Cbl-3; which function as E3 ubiquitin ligases and molecular adaptor to promote ubiquitin conjugation to the p85 regulatory subunit of PI3K. Cbl-b acts as a negative regulator of T-cell activation. Cbl-b KO mice spontaneously develop massive infiltration of activated T cells and B cells in multiple organs from 3 months of age, including the intestine, salivary glands, pancreas, liver, lung, heart, bladder, and connective tissues.[85]

XI. Blimp-1 KO Mice

B lymphocyte-induced maturation protein-1 (Blimp-1) is a zinc finger-containing transcriptional repressor that is necessary for the terminal differentiation of B cells into antibody-secreting plasma cells. In addition to B cells, Blimp-1 is expressed by several different cell lineages, including memory, but not naïve, T cells, myeloid lineage cells, and epithelial cells. Interestingly, T-cell-specific depletion of Blimp-1 (under control of Lck promoter) induces the spontaneous development of colitis.[86] Twenty percent of T-cell-specific Blimp-1 KO mice develop colitis at 4 weeks of age and the colitis is recognized in 83% of them by 21 weeks of age. This colitis is characterized by Th1 response. Treg cells expand normally, but the regulatory function is impaired by the absence of Blimp-1.

XII. A20 KO Mice

A20 is an ubiquitin-editing enzyme that is necessary for the termination of NFκB1 activation after stimulation with TNF or TLRLs. A20 KO mice develop inflammation in multiple organs such as liver, kidney, intestine, and joints at 3–6 weeks of age.[87] The colitis is independent of T and B cells as indicated by the development of inflammation in RAG1-deficient A20 double KO mice. Although A20 is critical for restricting TNF-induced NFκB1 signaling, both TNF-α and TNFR1 are dispensable for the development of this colitis. Epithelial cell-specific deletion of A20 does not elicit the spontaneous development of colitis,[88] suggesting that other cell types such as myeloid lineage cells may be primarily responsible for the colitis induced by A20 deficiency.

XIII. SHIP KO Mice

The Src homology 2 (SH2)-containing inositol-5 phosphatase (SHIP) is activated in response to various growth factors as well as TCR and BCR ligations. *SHIP* gene is located within an IBD association locus 2q37. Ninety-four percent of SHIP KO mice develop transmural, segmental ileitis at 6–8 weeks of age.[89] Granulomas are detectable in 28% of mice. Interestingly, inducible deletion of SHIP during adult life by administration of polyI/C also results in the spontaneous development of ileitis within 4–5 weeks after gene deletion. SHIP KO chimera mice, which are reconstituted with WT bone marrow, do not develop ileitis. In contrast, adoptive transfer of all splenocytes, but not $CD3^+$ T cells or NK cells, from SHIP KO mice induces the development of ileitis in WT mice at 1 month after cell transfer. Therefore, it is possible that granulocyte–monocyte lineage cells may be primarily responsible for the ileitis.

XIV. Gαi2 KO Mice

G-protein is a family of GTP-binding proteins that are involved in a wide variety of transmembrane signaling systems. 100% of Gαi2 KO mice on 129/sv background develop a lethal, diffuse colitis at 16–20 weeks of age, and adenocarcinoma is observed in 30% of them.[90] In contrast, Gαi2 KO mice on C57BL/6 background are relatively resistant to the colitis development. The colitis in 129/sv background is characterized by Th1, but not Th2, response and by reduction of B-cell subsets, including transitional type 2, marginal zone, and B1a B-cell subsets.[91] Reduced production of IL-10 by dendritic cells in response to CpG is observed in 129/sv Gαi2 KO, but not C57BL/6 Gαi2 KO, mice.[92]

XV. TNF(ARE) Mice

A regulatory sequence, which consists of adenosine–uracil multimers (AU-rich elements, ARE), is located in the 3′-untranslated region (3′-UTR) of transcripts encoding cytokines. ARE is responsible for the mRNA destabilization and translational silencing. Deletion of a 69 bp-nucleotide encompassing the TNF-ARE enhances the stabilization of TNF-α expressions and induces the spontaneous development of intestinal inflammation between 2 and 4 weeks of age.[93] The inflammation is localized primarily to the terminal ileum and occasionally to the proximal colon.

A series of double KO mouse systems that cross TNF(ARE) mice with RAG KO, TNFR1 KO, TNFR2 KO, CD4 KO, CD8 KO, TCRδ KO, IFN-γ KO, IL-12p40 KO, TNF-α, or MAPp38 KO mice suggest the pathogenic role of $CD8^+$ T cells, IL-12p40, TNF-α, and IFN-γ in this ileitis.[94] The ileitis of TNF (ARE) mice is associated with early increase in integrin $\alpha E\beta 7^+CD8\alpha\beta^+$ T cells producing IFN-γ and TNF-α in the periphery as well as the intraepithelial compartment.[95] Development of this ileitis is critically dependent on the integrin β7, whereas the function of CCL25/CCR9 is dispensable. In contrast to $CD8^+$ T cells, $CD4^+$ T cells play a protective role in this model. No obvious role of B cells or TCRγδ in this ileitis is recognized. MAPp38 is responsible for the suppression of ileitis, and absence of MAPp38 elicits the development of granulomas. Both, but not either, deficiencies of TNFR1 and TNFR2 inhibit the development of this ileitis.

XVI. LIGHT Tg Mice

A member of TNF super family LIGHT (TNFSF14), which is expressed by activated T cells, serves as a key component of the communication system that controls the response of T cells. LIGHT interacts with two cell-surface receptors, herpes virus entry mediator (HVEM), and lymphotixin (LT) β receptor, and the interaction is inhibited by soluble decoy receptor 3. T-cell-specific overexpressions (under a control of CD2 promoter) of human LIGHT induce the spontaneous development of hepatitis and small intestinal inflammation characterized by villus atrophy.[96] Adoptive transfer of total MLN cells from LIGHT Tg mice induces the development of not only ileitis but also colitis in recipient RAG1 KO mice.[97] In CD45RB model system, HVEM-deficient $CD4^+$ $CD45RB^{high}$ T cells induce milder form of colitis in recipient RAG KO mice than WT-derived $CD4^+$ $CD45RB^{high}$ T cells do, whereas the colitis is exacerbated when HVEM-deficient RAG double KO mice are used as recipients.[98]

XVII. TNFSF15 Tg Mice

TNFSF15 (also known as TL1A) is a member of the TNF super family that binds to death domain receptor 3 (DR3, TNFRSF25). TNFSF15 has been identified as an association gene of both CD and UC.[16–20] TNFSF15, which is produced primarily by lamina propria (LP) macrophages in CD patients, promotes the production of IFN-γ and IL-17 by the LP $CD4^+$ T cells.[99] In addition, the pathogenic role of TNFSF15 in colitis and its ability to enhance both Th1 and Th17 responses have been demonstrated using Gαi2 colitis model.[100]

Ectopic overexpressions of TNFSF15 in T cells (under control of CD2 promoter) causes spontaneous development of transmural inflammation in the small, but not large, intestine at 100% of penetration after 6 weeks of age.[101] This ileitis is associated with enhanced expressions of IL-13, IL-5, and IL-17. Interestingly, IL-13 is primarily responsible for the development of this ileitis. The development of ileitis is abrogated when TNFSF15 Tg mice are crossed with DR3 KO mice. In addition, overexpression of TNFSF15 in $CD11c^+$ cells (under control of CD11c promoter) also induces the spontaneous development of ileitis. However, the pathological change is milder as compared to T-cell-specific TNFSF Tg mice. Adoptive transfer of $CD4^+$ T cells from TNFSF Tg mice can induce both colitis and ileitis that is characterized by IL-13-independent fibrostenotic changes.

XVIII. IL-7 Tg Mice

IL-7 is essentially involved in T-cell homeostasis at various developmental stages. This cytokine is required for the development of mature T cells in the thymus and also prompts the survival of naïve and memory T cells in the periphery. IL-7 has been identified as a candidate gene associated with UC.[16] IL-7 Tg mice spontaneously develop colitis between 4 and 12 weeks of age depending on the copy number of transgenes.[102] Increase in IL-7 receptor expressions on mucosal lymphocytes is associated with this colitis. IL-7 also participates in the development of different types of chronic colitis seen in CD45RB model, TCRα KO mice, and IL-10 KO mice by enhancing the survival of effector memory $CD4^+$ T cells.[103,104] Interestingly, IL-7 contributes to the perpetuation of chronic colitis by generating long-lived colitogenic memory $CD4^+$ T cells that reside in the bone marrow.[105] Systemic, but not intestinal, IL-7 may be required for the persistence of chronic colitis.[106]

XIX. IL-15 Tg Mice

IL-2 and IL-15 bind to their heterotrimeric receptors that have two receptor subunits in common, but these two cytokines have contrasting roles in adaptive immune responses. IL-2 induces AICD to eliminate potentially self-reactive T cells, whereas IL-15 may play a crucial role for the maintenance of memory T-cell responses particularly to invading pathogens. Interestingly, overexpressions of IL-15 in the thymus and intestinal epithelial cells under a control of $T3^b$ promoter induce the spontaneous development of inflammation in proximal small intestine.[107] This inflammation starts from 3 months of age with 100% penetration to 6 months of age. The development of small intestinal inflammation is associated with an increase in $CD8\alpha\beta^+$ NK T cells and an enhanced production of IFN-γ and TNF-α.

XX. CD40L Tg Mice

CD40 ligand (CD40L, also known as gp39 or CD154) on T cells interacts with CD40 expressed on the potentially or professional antigen-presenting cells (APCs) such as B cells, dendritic cells, and macrophages. This interaction provides bidirectional costimulatory signals to both T cells and APCs to activate humoral as well as cell-mediated immune responses. T-cell-specific overexpressions of CD40L (under a control of Lck promoter) induce the development of lethal, transmural, granulomatous colitis between 3 and 6 weeks of age depending on the copy number of transgenes.[108] The T-cell-specific CD40L Tg mice exhibit an aberrant thymic selection and an enhanced Th1 response and also develop inflammation in the stomach, lung, and kidney.

B cells do not express CD40L (except some SLE patients). Interestingly, ectopic overexpression of CD40L on B cells (under control of VH promoter) induces spontaneous development of not only intestinal inflammation (colitis and ileitis) but also glomerulonephritis at 8–15 weeks of age.[109,110] The development of intestinal inflammation depends on CD40 as indicated by the absence of inflammation when B-cell-specific CD40L Tg mice are crossed with CD40 KO mice. Unexpectedly, specific overexpression of CD40L in skin epidermis (under control of keratin 8 promoter) induces the spontaneous development of not only dermatitis but also systemic autoimmune diseases.[111]

XXI. Soluble B7.2 Tg Mice

B7.2 (also known as CD86) is a costimulatory molecule expressed by APCs, which interacts with CD28 for T-cell activation and survival and with CTLA4 for immune regulation. Soluble B7.2 Ig Fc fusion protein, which is artificially generated, interacts with both CD28 and CTLA4. Interestingly, Tg mice, which are genetically engineered to express soluble B7.2 Ig Fc under the control of a liver-specific promoter, develop transmural colitis with extensive inflammatory-cell infiltration at 8–10 weeks of age.[112] This colitis is dependent on IFN-γ-producing T cells activated by CD28 as indicated by the absence of colitis in soluble B7.2 Tg mouse groups crossed with TCRβ KO mice, CD28 KO mice, or IFN-γ KO mice. B cells are not required for this colitis. Interestingly, soluble B7.2 Tg mice develop more exacerbated form of colitis in the absence of both B7.1 and B7.2 (soluble B7.2 Tg mice crossed with B7.1 × B7.2 double KO mice).

XXII. Integrin αV KO Mice

Integrins are dimeric cell-surface receptors composed of α and β subunits. Integrin αV is the most promiscuous α subunit, associated with β1, β3, β5, β6, and β8 and participating in many cellular processes such as cell adhesion and cell survival. Specific deletion of αV in endothelial and hematopoietic cells (under control of tie 2 promoter) elicits multiorgans inflammation from 14 weeks of age.[113] The organs affected include the colon, cecum, peritoneum, liver, nasal cavity, and respiratory tract, but not the small intestine. This inflammation is characterized by the enhancement of both IL-4 and IFN-γ productions and by the inability of gut-associated macrophages and dendritic cells to remove apoptotic cells. Apoptotic cells have previously been proposed to be a major source of self-antigens for the stimulation of colitogenic adaptive immunity.[63] Interestingly, Foxp3$^+$ Treg cells increase in the MLN but decrease in the colon of αV KO mice.

XXIII. Integrin β8 KO Mice

Integrin β8 is associated with integrin αV subunit to form αVβ8 that is capable of activating TGFβ. Mice lacking αV die before or shortly after birth from defects in brain vascular development. In contrast, mice with specific deletion of β8 in both CD4$^+$ T cells and dendritic cells (under control of Vav1 promoter) survive and begin to develop a progressive wasting disorder from 4 to 5 months of age.[114] All mice then develop colitis by 10 months of age. Interestingly, mice with deletion of β8 only in dendritic cells (under control of CD11c promoter) develop colitis at earlier time point (6–7 months of age), whereas mice with deletion of β8 only in CD4$^+$ T cells (under control of CD4 promoter) fail to develop colitis. Therefore, it is possible that integrin β8 plays different roles in colitis depending on the cell type involved, for example, protective role of β8 on dendritic cells versus pathogenic role of β8 on CD4$^+$ T cells. The colitis of β8 KO mice is characterized by significant increase in the production of autoantibodies against double stranded DNA and ribonuclear protein and by enhancement of both Th1 (IFN-γ) and Th2 (IL-4) responses.

XXIV. STAT4 Tg Mice

Signal transducer and activator of transcription (STAT)4 is a transcription factor that promotes Th1 development. Mice, which are genetically engineered to overexpress STAT4 when cytomegaro virus (CMV) promoter is activated, develop transmural colitis within 7–14 days after immunization with DNP-KLH/CFA for activation of CMV promoter.[115] The colonic CD4$^+$ T cells produce TNF-α and IFN-γ but not IL-4. Adoptive transfer of CD4$^+$ T cells from the inflamed colon of STAT4 Tg mice induces colitis in recipient SCID mice.

XXV. STAT3 KO Mice

STAT3 is a master transcriptional factor involved in a broad spectrum of adaptive and innate immune functions such as Th17 differentiation and epithelial regeneration. STAT3 is highlighted by recent human genome-wide association studies that identify STAT3 as a susceptibility gene of both CD and UC.[16–20] Alternatively, it is becoming increasingly apparent from IBD models that function of STAT3 in IBD is much more complicated than previously predicted. For example, absence of STAT3 in $CD4^+$ T cells contributes to the improvement of colitis observed in CD45RB model and IL-2 KO mice.[42,116] In contrast, conditional KO mice with myeloid cell-specific deletion (e.g., macrophages and neutrophils) of STAT3 spontaneously develop colitis at 20 weeks of age.[117] This development of colitis is associated with a decreased production of IL-10 by macrophages and an enhanced Th1 response. Interestingly, the conditional STAT3 KO mice develop no colitis when they are crossed with RAG1 KO mice lacking T and B cells,[118] suggesting the requirement of both intact STAT3 signaling in adaptive immunity and STAT3 deficiency in innate immunity for development of this colitis. In addition, specific deletion of STAT3 in epithelial cells increases the susceptibility of mice to DSS-induced colitis.[119] These findings suggest that STAT3 activation in innate versus adaptive immune responses plays distinctly different roles in the pathogenesis of colitis.

Interestingly, inducible deletion of STAT3 during adult life leads to rapid development of colitis.[120] Since the deletion is predicted to occur in all cell types, it is possible that the overall function of STAT3 in colitis may rely more on innate, rather than adaptive, immune response. Several cytokines and growth factors that are capable of activating STAT3 have been identified, including IL-6, IL-7, IL-9, IL-10, IL-11, IL-15, IL-22, IFN-α/β, leukemia inhibitory factor (LIF), hepatocyte growth factor (HGF), oncostatin M (OSM), leptin, and growth hormone (GH). Among them, IL-6 seems to serve as the strongest activator of STAT3 particularly in the $CD4^+$ T cells of IBD patients. Indeed, a pilot randomized trial shows the beneficial effect of humanized mAb to IL-6 receptor on active CD patients.[121] In contrast, activation of STAT3 in epithelial cells under intestinal inflammatory conditions is more effectively induced by IL-22 rather than IL-6.[71,119] Indeed, IL-22 plays a protective role of several different types of experimental colitis by enhancing STAT3-dependent epithelial barrier functions.[71,119,122]

XXVI. SOCS1 Tg Mice

Several cytokines activate JAK and STATs signaling pathway, which is in turn suppressed by suppressor of cytokine signaling (SOCS) protein that is characterized by the presence of a SH2 domain and a SOCS box. Approximately 40% of T-cell-specific SOCS1 Tg mice (under control of Lck promoter) develop colitis after 15 weeks of age when they are kept under conventional conditions.[123] Interestingly, the SOCS1 Tg mice do not develop colitis under specific pathogen-free condition. This colitis is characterized by enhanced Th1, but not Th2, response, and the colonic $CD4^+$ T cells exhibit high expression levels of CD28 and low expression levels of CTLA4.

XXVII. Gp130 KI Mice

IL-6 activates various cell types carrying the membrane-bound IL-6R/gp130 heterodimer receptors (classical IL-6 signaling) as well as $IL\text{-}6R^-$ $gp130^+$ cells via the soluble IL-6R (IL-6 *trans*-signaling). The gp130 KI mice, which carry a truncation mutation within *gp130* gene that deletes all STAT3-binding sites, spontaneously develop ulceration in the gastric pylorus and the anorectal region at 4 months of age when maintained under specific pathogen-free condition.[124] Approximately 40% of gp130 KI mice develops ulcerations in the cecum when they are kept under conventional conditions. Severe joint disease is also detectable in these mice.

XXVIII. NFκB1 KO Mice

Canonical NFκB (NFκB1) pathway is activated by tri-complex of IκB kinase (IKK)1, IKK2, and NEMO (also called IKKγ). Inhibition of NFκB1 has been shown to suppress T-cell-mediated colitis such as TNBS colitis.[125] In contrast, mice with specific deletion of NEMO in epithelial cells spontaneously develop colitis at 6 weeks of age.[126] This colitis, which is associated with extensive epithelial cell apoptosis and downregulation of β-defensin 3 expressions, fully depends on MyD88 and TNFR1 as indicated by the absence of colitis in MyD88 or TNFR1-deficient epithelial cell-specific NEMO double KO mice. In addition, absence of both, but not either, of IKK1 and IKK2 in epithelial cells also leads to spontaneous development of colitis.[126] These findings highlight the protective role of epithelial cell-specific canonical NFκB pathway in colitis.

In contrast to deficiency of IKK subunits, deficiency of an NFκB subunit p50 with heterozygous allele (+/−) of RelA does not result in the spontaneous development of colitis but makes mice more susceptible to *Helicobacter hepaticus*-induced colitis.[127] Alternatively, approximately 10–15% of mice with epithelial cell-specific deletion of RelA (under control of villin promoter) exhibit abdominal discoloration and diarrhea within 3 days after birth and die within 25 days.[128] Crypt-epithelial structure in the small intestine is completely lost in these mice. Interestingly, some epithelial cell-specific RelA KO mice somehow survive but they do not develop intestinal inflammation.

XXIX. Runx3 KO Mice

Runx3, which is expressed by CD4/CD8 double negative and single positive thymocytes and peripheral DCs, functions as context-dependent transcription factor involved in neurogenesis, thymopoiesis, and dendritic cell maturation. 100% of Runx3 KO mice develop colitis spontaneously by 4 weeks of age and the inflammation is expanded to the small intestine in 20% of them.[129] The colitis is characterized by the enhancement of both Th1 and Th2 (IL-4 not IL-5) responses and by decrease in IL-18 expression. Gastric mucosal hyperplasia is also observed in Runx3 KO mice at 8 months of age. Absence of Runx3 impairs full activation of Treg and $CD8^+$ T cells with immune regulatory ability.[130]

XXX. TLR5 KO Mice

Toll-like receptors (TLRs) are pattern recognition receptors that selectively recognize different microbial products. KO mouse strains, which are engineered to lack individual TLRs (TLR2, TLR4, TLR5, TLR7, and TLR9), have already been generated. Among them, only TLR5 KO mice have been shown to develop colitis spontaneously.[131] Approximately 30% of TLR5 KO mice may develop colitis characterized by rectal prolapse, bleeding, and body weight loss at 8–12 weeks of age. The inflammation is primarily observed in the cecum and proximal colon. Enhanced IL-1β activity due to lack of epithelial cell-derived soluble IL-1 receptor antagonists may be responsible to promote the colitis in TLR5 KO mice. Deficiency of TLR4 improves the colitis of TLR5 KO mice, whereas deficiency of IL-10 exacerbates it. However, the idea of spontaneous development of colitis in TLR5 KO mice has recently been challenged by another group showing no evidence of spontaneous inflammation or metabolic abnormalities in TLR5 KO mice.[132]

The function of TLR5 in IBD seems more complicated than previously predicted because bacterial flagellin (the ligand of TLR5) represents a dominant pathogenic antigen in human CD.[133] In addition, it has been well documented in flagellin-specific TCR transgenic mice (termed CBir1 Tg) that flagellin-specific $CD4^+$ T cells are able to induce colitis through two expansion steps: spontaneous T-cell proliferation driven by MyD88-dependent IL-6 (innate immune-mediated proliferation) and flagellin-specific T-cell proliferation (antigen-specific adaptive immune-mediated proliferation).[134]

XXXI. Enteric Glia KO

Tg mice, which are genetically engineered to express herpes simplex virus thymidine kinase under control of glial fibrillary acidic protein (GFAP), ablate GFAP-positive glial cells in the small intestine after administration of an antiviral agent ganciclovir for 14–17 days. Interestingly, the ablation of enteric glial cells induces fulminating and fatal ileitis characterized by necrosis, severe inflammation, and hemorrhage.[135] The enteric glia KO mice exhibit intestinal epithelial barrier dysfunction as indicated by an increase in epithelial permeability.[136] However, the necrotic ileitis is not improved by depletion of enteric bacteria through administration of broad spectrum of antibiotics.[135]

XXXII. XBP1 KO Mice

X-box-binding protein (XBP1) plays a critical role in unfolded protein response that is a primitive cellular pathway that is engaged in response to endoplasmic reticulum stress. Although the function of XBP1 to induce the terminal differentiation of B cells into plasma cells has been well known, the critical role of XBP1 in epithelial cell function is clearly highlighted by the fact that specific deletion of XBP1 in intestinal epithelial cells (under control of villin promoter) leads to the spontaneous development of small intestinal inflammation.[137] This ileitis is characterized by the absence of Paneth cells-secreting defensin. Interestingly, ileitis is observed in not only 60% of homozygous but also 30% of heterozygous mice, suggesting the importance of monoallelic expression of XBP1 in the pathogenesis of ileitis. Most notably, an association of XBP1 variants in both human CD and UC has been identified.[137]

XXXIII. Atg5 KO Mice

Autophagy is a cellular pathway involved in protein and organelle degradation. Although autophagy was thought to be a primary response to starvation, recently accumulating data indicate that it exerts diverse physiological functions ranging from microbial infections to MHC class II-dependent antigen presentation.[138] Importance of autophagy process in IBD is highlighted by the identification of autophagy-related gene (Atg) *16L1* and *IRGM*, a gene essential for autophagy, as major susceptibility genes of CD.[16–20] In order to identify the functional role of autophagy in IBD, two different mouse strains have been generated to lack functional Atg16L1.[139,140] One of the mouse strains is generated through gene trap-mediated disruption of Atg16L1. This mouse strain exhibits abnormality of Paneth cell function, particularly in the granule exocytosis process.[139] Another Atg16L1-deficient mouse strain is highly susceptible to DSS-induced colitis.[140] However, both strains fail to develop colitis or ileitis spontaneously, suggesting other factors in addition to Atg16L1 polymorphisms may be required for the induction of CD.

Neonatal lethality is seen in mice lacking another autophagy gene—*Atg5*. Therefore, Nedjic *et al.* engraft the thymi from neonatal Atg5 KO mice under the kidney capsule of athymic (nu/nu) mice to see the role of Atg5 in thymic selection. Unexpectedly, the nu/nu mice with implantation of Atg5 KO thymi develop multi-organ inflammation in the colon, liver, lung, and uterus.[141] Adoptive transfer of $CD4^+$ T cells from the Atg5 KO thymic-engrafted nu/nu mice into nu/nu recipients recapitulate essentially all aspects of inflammation as observed in the donor mice. Similarly, the donor $CD8^+$ T cells induce inflammation in the recipients.

XXXIV. mK8 KO Mice

Majority of single-layered epithelial cells express keratin 8 (mK8). K8-targeted mutation in mice on C57BL/129 background causes embryonic lethality (94% penetration). In contrast, K8-deficient mice on FVB/N background survive and develop colorectal, but not small intestinal, hyperplasia with rectal prolapse at 9 weeks of age.[142] The elongation of epithelial crypt is accompanied with inflammatory-cell infiltration extended to the submucosa. Of note, K8 deficiency confers resistance to colonocyte apoptosis in a microflora-dependent manner.[143]

XXXV. N-Cadherin Mutant Mice

N-cadherin is a transmembrane glycoprotein that mediates homophilic adhesive interactions between cells. Chimeric mice, which are generated from ES cells to carry dominant negative mutation of N-cadherin and express it only in small intestinal epithelial cells (under the control of a Fabp promoter), exhibit loss of endogenous E-cadherin that is localized to basolateral surfaces and apical junction complexes of epithelial cells. The N-cadherin mutant mice develop transmural inflammation in the jejunum by 3 months of age.[144]

XXXVI. Mdr1a KO Mice

Multiple drug resistance (Mdr) 1 belongs to a family of transporters known as ATP-binding cassette transporters, which are characterized by the ability to pump small amphiphilic and hydrophobic molecules across membranes in an ATP-dependent manner. Mdr1a is expressed by epithelial cells, some T cell subsets, hematopoietic cells, and blood–brain barrier cells. Interestingly, 20–25% of Mdr1a KO mice on FVB background develop a continuous colitis with significant increase in serum Ig levels by 1 year of age.[145] Depletion of enteric bacteria by administration of antibiotics improves this colitis. Treatment with CRX-526, an antagonist of TLR4 activity, also suppresses the colitis.[146] In addition, butyrophilin-like 2 (BTNL2), a B7-like molecule within butyroohilin gene family, may serve as a negative costimulatory molecule to suppress this colitis.[147]

XXXVII. GPX KO Mice

Glutathione peroxidases (GPX) are selenium-dependent hydroperoxidase-reducing enzymes that contribute in reducing H_2O_2 and fatty acid hydroperoxides. Mice deficient for both, but not either, GPX1 and GPX2 spontaneously develop perianal ulceration, diarrhea, and hypothermia at 14 days of age.[148] Histologically recognized inflammation in the colon and ileum can be detected at 24 days of age. Approximately 40% of GPX1 × GPX2 double KO mice die between 20 and 36 days after birth. One hundred and twenty-nine background mice are more susceptible to this colitis as compared to C57BL/6 background.[149] Cholesterol diet exacerbates this colitis presumably through reduction of unfolded protein (ER stress) responses.

XXXVIII. Muc2 KO Mice

There is a layer of mucus covering the surface of intestinal tract, which serves as a lubricant and a physiological barrier between luminal contents and mucosal surface. Mucins, which are a major component of intestinal mucus, are heavily *O*-glycosylated protein that are classified into two groups—secreted (e.g., Muc2, 5, and 6) and membrane-bound (e.g., Muc1, 3, 4, 12, 13, and 17) mucins. UC, which is characterized by a thin mucus layer in association with goblet cell depletion, exhibits strong expressions of Muc2 and Muc4 and low expressions of Muc1 and Muc3. Muc2 KO mice were initially established as a cancer model that spontaneously develops adenoma and adenocarcinoma in the colon at 6 months of age.[150] Interestingly, colitis, which is characterized by crypt hyperplasia, lack of goblet cells, mild inflammatory-cell infiltration, and increased TNF-α expression, is recognizable primarily in the distal colon of Muc2 KO mice from 5 weeks of age.[151] The colitis is further exacerbated (as indicated by 50% mortality) when Muc2 KO mice are crossed with IL-10 KO mice.[152]

XXXIX. C1galt1 KO Mice

Colonic mucus carries large numbers of *O*-glycan, which account for approximately 80% of the mass of mucus molecules. *O*-glycan synthesis is initiated from a primary core structure termed Tn antigen, which is formed by a glycosylation enzyme termed core 1β1,3-galactosyltransferase (C1galt1). After establishment of the core formation, sequential addition of saccharides is induced by other enzymes to elongate the *O*-glycan. Interestingly, epithelial cell-specific deletion of C1galt1 induces spontaneous development of colitis primarily in the distal part of colon with 100% penetration at 12 weeks of age.[153]

The epithelial cell-specific C1galt1 KO mice still develop colitis when crossed with RAG1 KO mice, suggesting no requirement of T or B cells in the induction of this colitis. Interestingly, neither Myd88 nor TLR4 deficiency protects the conditional C1galt1 KO mice from developing colitis, whereas administration of broad spectrum of antibiotics improves the colitis. Most notably, epithelial cell-specific and inducible C1galt1 KO mice, which lack C1galt1 expressions in epithelial cells during the adult life by administration of tamoxifen, rapidly develop colitis within 10 days after tamoxifen induction.[153] The importance of these findings is highlighted by that approximately 30% of UC patients lack *O*-glycan synthesis in epithelial cells due to somatic mutations within Cosmc; it serves as a chaperon for C1galt1-mediated *O*-glycan synthesis.[153]

Core 3 β1,3-*N*-acetylglucosaminyl transferase and core 2 β1,6-*N*-acetylglucosaminyltransferases are both involved in the elongation (maturation) of *O*-glycan after the primary core structure is formed. Absence of these enzymes increases the susceptibility of mice to DSS-induced colitis but fails to induce spontaneous development of colitis.[154,155]

XL. NFATc2/RAG DKO Mice

The nuclear factor of activated T cells (NFAT)c 2 is an immune-specific calcium-regulated NFTA family member that plays redundant function in gene regulation. NFATc1 and NFATc2 together are the most important NFAT isoforms in T-cell differentiation. NFATc2 KO mice do not develop colitis, but double KO mice deficient in both NFATc2 and RAG2 (lacking T and B cells) spontaneously develop colitis with rectal prolapse at 15 weeks of age.[156] Adoptive transfer of B cells, but not T cells, from WT mice can improve this colitis in an IL-10-independent manner. In addition, restoration of a monoassociated B cell population specific for hen egg lysosome (HEL) does not improve the colitis in NFATc2 × RAG2 double KO mice, suggesting that polyclonal activation of B cells is required for the suppression of this colitis.

XLI. T-bet/RAG DKO Mice

T-bet is a T-box transcriptional factor involved in the differentiation of Th1 T cells. Inhibition of T-bet in $CD4^+$ T cells contributes to the improvement of T-cell-mediated colitis in CD45RB model.[157] In contrast, although T-bet KO mice do not develop colitis, T-bet-deficient RAG2 double KO mice, which lack both T and B cells, spontaneously develop colitis from 4 weeks of age.[21] This colitis is characterized by continuous area of inflammation expanded from distal part of colon (homologue of human rectum) and by increased production of TNF-α but not any other cytokines, including IFN-γ, IL-1, IL-6, IL-10, IL-12, IL-13, or IL-23. Interestingly, the colitis in T-bet-deficient RAG2 double KO mice is both vertically (maternally from mother) and horizontally (from neighbors) transmissible to T-bet-sufficient mice such as WT mice. Two bacterial strains, *Klebsiella pneumonia* and *Proteus mirabilis*, are responsible for eliciting this colitis,[158] whereas *Bifidobacterium animalis* contributes to the suppression by increasing the abundance of select lactate-consuming and butyrate-producing bacteria.[159]

XLII. Anti-CD40mAb Model

Cognate interaction of T and B cells through CD154 (T cells) and CD40 (B cells) is implicated in the activation of adaptive immune responses. Interestingly, administration of agonistic CD40 mAbs induces acute colitis in RAG1 KO mice lacking T and B cells but not in WT mice.[160] The CD40mAb-treated RAG1 KO mice develop colitis within 4 days, and the colitis is observed until 3 weeks after injection. Unlike chronic colitis models, the colitis can be induced in germ-free environment and it is mediated by IL-23 and partially by TNF-α and IFN-γ.[160]

Agonistic CD40mAb is unable to induce colitis in common γ chain-deficient RAG1 double KO mice that lack all lymphoid lineage cells.[161] Recent studies identify two cell subsets that express NKp46 in the intestine—RORγT-independent NKp46$^+$ NK cells and RORγT-dependent NKp46$^+$ innate lymphoid cells. RORγT-dependent NKp46$^+$ innate lymphoid cells maintain epithelial homeostasis by producing IL-22, whereas RORγT-independent NKp46$^+$ NK cells contribute to the development of colitis in CD40mAb-treated RAG1 KO mice through IL-23-dependent production of IFN-γ.[161]

XLIII. C3H/HeJBir Mice

C3H/HeJBir mice are a substrain of mice, which was generated by a program of selected breeding of C3H/HeJ mice. C3H/HeJBir mice develop inflammation mainly in the cecum and right colon at 3–6 weeks of age, and the inflammation resolves spontaneously by 12 weeks of age.[162] A gene locus on chromosome 3, termed CdCs1, seems to make this mouse substrain susceptible to colitis.[163] Bacterial antigen-stimulated CD4$^+$ T cells from C3H/HeJBir mice have the ability to elicit the development of colitis in immune-deficient hosts.[164] Through a humoral screening system of cecal bacterial cDNA library, bacterial flagellins have been identified as the dominant bacterial antigen for stimulating the colitogenic $CD4^+$ T cells.[133] Importance of this finding is highlighted by the responsiveness of $CD4^+$ T cells from CD patients to bacterial flagellin antigen.[133]

XLIV. SAMP1/Yit Model

SAMP1 mouse strain is one of senescence-prone strains derived from 24 generations of siblings from a litter of AKR/J mice. SAMP1 mice show early signs of senescence such as amyloidosis, alopecia, and osteoporosis. There are two substrains of SAMP1 mice, termed SAMP1/Yit and SAMP1/YitFc. These substrains develop discontinuous, transmural inflammatory lesions in the terminal ileum with 100% penetrance by 30 weeks of age.[165,166] Unlike colitis models, enteric bacteria are required for the exacerbation, but not the induction, of this ileitis.[167] A series of bone marrow chimera mouse experiments indicates that the primary defect conferring ileitis in SAMP1/Yit mice originates from a nonhematopoietic source, presumably epithelial cells.[168] $CD4^+$ T cells producing IFN-γ and TNF-α are also responsible for the development of this ileitis.[166] Interestingly, the established ileitis is characterized by not only Th1 but also Th2 responses as indicated by significant increase in IL-5 and IL-13 expressions in the inflamed ileum. Indeed, the ileitis is treatable by either anti-IL-4 or anti-IL-5 neutralization.[169,170] The severity of ileitis is associated with an increase in the number of activated B cells (as indicated by IgA production) in the mesenteric lymph nodes, and they contribute to the development of ileitis by blocking $CD4^+CD25^+$ Treg.[171]

XLV. CD45RB Model

Adoptive transfer of naïve $CD4^+$ T cells ($CD4^+CD45RB^{high}$ T cells) from the spleen of WT mice induces transmural colitis in immune-deficient recipients at 6–8 weeks after cell transfer. The colitis cannot be induced when mixture of naïve and memory $CD4^+$ T cells (Treg) are used as donor cells. This model system, which was initially developed by Dr. Powrie,[7] has not only provided significant contributions for dissecting the fundamental mechanism of CD but also made a major breakthrough by bringing the concept of Treg to the fore.

The colitis of CD45RB model is mediated by Th1 responses associated with IFN-γ and TNF-α productions.[8] TNF-α is derived primarily by nonlymphoid cells in the recipient.[172] IL-6 *trans*-signaling is required for this colitis,[173] and

IL-4 may indirectly promote Th1-type immune responses to sustain the colitis.[174] IL-4 blocks the TGFβ-dependent development of $Foxp3^+$ Treg cells and induces a unique $CD4^+$ T cell population that plays a pathogenic role in the colitis through productions of both IL-9 and IL-10.[175] In order to be colitogenic, differentiation/expansion of donor-derived naïve T cells is induced even in the absence of gut-associated lymphoid tissues such as Peyer's patches and mesenteric lymph nodes.[176]

Several regulatory T cell subsets have been identified using CD45RB model. Colitis in this model is suppressed by TGFβ1-producing memory $CD4^+$ T cells (Th3 T cells)[177] and by an antigen OVA-specific IL-10-producing memory T cell subset termed Tr1.[178] Most importantly, numerous studies regarding $CD4^+CD25^+Foxp3^+$ Treg have been done using this model and, as shown in Table IV, this model has provided significant contributions to the identification of critical molecules involved in the inhibition versus promotion of Treg development/function. One of the important questions in IBD is: although Treg is accumulated in the inflamed colon of IBD patients, why is it not able to effectively suppress the inflammation presented. The clue to answer this question may be derived from the CD45RB model. One possibility is that Treg function may be impaired in IBD patients. This possibility is supported by that Treg cells are unable to maintain their Foxp3 expression and regulatory activity in the absence of paracrine IL-10.[179] Indeed, absence of IL-10-inducing STAT3 signaling in Treg cells abolishes their ability to inhibit Th17 and IFN-γ^+ Th17 cell responses and colitis development in CD45RB model.[180,181] Another possibility is that effector T cells cannot respond to Treg cells in IBD patients. This possibility is supported by that cotransfer of naïve $CD4^+$ T cells from Smad7 Tg mice with WT Treg cells induces colitis in recipient RAG1 KO mice; on the contrary, colitis is not observed when naïve $CD4^+$ T cells from WT mice were cotransferred with WT Treg cells.[182]

Interestingly, a unique $CD4^+$ T cell population, which expresses both Foxp3 and IL-17, is specifically found in the inflamed colon of CD, but not UC, patients.[183] Similarly, $CD4^+$ $Foxp3^+$ IL-17^+ T cells develop in the colon of CD45RB model. Interestingly, the $CD4^+$ $Foxp3^+$ IL-17^+ T cells are converted from $Foxp3^+$ Treg in an IL-12p70-dependent manner. Although they express IL-17, they still maintain the regulatory ability to suppress this colitis.[184] In addition to $CD4^+$ T cells, other cell populations with potential ability to suppress colitis has been found in CD45RB model, including $CD8^+CD28^-$ T cells, $CD8^+CD122^+$ T cells, $CD8^+CD11c^+$ T cells, $CD8^+$ intraepithelial lymphocytes (IELs), B cells, NK cells, NKT cells, and $CD103^+$ DCs (Table IV).

IL-23 receptor is a major IBD susceptibility gene—the polymorphisms are negatively associated with the development of IBD.[16–20] The pathogenic role of IL-23 and its receptor has been well documented in CD45RB model.[185,186]

TABLE IV
PATHOGENIC VERSUS PROTECTIVE FACTORS IN THE COLITIS OF CD45RB MODEL

Pathogenic	Protective
Cell types	
$CD4^+CD45RB^{high}$ naïve T cells, $CD4^+$ $CD62L^+$ naïve T cells, IL-9^+ IL-10^+ $CD4^+$ T cells, integrin $\alpha4/\beta7^+$ T cells, flagellin-specific memory $CD4^+$ T cells, E-cadherin$^+$ DCs, $CD11b^+$ cells	$CD4^+$ $Foxp3^+$ $CD25^+$ T cells, $CD103^+$ $CD25^-$ $PD1^+$ T cells, dual TCR-expressing $CD4^+$ T cells, IL-17-producing $Foxp3^+$ $CD4^+$ T cells, IL-10-producing Tr1, $CD4^+$ $CD8\alpha\alpha^+$ IELs, $CD8^+$ $CD28^-$ T cells, $CD8^+$ $CD122^+$ T cells, NK T cells, small intestinal $CD11c^+$ $CD8^+$ T cells, $CD103^+$ DCs, $CD11c^+$ $CD8^+$ DCs, B cells, NK cells, neutrophils
Endogenous factors	
Cell-specific:	*Cell-specific:*
CD27 on CD4, CD48 in recipient, memory CD4-derived IL-22, cMaf in memory T cells, cRel in CD4, CX3CR1 on DCs, HVEM (herpes virus entry mediator) on CD4, IRF1 (IFN regulatory factor) in CD4, IRF4 in CD4, MyD88 in CD4, PDK1 in CD4, PKCθ (protein kinase C) in CD4, STAT4 in CD4, T-bet in CD4, T cell-derived IL-2	A(2A) adenosine receptor on Treg, CCR4 on Treg, naïve CD4-derived IL-22, cMaf in naïve T cells, CD80/CD86 in recipient, GITR on Treg, latency-associated TGFβ1 on Treg, NK cell-derived IL-22, PDK1 in Treg, PI3Kp110δ in Treg, PTEN in Treg, Smad7 in T cells, TNFR2 on CD4, WASP in Treg, HVEM in recipient
Others:	*Others:*
Carbohydrate modification of *N*-glycan, B7-H1, CD70, CD80, CD154, CD161, IFN-γ, IL-4, IL-6, IL-12p40, IL-18, IL-21, IL-23, IL-27, leptin, LFA-1, neurokinin 1, NKG2D, systemic but not local IL-7, TNF-α, triggering receptor expressed on myeloid cell-1 (TREM-1), VEGF	4-1BB, CCR4, CTLA-4, IL-10, IL-18, IL-35, OX40, PPARγ, RANKL, TCRα chain, TGFβ1, TLR9
Exogenous factors	
Segmented flagmentous bacteria (SFB) Bacterial-derived ATP	Anti-Fas L antibody Dietary *n*-3 polysaturated fatty acids Filamentous hemagglutinin, FTY720 Germinated barley food

The pathogenic mechanism of IL-23 is mediated by the ability to not only promote T cell proliferation but also inhibit the differentiation of Treg.[186] The IL-23 in CD45RB model is produced primarily by E-cadherin$^+$ $CD103^-$ $CD11b^+$ $CD8\alpha^-$ $Gr1^-$ DC subset, which also produces IL-6.[187] Interestingly, expansion of this DC subset is specifically observed in adaptive immune-mediated, but not innate-mediated, colitis.

IL-23 serves as a key determinant of maintaining Th17 cells ($CD4^+$ T cells producing IL-17). Many novel insights into the Th17 biology have been drawn from CD45RB model. Enteric bacteria-derived adenosine triphosphate (ATP) promotes the colitis of CD45RB model by inducing the development of intestinal Th17 cells through stimulation of $CD70^+CD11c^{low}$ unique APCs.[188] *Enterococcus gallinarum*, which represents vancomycin-resistant gram (+), is recently identified as a potential ATP-secreting bacterium.[189] Segmented filamentous bacterium (SFB), which is susceptible to vancomycin, also induces the development of Th17 cells.[190] In addition, IL-17-producing TCRγδ T cells contribute to the pathogenesis of colitis in CD45RB model by promoting the development of Th17 cells.[191] Interestingly, a previously unappreciated "innate lymphoid cell" population, which is characterized by Lin^- $s\text{-}kit^-$ Sca^+ $Thy1^+$ phenotype, exists in the colon.[192] The innate lymphoid cells produce IL-17 in response to IL-23 and also promote the development of Th17 response.

Although the findings described above tend to suggest the pathogenic ability of Th17 cells, the IL-23-dependent colitis of CD45RB model does not require IL-17 secretion by $CD4^+$ T cells.[193] Indeed, adoptive transfer of $CD4^+$ T cells from IL-17F KO mice or from IL-17A × IL-17F double KO mice still induces severe colitis in recipient RAG KO mice.[194,195] In addition, protective role of $CD4^+$ T-cell-derived IL-17A in this colitis has been demonstrated.[196] Alternatively, adoptive transfer of RORγt-deficient $CD4^+$ T cells that are unable to differentiate into Th17 cells fails to induce colitis in the recipient RAG KO mice.[195] Therefore, it is possible that other cytokines such as IL-22 and IL-26 rather than IL-17A and IL-17F, which are produced by Th17 cells, cause the colitis in CD45RB model. However, IL-22-deficient $CD4^+CD45RB^{high}$ T cells can still induce colitis in CD45RB model.[195] In addition, the colitis is rather exacerbated when $CD4^+CD45RB^{high}$ T cells from IL-22 KO mice are transferred into IL-22-deficient RAG double KO mice,[122] suggesting the protective role of both naïve $CD4^+$ T cell- and non-T cell-derived IL-22 in this colitis. In contrast, a recent study suggests that adoptive transfer of $CD25^+$ (Treg)-depleted $CD4^+CD45RB^{low}$ memory T cells induces colitis in RAG1 KO mice in an IL-22-dependent manner.[197] Therefore, it is possible that IL-22 somehow plays dual roles in the colitis of CD45RB model depending on the producers (naïve $CD4^+$ T cells and non-T cells versus memory $CD4^+$ T cells).

Unique $CD4^+$ T-cell subset expressing both IL-17 and IFN-γ can be specifically detected in the inflamed colon of CD, but not UC, patients.[198] Studies using a reporter mouse system have recently identified the $IL\text{-}17^+$ $IFN\text{-}\gamma^+$ $CD4^+$ T cells and $IL\text{-}17^+$ $IFN\text{-}\gamma^-$ $CD4^+$ T cells as an immediate precursor of IFN-γ-producing Th1 T cells.[134,199] Therefore, it is possible that the consequent mediator of colitis in CD45RB model is Th1, rather than Th17, T cells. However, given the fact that anti-IFN-γ therapy has failed in human CD,[200] there is a call for extensive research in the hopes to learn more about this highly complicated mechanism that underlies the CD45RB model and CD. In this regard, a recent study identifies that deficiencies of all, but not either, of three transcription factors (T-bet, STAT6, and STAT4) in donor $CD4^+$ T cells are required to fully abolish the development of colitis in CD45RB model.[201]

XLVI. Human CD3ε Model

Overexpression of human CD3ε26 disrupts the normal thymic selection in mice. Transplantation of T cell-depleted bone marrow cells from WT mice induces the development of colitis in the CD3ε Tg mice at 5–8 weeks after transplantation, and this colitis can be inhibited by preimplantation of WT thymi under the recipient kidney capsule.[202] Both TCRαβ T cells and TCRγδ T cells mediate this colitis,[203] and enteric bacteria are required for eliciting the pathogenic ability of these T cells, particularly $CD4^+$ T cells.[204] Interestingly, absence of IFN-γ still induces the development of severe colitis, whereas absence of STAT4, IL-12p40, or TNF-α improves (but not abolishes) the colitis in this model.[205] Therefore, it is possible that pathogenic effects of IL-12/STAT4 or other Th1-promoting pathways are not limited to the induction of IFN-γ-producing T cells in this model. This colitis is ameliorated when CD48, receptor of CD2 and 2B4, is absent from both, but not either, T cells and APCs.[206] In addition, continuous production of macrophage-migration inhibitory factor (MIF) by innate immune system is required for the development of this colitis.[207]

XLVII. CD8-Transfer Models

Adoptive transfer of heat shock protein (hsp) 60-reactive $CD8^+$ T cell clones, which coexpress Vβ8.1 and Vα8 and possess Fas-dependent cytolytic activity, induces inflammation in the small intestine of recipient TCRβ KO mice.[208] This ileitis occurs in the absence of enteric bacteria.

Adoptive transfer of ovalbumin (OVA)-specific $CD8^+$ T cells cannot induce inflammation in the recipient mice engineered to express the exogenous antigen OVA in the intestinal epithelial cells. However, infection with a virus (VSV) to induce temporary damage on the epithelial cells leads to development of chronic inflammation in the small intestine.[209] This model provided the first evidence to link innate and adaptive immune responses together in the pathogenesis of IBD.

Like CD45RB model, a single transfer of $CD44^{low}$ $CD62L^+$ naïve $CD8^+$ T cells from WT mice can induce colitis in recipient RAG2 KO mice. This colitis is IL-6 dependent and mediated by IL-17-producing $CD8^+$ T cells.[210]

XLVIII. ECOVA Model

SCID mice reconstituted with $CD4^+$ T cells from Tg mice bearing chicken OVA-specific TCR do not develop inflammation. Alternatively, preinoculation of *Escherichia coli* engineered to express OVA allows OVA-specific $CD4^+$ T cells to induce transmural colitis in the recipient SCID mice within 12–20 days after $CD4^+$ T-cell transfer.[211,212] A technical impediment confronted at the outset of this system is the difficulty in the development of laboratory *E. coli* that could stably express OVA in the presence of host enteric flora *in vivo*. To overcome the problem, one system uses the treatment of hosts with antibiotics to delete the enteric bacteria[212] and another system elegantly employs adoption of the prir15 promoter in the laboratory *E. coli* strain.[211] The colitis of this model can be induced not only by OVA-specific Th1 but also by OVA-specific Th2 T-cell subsets.[211] In addition, this colitis is mediated by TLR2 activation, and absence of NOD2 makes mice highly susceptible to this colitis by enhancing the ability of APCs to stimulate Th1 T cells.[213]

XLIX. TNBS Model

Acute colitis in rats, mice, rabbits, and pig can be induced by an enema containing a contact-sensitizing allergen, trinitrobenzene sulfonic acid (TNBS), in 50% ethanol solution.[214] The TNBS initially causes acute discrete foci of necrosis and inflammation, and this change is believed to be followed by immune responses to "hapten-modified self-antigen."[4] In mice, TNBS colitis is highly dependent on the strains. SJL/J, C3HeJ, and Balb/c are susceptible strains, whereas C57BL/6 and DBA/2 are highly resistant strains.[215]

The relevance of this model to CD is suggested by the involvement of NOD2 (a key CD susceptibility gene) in the pathogenesis of TNBS colitis. Administration of plasmid carrying intact NOD2, but not plasmid carrying CD-associated frame-shift-mutated NOD2, makes mice more resistant to this colitis.[216] In addition, administration of muramyl dipeptide (MDP), a ligand of NOD2, protects mice from TNBS colitis by reducing cytokine responses against multiple TLRs ligands in an IFN regulatory factor (IRF) 4-dependent manner.[217] In addition to the relevance to CD, TNBS colitis is an economical and time saving model. Therefore, many studies have been conducted in this model to dissect the mechanism of CD and to evaluate the therapeutic potential of exogenous compounds on CD (Table V). Interestingly, many studies regarding complementary and alternative medicine in IBD have been done using this model. In some cases, dinitrosulfonic acid (DNS) is used as a mimic of TNBS, but the DNS-induced inflammation may more closely reflect irritable bowel syndrome or food allergy rather than CD.[218]

TABLE V
PATHOGENIC VERSUS PROTECTIVE FACTORS IN MURINE TNBS COLITIS

Pathogenic endogenous factors

Soluble molecules:

Catecholamines, LIGHT, melanin-concentrating hormone (MCH), complement C5a, IL-6, IL-12p70, IL-16, IL-18, macrophage inhibitory factor (MIF), macrophage inflammatory protein 1α (MIP1α), MIP3α, neurotensin, Smad7, TNF-α, TNF-like weak inducer of apoptosis (TWEAK)

Enzymes:

Angiotensinogen, glycolipid biosynthesis, inducible nitric oxide synthase (iNOS), leptin, mast cell protease 6, matrix metalloproteinase 2 (MMP2), MMP9, poly(AID-ribose) polymerase-1 (PARP1), poly(AID-ribose) synthetase (PARS), vanin1 (epithelial ectoenzyme)

Receptors:

CD11b/CD18, CD30/CD30L, CD39, CD44v7, CD80, CTLA4, complement receptor 3 (CR3), integrin α1, IL-17R, meurokinin-1 receptor, protease-activated receptor 2 (PAR2), transient receptor potential ankyrin 1 (TRPA1)

Signaling molecules:

IFN regulatory factor 4 (IRF4), MAPp38α in myeloid cells, NFκBp65, p53-upregulated modulator of apoptosis (PUMA), SOCS1, STAT3

Pathogenic exogenous factors

Bilateral adrenalectomy, *Klebsiella pneumoniae*, stress

Protective endogenous factors

Cells and function:

Liver environment for immune balance, NK cells, NKT cells, oral tolerance, probiotics-treated DCs, TCRγδT cells

Soluble molecules:

A33 antigen, allograft inflammatory factor 1 (AIF1), basic fibrocyte growth factor (bFGF), galecin-1, G-CSF, ghrelin, growth hormone, hepatocyte growth factor (HGF), hypoxia-inducible factor (HIF), IL-10, IL-11, IL-23, IL-25, IL-27, intestinal trefoil factor, placental growth hormone, resistin-like molecule β (RELMβ), T cell Ig mucin 3 (Tim3), TGFβ1, vasoactive intestinal peptide (VIP)

Enzymes:

Elafin (endogenous serine protease inhibitor), indoleamine 2,3-dioxygenase (IDO), superoxide dismutase

Receptors:

4-1BB, canabinoid receptors, CCL2, CD73, glucocorticoid-induced TNF receptor (GITR), multi drug resistance protein 1 (MDR1), muopioid receptor (MOR), peroxisome proliferator-activated receptor γ (PPAR γ), proteinase-activated receptor 2 (PAR2), retinoid X receptor (RXR), serotonin reuptake transporter, TNFR1

Signaling molecules:

IRF4, forkhead box transcription factor 4 (Foxo4), MAPp38α in epithelial cells, NOD2, STAT5b

Exogenous agents with therapeutic potential

Biological therapy:

2,3,7,8-Tetrachlorodibenzo-*p*-dioxin (TCDD, AhR agonist), 2′4′6′-tris chalcone (TMMC), 3,3′-diindolylmethane (DIM), 4sc-101 (inhibitor of dihydroorotate dehydrogenase), 4AZA1378 (pteridine derivative), 5-aminoimidazole-4-carboxamide ribonucleoside (AICAR), 15-deoxyspergualin, adrenomedullin, AS605240 (PI3Kγ inhibitor), atorvastatin, bindarit (inhibitor of MCP1 synthesis), bone marrow stromal cell therapy, geranyl geranylacetone,

(*Continues*)

TABLE V (*Continued*)

glatiramer acetate, indomethacin heptyl ester, INT747 (nuclear famesoid X receptor agonist), follistatin (natural antagonist of activin), FTY720 (synthetic sphingosine analog of myriocin), glatiramer acetate, glucagon-like peptide 2 (nutritional absorptive factor), KPV (tripeptide Lys-Pro-Val), O-1602 (agonist at the GPR55 cannabinoid receptor), recombinant spondin 1 (epithelial mitogen), rifaxlmin (pregnane X receptor activator), SB225002 (CXCR2 antagonist), simvastatin, tacrolimus (immune suppressant), tanshinone IIA (antioxidant), ZK-192 (beta-oxidation-resistant 3-oxa-ATL analog)
Pro/prebiotics bacteria:
Bacillus polyfermenticus, *Bacillus subtilis* PB6, *Bifidobacterium longum*, *Faecalibacterium prausnitzii*, *Heligmosomoides polygyrus*, *Lactobacillus fermentus*, probiotix mixture #2, *Schistosoma japonicum* egg, *Schistosoma mansoni*, *Yersinia pseudotuberculosis*
Bacterial products and antibiotics:
Ciprofloxacin, clotrimazole (antifungal imidazole), food grade yeasts, *Lactobacillus* peptidoglycan, minocycline (antibiotics), recombinant cholera toxin B, rifaximin, *Trichinella spiralis* 53 kDa protein
Complementary and alternative medicine:
Berberine (*Coptidis japonica*), Bojanggunbi-Tang (Korean herbal medicine), bolinaquinone (marine product), carbon monoxide (smoking), catalposide, cirsilineol (flower oil from *Artemisia vestita*), crocetin (flower product), curcumin (curry compound), Daikin-Chu-To (Japanese herbal medicine), electric shock, epigallo catechin-3-gallate (green tea), glutamine, guggulsterone (plant-derived steroid), hawthornfruit, lancemaside A (*Codonopsis lanceolata*), matrine (alkaloid), Paenol enema (Chinese herbal), pentacyclic triterpenes, petrosaspongiolide M (marine product), polygalae roots, Si-Ni-San (Chinese herbal medicine), sinomenine (alkaloid), soyasaponin1 (soybean), taurine, theaflavin-3-3′-digallate (black tea), thearubigin (polyphenol), *trans*-retinoic acid, triterpenoid saponin, vanillin (a flavor compound in food), vitamin D analog 22-ene-25-oxa-vitamin D, yogurt

Beneficial effect of anti-IL-12p40 mAbs on colitis was initially proposed in TNBS model.[219] This discovery has prompted physicians to conduct human trials of anti-IL-12p40 mAb therapy in active CD patients, which have demonstrated primary evidence of efficacy.[220] IL-12p40 is associated with IL-12p35 to form IL-12p70 and with p19 to form IL-23. IL-23 receptor (IL-23R) is a major IBD susceptibility gene,[16–20] and pathogenic role of IL-23, but not IL-12p70, is becoming increasingly apparent in many IBD models. In contrast, the colitis of TNBS model may be mediated by IL-12p70 and be protected by IL-23, as indicated by an observation that TNBS colitis is exacerbated in IL-23 (p19) KO mice.[221] The exacerbation of colitis is mediated by enhanced IL-12p70 production. Similarly, IL-12p35 KO mice lacking IL-12p70 develop mild TNBS colitis, whereas IL-12p40 KO mice lacking both IL-12p70 and IL-23 develop more exacerbated colitis.[222] Administration of anti-IL-12p40 mAbs significantly exacerbates the TNBS colitis of IL-12p35 KO mice. Alternatively, IL-23 serves as the key factor of maintaining Th17 cells and inducing IL-17-producing innate lymphoid cells.[192] IL-17 receptor KO mice are shown to resist against TNBS colitis.[223]

IL-12p70 is a well-known cytokine to promote the development of IFN-γ-producing $CD4^+$ (Th1) T cells. Consistent with the dependence of TNBS colitis on IL-12p70, this colitis is characterized by enhanced Th1 response.[219,224] Therefore, dependence of TNBS colitis on Th1 TCRαβ T cells has generally been believed. However, this concept is challenged by a series of recent findings. A study suggests that TCRγδ T cells are protective but TCRαβ T cells do not significantly influence the initiation or perpetuation of this colitis.[225] Indeed, TNBS colitis can be induced in RAG1 KO mice that lack both T and B cells.[226] In addition, WT and IFN-γ receptor KO mice exhibit the comparable susceptibility against TNBS colitis.[227] Similarly, administration of anti-IFN-γ mAbs has no effect on this colitis.[228] Interestingly, TNBS colitis can be induced in IFN-γ KO mice of Balb/c background, and the colitis is mediated by IL-4.[229]

L. Oxazolone Model

Oxazolone (4-ethoxylmethylene-2-phenyloxazol-5-one) is a classical haptenating agent that has historically been used for studying delayed-type hypersensitive responses in skin. An intrarectal administration of high dose (6 mg/mouse) of oxazolone resolved in ethanol induces rapid onset of inflammation in the distal part of colon in an autoimmune prone mouse strain SJL/J by 2 days after administration.[230] This colitis is characterized by hemorrhagic inflammation and severe submucosal edema, and it is mediated by IL-4 and IL-13-producing NKT cells.[230,231] Human UC also exhibits the expansion of IL-13-producing $CD161^+$ NKT cells that do not express an invariant TCRα chain (TCRα24, a marker of conventional NKT cells) or is not activated by a classical NKT cell ligand α-GalCer.[232] IL-12p40 plays no role in the oxazolone colitis of SJL/J mice.

Although majority of conventional mouse strains were resistant to oxazolone-induced colitis, presensitization of skin with oxazolone before the intrarectal administration has been found to induce colitis reproducibly in C57BL/6.[233] The colitis, which is induced in C57BL/6 mice by low dose (3% for skin sensitization and 1% for intrarectal administration) of oxazolone, is mediated by both Th1 and Th2 responses.[233] Epstein–Barr virus-induced gene 3 (*EBI3*), which is associated with IL-27p28 to form IL-27 and with IL-12p35 to form IL-35, contributes to the development of oxazolone colitis by enhancing Th2 immune responses.[234] In addition, IL-6, rather than IL-23 and IL-17, are essential for this colitis, and the IL-6 production depends on NFATc2 that has been found to increase in UC patients.[235] TNF is generally believed to exert proinflammatory activities, but there is increasing evidence for anti-inflammatory action of TNF in some inflammatory conditions. Interestingly, administration of TNF ameliorates oxazolone colitis by inducing extra-adrenal glucocorticoid synthesis.[236]

LI. DSS Model

Dextran sulfate sodium (DSS) is polymer of sulfated polysaccharide. Continuous administration of 1–5% DSS in drinking water induces acute intestinal injury characterized by bloody diarrhea, ulcerations, intestinal inflammation, body weight loss, and shortening of colon length. This injury seems to be induced by direct hyperosmotic damage to epithelial cells[237], and it can be induced in immune-deficient mice that lack T and B cells.[238] C3H/HeJ mice are highly susceptible to this injury model as compared to other strains of mice such as C57BL/6. In addition, male tends to be more susceptible than female.[238]

Three kinds of regimens have been used in this model system. First regimen is a continuous administration of DSS to study the induction (acute) phase of intestinal injury. In contrast to majority of IBD models, DSS model recovers spontaneously after termination of DSS treatment, allowing us to more closely examine the mechanism of recovery processes from intestinal inflammation such as epithelial healing and development of regulatory immune cells with ability to enhance the wound healing.[119,239] In this regard, second regimen is to have mice treated with DSS for a few days. The recovery phase is then examined after termination of treatment by changing DSS to normal drinking water. Third regimen is to induce chronic colitis. To do so, mice are treated with repeated administrations of DSS with each interval.

A distinctly different aspect with DSS model from the majority of other IBD models is that enteric bacteria contribute to the suppression of the acute colitis. Mice kept under germ-free environment develop lethal colitis with massive intestinal bleeding in response to DSS.[240] Similarly, treatment of mice with wide spectrum of antibiotics causes lethal colitis with massive intestinal bleeding after DSS administration.[241] In addition, DSS treatment significantly increases the mortality rate in mice deficient in MyD88 or TLR4 as compared to WT mice.[241–243] TLR9 also protects mice from DSS-induced colitis by enhancing the production of type I IFN by epithelial cells.[244]

Since DSS colitis is easily used and time- and cost-saving, this model has been most widely used (Tables VI and VII). Indeed, many genetically engineered mice have been subjected to this system (Table VI), and therapeutic potential of many compounds including the field of complementary and alternative medicine have been evaluated using this model (Table VII). However, it should be important to mention that DSS model has several concerns

TABLE VI
Resistant Versus Susceptible Mice Against DSS Colitis

Susceptible mice to DSS colitis	Resistant mice to DSS colitis
Cytokines and hormones	**Cytokines and hormones**
GM-CSF KO mice	GH Tg mice (growth hormone)
HGFA KO mice (hepatocyte growth factor activator)	gp130 KO mice
IL-1 receptor 1 KO mice	gp130 [myeloid cell-specific] KO mice
IL-6 KO mice	IFN-γ KO mice
IL-10 KO mice	*IL-6 KO mice*
IL-15 KO mice	IL-12p35 KO mice
IL-18 Tg mice	*IL-15 KO mice*
IL-18 KO mice	IL-17A KO mice
IL-18R KO mice	IL-21 KO mice
IL-19 KO mice	IL-27R KO mice
IL-22 KO mice	IL-32R Tg mice (at chronic phase)
IL-32R Tg mice (at induction phase)	LIGHT KO mice
KGF KO mice (keratinocyte growth factor)	*OPN KO mice*
MC1Re/e mice (carrying a frame-shift mutation in α-melanocyte stimulating hormone gene)	OPN KO mice at chronic phase
MFG-E8 KO mice (milk fat globule-EGF factor 8)	TGFα Tg mice
OPN KO mice (osteopontin)	VEGFR1 Tg mice
OPN KO mice at induction phase	WSX-1 KO mice (a component of IL-27 receptor)
PLGF KO mice (placental growth factor)	
TAK1 KO [epi-specific] mice (TGFβ-activated kinase 1)	
TGFα KO mice	
TGFβDNRII [CD68-specific] Tg mice	
TGFβDNRII [epi-specific] Tg mice	
TNF-α KO mice	
VEGF-A Tg mice (vascular endothelial growth factor-A)	
Chemokines and homing	**Chemokines and homing**
AnxA1 KO mice (annexin A1)	CCR2 KO mice
CCR9 KO mice	CCR5 KO mice
CXCL1 KO mice	CX3CR1 KO mice
MIF Tg mice (macrophage-migration inhibitory factor)	CXCL16 KO mice
MIP2 [epi-specific] Tg mice (macrophage inflammatory protein-2)	CXCR2 KO mice
PSGL1 KO mice (P-selectin glycoprotein ligand-1)	Integrin α1β1 KO mice
	MIF KO mice
	PHIL Tg mice (that are devoid of eosinophils)
	FucTVII KO mice (fucosyltransferase 7)
	SLC11a1 KO mice (natural resistance-associated macrophage protein 1)
Receptors and costimulation	**Receptors and costimulation**
A(2B)AR KO mice (adenosine 2B receptor)	*A(2B)AR KO mice*
AhR KO mice (aryl hydrocarbon receptor)	C3 KO mice (complement C3)
C5 KO mice (complement 5)	CD5 KO mice

(*Continues*)

TABLE VI (*Continued*)

Susceptible mice to DSS colitis	Resistant mice to DSS colitis
CBIR KO mice (cannabinoid receptor 1)	CD11a KO mice
CD39 KO mice	CD18 KO mice
CD95 KO mice (Fas ligand)	CD30 L KO mice
CD95 [epi-specific] KO mice	CD34 KO mice
PIRB KO mice (paired immunoglobulin-like receptor B)	CD40 KO mice
RAGE KO mice (receptor for advanced glycation end products)	CD40L KO mice
Syndecan 1 KO mice	CD54 KO mice
TCRδ KO mice (absence of TCRγδ T cells)	Galectin-3 KO mice
TNFR1-deficient RAG double KO mice	Necl-5 KO mice (CD155)
VDR KO mice (vitamin D receptor)	RAG1 KO mice (absence of both T and B cells)
Waved-2 mice (defect of EGF-R)	TR2 KO mice (TNFR-related 2)
	VPAC1 KO mice (T cell G-protein-coupled receptor for vasoactive intestinal peptide)
Signaling molecules	**Signaling molecules**
ASK KO mice (apoptosis signal-regulating kinase 1)	Fra1 Tg mice (component of the dimeric transcription factor AP1)
BLT2 KO mice (low-affinity leukotriene B4 receptor)	GPR43 KO mice (G-protein-coupled receptor 43)
Cdx2 heterozygous mice (intestine-specific transcription factor)	HDC KO mice (histidine decarboxylase)
Cux1HD KO mice (ubiquitous transcription factor)	MAPp38α [myeloid cell-specific] KO mice
Fos KO mice	PAR2 KO mice (proteinase-activated receptor)
Gly96 KO mice (immediate early gene-X1)	PI3Kγ mutant mice
GRK6 KO mice (G-protein-coupled receptor kinase 6)	SK1 KO mice (aphingosine kinase 1)
HNF4α [epi-specific] KO mice (hepatocyte nuclear factor 4α)	Smad3 heterozygous KO mice
IRF1 KO mice (IFN regulatory factor)	XID mice (absence of Bruton tyrosine kinase)
IRF4 KO mice	
JNK2 KO mice (c-Jun N-terminal kinase)	
MaFIA Tg mice (macrophage Fas-induced apoptosis)	
MAPp38α [epi-specific] KO mice	
Nrf2 KO mice (nuclear factor—erythroid 2-related factor)	
PKC lota KO [epi-specific] mice (protein kinase C lota)	
Raf1 [epi-specific] KO mice (raf-1 kinase)	
RelA [epi-specific] KO mice (NFκB subunit)	
Smad5 [epi-specific] KO mice	
SOCS1 [T-cell-specific] KO mice (suppressor of cytokine signaling-1)	
STAT3 [epi-specific] KO mice	
STAT6 KO mice	
TC-PTP heterozygous mice (T cell protein tyrosine phosphatase)	

(*Continues*)

TABLE VI (*Continued*)

Susceptible mice to DSS colitis	Resistant mice to DSS colitis
Inflammasome and oxidase stress	**Inflammasome and oxidase stress**
AGP Tg mice (α1-acid glycoprotein acute phase protein)	Casp12 KO mice (caspase-12)
Casp1 KO mice (caspase-1)	Cu/Zn-SOD Tg mice (Cu/Zn-superoxide dismutase)
Cox2 [myeloid cell-specific] KO mice	LMP2 KO mice (low molecular mass polypeptide 2)
Cox2 [endothelial cell-specific] KO mice	*NLRP3 KO mice*
EP4 KO mice (prostaglandin receptor EP4)	*iNOS KO mice*
Gp91 KO mice (neutrophil oxidative burst mediator)	nNOS KO mice (neuronal nitric oxide synthase)
eNOS KO mice (endothelial nitric oxide synthase)	Perforin KO mice
iNOS KO mice (inducible nitric oxide synthase)	TRX1 Tg mice (thioredoxin-1)
NLRP3 KO mice (nucleotide-binding domain, leucin-rich-containing protein)	
PPARβ/δ KO mice (peroxisome proliferator-activated receptor)	
PPARγ [epi-specific] KO mice	
PPARγ [macrophage-specific] KO mice	
Pycard KO mice (an inflammasome adaptor protein)	
SOD Tg mice (superoxide dismutase)	
Epithelial defense and functions	**Epithelial defense and functions**
C2GnT2 KO mice (core 2 β1,6-*N*-acetylglucosaminyltransferases)	Muc1 KO mice (mucin 1)
C3GnT KO mice (core 3β1,3-*N*-acetylglucosaminyl transferase)	*RELMα KO mice*
CLCN5 KO mice (intracellular Cl(−)H(+) exchanger)	RELMβ KO mice
CD98 [epi-specific] Tg mice	TRPV1 KO mice (transient receptor potential vanilloid I)
JAM-A KO mice (junctional adhesion molecule A)	Vanin1 KO mice (epithelial pantetheinase)
M3 KO mice (M3 muscarinic receptor)	
Mdr1 KO mice (multiple drug resistant)	
Mtgr1 KO mice (myeloid translocation gene-related 1)	
Nas1 KO mice (Nas1 sulfate transporter)	
NHE3 KO mice (sodium-hydrogen antiporter 3)	
Reduced DKK1 mice (antagonist of Wnt-βcatenin signal)	
RELMα KO mice (resistin-like molecule)	
TFF2 KO mice (trefoil factor 2)	
TFF3 KO mice (trefoil factor 3)	
Stress protein and hypoxia	**Stress protein and hypoxia**
Atg19L1 KO mice (autophagy 16L1)	HSF1 Tg mice
HSF1 KO mice (heat shock factor 1)	Hsp70 [epi-specific] Tg mice (heat shock protein 70)
IRE1β KO mice (protein sensing and responding to ER stress)	PHD1 KO mice (Hypoxia inducible factor-prolyl hydroxylase)
Mbtps1 mutant mice (membrane-bound transcription factor peptidase site 1)	

(*Continues*)

TABLE VI (*Continued*)

Susceptible mice to DSS colitis	Resistant mice to DSS colitis
Metabolisms and mesenchimal structure	**Metabolisms and mesenchimal structure**
ADAM17 KO mice (a metalloprotease)	A3AR KO mice (adenosine A3 receptor)
ApoA IV KO mice (apolipoprotein A-IV)	*APN KO mice*
APN KO mice (adiponectin)	Elafin Tg mice (serine protease inhibitor)
IAP KO mice (intestinal alkaline phosphatase)	Epimorphin KO mice
Meprin A/B double KO mice (meprin metalloproteinase)	Fat1 Tg mice (enhancement of biosynthesis of omega-3 polyunsaturated fatty acids)
MMP2 KO mice (matrix metalloproteinase)	INS-GAS Tg mice (amidated gastrin)
SCD1KO mice (stearoyl-CoA desaturase 1)	MCP6 KO mice (mast cell protease)
Serum amyloid A1/B1 double KO mice	MMP2 × MMP9 double KO mice
TKTL1 KO mice (transketolase-like 1)	*MMP2 KO mice*
	MMP-9 KO mice
	MT KO mice (metallothionein binding protein to zinc)
	NE KO mice (protease neutrophil elastase)
	Ob/Ob mice (leptin-deficient)
	Op/Op mice (osteopetrotic mice)
	PR3 KO mice (proteinase 3)
	S/S mice (mutation in leptin receptor)
	W/W mutant mice (absence of Cajal cells)
Oncogenic	**Oncogenic**
Atm KO mice (telangiectasia mutated, a DNA double strand break recognition and response protein)	PUMA KO mice (P53-upregulated modulator of apoptosis)
Foxo3 KO mice (a tumor suppressor gene)	
GLI1 KO mice (glioma-associated oncogene homolog 1)	
TP53INP1 KO mice (tumor p53-induced nuclear protein 1)	
VHL [epi-specific] KO mice (von Hippel–Lindau tumor suppressor protein)	
Kruppel-like factor heterozygous mice	
Neurogenic and angiogenic	**Neurogenic and angiogenic**
Endoglin heterozygous mice (molecule essential for physiological angiogenesis)	AtiaR KO mice (angiotensin II type 1a receptor)
PACAP KO mice (pituitary adenylate cyclase-activating polypeptide)	LMP7 KO mice (an immunoproteasome subunit)
TSP1 KO mice (thrombospondin-1)	NOR KO mice (nociceptin receptor)
	NPY KO mice (neuropeptide Y)
	TPH1 KO mice (tryptophan hydroxylase 1)

(*Continues*)

TABLE VI (*Continued*)

Susceptible mice to DSS colitis	Resistant mice to DSS colitis
Mice that were shown to develop comparable grade of DSS colitis as compared to WT	
BLT1 KO mice (low-affinity leukotriene B(4) receptor), bradykinin 2 receptor KO mice, CXCR4 KO mice, dipeptidyl peptidase IV KO mice, Gal1R KO mice (galanin-1receptor), IL-1R KO mice, IL-5 KO mice, endothelial-specific JAM-A KO mice, mast cell-deficient mice, Min KO mice, MT KO mice (metallothionein), MT Tg mice, NIRP4 KO mice (nucleotide-binding domain, leucin-rich-containing component), *NOD2 KO mice*, p47phox KO mice (p47phox subunit of NADPH oxidase), stearoyl-CoA desaturase 1 KO mice, *TLR2 KO mice*, *TNF-α KO mice*	

KO, knockout; Tg, transgenic.

Cell-specific conditional KO/Tg mice are indicated by [ ] in which the cell targeted is mentioned.

As indicated by *italic*, both protection and exacerbation of DSS colitis have been described in some KO/Tg mice.

particularly when used for the evaluation of such therapeutic agents. For example, the severity of inflammation is well known to vary considerably from mouse to mouse as well as from region to region even in an individual colon, making the assessment of inflammation more difficult as compared to other models. Different protocols with inconsistent DSS doses (1–5%) and treatment period (3–7 days) also affect the outcome. In addition, genetic background is a key factor of determining the susceptibility of mice to DSS colitis,[4] which would be very critical when mixed background mice are used. Further, the instability of DSS activity under acidic conditions[245] tends to raise a concern when acidified water is applied to mice. Indeed, as seen in Table VI, DSS model has made several discrepancies in data. Therefore, careful attention should be made when DSS model is used particularly for evaluating the therapeutic effect of some candidate agents on IBD.[246]

LII. Conclusion

There are currently many murine IBD models available to use. As proposed in Table VIII, each model has specific advantage(s) over other models. For example, a series of meritorious information to understand the adaptive immune mechanism involved in the pathogenesis of IBD have been brought from CD45RB model. TNBS model has provided significant impact on developing anti-IL-12p40 therapy that is currently applying to human CD. IL-10 KO model has provided significant contributions to the understanding of probiotics mechanism in IBD. TCRα KO model has been shown to share unique etiological characteristics (e.g., appendectomy and smoking) with human UC.

TABLE VII

EXOGENOUS FACTORS THAT IMPROVES OR EXACERBATES DSS COLITIS

Exogenous factors with therapeutic effect on DSS colitis

Biological therapy:

2,3,7,8-Tetrachlorodibenzo-*p*-dioxin (TCDD, AhR activator), 4,4′-dibromodibenzylmetane (ER stress inhibitor), 4-OHDHA (PPARγ agonist), 4SC-101 (immune suppressant), abscisic acid, adrenomedullin, AGN205203 (E prostanoid receptor 4 agonist), AMD3100 (CXCR4 antagonist), AS605240 (PI3Kγ inhibitor), ATL-801 (adenosine A2B receptor antagonist), anti-CD3 Ab, anti-CD44v7 Ab, anti-CSF1 Ab, anti-high mobility group box 1 (HMGB1) Ab, anti-IL-1β Ab, anti-IL-12p40 Ab, anti-PECAM1 Ab, antithrombin agent, anti-TLR4 Ab, anti-TNFSF15 Ab, balsalazide (5-ASA prodrug), bortezomib (proteasome inhibitor), bosentan (endothelin receptor antagonist), CA-074 (cathepsin B inhibitor), chloramidine (protein arginine deiminase inhibitor), C10 (phenyl methimazole), COG112 (apolipoprotein E-mimetic peptide), cyclosporine A, dibenzazepine (Notch pathway blocker), endostatin, ENalaprilat (angiotensin-converting enzyme inhibitor), ER-319711 (dipeptidyl peptidase IV inhibitor), enoxaparin (syndecan 1 analog heparin), FGF-20, fibronectin, fibratide (synthetic FGF mimetic peptide), fluoxetine (serotonin reuptake inhibitor), follistatin (activin antagonist), FTY720 (sphingosine-1 phosphate receptor modulator), geranylgeranylacetone (hsp70 inducer), glatiramer acetate, GI270384X (inhibitor of adhesion molecule expression), GL1001 (angiotensin-converting enzyme), GM-CSF therapy, imidazoacridinone derivate (Flt3 receptor tyrosine kinase inhibitor), INT-747 (nuclear farnesoid X receptor agonist), ISO-F (inhibitor targeting the catalytic pocket of MIF), ITF2357 (histone deacetylase inhibitor), KPV (tripeptide Lys-Pro-Val), lyprinol (lipids from New Zealand green-lipped mussel), mesalazine, nafamostat mesilate (tryptase inhibitor), LPS through rectal administration, melatonin, melphalan (anticancer drug), naltrexone (opioid antagonist), NEMO-binding domain peptide, non-anaphylatogenic anti-IgE Ab, O-1602 (cannabinoid receptor GPR55 agonist), OCH (α-galactosyl ceramide), OLL1181 (activator of AhR), ONO-5046 (neutrophil elastase inhibitor), P2281 (mTOR inhibitor), P59/99 (dipeptidyl peptidase inhibitor), papaverine, PDTC (NFκB inhibitor), pepstatin A (cathepsin D inhibitor), R-spondin 1 (epithelial mitogen), polaprezinc (anti-ulcer drug), Raf-1 kinase, RDP58 (TNF synthesis inhibitor), recombinant IL-25, recombinant trefoil factor 3, repifermin (truncated form of keratinocyte growth factor 2), resolvin E1, rifaximin (pregnane X receptor activator), rofecoxib (cox2 inhibitor), SAHA (histone deacetylase inhibitor), SB216763 (glycogen synthase kinase 3β inhibitor), serine-threonine protein kinase AKT inhibitor, SM-19712 (endothelin-converting enzyme), SMA-7 (difluoromethylene analogs of sphingomyelin), SP600125 (JNK inhibitor), thioredoxin-1, TNJ-419 (synthetic compound from acteoside), tranilast through enema, TRPV1 (vanilloid receptor type I antagonist), TV-5010 (high molecular weight of glatiramer acetate), TZD (PPARγ agonist), valsartan (angiotensin II antagonist), vapiprost (thromboxane-prostanoid receptor antagonist), Wartmann (PI3K inhibitor), ZK1916784 (calcitriol analog), UR-1505 (chemical related to salicylic acid), urocanic acid

Gene and cell therapy:

Appendectomy, adenoviral delivered glial-derived neurotrophic factor, administration of hypertonic saline, adoptive transfer of IL-10-producing B10 B cells, AP1 decoy oligodeoxynucleotides, *Bifidobacterium* engineered to secrete IL-10, colon-specific lipid based gene delivery of IL-22, feed with autologous inflamed colon, fusion protein of IL-2 linked together with caspase-3, immortalized CD34$^-$ stem cells, insulin-like growth factor-1-coated sutures, intracolonic administration of ENA-78 antisense oligonucleotide, mesenchymal stem cell therapy, *Lactobacillus casei* BL23 engineered to produce superoxide dismutase, *Lactobacillus* engineered to secrete trefoil factor, nanoparticles loaded with cyclin D1-siRNA, NFκB decoy oligonucleotides, *Salmonella typhimurium* SL7207 carrying plasmid encoding Cu/Zn-superoxide dismutase

(*Continues*)

TABLE VII (*Continued*)

Microorganisms and products:
Anisakis simplex, *Bacillus polyfermenticus*, *Bacteroides ovatus*, *Bifidobacterium lactis*, cathelicidin (anti-bacterial peptide), colifagina (soluble bacterial antigen), *E. coli* M-17, *E. coli* Nissle 1917, *Enterococcus durans*, *Enterococcus faecalis*, heat-killed *lactobacillus brevis*, *Helicbacter pylori* DNA, *Inonotus obliquus* (fungus), *Lactobacillus reuteri*, *Lactobacillus rhamnosus*, *Lactococcus lactis*, minocycline, parabacteroides distasonis antigen, PCSK (TLR2 ligand), poly(IC) (TLR3 ligand), probiotics mixture VSL#3, *Saccharomyces boulardii*, *Schistosoma mansoni* infection
Complementary and alternative medicine:
α-Phenyl-*N*-*tert*-butylnitrone, 2-(*RS*)-*n*-propylthiazolidine-4(*R*)-carboxylic acid, 4-methoxy-5-hydroxy-canthin-6-one, 7-hydroxy-frullanolide, acanthopanax (Korean herb product), aglycones, Aloe, American ginseng, apple procyanidins, *Arctium lappa*, β-caryophyllene (plant metabolite), blueberry, Bojanggunbi-tang (Korean herbal medicine), bovine milk, Brassica food plant, brown rice, butyrate, caffeic acid, cellobiose, cheese whey protein, chrysin, coenzyme Q10, conjugated linoleic acid, curcumin, DA-6034 (flavonoid), dietary fish oil, D-ribose-L-cysteine, docosahexaenoic acid, dosmalfate, eicosanoid lipoxin A4, embelin, freeze-dried black raspberry, glabridin (liquor rice), goat milk, grain food processed, guggulsterone, gum resin, hen egg lysozyme, hesperidin (citrus fruit), hookworm product, hyaluronic acid, hydrogen gas, IDS30 (stinging nettle leaf extract), Kurozu (Japanese black vinegar), L-cysteine, L-tryptophan, *Mangifera indica*, mushroom, *N*-acetylcysteine, okadaic acid through enema, partially hydrolyzed guar gum (water soluble dietary fiber), *Patrinia scabicsaefolia*, piceatannol (grape compound), phenylethanoid acteoside, pomegranate, puberty, resveratrol, roquinimex, rutin, sialyl(α2,3)lactose-deficient breast milk, smoking, *Solanum nigrum* linne fruit, sophoridine, tacrolimus, taurine, tea catechin, tetrandrine (Chinese herb component), tryptanthrin, vanillic acid, xanthorrhizol, *Zanthoxylum piperitum* fruit, zerumbone (condiment in Southeast Asia), zinc, ziziphora clinopoides (Iranian herbal medicine)
Exogenous factors that exacerbate DSS colitis
Biological therapy:
Anti-CD98 Ab, antiflagellin IgG, anti-Gr1 Ab, anti-IL-17 Ab, AR-R17779 (α7-nicotinic acetylcholine receptor agonist), basic fibroblast growth factor, cox1/2 inhibitor, FR167653 (MAPp38 inhibitor), fructo-oligosaccharide, pseudoephedrine, *sn*-protoporphyrin (heme oxygenase-1 inhibitor)
Environment:
Adhesive invasive *E. coli* (AIEC), *Bacteroides fragillis*, *Candida albicans*, CpG (TLR9 ligand), *Cytomegalovirus* infection, depressive-like behavior, *E. coli* O6K13, flagellin, germ-free environment, repeated exposure to chronic psychosocial stress, sleep deprivation, vagotomy, water avoidance stress
Complementary and alternative medicine:
Dietary iron supplement, high fat diet, low Ca^+ diet, omega-3 fatty acids, potato skin, soy bean, vitamin D-deficiency, vitamin B, Western style diet

DSS model has been useful for dissecting the mechanism of inflammation-associated epithelial homeostasis and wound healing process. In addition, many new IBD models employing cutting-edge genetic technologies have been established increasingly, and they have brought several novel concepts on IBD pathogenesis to the fore. Since none of models can fully reflect human IBD and each model has some demerits, further accumulation of data from each model systems (but not from one model system) would compensate the

TABLE VIII
SPECIFIC CHARACTERISTICS OF EACH IBD MODEL

Models	Useful for studying
Anti-CD40	Innate lymphocyte cells
C3HeJ/Bir	Flagellin-specific $CD4^+$ T cell response
CD3εTg	Thymic selection, IL-12p40-dependent non-Th1 pathway
CD45RB	Treg and other regulatory T cells, pathogenic IL-23, $IL\text{-}17^+IFN\text{-}\gamma^+$ T cells
Clgalt1 KO	Epithelial defense
DSS	Epithelial homeostasis/regeneration, wound healing process
IL-2 KO	Autoimmune response, activation of $CD4^+$ T cells, maintenance of Treg
IL-10 KO	IBD susceptibility gene, probiotics, *Helicobacter*, NSAID
Oxazolone	IL-13, NKT cells, steroidogenesis, fibrosis
PDK1 KO	Pathogenic TCRγδ T cells
SAMP1	Ileitis, cell homing, pathogenic B cells
SHIP KO	Granulocytes and monocytes
STAT3 KO	IBD susceptibility gene, innate and adaptive immune linkage
T-bet/RAG	Microbiota community
TCRα KO	Appendectomy, CO (smoking), regulatory B cells, IL-22-defense cascade
TGFβ KO	$CD4^+$ T cell differentiation
TNBS	NOD2 function, pathogenic IL-12p70, and protective IL-23 functions
TNF(ARE)	TNF signaling, pathogenic $CD8^+$ T cells
WASP KO	T-cell signaling, Treg function
XBP1 KO	IBD susceptibility gene, ER stress

pitfalls of each other and provide important interventions for developing new therapeutic measures such as cell-specific therapy and personalized therapy to save the lives of patients with IBD more safely and more effectively.[15,200]

ACKNOWLEDGMENTS

We greatly thank Cindy W. Lau for her excellent editorial help and NIH (RO1AI081807, RC1DK086242), Crohn's and colitis foundation of America, and Broad Foundation for supporting our projects.

REFERENCES

1. Xavier RJ, Podilsky DK. Unravelling the pathogenesis of inflammatory bowel disease. *Nature* 2007;**448**:427–34.
2. Kaser A, Zeissig S, Blumberg RS. Inflammatory bowel disease. *Annu Rev Immunol* 2010;**28**:573–621.
3. Kirsner JB, Elchelepp J. The production of an experimental ulcerative colitis in rabbits. *Trans Assoc Am Physicians* 1957;**70**:102–19.

4. Elson CO, Sartor RB, Tennyson GS, Riddell RH. Experimental models of inflammatory bowel disease. *Gastroenterology* 1995;**109**:1344–67.
5. Chalifoux LV, Bronson RT. Colonic adenocarcinoma associated with chronic colitis in cotton top marmosets, Saguinus oedipus. *Gastroenterology* 1981;**80**:942–6.
6. Podolsky DK, Lobb R, King N, Benjamin CD, Pepinsky B, Sehgal P, et al. Attenuation of colitis in the cotton-top tamarin by anti-alpha 4 integrin monoclonal antibody. *J Clin Invest* 1993;**92**:372–80.
7. Powrie F, Mason D. OX-22high CD4 + T cells induce wasting disease with multiple organ pathology: prevention by the OX-22low subset. *J Exp Med* 1990;**172**:1701–8.
8. Powrie F, Leach MW, Mauze S, Menon S, Caddle LB, Coffman RL. Inhibition of Th1 responses prevents inflammatory bowel disease in scid mice reconstituted with CD45RBhi CD4 + T cells. *Immunity* 1994;**1**:553–62.
9. Hammer RE, Maika SD, Richardson JA, Tang JP, Taurog JD. Spontaneous inflammatory disease in transgenic rats expressing HLA-B27 and human beta 2m: an animal model of HLA-B27-associated human disorders. *Cell* 1990;**63**:1099–112.
10. Sadlack B, Merz H, Schorle H, Schimpl A, Feller AC, Horak I. Ulcerative colitis-like disease in mice with a disrupted interleukin-2 gene. *Cell* 1993;**75**:253–61.
11. Kühn R, Löhler J, Rennick D, Rajewsky K, Müller W. Interleukin-10-deficient mice develop chronic enterocolitis. *Cell* 1993;**75**:263–74.
12. Mombaerts P, Mizoguchi E, Grusby MJ, Glimcher LH, Bhan AK, Tonegawa S. Spontaneous development of inflammatory bowel disease in T cell receptor mutant mice. *Cell* 1993;**75**:274–82.
13. Strober W, Fuss I, Mannon P. The fundamental basis of inflammatory bowel disease. *J Clin Invest* 2007;**117**:514–21.
14. Elson CO, Cong Y, McCracken VJ, Dimmitt RA, Lorenz RG, Weaver CT. Experimental models of inflammatory bowel disease reveal innate, adaptive, and regulatory mechanisms of host dialogue with the microbiota. *Immunol Rev* 2005;**206**:260–76.
15. Mizoguchi A, Mizoguchi E. Animal models of IBD: linkage to human disease. *Curr Opin Pharmacol* 2010;**10**:578–87.
16. Anderson CA, Boucher G, Lees CW, Franke A, D'Amato M, Taylor KD, et al. Meta-analysis identifies 29 additional ulcerative colitis risk loci, increasing the number of confirmed associations to 47. *Nat Genet* 2011;**43**:246–52.
17. Franke A, McGovern DP, Barrett JC, Wang K, Radford-Smith GL, Ahmad T, et al. Genome-wide meta-analysis increases to 71 the number of confirmed Crohn's disease susceptibility loci. *Nat Genet* 2010;**42**:1118–25.
18. Abraham C, Cho JH. Inflammatory bowel disease. *N Engl J Med* 2009;**361**:2066–78.
19. Lees CW, Barrett JC, Parkes M. New IBD genetics: common pathways with other diseases. *Gut* 2011; PMID:21300624 [Epub ahead of print].
20. Khor B, Gardet A, Xavier RJ. Genetics and pathogenesis of inflammatory bowel disease. *Nature* 2011;**474**:307–17.
21. Garrett WS, Lord GM, Punit S, Lugo-Villarino G, Mazmanian SK, Ito S, et al. Communicable ulcerative colitis induced by T-bet deficiency in the innate immune system. *Cell* 2007;**131**:33–45.
22. Matharu KS, Mizoguchi E, Cotoner CA, Nguyen DD, Mingle B, Iweala OI, et al. Toll-like receptor 4-mediated regulation of spontaneous Helicobacter-dependent colitis in IL-10-deficient mice. *Gastroenterology* 2009;**137**:1380–90.
23. Hoentjen F, Harmsen HJ, Braat H, Torrice CD, Mann BA, Sartor RB, et al. Antibiotics with a selective aerobic or anaerobic spectrum have different therapeutic activities in various regions of the colon in interleukin 10 gene deficient mice. *Gut* 2003;**52**:1721–7.

24. González-Navajas JM, Fine S, Law J, Datta SK, Nguyen KP, Yu M, et al. TLR4 signaling in effector CD4+ T cells regulates TCR activation and experimental colitis in mice. *J Clin Invest* 2010;**120**:570–81.
25. Rakoff-Nahoum S, Hao L, Medzhitov R. Role of toll-like receptors in spontaneous commensal-dependent colitis. *Immunity* 2006;**25**:319–29.
26. Biswas A, Wilmanski J, Forsman H, Hrncir T, Hao L, Tlaskalova-Hogenova H, et al. Negative regulation of Toll-like receptor signaling plays an essential role in homeostasis of the intestine. *Eur J Immunol* 2011;**41**:182–94.
27. Berg DJ, Zhang J, Weinstock JV, Ismail HF, Earle KA, Alila H, et al. Rapid development of colitis in NSAID-treated IL-10-deficient mice. *Gastroenterology* 2002;**123**:1527–42.
28. Wilson MS, Ramalingam TR, Rivollier A, Shenderov K, Mentink-Kane MM, Madala SK, et al. Colitis and intestinal inflammation in IL10−/− mice results from IL-13Rα2-mediated attenuation of IL-13 activity. *Gastroenterology* 2011;**140**:254–64.
29. Davidson NJ, Leach MW, Fort MM, Thompson-Snipes L, Kühn R, Müller W, et al. T helper cell 1-type CD4+ T cells, but not B cells, mediate colitis in interleukin 10-deficient mice. *J Exp Med* 1996;**184**:241–51.
30. Read S, Malmström V, Powrie F. Cytotoxic T lymphocyte-associated antigen 4 plays an essential role in the function of CD25(+)CD4(+) regulatory cells that control intestinal inflammation. *J Exp Med* 2000;**192**:295–302.
31. Davidson NJ, Hudak SA, Lesley RE, Menon S, Leach MW, Rennick DM. IL-12, but not IFN-gamma, plays a major role in sustaining the chronic phase of colitis in IL-10-deficient mice. *J Immunol* 1998;**161**:3143–9.
32. Sheikh SZ, Matsuoka K, Kobayashi T, Li F, Rubinas T, Plevy SE. IFN-gamma is a negative regulator of IL-23 in murine macrophages and experimental colitis. *J Immunol* 2010;**184**:4069–73.
33. Yen D, Cheung J, Scheerens H, Poulet F, McClanahan T, McKenzie B, et al. IL-23 is essential for T cell-mediated colitis and promotes inflammation via IL-17 and IL-6. *J Clin Invest* 2006;**116**:1310–6.
34. Willerford DM, Chen J, Ferry JA, Davidson L, Ma A, Alt FW. Interleukin-2 receptor alpha chain regulates the size and content of the peripheral lymphoid compartment. *Immunity* 1995;**3**:521–30.
35. Murata Y, Yamashita A, Saito T, Sugamura K, Hamuro J. The conversion of redox status of peritoneal macrophages during pathological progression of spontaneous inflammatory bowel disease in Janus family tyrosine kinase 3(−/−) and IL-2 receptor gamma(−/−) mice. *Int Immunol* 2002;**14**:627–36.
36. Poussier P, Ning T, Chen J, Banerjee D, Julius M. Intestinal inflammation observed in IL-2R/IL-2 mutant mice is associated with impaired intestinal T lymphopoiesis. *Gastroenterology* 2000;**118**:880–91.
37. Contractor NV, Bassiri H, Reya T, Park AY, Baumgart DC, Wasik MA, et al. Lymphoid hyperplasia, autoimmunity, and compromised intestinal intraepithelial lymphocyte development in colitis-free gnotobiotic IL-2-deficient mice. *J Immunol* 1998;**160**:385–94.
38. Ma A, Datta M, Margosian E, Chen J, Horak I. T cells, but not B cells, are required for bowel inflammation in interleukin 2-deficient mice. *J Exp Med* 1995;**182**:1567–72.
39. Nagahama K, Ogawa A, Shirane K, Shimomura Y, Sugimoto K, Mizoguchi A. Protein kinase C theta plays a fundamental role in different types of chronic colitis. *Gastroenterology* 2008;**134**:459–69.
40. Simpson SJ, Mizoguchi E, Allen D, Bhan AK, Terhorst C. Evidence that CD4+, but not CD8+ T cells are responsible for murine interleukin-2-deficient colitis. *Eur J Immunol* 1995;**25**:2618–25.

41. Ashcroft AJ, Cruickshank SM, Croucher PI, Perry MJ, Rollinson S, Lippitt JM, et al. Colonic dendritic cells, intestinal inflammation, and T cell-mediated bone destruction are modulated by recombinant osteoprotegerin. *Immunity* 2003;**19**:849–61.
42. Yang XP, Ghoreschi K, Steward-Tharp SM, Rodriguez-Canales J, Zhu J, Grainger JR, et al. Opposing regulation of the locus encoding IL-17 through direct, reciprocal actions of STAT3 and STAT5. *Nat Immunol* 2011;**12**:247–54.
43. Burchill MA, Yang J, Vang KB, Moon JJ, Chu HH, Lio CW, et al. Linked T cell receptor and cytokine signaling govern the development of the regulatory T cell repertoire. *Immunity* 2008;**28**:112–21.
44. Davidson TS, DiPaolo RJ, Andersson J, Shevach EM. IL-2 is essential for TGF-beta-mediated induction of Foxp3+ T regulatory cells. *J Immunol* 2007;**178**:4022–6.
45. Furtado GC, Curotto de Lafaille MA, Kutchukhidze N, Lafaille JJ. Interleukin 2 signaling is required for CD4(+) regulatory T cell function. *J Exp Med* 2002;**196**:851–7.
46. Bhan AK, Mizoguchi E, Smith RN, Mizoguchi A. Colitis in transgenic and knockout animals as models of human inflammatory bowel disease. *Immunol Rev* 1999;**169**:195–207.
47. Mizoguchi A, Mizoguchi E, Saubermann LJ, Higaki K, Blumberg RS, Bhan AK. Limited CD4 T-cell diversity associated with colitis in T-cell receptor alpha mutant mice requires a T helper 2 environment. *Gastroenterology* 2000;**119**:983–95.
48. Kishi D, Takahashi I, Kai Y, Tamagawa H, Iijima H, Obunai S, et al. Alteration of V beta usage and cytokine production of CD4+ TCR beta beta homodimer T cells by elimination of Bacteroides vulgatus prevents colitis in TCR alpha-chain-deficient mice. *J Immunol* 2000;**165**:5891–9.
49. Mizoguchi A, Mizoguchi E, Chiba C, Spiekermann GM, Tonegawa S, Nagler-Anderson C, et al. Cytokine imbalance and autoantibody production in T cell receptor-alpha mutant mice with inflammatory bowel disease. *J Exp Med* 1996;**183**:847–56.
50. Mizoguchi A, Mizoguchi E, Bhan AK. The critical role of interleukin 4 but not interferon gamma in the pathogenesis of colitis in T-cell receptor alpha mutant mice. *Gastroenterology* 1999;**116**:320–6.
51. Iijima H, Takahashi I, Kishi D, Kim JK, Kawano S, Hori M, et al. Alteration of interleukin 4 production results in the inhibition of T helper type 2 cell-dominated inflammatory bowel disease in T cell receptor alpha chain-deficient mice. *J Exp Med* 1999;**190**:607–15.
52. Dash P, McClaren JL, Oguin 3rd TH, Rothwell W, Todd B, Morris MY, et al. Paired analysis of TCRα and TCRβ chains at the single-cell level in mice. *J Clin Invest* 2011;**121**:288–95.
53. Morgan NV, Goddard S, Cardno TS, McDonald D, Rahman F, Barge D, et al. Mutation in the TCRα subunit constant gene (TRAC) leads to a human immunodeficiency disorder characterized by a lack of TCRαβ+ T cells. *J Clin Invest* 2011;**121**:695–702.
54. Dianda L, Hanby AM, Wright NA, Sebesteny A, Hayday AC, Owen MJ. T cell receptor-alpha beta-deficient mice fail to develop colitis in the absence of a microbial environment. *Am J Pathol* 1997;**150**:91–7.
55. Chin EY, Dangler CA, Fox JG, Schauer DB. Helicobacter hepaticus infection triggers inflammatory bowel disease in T cell receptor alphabeta mutant mice. *Comp Med* 2000;**50**:586–94.
56. Burich A, Hershberg R, Waggie K, Zeng W, Brabb T, Westrich G, et al. Helicobacter-induced inflammatory bowel disease in IL-10- and T cell-deficient mice. *Am J Physiol Gastrointest Liver Physiol* 2001;**281**:G764–78.
57. Sydora BC, Tavernini MM, Doyle J, Fedorak RN. A defect in epithelial barrier integrity is not required for a systemic response to bacterial antigens or intestinal injury in T cell receptor-alpha gene-deficient mice. *Inflamm Bowel Dis* 2006;**12**:750–7.

58. Gaskins HR, Vondrak-Juergens GL, McCracken BA, Woolsey JH. Specific-pathogen-free conditions enhance inflammatory bowel disease in T-cell receptor knockout, but not C3H/HeJBir mice. *Lab Anim Sci* 1997;**47**:650–5.
59. Shimomura Y, Mizoguchi E, Sugimoto K, Kibe R, Benno Y, Mizoguchi A, et al. Regulatory role of B-1 B cells in chronic colitis. *Int Immunol* 2008;**20**:729–37.
60. López-Serrano P, Pérez-Calle JL, Pérez-Fernández MT, Fernández-Font JM, Boixeda de Miguel D, Fernández-Rodríguez CM. Environmental risk factors in inflammatory bowel diseases. Investigating the hygiene hypothesis: a Spanish case–control study. *Scand J Gastroenterol* 2010;**45**:1464–71.
61. Mizoguchi A, Mizoguchi E, Chiba C, Bhan AK. Role of appendix in the development of inflammatory bowel disease in TCR-alpha mutant mice. *J Exp Med* 1996;**184**:707–15.
62. Sheikh SZ, Hegazi RA, Kobayashi T, Onyiah JC, Russo SM, Matsuoka K, et al. An anti-inflammatory role for carbon monoxide and heme oxygenase-1 in chronic th2-mediated murine colitis. *J Immunol* 2011;**186**:5506–13.
63. Mizoguchi A, Mizoguchi E, Smith RN, Preffer FI, Bhan AK. Suppressive role of B cells in chronic colitis of T cell receptor alpha mutant mice. *J Exp Med* 1997;**186**:1749–56.
64. Mizoguchi A, Bhan AK. A case for regulatory B cells. *J Immunol* 2006;**176**:705–10.
65. Mizoguchi A, Mizoguchi E, Takedatsu H, Blumberg RS, Bhan AK. Chronic intestinal inflammatory condition generates IL-10-producing regulatory B cell subset characterized by CD1d upregulation. *Immunity* 2002;**16**:219–30.
66. Sugimoto K, Ogawa A, Shimomura Y, Nagahama K, Mizoguchi A, Bhan AK. Inducible IL-12-producing B cells regulate Th2-mediated intestinal inflammation. *Gastroenterology* 2007;**133**:124–36.
67. Leiper K, Martin K, Ellis A, Subramanian S, Watson AJ, Christmas SE, et al. Randomised placebo-controlled trial of rituximab (anti-CD20) in active ulcerative colitis. *Gut* 2011;**60**:1520–6.
68. Goetz M, Atreya R, Ghalibafian M, Galle PR, Neurath MF. Exacerbation of ulcerative colitis after rituximab salvage therapy. *Inflamm Bowel Dis* 2007;**13**:1365–8.
69. Ardelean DS, Gonska T, Wires S, Cutz E, Griffiths A, Harvey E, et al. Severe ulcerative colitis after rituximab therapy. *Pediatrics* 2010;**126**:e243–6.
70. El Fassi D, Nielsen CH, Junker P, Hasselbalch HC, Hegedüs L. Systemic adverse events following rituximab therapy in patients with Graves' disease. *J Endocrinol Invest* 2011;**34**:e163–7.
71. Sugimoto K, Ogawa A, Mizoguchi E, Shimomura Y, Andoh A, Bhan AK, et al. IL-22 ameliorates intestinal inflammation in a mouse model of ulcerative colitis. *J Clin Invest* 2008;**118**:534–44.
72. Shull MM, Ormsby I, Kier AB, Pawlowski S, Diebold RJ, Yin M, et al. Targeted disruption of the mouse transforming growth factor-beta 1 gene results in multifocal inflammatory disease. *Nature* 1992;**359**:693–9.
73. Gorelik L, Flavell RA. Abrogation of TGFbeta signaling in T cells leads to spontaneous T cell differentiation and autoimmune disease. *Immunity* 2000;**12**:171–81.
74. Moritoki Y, Lian ZX, Lindor K, Tuscano J, Tsuneyama K, Zhang W, et al. B-cell depletion with anti-CD20 ameliorates autoimmune cholangitis but exacerbates colitis in transforming growth factor-beta receptor II dominant negative mice. *Hepatology* 2009;**50**:1893–903.
75. Beck PL, Rosenberg IM, Xavier RJ, Koh T, Wong JF, Podolsky DK. Transforming growth factor-beta mediates intestinal healing and susceptibility to injury in vitro and in vivo through epithelial cells. *Am J Pathol* 2003;**162**:597–608.
76. Li MO, Wan YY, Flavell RA. T cell-produced transforming growth factor-beta1 controls T cell tolerance and regulates Th1- and Th17-cell differentiation. *Immunity* 2007;**26**:579–91.

77. Kajino-Sakamoto R, Inagaki M, Lippert E, Akira S, Robine S, Matsumoto K, et al. Enterocyte-derived TAK1 signaling prevents epithelium apoptosis and the development of ileitis and colitis. *J Immunol* 2008;**181**:1143–52.
78. Snapper SB, Rosen FS, Mizoguchi E, Cohen P, Khan W, Liu CH, et al. Wiskott-Aldrich syndrome protein-deficient mice reveal a role for WASP in T but not B cell activation. *Immunity* 1998;**9**:81–91.
79. Nguyen DD, Maillard MH, Cotta-de-Almeida V, Mizoguchi E, Klein C, Fuss I, et al. Lymphocyte-dependent and Th2 cytokine-associated colitis in mice deficient in Wiskott-Aldrich syndrome protein. *Gastroenterology* 2007;**133**:1188–97.
80. Maillard MH, Cotta-de-Almeida V, Takeshima F, Nguyen DD, Michetti P, Nagler C, et al. The Wiskott-Aldrich syndrome protein is required for the function of CD4(+)CD25(+)Foxp3(+) regulatory T cells. *J Exp Med* 2007;**204**:381–91.
81. Okkenhaug K, Bilancio A, Farjot G, Priddle H, Sancho S, Peskett E, et al. Impaired B and T cell antigen receptor signaling in p110delta PI 3-kinase mutant mice. *Science* 2002;**297**:1031–4.
82. Patton DT, Garden OA, Pearce WP, Clough LE, Monk CR, Leung E, et al. The phosphoinositide 3-kinase p110 delta is critical for the function of CD4+CD25+Foxp3 + regulatory T cells. *J Immunol* 2006;**177**:6598–602.
83. Uno JK, Rao KN, Matsuoka K, Sheikh SZ, Kobayashi T, Li F, et al. Altered macrophage function contributes to colitis in mice defective in the phosphoinositide-3 kinase subunit p110δ. *Gastroenterology* 2010;**139**:1642–53.
84. Park SG, Mathur R, Long M, Hosh N, Hao L, Hayden MS, et al. T regulatory cells maintain intestinal homeostasis by suppressing γδ T cells. *Immunity* 2010;**33**:791–803.
85. Bachmaier K, Krawczyk C, Kozieradzki I, Kong YY, Sasaki T, Oliveira-dos-Santos A, et al. Negative regulation of lymphocyte activation and autoimmunity by the molecular adaptor Cbl-b. *Nature* 2000;**403**:211–6.
86. Martins GA, Cimmino L, Shapiro-Shelef M, Szabolcs M, Herron A, Magnusdottir E, et al. Transcriptional repressor Blimp-1 regulates T cell homeostasis and function. *Nat Immunol* 2006;**7**:457–65.
87. Lee EG, Boone DL, Chai S, Libby SL, Chien M, Lodolce JP, et al. Failure to regulate TNF-induced NF-kappaB and cell death responses in A20-deficient mice. *Science* 2000;**289**:2350–4.
88. Vereecke L, Sze M, Mc Guire C, Rogiers B, Chu Y, Schmidt-Supprian M, et al. Enterocyte-specific A20 deficiency sensitizes to tumor necrosis factor-induced toxicity and experimental colitis. *J Exp Med* 2010;**207**:1513–23.
89. Kerr WG, Park MY, Maubert M, Engelman RW. SHIP deficiency causes Crohn's disease-like ileitis. *Gut* 2011;**60**:177–88.
90. Rudolph U, Finegold MJ, Rich SS, Harriman GR, Srinivasan Y, Brabet P, et al. Ulcerative colitis and adenocarcinoma of the colon in G alpha i2-deficient mice. *Nat Genet* 1995;**10**:143–50.
91. Dalwadi H, Wei B, Schrage M, Spicher K, Su TT, Birnbaumer L, et al. B cell developmental requirement for the G alpha i2 gene. *J Immunol* 2003;**170**:1707–15.
92. Peña JA, Thompson-Snipes L, Calkins PR, Tatevian N, Puppi M, Finegold MJ. Alterations in myeloid dendritic cell innate immune responses in the Galphai2-deficient mouse model of colitis. *Inflamm Bowel Dis* 2009;**15**:248–60.
93. Kontoyiannis D, Pasparakis M, Pizarro TT, Cominelli F, Kollias G. Impaired on/off regulation of TNF biosynthesis in mice lacking TNF AU-rich elements: implications for joint and gut-associated immunopathologies. *Immunity* 1999;**10**:387–98.
94. Kontoyiannis D, Boulougouris G, Manoloukos M, Armaka M, Apostolaki M, Pizarro T, et al. Genetic dissection of the cellular pathways and signaling mechanisms in modeled tumor

necrosis factor-induced Crohn's-like inflammatory bowel disease. *J Exp Med* 2002;**196**:1563–74.
95. Apostolaki M, Manoloukos M, Roulis M, Wurbel MA, Müller W, Papadakis KA, et al. Role of beta7 integrin and the chemokine/chemokine receptor pair CCL25/CCR9 in modeled TNF-dependent Crohn's disease. *Gastroenterology* 2008;**134**:2025–35.
96. Shaikh RB, Santee S, Granger SW, Butrovich K, Cheung T, Kronenberg M, et al. Constitutive expression of LIGHT on T cells leads to lymphocyte activation, inflammation, and tissue destruction. *J Immunol* 2001;**167**:6330–7.
97. Wang J, Anders RA, Wang Y, Turner JR, Abraham C, Pfeffer K, et al. The critical role of LIGHT in promoting intestinal inflammation and Crohn's disease. *J Immunol* 2005;**174**:8173–82.
98. Steinberg MW, Turovskaya O, Shaikh RB, Kim G, McCole DF, Pfeffer K, et al. A crucial role for HVEM and BTLA in preventing intestinal inflammation. *J Exp Med* 2008;**205**:1463–76.
99. Kamada N, Hisamatsu T, Honda H, Kobayashi T, Chinen H, Takayama T, et al. TL1A produced by lamina propria macrophages induces Th1 and Th17 immune responses in cooperation with IL-23 in patients with Crohn's disease. *Inflamm Bowel Dis* 2010;**16**:568–75.
100. Takedatsu H, Michelsen KS, Wei B, Landers CJ, Thomas LS, Dhall D, et al. TL1A (TNFSF15) regulates the development of chronic colitis by modulating both T-helper 1 and T-helper 17 activation. *Gastroenterology* 2008;**135**:552–67.
101. Meylan F, Song YJ, Fuss I, Villarreal S, Kahle E, Malm IJ, et al. The TNF-family cytokine TL1A drives IL-13-dependent small intestinal inflammation. *Mucosal Immunol* 2011;**4**:172–85.
102. Watanabe M, Ueno Y, Yajima T, Okamoto S, Hayashi T, Yamazaki M, et al. Interleukin 7 transgenic mice develop chronic colitis with decreased interleukin 7 protein accumulation in the colonic mucosa. *J Exp Med* 1998;**187**:389–402.
103. Yamazaki M, Yajima T, Tanabe M, Fukui K, Okada E, Okamoto R, et al. Mucosal T cells expressing high levels of IL-7 receptor are potential targets for treatment of chronic colitis. *J Immunol* 2003;**171**:1556–63.
104. Nemoto Y, Kanai T, Makita S, Okamoto R, Totsuka T, Takeda K, et al. Bone marrow retaining colitogenic CD4+ T cells may be a pathogenic reservoir for chronic colitis. *Gastroenterology* 2007;**132**:176–89.
105. Nemoto Y, Kanai T, Kameyama K, Shinohara T, Sakamoto N, Totsuka T, et al. Long-lived colitogenic CD4+ memory T cells residing outside the intestine participate in the perpetuation of chronic colitis. *J Immunol* 2009;**183**:5059–68.
106. Tomita T, Kanai T, Nemoto Y, Totsuka T, Okamoto R, Tsuchiya K, et al. Systemic, but not intestinal, IL-7 is essential for the persistence of chronic colitis. *J Immunol* 2008;**180**:383–90.
107. Ohta N, Hiroi T, Kweon MN, Kinoshita N, Jang MH, Mashimo T, et al. IL-15-dependent activation-induced cell death-resistant Th1 type CD8 alpha beta+NK1.1+ T cells for the development of small intestinal inflammation. *J Immunol* 2002;**169**:460–8.
108. Clegg CH, Rulffes JT, Haugen HS, Hoggatt IH, Aruffo A, Durham SK, et al. Thymus dysfunction and chronic inflammatory disease in gp39 transgenic mice. *Int Immunol* 1997;**9**:1111–22.
109. Kawamura T, Kanai T, Dohi T, Uraushihara K, Totsuka T, Iiyama R, et al. Ectopic CD40 ligand expression on B cells triggers intestinal inflammation. *J Immunol* 2004;**172**:6388–97.
110. Higuchi T, Aiba Y, Nomura T, Matsuda J, Mochida K, Suzuki M, et al. Ectopic expression of CD40 ligand on B cells induces lupus-like autoimmune disease. *J Immunol* 2002;**168**:9–12.
111. Mehling A, Loser K, Varga G, Metze D, Luger TA, Schwarz T, et al. Overexpression of CD40 ligand in murine epidermis results in chronic skin inflammation and systemic autoimmunity. *J Exp Med* 2001;**194**:615–28.

112. Kim G, Turovskaya O, Levin M, Byrne FR, Whoriskey JS, McCabe JG, et al. Spontaneous colitis occurrence in transgenic mice with altered B7-mediated costimulation. *J Immunol* 2008;**181**:5278–88.
113. Lacy-Hulbert A, Smith AM, Tissire H, Barry M, Crowley D, Bronson RT, et al. Ulcerative colitis and autoimmunity induced by loss of myeloid alphav integrins. *Proc Natl Acad Sci USA* 2007;**104**:15823–8.
114. Travis MA, Reizis B, Melton AC, Masteller E, Tang Q, Proctor JM, et al. Loss of integrin alpha (v)beta8 on dendritic cells causes autoimmunity and colitis in mice. *Nature* 2007;**449**:361–5.
115. Wirtz S, Finotto S, Kanzler S, Lohse AW, Blessing M, Lehr HA, et al. Chronic intestinal inflammation in STAT-4 transgenic mice: characterization of disease and adoptive transfer by TNF- plus IFN-gamma-producing CD4+ T cells that respond to bacterial antigens. *J Immunol* 1999;**162**:1884–8.
116. Durant L, Watford WT, Ramos HL, Laurence A, Vahedi G, Wei L, et al. Diverse targets of the transcription factor STAT3 contribute to T cell pathogenicity and homeostasis. *Immunity* 2010;**32**:605–15.
117. Takeda K, Clausen BE, Kaisho T, Tsujimura T, Terada N, Förster I, et al. Enhanced Th1 activity and development of chronic enterocolitis in mice devoid of Stat3 in macrophages and neutrophils. *Immunity* 1999;**10**:39–49.
118. Reindl W, Weiss S, Lehr HA, Förster I. Essential crosstalk between myeloid and lymphoid cells for development of chronic colitis in myeloid-specific signal transducer and activator of transcription 3-deficient mice. *Immunology* 2007;**120**:19–27.
119. Pickert G, Neufert C, Leppkes M, Zheng Y, Wittkopf N, Warntjen M, et al. STAT3 links IL-22 signaling in intestinal epithelial cells to mucosal wound healing. *J Exp Med* 2009;**206**:1465–72.
120. Alonzi T, Newton IP, Bryce PJ, Di Carlo E, Lattanzio G, Tripodi M, et al. Induced somatic inactivation of STAT3 in mice triggers the development of a fulminant form of enterocolitis. *Cytokine* 2004;**26**:45–56.
121. Ito H, Takazoe M, Fukuda Y, Hibi T, Kusugami K, Andoh A, et al. A pilot randomized trial of a human anti-interleukin-6 receptor monoclonal antibody in active Crohn's disease. *Gastroenterology* 2004;**126**:989–96.
122. Zenewicz LA, Yancopoulos GD, Valenzuela DM, Murphy AJ, Stevens S, Flavell RA. Innate and adaptive interleukin-22 protects mice from inflammatory bowel disease. *Immunity* 2008;**29**:947–57.
123. Inagaki-Ohara K, Sasaki A, Matsuzaki G, Ikeda T, Hotokezaka M, Chijiiwa K, et al. Suppressor of cytokine signalling 1 in lymphocytes regulates the development of intestinal inflammation in mice. *Gut* 2006;**55**:212–9.
124. Ernst M, Inglese M, Waring P, Campbell IK, Bao S, Clay FJ, et al. Defective gp130-mediated signal transducer and activator of transcription (STAT) signaling results in degenerative joint disease, gastrointestinal ulceration, and failure of uterine implantation. *J Exp Med* 2001;**194**:189–203.
125. Neurath MF, Pettersson S, Meyer zum Büschenfelde KH, Strober W. Local administration of antisense phosphorothioate oligonucleotides to the p65 subunit of NF-kappa B abrogates established experimental colitis in mice. *Nat Med* 1996;**2**:998–1004.
126. Nenci A, Becker C, Wullaert A, Gareus R, van Loo G, Danese S, et al. Epithelial NEMO links innate immunity to chronic intestinal inflammation. *Nature* 2007;**446**:557–61.
127. Tomczak MF, Erdman SE, Poutahidis T, Rogers AB, Holcombe H, Plank B, et al. NF-kappa B is required within the innate immune system to inhibit microflora-induced colitis and expression of IL-12 p40. *J Immunol* 2003;**171**:1484–92.
128. Steinbrecher KA, Harmel-Laws E, Sitcheran R, Baldwin AS. Loss of epithelial RelA results in deregulated intestinal proliferative/apoptotic homeostasis and susceptibility to inflammation. *J Immunol* 2008;**180**:2588–99.

129. Brenner O, Levanon D, Negreanu V, Golubkov O, Fainaru O, Woolf E, et al. Loss of Runx3 function in leukocytes is associated with spontaneously developed colitis and gastric mucosal hyperplasia. *Proc Natl Acad Sci USA* 2004;**101**:16016–21.
130. Sugai M, Aoki K, Osato M, Nambu Y, Ito K, Taketo MM, et al. Runx3 is required for full activation of regulatory T cells to prevent colitis-associated tumor formation. *J Immunol* 2011;**186**:6515–20.
131. Vijay-Kumar M, Sanders CJ, Taylor RT, Kumar A, Aitken JD, Sitaraman SV, et al. Deletion of TLR5 results in spontaneous colitis in mice. *J Clin Invest* 2007;**117**:3909–21.
132. Letran SE, Lee SJ, Atif SM, Flores-Langarica A, Uematsu S, Akira S, et al. TLR5-deficient mice lack basal inflammatory and metabolic defects but exhibit impaired CD4 T cell responses to a flagellated pathogen. *J Immunol* 2011;**186**:5406–12.
133. Lodes MJ, Cong Y, Elson CO, Mohamath R, Landers CJ, Targan SR, et al. Bacterial flagellin is a dominant antigen in Crohn disease. *J Clin Invest* 2004;**113**:1296–306.
134. Feng T, Wang L, Schoeb TR, Elson CO, Cong Y. Microbiota innate stimulation is a prerequisite for T cell spontaneous proliferation and induction of experimental colitis. *J Exp Med* 2010;**207**:1321–32.
135. Bush TG, Savidge TC, Freeman TC, Cox HJ, Campbell EA, Mucke L, et al. Fulminant jejuno-ileitis following ablation of enteric glia in adult transgenic mice. *Cell* 1998;**93**:189–201.
136. Savidge TC, Newman P, Pothoulakis C, Ruhl A, Neunlist M, Bourreille A, et al. Enteric glia regulate intestinal barrier function and inflammation via release of S-nitrosoglutathione. *Gastroenterology* 2007;**132**:1344–58.
137. Kaser A, Lee AH, Franke A, Glickman JN, Zeissig S, Tilg H, et al. XBP1 links ER stress to intestinal inflammation and confers genetic risk for human inflammatory bowel disease. *Cell* 2008;**134**:743–56.
138. Stappenbeck TS, Rioux JD, Mizoguchi A, Saitoh T, Huett A, Darfeuille-Michaud A, et al. Crohn disease: a current perspective on genetics, autophagy and immunity. *Autophagy* 2010;**7**:335–74.
139. Cadwell K, Liu JY, Brown SL, Miyoshi H, Loh J, Lennerz JK, et al. A key role for autophagy and the autophagy gene Atg16l1 in mouse and human intestinal Paneth cells. *Nature* 2008;**456**:259–63.
140. Saitoh T, Fujita N, Jang MH, Uematsu S, Yang BG, Satoh T, et al. Loss of the autophagy protein Atg16L1 enhances endotoxin-induced IL-1beta production. *Nature* 2008;**456**:264–8.
141. Nedjic J, Aichinger M, Emmerich J, Mizushima N, Klein L. Autophagy in thymic epithelium shapes the T-cell repertoire and is essential for tolerance. *Nature* 2008;**455**:396–400.
142. Baribault H, Penner J, Iozzo RV, Wilson-Heiner M. Colorectal hyperplasia and inflammation in keratin 8-deficient FVB/N mice. *Genes Dev* 1994;**8**:2964–73.
143. Habtezion A, Toivola DM, Asghar MN, Kronmal GS, Brooks JD, Butcher EC, et al. Absence of keratin 8 confers a paradoxical microflora-dependent resistance to apoptosis in the colon. *Proc Natl Acad Sci USA* 2011;**108**:1445–50.
144. Hermiston ML, Gordon JI. Inflammatory bowel disease and adenomas in mice expressing a dominant negative N-cadherin. *Science* 1995;**270**:1203–7.
145. Panwala CM, Jones JC, Viney JL. A novel model of inflammatory bowel disease: mice deficient for the multiple drug resistance gene, mdr1a, spontaneously develop colitis. *J Immunol* 1998;**161**:5733–44.
146. Fort MM, Mozaffarian A, Stöver AG, Correia Jda S, Johnson DA, Crane RT, et al. A synthetic TLR4 antagonist has anti-inflammatory effects in two murine models of inflammatory bowel disease. *J Immunol* 2005;**174**:6416–23.
147. Arnett HA, Escobar SS, Gonzalez-Suarez E, Budelsky AL, Steffen LA, Boiani N, et al. BTNL2, a butyrophilin/B7-like molecule, is a negative costimulatory molecule modulated in intestinal inflammation. *J Immunol* 2007;**178**:1523–33.

148. Esworthy RS, Aranda R, Martín MG, Doroshow JH, Binder SW, Chu FF. Mice with combined disruption of Gpx1 and Gpx2 genes have colitis. *Am J Physiol Gastrointest Liver Physiol* 2001;**281**:G848–55.
149. Esworthy RS, Kim BW, Larson GP, Yip ML, Smith DD, Li M, et al. Colitis locus on chromosome 2 impacting the severity of early-onset disease in mice deficient in GPX1 and GPX2. *Inflamm Bowel Dis* 2010;**17**:1373–86.
150. Velcich A, Yang W, Heyer J, Fragale A, Nicholas C, Viani S, et al. Colorectal cancer in mice genetically deficient in the mucin Muc2. *Science* 2002;**295**:1726–9.
151. Van der Sluis M, De Koning BA, De Bruijn AC, Velcich A, Meijerink JP, Van Goudoever JB, et al. Muc2-deficient mice spontaneously develop colitis, indicating that MUC2 is critical for colonic protection. *Gastroenterology* 2006;**131**:117–29.
152. van der Sluis M, Bouma J, Vincent A, Velcich A, Carraway KL, Büller HA, et al. Combined defects in epithelial and immunoregulatory factors exacerbate the pathogenesis of inflammation: mucin 2-interleukin 10-deficient mice. *Lab Invest* 2008;**88**:634–42.
153. Fu J, Wei B, Wen T, Johansson ME, Liu X, Bradford E, et al. Loss of intestinal core 1-derived O-glycans causes spontaneous colitis in mice. *J Clin Invest* 2011;**121**:1657–66.
154. An G, Wei B, Xia B, McDaniel JM, Ju T, Cummings RD, et al. Increased susceptibility to colitis and colorectal tumors in mice lacking core 3-derived O-glycans. *J Exp Med* 2007;**204**:1417–29.
155. Stone EL, Ismail MN, Lee SH, Luu Y, Ramirez K, Haslam SM, et al. Glycosyltransferase function in core 2-type protein O glycosylation. *Mol Cell Biol* 2009;**29**:3770–82.
156. Gerth AJ, Lin L, Neurath MF, Glimcher LH, Peng SL. An innate cell-mediated, murine ulcerative colitis-like syndrome in the absence of nuclear factor of activated T cells. *Gastroenterology* 2004;**126**:1115–21.
157. Neurath MF, Weigmann B, Finotto S, Glickman J, Nieuwenhuis E, Iijima H, et al. The transcription factor T-bet regulates mucosal T cell activation in experimental colitis and Crohn's disease. *J Exp Med* 2002;**195**:1129–43.
158. Garrett WS, Gallini CA, Yatsunenko T, Michaud M, DuBois A, Delaney ML, et al. Enterobacteriaceae act in concert with the gut microbiota to induce spontaneous and maternally transmitted colitis. *Cell Host Microbe* 2010;**8**:292–300.
159. Veiga P, Gallini CA, Beal C, Michaud M, Delaney ML, DuBois A, et al. *Bifidobacterium animalis* subsp. lactis fermented milk product reduces inflammation by altering a niche for colitogenic microbes. *Proc Natl Acad Sci USA* 2010;**107**:18132–7.
160. Uhlig HH, McKenzie BS, Hue S, Thompson C, Joyce-Shaikh B, Stepankova R, et al. Differential activity of IL-12 and IL-23 in mucosal and systemic innate immune pathology. *Immunity* 2006;**25**:309–18.
161. Vonarbourg C, Mortha A, Bui VL, Hernandez PP, Kiss EA, Hoyler T, et al. Regulated expression of nuclear receptor RORγt confers distinct functional fates to NK cell receptor-expressing RORγt(+) innate lymphocytes. *Immunity* 2010;**33**:736–51.
162. Sundberg JP, Elson CO, Bedigian H, Birkenmeier EH. Spontaneous, heritable colitis in a new substrain of C3H/HeJ mice. *Gastroenterology* 1994;**107**:1726–35.
163. Beckwith J, Cong Y, Sundberg JP, Elson CO, Leiter EH. Cdcs1, a major colitogenic locus in mice, regulates innate and adaptive immune response to enteric bacterial antigens. *Gastroenterology* 2005;**129**:1473–84.
164. Cong Y, Brandwein SL, McCabe RP, Lazenby A, Birkenmeier EH, Sundberg JP, et al. CD4+ T cells reactive to enteric bacterial antigens in spontaneously colitic C3H/HeJBir mice: increased T helper cell type 1 response and ability to transfer disease. *J Exp Med* 1998;**187**:855–64.

165. Matsumoto S, Okabe Y, Setoyama H, Takayama K, Ohtsuka J, Funahashi H, et al. Inflammatory bowel disease-like enteritis and caecitis in a senescence accelerated mouse P1/Yit strain. *Gut* 1998;**43**:71–8.
166. Kosiewicz MM, Nast CC, Krishnan A, Rivera-Nieves J, Moskaluk CA, Matsumoto S, et al. Th1-type responses mediate spontaneous ileitis in a novel murine model of Crohn's disease. *J Clin Invest* 2001;**107**:695–702.
167. Bamias G, Okazawa A, Rivera-Nieves J, Arseneau KO, De La Rue SA, Pizarro TT, et al. Commensal bacteria exacerbate intestinal inflammation but are not essential for the development of murine ileitis. *J Immunol* 2007;**178**:1809–18.
168. Olson TS, Reuter BK, Scott KG, Morris MA, Wang XM, Hancock LN, et al. The primary defect in experimental ileitis originates from a nonhematopoietic source. *J Exp Med* 2006;**203**:541–52.
169. Takedatsu H, Mitsuyama K, Matsumoto S, Handa K, Suzuki A, Takedatsu H, et al. Interleukin-5 participates in the pathogenesis of ileitis in SAMP1/Yit mice. *Eur J Immunol* 2004;**34**:1561–9.
170. Bamias G, Martin C, Mishina M, Ross WG, Rivera-Nieves J, Marini M, et al. Proinflammatory effects of TH2 cytokines in a murine model of chronic small intestinal inflammation. *Gastroenterology* 2005;**128**:654–66.
171. Olson TS, Bamias G, Naganuma M, Rivera-Nieves J, Burcin TL, Ross W, et al. Expanded B cell population blocks regulatory T cells and exacerbates ileitis in a murine model of Crohn disease. *J Clin Invest* 2004;**114**:389–98.
172. Corazza N, Eichenberger S, Eugster HP, Mueller C. Nonlymphocyte-derived tumor necrosis factor is required for induction of colitis in recombination activating gene (RAG)2(−/−) mice upon transfer of CD4(+)CD45RB(hi) T cells. *J Exp Med* 1999;**190**:1479–92.
173. Atreya R, Mudter J, Finotto S, Müllberg J, Jostock T, Wirtz S, et al. Blockade of interleukin 6 trans signaling suppresses T-cell resistance against apoptosis in chronic intestinal inflammation: evidence in crohn disease and experimental colitis in vivo. *Nat Med* 2000;**6**:583–8.
174. Fort MM, Leach MW, Rennick DM. A role for NK cells as regulators of CD4+ T cells in a transfer model of colitis. *J Immunol* 1998;**161**:3256–61.
175. Dardalhon V, Awasthi A, Kwon H, Galileos G, Gao W, Sobel RA, et al. IL-4 inhibits TGF-beta-induced Foxp3+ T cells and, together with TGF-beta, generates IL-9+ IL-10+ Foxp3(−) effector T cells. *Nat Immunol* 2008;**9**:1347–55.
176. Takebayashi K, Koboziev I, Ostanin DV, Gray L, Karlsson F, Robinson-Jackson SA, et al. Role of the gut-associated and secondary lymphoid tissue in the induction of chronic colitis. *Inflamm Bowel Dis* 2011;**17**:268–78.
177. Powrie F, Carlino J, Leach MW, Mauze S, Coffman RL. A critical role for transforming growth factor-beta but not interleukin 4 in the suppression of T helper type 1-mediated colitis by CD45RB(low) CD4+ T cells. *J Exp Med* 1996;**183**:2669–74.
178. Groux H, O'Garra A, Bigler M, Rouleau M, Antonenko S, de Vries JE, et al. A CD4+ T-cell subset inhibits antigen-specific T-cell responses and prevents colitis. *Nature* 1997;**389**:737–42.
179. Murai M, Turovskaya O, Kim G, Madan R, Karp CL, Cheroutre H, et al. Interleukin 10 acts on regulatory T cells to maintain expression of the transcription factor Foxp3 and suppressive function in mice with colitis. *Nat Immunol* 2009;**10**:1178–84.
180. Chaudhry A, Samstein RM, Treuting P, Liang Y, Pils MC, Heinrich JM, et al. Interleukin-10 signaling in regulatory T cells is required for suppression of th17 cell-mediated inflammation. *Immunity* 2011;**34**:566–78.
181. Huber S, Gagliani N, Esplugues E, O'Connor Jr. W, Huber FJ, Chaudhry A, et al. Th17 cells express interleukin-10 receptor and are controlled by Foxp3(−) and Foxp3(+) regulatory CD4(+) T cells in an interleukin-10-dependent manner. *Immunity* 2011;**34**:554–65.

182. Fantini MC, Rizzo A, Fina D, Caruso R, Sarra M, Stolfi C, et al. Smad7 controls resistance of colitogenic T cells to regulatory T cell-mediated suppression. *Gastroenterology* 2009;**136**:1308–16.
183. Hovhannisyan Z, Treatman J, Littman DR, Mayer L. Characterization of interleukin-17-producing regulatory T cells in inflamed intestinal mucosa from patients with inflammatory bowel diseases. *Gastroenterology* 2011;**140**:957–65.
184. Feng T, Cao AT, Weaver CT, Elson CO, Cong Y. IL-12 converts Foxp3 + regulatory T cells to Foxp3 + IFN-γ + T cells with inhibitory functions during induction of colitis. *Gastroenterology* 2011;**140**:2031–43.
185. Kullberg MC, Jankovic D, Feng CG, Hue S, Gorelick PL, McKenzie BS, et al. IL-23 plays a key role in Helicobacter hepaticus-induced T cell-dependent colitis. *J Exp Med* 2006;**203**:2485–94.
186. Ahern PP, Schiering C, Buonocore S, McGeachy MJ, Cua DJ, Maloy KJ, et al. Interleukin-23 drives intestinal inflammation through direct activity on T cells. *Immunity* 2010;**33**:279–88.
187. Siddiqui KR, Laffont S, Powrie F. E-cadherin marks a subset of inflammatory dendritic cells that promote T cell-mediated colitis. *Immunity* 2010;**32**:557–67.
188. Atarashi K, Nishimura J, Shima T, Umesaki Y, Yamamoto M, Onoue M, et al. ATP drives lamina propria T(H)17 cell differentiation. *Nature* 2008;**455**:808–12.
189. Iwase T, Shinji H, Tajima A, Sato F, Tamura T, Iwamoto T, et al. Isolation and identification of ATP-secreting bacteria from mice and humans. *J Clin Microbiol* 2010;**48**:1949–51.
190. Ivanov II, Atarashi K, Manel N, Brodie EL, Shima T, Karaoz U, et al. Induction of intestinal Th17 cells by segmented filamentous bacteria. *Cell* 2009;**139**:485–98.
191. Do JS, Visperas A, Dong C, Baldwin 3rd WM, Min B. Generation of colitogenic Th17 CD4 T cells is enhanced by IL-17 + {gamma}{delta} T cells. *J Immunol* 2011;**186**:4546–50.
192. Buonocore S, Ahern PP, Uhlig HH, Ivanov II, Littman DR, Maloy KJ, et al. Innate lymphoid cells drive interleukin-23-dependent innate intestinal pathology. *Nature* 2010;**464**:1371–5.
193. Izcue A, Hue S, Buonocore S, Arancibia-Cárcamo CV, Ahern PP, Iwakura Y, et al. Interleukin-23 restrains regulatory T cell activity to drive T cell-dependent colitis. *Immunity* 2008;**28**:559–70.
194. Noguchi D, Wakita D, Tajima M, Ashino S, Iwakura Y, Zhang Y, et al. Blocking of IL-6 signaling pathway prevents CD4 + T cell-mediated colitis in a T(h)17-independent manner. *Int Immunol* 2007;**19**:1431–40.
195. Leppkes M, Becker C, Ivanov II, Hirth S, Wirtz S, Neufert C, et al. RORgamma-expressing Th17 cells induce murine chronic intestinal inflammation via redundant effects of IL-17A and IL-17F. *Gastroenterology* 2009;**136**:257–67.
196. O'Connor Jr. W, Kamanaka M, Booth CJ, Town T, Nakae S, Iwakura Y, et al. A protective function for interleukin 17A in T cell-mediated intestinal inflammation. *Nat Immunol* 2009;**10**:603–9.
197. Kamanaka M, Huber S, Zenewicz LA, Gagliani N, Rathinam C, O'Connor Jr. W, et al. Memory/effector (CD45RBlo) CD4 T cells are controlled directly by IL-10 and cause IL-22-dependent intestinal pathology. *J Exp Med* 2011;**208**:1027–40.
198. Annunziato F, Cosmi L, Santarlasci V, Maggi L, Liotta F, Mazzinghi B, et al. Phenotypic and functional features of human Th17 cells. *J Exp Med* 2007;**204**:1849–61.
199. Hirota K, Duarte JH, Veldhoen M, Hornsby E, Li Y, Cua DJ, et al. Fate mapping of IL-17-producing T cells in inflammatory responses. *Nat Immunol* 2011;**12**:255–63.
200. Plevy SE, Targan SR. Future therapeutic approaches for inflammatory bowel diseases. *Gastroenterology* 2011;**140**:1838–46.
201. Xu J, Yang Y, Qiu G, Lal G, Yin N, Wu Z, et al. Stat4 is critical for the balance between Th17 cells and regulatory T cells in colitis. *J Immunol* 2011;**186**:6597–606.

202. Holländer GA, Simpson SJ, Mizoguchi E, Nichogiannopoulou A, She J, Gutierrez-Ramos JC, et al. Severe colitis in mice with aberrant thymic selection. *Immunity* 1995;**3**:27–38.
203. Simpson SJ, Holländer GA, Mizoguchi E, Allen D, Bhan AK, Wang B, et al. Expression of pro-inflammatory cytokines by TCR alpha beta+ and TCR gamma delta+ T cells in an experimental model of colitis. *Eur J Immunol* 1997;**27**:17–25.
204. Veltkamp C, Tonkonogy SL, De Jong YP, Albright C, Grenther WB, Balish E, et al. Continuous stimulation by normal luminal bacteria is essential for the development and perpetuation of colitis in Tg(epsilon26) mice. *Gastroenterology* 2001;**120**:900–13.
205. Simpson SJ, Shah S, Comiskey M, de Jong YP, Wang B, Mizoguchi E, et al. T cell-mediated pathology in two models of experimental colitis depends predominantly on the interleukin 12/Signal transducer and activator of transcription (Stat)-4 pathway, but is not conditional on interferon gamma expression by T cells. *J Exp Med* 1998;**187**:1225–34.
206. Abadía-Molina AC, Ji H, Faubion WA, Julien A, Latchman Y, Yagita H, et al. CD48 controls T-cell and antigen-presenting cell functions in experimental colitis. *Gastroenterology* 2006;**130**:424–34.
207. de Jong YP, Abadia-Molina AC, Satoskar AR, Clarke K, Rietdijk ST, Faubion WA, et al. Development of chronic colitis is dependent on the cytokine MIF. *Nat Immunol* 2001;**2**:1061–6.
208. Steinhoff U, Brinkmann V, Klemm U, Aichele P, Seiler P, Brandt U, et al. Autoimmune intestinal pathology induced by hsp60-specific CD8 T cells. *Immunity* 1999;**11**:349–58.
209. Vezys V, Olson S, Lefrançois L. Expression of intestine-specific antigen reveals novel pathways of CD8 T cell tolerance induction. *Immunity* 2000;**12**:505–14.
210. Tajima M, Wakita D, Noguchi D, Chamoto K, Yue Z, Fugo K, et al. IL-6-dependent spontaneous proliferation is required for the induction of colitogenic IL-17-producing CD8+ T cells. *J Exp Med* 2008;**205**:1019–27.
211. Iqbal N, Oliver JR, Wagner FH, Lazenby AS, Elson CO, Weaver CT. T helper 1 and T helper 2 cells are pathogenic in an antigen-specific model of colitis. *J Exp Med* 2002;**195**:71–84.
212. Yoshida M, Shirai Y, Watanabe T, Yamori M, Iwakura Y, Chiba T, et al. Differential localization of colitogenic Th1 and Th2 cells monospecific to a microflora-associated antigen in mice. *Gastroenterology* 2002;**123**:1949–61.
213. Watanabe T, Kitani A, Murray PJ, Wakatsuki Y, Fuss IJ, Strober W. Nucleotide binding oligomerization domain 2 deficiency leads to dysregulated TLR2 signaling and induction of antigen-specific colitis. *Immunity* 2006;**25**:473–85.
214. Morris GP, Beck PL, Herridge MS, Depew WT, Szewczuk MR, Wallace JL. Hapten-induced model of chronic inflammation and ulceration in the rat colon. *Gastroenterology* 1989;**96**:795–803.
215. Elson CO, Beagley KW, Sharmanov AT, Fujihashi K, Kiyono H, Tennyson GS, et al. Hapten-induced model of murine inflammatory bowel disease: mucosa immune responses and protection by tolerance. *J Immunol* 1996;**157**:2174–85.
216. Yang Z, Fuss IJ, Watanabe T, Asano N, Davey MP, Rosenbaum JT, et al. NOD2 transgenic mice exhibit enhanced MDP-mediated down-regulation of TLR2 responses and resistance to colitis induction. *Gastroenterology* 2007;**133**:1510–21.
217. Watanabe T, Asano N, Murray PJ, Ozato K, Tailor P, Fuss IJ, et al. Muramyl dipeptide activation of nucleotide-binding oligomerization domain 2 protects mice from experimental colitis. *J Clin Invest* 2008;**118**:545–59.
218. Bailón E, Cueto-Sola M, Utrilla P, Nieto A, Garrido-Mesa N, Celada A, et al. DNFB-DNS hapten-induced colitis in mice should not be considered a model of inflammatory bowel disease. *Inflamm Bowel Dis* 2011;**17**:2087–101.
219. Neurath MF, Fuss I, Kelsall BL, Stüber E, Strober W. Antibodies to interleukin 12 abrogate established experimental colitis in mice. *J Exp Med* 1995;**182**:1281–90.

220. Mannon PJ, Fuss IJ, Mayer L, Elson CO, Sandborn WJ, Present D, et al. Anti-IL-12 Crohn's Disease Study Group. Anti-interleukin-12 antibody for active Crohn's disease. *N Engl J Med* 2004;**351**:2069–79.
221. Becker C, Dornhoff H, Neufert C, Fantini MC, Wirtz S, Huebner S, et al. IL-23 cross-regulates IL-12 production in T cell-dependent experimental colitis. *J Immunol* 2006;**177**:2760–4.
222. Camoglio L, te Velde AA, de Boer A, ten Kate FJ, Kopf M, van Deventer SJ. Hapten-induced colitis associated with maintained Th1 and inflammatory responses in IFN-gamma receptor-deficient mice. *Eur J Immunol* 2000;**30**:1486–95.
223. Zhang Z, Zheng M, Bindas J, Schwarzenberger P, Kolls JK. Critical role of IL-17 receptor signaling in acute TNBS-induced colitis. *Inflamm Bowel Dis* 2006;**12**:382–8.
224. Boirivant M, Fuss IJ, Ferroni L, De Pascale M, Strober W. Oral administration of recombinant cholera toxin subunit B inhibits IL-12-mediated murine experimental (trinitrobenzene sulfonic acid) colitis. *J Immunol* 2001;**166**:3522–32.
225. Hoffmann JC, Peters K, Henschke S, Herrmann B, Pfister K, Westermann J, et al. Role of T lymphocytes in rat 2,4,6-trinitrobenzene sulphonic acid (TNBS) induced colitis: increased mortality after gammadelta T cell depletion and no effect of alphabeta T cell depletion. *Gut* 2001;**48**:489–95.
226. Fiorucci S, Mencarelli A, Palazzetti B, Sprague AG, Distrutti E, Morelli A, et al. Importance of innate immunity and collagen binding integrin alpha1beta1 in TNBS-induced colitis. *Immunity* 2002;**17**:769–80.
227. Camoglio L, Juffermans NP, Peppelenbosch M, te Velde AA, ten Kate FJ, van Deventer SJ, et al. Contrasting roles of IL-12p40 and IL-12p35 in the development of hapten-induced colitis. *Eur J Immunol* 2002;**32**:261–9.
228. Tozawa K, Hanai H, Sugimoto K, Baba S, Sugimura H, Aoshi T, et al. Evidence for the critical role of interleukin-12 but not interferon-gamma in the pathogenesis of experimental colitis in mice. *J Gastroenterol Hepatol* 2003;**18**:578–87.
229. Dohi T, Fujihashi K, Rennert PD, Iwatani K, Kiyono H, McGhee JR. Hapten-induced colitis is associated with colonic patch hypertrophy and T helper cell 2-type responses. *J Exp Med* 1999;**189**:1169–80.
230. Boirivant M, Fuss IJ, Chu A, Strober W. Oxazolone colitis: a murine model of T helper cell type 2 colitis treatable with antibodies to interleukin 4. *J Exp Med* 1998;**188**:1929–39.
231. Heller F, Fuss IJ, Nieuwenhuis EE, Blumberg RS, Strober W. Oxazolone colitis, a Th2 colitis model resembling ulcerative colitis, is mediated by IL-13-producing NK-T cells. *Immunity* 2002;**17**:629–38.
232. Fuss IJ, Heller F, Boirivant M, Leon F, Yoshida M, Fichtner-Feigl S, et al. Nonclassical CD1d-restricted NK T cells that produce IL-13 characterize an atypical Th2 response in ulcerative colitis. *J Clin Invest* 2004;**113**:1490–7.
233. Iijima H, Neurath MF, Nagaishi T, Glickman JN, Nieuwenhuis EE, Nakajima A, et al. Specific regulation of T helper cell 1-mediated murine colitis by CEACAM1. *J Exp Med* 2004;**199**:471–82.
234. Nieuwenhuis EE, Neurath MF, Corazza N, Iijima H, Trgovcich J, Wirtz S, et al. Disruption of T helper 2-immune responses in Epstein-Barr virus-induced gene 3-deficient mice. *Proc Natl Acad Sci USA* 2002;**99**:16951–6.
235. Weigmann B, Lehr HA, Yancopoulos G, Valenzuela D, Murphy A, Stevens S, et al. The transcription factor NFATc2 controls IL-6-dependent T cell activation in experimental colitis. *J Exp Med* 2008;**205**:2099–110.
236. Noti M, Corazza N, Mueller C, Berger B, Brunner T. TNF suppresses acute intestinal inflammation by inducing local glucocorticoid synthesis. *J Exp Med* 2010;**207**:1057–66.

237. Schwartz L, Abolhassani M, Pooya M, Steyaert JM, Wertz X, Israël M, et al. Hyperosmotic stress contributes to mouse colonic inflammation through the methylation of protein phosphatase 2A. *Am J Physiol Gastrointest Liver Physiol* 2008;**295**:G934–41.
238. Dieleman LA, Ridwan BU, Tennyson GS, Beagley KW, Bucy RP, Elson CO. Dextran sulfate sodium-induced colitis occurs in severe combined immunodeficient mice. *Gastroenterology* 1994;**107**:1643–52.
239. Shimomura Y, Ogawa A, Kawada M, Sugimoto K, Mizoguchi E, Shi HN, et al. A unique B2 B cell subset in the intestine. *J Exp Med* 2008;**205**:1343–55.
240. Kitajima S, Morimoto M, Sagara E, Shimizu C, Ikeda Y. Dextran sodium sulfate-induced colitis in germ-free IQI/Jic mice. *Exp Anim* 2001;**50**:387–95.
241. Rakoff-Nahoum S, Paglino J, Eslami-Varzaneh F, Edberg S, Medzhitov R. Recognition of commensal microflora by toll-like receptors is required for intestinal homeostasis. *Cell* 2004;**118**:229–41.
242. Araki A, Kanai T, Ishikura T, Makita S, Uraushihara K, Iiyama R, et al. MyD88-deficient mice develop severe intestinal inflammation in dextran sodium sulfate colitis. *J Gastroenterol* 2005;**40**:16–23.
243. Cario E, Gerken G, Podolsky DK. Toll-like receptor 2 controls mucosal inflammation by regulating epithelial barrier function. *Gastroenterology* 2007;**132**:1359–74.
244. Katakura K, Lee J, Rachmilewitz D, Li G, Eckmann L, Raz E. Toll-like receptor 9-induced type I IFN protects mice from experimental colitis. *J Clin Invest* 2005;**115**:695–702.
245. Araki Y, Mukaisyo K, Sugihara H, Hattori T. Decomposition of dextran sulfate sodium under alkaline, acidic, high temperature and high pressure conditions. *Oncol Rep* 2008;**20**:147–9.
246. Mizoguchi A, Mizoguchi E. Inflammatory bowel disease, past, present and future: lessons from animal models. *J Gastroenterol* 2008;**43**:1–17.

Animal Models of Molecular Pathology: Systemic Lupus Erythematosus

ALLISON SANG, YIMING YIN,
YING-YI ZHENG, AND
LAURENCE MOREL

Department of Pathology, Immunology, and Laboratory Medicine, University of Florida, Gainesville, Florida, USA

Systemic lupus erythematosus (SLE) is an autoimmune disease that affects multiple organ systems. A hallmark of SLE is the production of antinuclear antibodies against nuclear antigens such as chromatin and DNA. High levels of autoAbs promote the formation of immune complexes which can lead to the development of glomerulonephritis and progress to end-stage renal failure.[1] Although the exact etiology of SLE is unknown, it is thought to be multifactorial in nature. A combination of environmental, hormonal, and a predisposed genetic background lead to the development of this disorder.[2] Here, we review the various mouse models that have been used to study SLE and discuss how

Progress in Molecular Biology
and Translational Science, Vol. 105
DOI: 10.1016/B978-0-12-394596-9.00010-X

1877-1173/12 $35.00

their study has led to a better understanding of the genetic and cellular factors involved in the development of systemic autoimmunity and lupus-like clinical symptoms. We also review the mouse studies that have explored the molecular pathways that are altered in this disease and the investigation of their therapeutic potentials.

I. Murine Models of Lupus

Mouse models have greatly contributed to our understanding of the genetics and etiopathology of systemic lupus erythematosus (SLE). They are also an ideal platform for the test of novel drugs and therapies. In general, mouse models of SLE can be divided to three groups: spontaneous models, induced models, and genetically modified models. The most widely used spontaneous models are the F1 hybrid between the New Zealand Black (NZB) and New Zealand White (NZW) strains (NZB/W F1) and its derivatives, the MRL/*lpr* and BXSB/Yaa strains. The most commonly used induced models are the pristane-induced lupus model and the chronic graft-versus-host disease models (cGVHD). The classical mouse models of spontaneous and induced SLE are summarized in Table I.

There is a plethora of genetically modified mouse models of SLE in which one or more genes are disrupted. All of these models share some of the symptoms of human SLE, such as autoAb production, lymphoid activation and hyperplasia, and lupus nephritis. Notably, many of the genetically modified models were generated on a C57BL6/J (B6) × 129/Sv (129) mixed background, which is known to induce by itself spontaneous systemic autoimmunity.[9,13,14] The contribution of these genetically modified models to our understanding of lupus pathogenesis has been discussed in excellent reviews.[15–18]

A. Spontaneous Models of Lupus

1. NZB/W F1 and NZM

The NZB/W F1 model of lupus is generated by the F1 hybrid between the NZB and NZW strains.[3] Although autoimmune manifestations in NZB mice are mostly limited to hemolytic autoimmune anemia and NZW mice are essentially non-autoimmune, NZB/W F1 hybrids develop severe lupus-like phenotypes comparable to that of lupus patients, including a strong female bias.[15] An accidental backcross between NZB/W F1 and NZW followed by brother–sister matings generated 27 different recombinant inbred strains of New Zealand Mixed (NZM) mice.[4] Among them, NZM2328 and NZM2410 strains exhibit highly penetrant lupus-like symptoms and have been used as

TABLE I

CLASSIC MOUSE MODELS OF SPONTANEOUS AND INDUCED LUPUS

Name	Generation	Lupus-like symptoms	Gender bias
Spontaneous models			
NZB/W F1[3]	F1 hybrid between the NZB and NZW strains	Lymphadenopathy, splenomegaly, hypergammaglobulinemia ANA (mainly IgG2a and IgG3), IC-mediated GN	Strongly biased in favor of females
NZM2410[4,5]	A backcross between NZB/W F1 and NZW followed by brother–sister matings	Similar to NZB/W F1 with high penetrance	Biased in favor of females, with about a month delay in males
NZM2328[4,6]	A backcross between NZB/W F1 and NZW followed by brother–sister matings	Similar to NZB/W F1 with high penetrance	Strongly biased in favor of females, greater than in NZM2410
MRL/*lpr*[7]	Lymphoproliferation (*lpr*) mutation in the *Fas* gene on the MRL background	Lymphadenopathy due to an accumulation of double negative ($CD4^{-}CD8^{-}$) and $B220^{+}$ T-cells, hypergammaglobulinemia, high titers of autoAbs (ANA, rheumatoid factors) and circulating ICs. Most severe clinical disease of all spontaneous models	Both males and females are significantly affected
BXSB/Yaa[8]	An inbreed strain generated by (B6 × SB/Le)F1 backcrossed to SB/Le	Secondary lymphoid tissue hyperplasia, IC-mediated GN, hypergammaglobulinemia, ANA, and high serum antiretroviral gp70 IgG titers, monocytosis	Much more severe disease and earlier onset in males
B6 × 129[9]	1. F1 Hybrid between the B6 and 129 strains 2. Genetically modified mice generated using 129 embryonic stem (ES) cells from strain and B6 blastocytes	AutoAb production, mild GN, hyperactive T cells	Studies performed mostly on females
Induced models			
Pristane-induced lupus[10]	Intraperitoneal injection of pristane	AutoAb production, IC-mediated GN, hemorrhagic pulmonary capillaritis, arthritis	Disease is more severe in females than in males in the SJL/J strain
cGVHD[11,12]	Allograft response generated by allogenic $CD4^{+}$ T cells	Polyclonal B-cell activation, autoAb production, GN	More severe in females in the parent into F1 (P → F1) model

lupus models.[5–7] Different from NZB/W F1, the NZM strains have homozygous genomes and are enriched for lupus-susceptibility loci, which have greatly facilitated genetic analyses of SLE.

2. MRL/*lpr*

The MRL strain was developed from several crosses of inbred strains including LG/J, C3H/Di, B6, and AKR/J.[15] One MRL substrain, termed MRL/*lpr*, develops an SLE-like phenotype.[7] Unlike the NZB/W F1 mice, both male and female MRL/*lpr* mice are significantly affected. This phenotype was attributed to a recessive mutation termed lymphoproliferation (*lpr*), which leads to aberrant, nonfunctional transcripts of the *Fas* gene.[19–21] Fas is a surface-bound receptor belonging to the tumor necrosis factor receptor (TNF-R) family that induces apoptosis upon interacting with its ligand, FasL.[22] Expectedly, a spontaneous mutation in the gene encoding for FasL gene, termed generalized lymphoproliferative disease (*gld*), was shown to cause an autoimmune disorder similar to the *lpr* mutation.[23,24] Interestingly, the MRL genetic background plays an important role in the disease development. Expression of the *lpr* mutation in either the C3H/HeJ or the B6 strains did not result in nephritis.[25] In contrast, Fas-sufficient, MRL$^{+/+}$ mice displayed the pathological effects of nephritis; though onset of symptoms was significantly delayed when compared to MRL/*lpr* mice.

3. BXSB/*Yaa*

A recombinant inbred strain derived from the backcross of (B6 × SB/Le) F1 to SB/Le, termed BXSB/Mp (BXSB/*Yaa*), develops a highly penetrant lupus-like disease. The symptoms are much more severe with earlier onset in males.[7,8,26] Mean survival is roughly 5 months in males and 14 months in females, and the major cause of death is proliferative glomerulonephritis (GN).[7,8] The disease acceleration in BXSB males is not due to a hormonal effect, as orchiectomy did not delay onset or severity in BXSB males, whereas ovariectomy did not exacerbate disease in females.[27] BXSB-derived Y chromosome exacerbated disease when combined with either the NZW or MRL genetic backgrounds,[28,29] indicating that an element, termed Y-linked autoimmune accelerator (*Yaa*), is responsible for the acute disease in BXSB males. It is now known that *Yaa* is due to a translocation of the telomeric end of the X chromosome to the Y chromosome. This translocation results in a duplication and an approximately twofold greater expression of at least 16 genes.[30,31] Among these genes is Toll-like Receptor 7 (*Tlr7*), which encodes an innate pattern recognition receptor found in B cells and antigen-presenting cells (APCs). Deletion of the endogenous copy of *Tlr7* from the X chromosome abrogates *Yaa*-induced lupus phenotypes,[32–34] demonstrating that the *Tlr7* gene duplication is the major functional contributor to the *Yaa* phenotype.

Other genes in the translocated *Yaa* interval may also contribute, since lupus-prone $Yaa^{+}.Tlr7^{-}$ mice still display some humoral abnormalities, increased autoAb production, and a decreased marginal zone (MZ) B-cell compartment as compared to lupus-prone $Yaa^{-}.Tlr7^{+}$ mice.[32,34]

4. B6 × 129/Sv

Gene knockout (KO) models are widely used to study the function of genes, including their role in SLE. In KO technology, a gene is usually first disrupted in the embryonic stem (ES) cells by homologous recombination resulting in the disruption or removal of a piece of DNA within this gene. After that, the genetically modified ES cells are introduced to the inner cavity of a blastocyte, creating chimeric embryos that are put back into female mice. ES cell lines from 129 strain and blastocytes from B6 strain have been usually employed because of their efficiency and reliability. Consequently, newly generated KO mice are usually of a B6 and 129 mixed background. After that the disrupted gene can be bred onto a B6 background to dilute the contribution of the 129 genome, but at least a large region flanking the *KO* gene remains of 129 origin, unless extreme measures are taken to select for recombination between tightly linked markers.[14] Intriguingly, lupus-like disease is one of the most common outcomes of a genetic KO in mice and the lupus phenotypes of KO mice are largely affected by the amount of 129 genome they carry. C1q deficiency led to full expression of lupus-like disease in B6 × 129 background, which almost completely disappeared after a seven generation backcross onto B6.[35,36] Similar results were sporadically reported for other KO, such as the MFG-E8,[37,38] indicating that the B6 × 129 background predisposes mice to the development of SLE. The analysis of a cross between the B6 and 129 strains has identified a number of susceptibility loci contributed from both strains,[9,39–42] the strongest of which is located on the telomeric chromosome 1 of the 129 strain. Therefore, the interaction between B6 and 129 genomes may be responsible for some of the autoimmune traits seen in KO models of a B6 × 129 background, and 129 ES cells have been largely replaced by B6-derived ES cells to target genes with suspected autoimmune contributions.

B. Induced Models of Lupus

1. The Pristane-Induced Lupus Model

Like humans, mice can develop a lupus-like disease after exposure to certain environmental triggers. Intraperitoneal injections of pristane (2,6,10,14-tetramethylpentadecane, TMPD), an isoprenoid alkane found at high concentration in mineral oil, is a standard method to obtain ascitic fluid enriched in monoclonal Abs. However, pristane injections in non-autoimmune BALB/c mice also induce lupus symptoms.[10] Almost all mouse strains are susceptible, to a variable extent, to the pristane-induced production of autoAbs and lupus manifestations,[43] confirming that the environment has a considerable importance in lupus. Pristane-induced lupus is more severe in females than in males, at least in the SJL/J strain.[44] Recent findings have shown that pristane-induced lupus is highly dependent on the production of type 1 interferons (IFNs),[45,46] making this model highly relevant to human SLE (see below).

2. Induced Chronic Graft-Versus-Host Disease

A single injection of allogenic donor splenocytes can induce a lupus-like syndrome in the host with a severity correlating to the number of injected cells.[47,48] Because this method induces a rapid disease onset from a known time point, it has greatly facilitated the study of lupus induction as compared to the spontaneous models. As an example, autoAbs are detectable as early as 10 days after induction, whereas it can be several weeks to months after birth for the same phenotype to develop in spontaneous lupus models.

In the parent into F1 (P → F1) model, lymphocytes are transferred into a semiallogenic recipient such that the donor cells react to the host, and the host is tolerant to the donor allograft.[11] Depending on the model used, an acute or a chronic GVHD (aGVHD or cGVHD) ensues. In most models of P → F1, such as B6 into (B6 × DBA)F1, aGVHD is induced by donor $CD8^+$ T cells, which manifests as severe lymphopenia of host cells as well as strain-specific connective tissue diseases. Host B cells are most sensitive to this cytotoxic T lymphocyte (CTL) killing and may be almost completely ablated after 2 weeks, followed by host APCs, then $CD4^+$ T cells, and finally host $CD8^+$ T cells. Conversely, DBA into (B6 × DBA)F1 (DBA → BDF1) results in cGVHD, a donor $CD4^+$-mediated lupus-like response. The lupus pathogenesis in DBA → BDF1 model is more severe in females than males.[49,50] The induction of a cGVHD instead of aGVDH response is caused by a diminished capacity for DBA $CD8^+$ T cells to produce a CTL response to the F1 host. Correspondingly, B6 into (B6 × DBA)F1 will result in cGVHD if the donor cells are depleted of $CD8^+$ T cells.

In the *B6.H-2^{bm12}* ↔ B6 cGVHD model, B6.H-2^{bm12} lymphocytes, whose MHC-II locus confers a three-amino acid substitution in H-2^{b}, are transferred into B6 hosts with the wild-type H-2^{b} locus, and cGVHD develops.[12] Since the major histocompatibility complex (MHC) disparity between donor and host is limited to class II in this model, $CD8^+$ T-cell depletion of donor lymphocytes is not required. Also, since this model works equally well with B6.H-2^{bm12} → B6 and B6 → B6.H-2^{bm12}, donor and host gene modifications can be examined by the use of KO and transgenic (Tg) mice.

II. Role of the Major Histocompatibility Complex

As for many other autoimmune diseases, the MHC plays a preeminent role in lupus pathogenesis. Class II Aβ deficiency[51] and β2 microglobulin deficiency (which results into class I MHC deficiency)[52] protect MRL/*lpr* mice from lupus nephritis. These mice were, however, deficient in $CD4^+$ and $CD8^+$ T cells, respectively, and it was therefore impossible to tease apart the respective roles of the MHC and T cells from these studies. Genetics studies have, however, unequivocally associated the MHC and lupus. Meta-analyses of genome-wide association studies (GWAS) have revealed that the strongest association with human lupus lies in the class II *DRB1* gene, with two alleles, DR3 and DR2, reaching the accepted genome-wide significance ($p < 5 \times 10^{-8}$) in at least one study.[53] However, the MHC is a gene-rich region with a high level of linkage disequilibrium. This complexity has made the identification of the causative alleles even more difficult than for the other susceptibility loci. It is highly likely that variations in the *DRB1* gene affect Ag presentation and the shape of the $CD4^+$ T-cell repertoire, and consequently susceptibility to lupus. In addition to DRB1, additional studies, including deep-sequencing of the MHC region in a large number of lupus patients and controls, will be necessary to determine whether other closely linked genes are also involved. Significant associations with both *TAP1* and *TAP2* genes with SLE have been recently reported, not surprisingly in interaction with DRB1 haplotypes.[54] As it will be detailed below, studies in mouse models of lupus strongly suggest the existence of both class II MHC genes and non-class I, non-class II MHC genes contributing to the disease.

The first report of an association between the MHC and lupus in the mouse occurred in the NZB/W F1 model in which the major contribution from the NZW genome was attributed to the $H\text{-}2^z$ allele.[55,56] The heterozygous $H\text{-}2^{d/z}$ genotype was strongly associated with disease as compared to either homozygous $H\text{-}2^d$ or $H2^z$. This strong association between heterozygous $H\text{-}2^z$ and lupus was further confirmed by a more comprehensive linkage analyses of a (NZB × NZW)F2 cross.[57] A backcross between NZM2410 and B6 revealed that the heterozygous $H\text{-}2^{b/z}$ genotype was also strongly associated with disease.[5] Congenic analysis showed that the $H\text{-}2^{z/z}$ homozygous genotype resulted in a strong suppression of lupus phenotypes on either NZW or B6 genetic backgrounds.[58] This suppressor locus, named *Sles1*, has been mapped to a less than 1 Mb region within the MHC that contains the class II genes, but excludes TNFα.[59] Interestingly, the suppressive function of *Sles1* is shared between the NZW and 129/SvJ alleles, with both strains showing strong lupus-susceptibility loci on a B6 background.[9,59] In addition to the unusual association of susceptibility with a heterozygous genotype, another level of complexity of the association between H-2 and lupus resides in its dependence on the genetic background, with H-$2^{d/z}$ being strongly associated with disease on an (NZB × B6) F1 but not on an (NZB × BALB/c)F1 background.[60] The context-dependence of the association between H-2 and lupus was confirmed by the lack of strong linkage between H-2 and clinical lupus in the other two major murine models, MRL/*lpr*[20] and BXSB/*Yaa*.[61] The H-2 haplotype is, however, involved in the production of autoAbs with MRL/*lpr* mice congenic for H-2^b producing a reduced repertoire of autoAbs with less IgG3 class-switching.[62] Moreover, H-2^b expression was sufficient to induce non-autoimmune BALB/c (H-2^d) mice to produce antinuclear class-switched autoAbs.[63] Overall, these results stress the importance of the interaction between MHC haplotypes and the genetic background.

The specific role of the MHC class II genes in SLE was investigated directly by transgenesis. The H-2^b haplotype, which is I-E-deficient, has been shown to be more disease-prone than either the H-2^d or H-2^k haplotypes, which express I-E, and this is due to the protective effect of the Ea molecule.[64,65] It has been proposed that the Ea^d gene is protective due to its production of peptides with a high affinity for the I-A molecules that are thereby interfering with the presentation of autoAgs.[66] This hypothesis was strengthened by the observation that Ea^d is protective in the absence of Eb expression, precluding protection through the expression of a functional Ea/Eb Ag-presenting dimer.[67] The protective role of Ea^d was confirmed in NZB/W F1 mice homozygous for A^d but deficient for E^d, which develop a disease that was comparable to that of control NZB/W F1 mice carrying the H-$2^{d/z}$ genotype.[68] Ea^d suppressive effect was, however, trumped by the expression of H-2^z. The role of the class II H-2^z genes in lupus was also investigated with Tg mice, and it was shown that on an NZB background, both the Ea^z and Eb^z[69] and the Aa^z and Ab^z[70] Tg segregated independently from disease, therefore failing to identify a causative relationship between these genes and SLE. The fact that E^z/E^b heterozygosity strongly correlated with disease in an NZM2410 cross[71] could therefore be interpreted as the involvement of a closely linked gene. A direct involvement of the class II *Aβ* gene in autoAb production was, however, shown in B6.*Sle1*.H-$2Ab^{+/-}$ mice, a congenic strain derived from NZM2410, in which the elimination of one copy of the $A\beta^b$ gene resulted in significantly enhanced autoAb production.[72] How $A\beta^b$ hemizygosity and A^z heterozygosity both amplify the autoimmune response is still open for speculations. The contribution of the human HLA-DR2 (DRB1*1502) allele, but not DR3, of the class II *DRβ* gene has been directly demonstrated as a Tg in the NZM2410 model.[71] Interestingly, only an autoAb shift to a more pathogenic profile and not clinical disease has been associated with segregation of the DRB1*1502 Tg in NZM2410 mice. This result was at odd with the very strong association of both DR2 and DR3 with human SLE.

In addition to the MHC class II genes, the NZW allele of TNFα, a gene located in the MHC locus, has been associated with low production of this cytokine and TNFα treatment has been shown to delay the development of disease in NZB/W F1 mice.[73–75] Congenic mice in which the origin of the TNFα and MHC class II genes segregated independently demonstrated an independent contribution of both genes.[76] Importantly, a low TNFα serum level is associated with increased disease severity in lupus patients.[77] To this date, the identity of the MHC genes that contribute to lupus in the mouse, except for TNFα, has not been identified. Genomic sequencing and haplotype-mapping across the small interval corresponding to the *Sles1* locus[59] will most likely provide answers to this puzzle that has now endured for over 20 years.

III. Role of Cell Signaling

Signaling pathways are composed of multiple molecules that initiate, promote, or dampen an immune response. Activated autoreactive B and T cells promote the development of symptoms characteristic of SLE. Abnormalities in signaling components have been found in SLE patients.[78] The immune system is dysregulated by changes in cell signaling thresholds as well as by defects in inhibitory molecules, resulting in a breakdown of tolerance, and the production of autoreactive B and T cells.

A. B-Cell Signaling

SLE has been coined as a disease of B-cell hyperactivity,[79] and not surprisingly, abnormalities with B-cell signaling are associated with SLE. For instance, stimulation of B cells from SLE patients with anti-IgM showed higher levels of calcium flux and increased phosphorylation of tyrosine residues.[80] Mouse models have been used extensively to investigate the mechanisms responsible for intrinsic B-cell activation. B cells from various lupus-prone mouse strains share the hyperactivation of multiple signaling pathways, with the AKT/mTOR axis playing a leading role.[81] Gene targeting of individual B-cell signaling molecules have identified two inhibitory receptors, Lyn and CD22, as key players in regulation B-cell lupus phenotypes. In addition, natural polymorphisms and gene targeting have established that *FCγRIIB*, another inhibitory receptor, represents another key molecule in preventing the development of lupus phenotypes in B cells.

1. Lyn

Decreased levels of Lyn have been found in both resting and activated B cells of SLE patients.[82] Lyn is a Src family protein tyrosine kinase that is involved in both positive and negative regulation of multiple signaling pathways in the immune response. Upon B-cell receptor (BCR) crosslinking, Lyn is rapidly activated to phosphorylate other BCR complex molecules such as CD19 and the Igα/Igβ subunits. However, Lyn also plays a major role in the negative regulation of the B-cell response, by phosphorylating inhibitory receptors such as CD22 and FcγRIIB, as well as by recruiting tyrosine phosphatases such as SHP-1 and SHP-2. The significance of Lyn in negative regulation of B cells first became apparent when Lyn$^{-/-}$ mice were shown to develop a lupus-like disease with antinuclear autoAbs, elevated serum IgM, and GN.[83] Despite this autoimmune phenotype, Lyn$^{-/-}$ mice have decreased levels of recirculating B cells as well as defective BCR-mediated signaling.[84,85] The autoimmune phenotype of Lyn$^{-/-}$ mice results from the abrogated Lyn-mediated inhibition of Bruton's tyrosine kinase (Btk) as Lyn and Btk double-deficiency did not result in autoimmunity.[86] Lyn$^{-/-}$ mice displayed a Btk-mediated accumulation of splenic plasmablasts and plasma cells. More recent studies have shown that IL-6 is required for the development of pathogenic IgG autoAbs in Lyn$^{-/-}$ mice.[87,88] While it is clear that Lyn deficiency leads to systemic autoimmunity, the reasons for its decreased expression in the B cells of SLE patients are not well understood. Several mechanisms are likely to be involved such as decreased gene expression,[82] increased ubiquination, and aberrant recruitment to lipid rafts.[89] In addition to deciphering the mechanisms regulating Lyn expression, a better characterization of the signaling pathways regulated by Lyn is needed for a better understanding of Lyn's role in SLE.

2. CD22

CD22 is a B-cell-specific glycoprotein that associates with membrane IgM and binds α2,6-linked sialic acid ligands expressed on various cell types. Upon BCR stimulation, Lyn phosphorylates CD22, which allows for the recruitment of SHP-1 phosphatase by CD22. Mice deficient in CD22 display dysregulation in B-cell functions, resulting in the production of somatically mutated, high-affinity autoAbs against dsDNA, but these autoimmune manifestations were not sufficient to induce a clinical disease.[90] Furthermore, B cells from CD22$^{-/-}$ mice were hyperresponsive to Ag stimulation and exhibited elevated calcium response.[91,92] Although no significant polymorphisms of CD22 are associated with human SLE, there is a candidate CD22 allele in murine lupus models. Several lupus-prone strains express the CD22^{a} allele, which results in aberrant CD22 mRNA transcripts due to a 794-bp nucleotide insertion containing short interspersed nucleotide elements. CD22^{a}-expressing B cells have decreased ligand-binding capacity versus CD22^{b}.[93,94] Therefore, a decrease in function in CD22 could potentially promote autoimmunity as it removes an important negative regulator in B-cell signaling. In addition, CD22 is part of the sialic acid acetylesterase pathway, which is a major contributor to B-cell peripheral tolerance.[95] Sialate:O-acetyl esterase encoded by the *Siae* gene is an enzyme that specifically removes acetyl moieties from the 9-OH position of α2-6-linked sialic acid, and therefore regulates CD22 function. *Siae*-deficient mice develop enhanced BCR activation, antinuclear autoAbs, and renal immune complex (IC) deposition.[96]

CD22 expression, unlike Lyn, is similar between SLE patients and healthy controls.[97] A biologic, Epratuzumab, targeting B cell by binding CD22 has, however, shown promising results for the treatment of SLE.[98] Besides the fact that CD22 is highly expressed on B cells, which may turn Epratuzumab into a B-cell-depleting agent, the mechanisms by which it acts is unclear at the present.[99] Interestingly, rare loss-of-function variants of *SIAE* have been associated with various autoimmune clinical manifestations,[100] establishing the involvement of the CD22/sialic acid acetylesterase pathway in maintaining B-cell tolerance in humans.

3. Fc Gamma Receptors

The Fc gamma receptors (FcγR) constitute a complex family of molecules that bind the Fc portion of IgG Abs with varying affinity and isotype specificity. Their binding of Ab/Ag ICs results in either activation or inhibitory signals.[101] Both activating and inhibitory FcγR have been associated with human SLE through functional and genetic studies.[18] In the mouse, studies have focused on *FcγRIIB*, an inhibitory low-affinity Fc receptor which contains an immunoreceptor tyrosine-based inhibitory motif (ITIM) in its cytoplasmic domain. There are two isoforms of *FcγRIIB*: *FcγRIIB1* is expressed in B cells and, when crosslinked to the BCR, results in an increased B-cell activation threshold, decreased Ag presentation, and Ab production. *FcγRIIB2* is mainly expressed in myeloid-derived cells, in which its binding to ICs inhibits Ag internalization and presentation, resulting in decreased T-cell priming.[102,103] The link between the defective expression of *FcγRIIB1* in B cells and lupus has been studied extensively in mouse models.

Reduced expression of *FcγRIIB* was observed in the lupus-prone strains NZB, BXSB, and MRL/*lpr*, due to unique polymorphisms in the *FcγRIIB* gene.[104] Additionally, polymorphisms in the NZM2410/NZW allele of *FcγRIIB* were associated with decreased expression of *FcγRIIB* in germinal center (GC) B cells and elevated class-switched plasma cells, although autoimmunity did not develop in the absence of additional lupus-susceptibility loci.[105] *FcγRIIB*-deficient mice break B-cell tolerance and develop nuclear autoAbs and GN in the B6 but not BALB/c background.[106,107] Additional studies with B6. *FcγRIIB*$^{-/-}$ mice have shown that FcγRIIB prevents the development of autoreactive B cells at a GC/early plasmablast checkpoint.[108]

These observations associating a reduced *FcγRIIB* expression with the presence of autoreactive B cells in mice have been validated by a meta-analysis that showed that *FCγRIIB* polymorphisms confer susceptibility to human SLE in specific ethnic backgrounds.[109] In addition, memory B cells and plasma cells from SLE patients failed to upregulate *FCγRIIB* expression as seen in healthy controls.[110,111] Therefore, *FcγRIIB* plays an important role in preventing systemic autoimmunity, specifically by inhibiting autoreactive B cells in the terminal phases of their differentiation as GC B cells and plasma cells.

B. T-Cell Signaling

Multiple alterations have been identified in the T-cell signaling pathway of SLE patients.[112] Initial studies have demonstrated that T-cell receptor (TCR) stimulation induced higher levels of intracellular calcium in T cells from SLE patients than from healthy controls. This is due to decreased levels of CD3ζ, a critical TCR-signaling component,[113] which is compensated with the increased expression of the analogous molecule FcεRγ (the common γ chain of the Fcε receptor).[114] FcεRγ, unlike CD3ζ, does not bind ZAP-70 (ζ-chain-associated protein kinase 70) but instead associates with the spleen tyrosine kinase (Syk) which results in the elevated levels of calcium after TCR engagement.[115] Syk inhibition in lupus-prone MRL/*lpr* and BAK/BAX double KO mice resulted in decreased spleen and lymph node size and suppressed the development of renal disease.[116] In NZB/W F1 mice, anti-Syk treatments reduced the levels of activated T cells and prolonged survival despite having no effect in the production of autoAbs.[117] Thus, altering the signaling pathway of T cells leads to a change in the strength of the autoimmune response and to clinical improvement in several mouse models of SLE.

Lipid rafts are cholesterol-rich zones containing signaling molecules that cluster after TCR stimulation. T cells from MRL/*lpr* mice contain clustered lipid rafts which increased with age.[118] These clusters contained adhesion, signaling, and costimulatory molecules which play a role in T-cell signaling. Disruption of the aggregated lipid rafts increased the survival of MRL/*lpr* mice. Furthermore, T cells from SLE patients were found to contain a higher number of lipid rafts which preclustered in the absence of TCR stimulation.[119] Therefore, in human and murine SLE, preclustered lipid rafts bring together diverse components of the T-cell signaling complex which promote aberrant T-cell activation.

In addition to changes in molecules involved in the activation and signal transduction of T cells, there is one well-known T-cell inhibitory molecule, cytotoxic T lymphocyte antigen-4 (CTLA-4), that has been implicated in SLE pathogenesis. CTLA-4 functions as an inhibitory receptor by binding CD80/86 expressed on the surface of APCs with higher avidity than the activating receptor CD28.[120] CTLA-4 plays a significant role in regulating the immune system as mice deficient in CTLA-4 suffered from uncontrolled T-cell activation leading to splenomegaly, lymphadenopathy, and lethal multiorgan T-cell infiltrates by 3–4 weeks of age.[121] In SLE, disease symptoms are not caused by

CTLA-4 downregulation as the CTLA-4 mRNA levels are similar between patients and healthy controls.[122] On the contrary, the characteristic high level of T-cell activation is associated with a greater percentage of T cells expressing CTLA-4 in SLE patients. It has been suggested, however, that CTLA-4 functions are defective in SLE as one study showed that CTLA-4 in SLE T-cells was excluded from lipid rafts.[123] Although significant associations between CTLA-4 polymorphisms and SLE have not been identified, a meta-analysis showed that the association between SLE and some CTLA-4 polymorphisms are race specific.[124] The strong CTLA-4 inhibition of T-cell function has suggested that it may be used as a treatment for SLE, with soluble receptor CTLA-4Ig preventing CD28 interaction with CD80/86. CTLA-4Ig treatments in $CTLA4^{-/-}$ mice prevented lymphoproliferation and multiorgan tissue damage.[125] Furthermore, CTLA-4Ig treatment prevents disease in several lupus-prone strains, and can revert lupus nephritis in combination with cyclophosphamide.[126] Abatacept, a humanized soluble CTLA-4 fusion molecule was recently tested in patients with non-life-threatening SLE, but the Phase II clinical trial did not meet primary and secondary endpoints.[127] A potential explanation for this failure was provided recently by a mouse study that showed that high levels of IFNα (which are a hallmark of human SLE, see below) not only accelerated disease in NZB/W F1 mice but also significantly reduced the efficacy of CTLA-4Ig treatment.[128]

C. Toll-Like Receptors

Immune cells express innate receptors such as Toll-like receptors (TLRs) that recognize conserved motifs shared by bacteria, virus, and fungi.[129] In addition to pathogen-derived ligands, two endosomal TLRs also recognize mammalian-derived cellular Ags that are relevant to SLE. TLR7 binds single-stranded RNA, and TLR9 binds unmethylated cytosine phosphate guanine (CpG) DNA[130] and mediates intracellular signaling through the myeloid differentiation primary response gene 88 (MyD88)-dependent pathway.[130] MyD88 associates with the cytoplasmic TIR (Toll/interleukin-1 receptor/resistance) domain of TLR 7 or TLR9, which activates IRAKs (IL-1 receptor-associated kinases) and TRAF6 (tumor necrosis factor receptor-associated factor 6), leading to NF-κβ translocation from the cytosol to the nucleus to induce the transcription of pro-inflammatory genes, including type I IFNs.[130] Because SLE is characterized by the production of autoAbs largely directed against DNA and RNA-containing complexes, numerous studies have investigated the role of TLR7 and TLR9 in the activation of autoreactive B cells. The role of TLR7 in murine SLE has been directly demonstrated with the Yaa locus (see above). Moreover, a polymorphism associated with an increased expression of TLR7 is associated with human SLE and the presence of an IFN signature, especially in males.[131] Genetic evidence for a role of TLR9 in lupus is weaker, with associations between TLR9 polymorphisms and SLE found in some studies[132] but not others.[133]

Considering the intracellular location of TLR7 and TLR9, their ligation in autoreactive B cells requires delivery of extracellular autoAgs to the endosome. The MRL/*lpr* AM14 BCR-Tg model (see below) showed that activation of AM14 autoreactive B cells by their autoAg IgG2a–chromatin ICs required an initial interaction between the IgG2a component of the IC with the BCR to internalize the IC, which is then followed by association of the internalized IC with TLR9.[134] Experiments using anti-DNA B cells from the 3H9 model confirmed that TLR9 recognizes the IC DNA component brought inside the cell.[135] While B cells bring in DNA-IC through their BCR, dendritic cells (DCs) use FcγR to bring in DNA-IC (see FcγR section). Briefly, mice deficient in FcγRIII, MyD88, or TLR9 demonstrated that murine DCs use FcγRIII to internalize chromatin-IC, which then ligate intracellular TLR9 and initiates MyD88 signaling, which induces the expression of pro-inflammatory cytokine TNFα.[136] In humans, DNA-IC is internalized by plasmacytoid DCs (pDCs) via binding to FcγRIIa, then activation of TLR9 leads to production of IFNα.[137] These studies clarified the mechanisms by which DNA-IC is brought to TLR9 to activate immune cells which can then contribute to the pathogenesis of lupus.

While uptake of DNA-IC by DCs leads to activation of TLR9, uptake of ribonucleoproteins by DCs leads to activation of TLR7. TLR7 in DCs recognizes the RNA component of ribonucleoproteins and upon ligation, stimulates the MyD88-dependent pathway which leads to production of type I IFN.[138] TLR7-dependent activation of autoreactive B cells by RNA-containing ICs has also been demonstrated with the AM14 Tg model.[139] Interestingly, this TLR7 activation of autoreactive B cells was dependent on an IFN-β feedback loop, confirming the key role that TLR7 plays in the type I IFN pathway in SLE. Because both TLR7 and TLR9 contribute to lupus by activating antinuclear autoreactive B cells or inducing type I IFN production by DCs, removal of these pathogenic TLR from a lupus-prone murine background should alleviate disease. TLR9-deficiency in MRL/*lpr* mice is associated with a decreased level of anti-DNA autoAbs, but with a more severe disease.[140,141] To the contrary, TRL7-deficiency not only prevented the production of autoAb directed to RNA-ICs but also protected from disease development.[140] Additional studies comparing MRL/*lpr* mice deficient for both TLR7 and TLR9 to single deficient mice showed a complex relationship between TLR7 and TLR9, with TLR9 regulating TLR-7 and MyD88-dependent autoAb production and pathology.[142] TLR7/9 activation of pDCs, which are the main producers of type 1 IFNs, is responsible for these pDCs to be intrinsically resistant to glucocorticoid-induced cell death in both humans and lupus-prone mice.[143] This suggested that the inhibition of the TLR7/9 pathways could lead to significant improvements of glucocorticoid therapy, by decreasing doses and improving efficacy. Moreover, the same bifunctional TLR7/9 inhibitor that restored pDCs sensitivity to glucocorticoids prevented and reversed autoimmune skin lesions in the NZB/W F1 model.[144] These are very promising therapeutic venues which are being evaluated in SLE patients.[145] However, considering the protective role of TLR9, therapies targeting only TLR7 should be also considered.

D. Apoptosis

Two types of apoptosis defects have been associated with lupus. First, the inability of immune cells to undergo programmed cell death, largely due to defects in the Fas pathway, results in the production of autoreactive lymphocytes and the development of lupus. Second, impaired removal of apoptotic debris induces inflammation, including the activation of the TRL7 and TLR9 pathways that also results in the development of lupus.

1. Fas Pathway

Interaction between surface-bound Fas receptor with its ligand (FasL) activates caspase-8, which in turn activates downstream effector caspases that are necessary for apoptosis. Fas, which belongs to the TNF-R family, is expressed in lymphoid, myeloid, and nonhematopoietic cell types, while FasL is primarily expressed in $CD8^+$ or activated $CD4^+$ T cells.[22] The *lpr* mutation severely impairs Fas expression, and results in a lymphoproliferative disease and a severe lupus-like phenotype, especially in the MRL/*lpr* mouse strain.[21] *lpr* expression (i.e., Fas deficiency) is necessary in both B and T cells in MRL/*lpr* mice as the autoimmune phenotype was decreased in the absence of either cell type.[146,147] Conditional inactivation of Fas in T cells in (B6 × MRL)F1 mice resulted in lymphopenia as the activated T cells upregulated FasL, which in turn resulted in the apoptosis of Fas-expressing peripheral lymphocytes. However, conditional deletion of Fas in either T or B cells, or both, in non-autoimmune B6 mice eliminated lymphoproliferation, but preserved autoAb production, indicating that Fas deficiency had to be present in both lymphoid and nonlymphoid cells to reconstitute the autoimmune phenotype of *lpr* mice.[148] Furthermore, deletion of Fas in DCs resulted in elevated antinuclear antibodies (ANAs), splenomegaly, and accumulation of mature DCs.[149] These results demonstrate a complex role of Fas in maintaining systemic immune tolerance that involves all the major immune cell types.

The role of Fas-mediated apoptosis in lupus was further demonstrated by a recessive mutation in the gene encoding for FasL, known as generalized lymphoproliferative disease (*gld*), which resulted in a lupus phenotype similar to that seen with the *lpr* mutation.[23,24] In humans, deficiency in Fas signaling leads to the development of autoimmune proliferative syndrome (ALPS).

ALPS shares many symptoms with SLE, but GN is uncommon in ALPS patients.[150] Therefore, deficiencies in the Fas pathway lead to lupus-like symptoms in both humans and murine models.

2. Clearance of Dead Cells

Most SLE autoAbs target intracellular Ags, such as chromatin and RNA complex. How the intracellular materials become antigenic remains a major question. Billions of cells die in an adult body every day, mainly through apoptotic programmed cell death.[151] It has been proposed that dead cells are a major source of SLE autoAgs, which are expressed on the surface of apoptotic cells.[152] This was demonstrated when non-autoimmune mice repeatedly injected with synergic apoptotic thymocytes developed autoAbs and IgG deposition in the glomeruli.[153]

An unique feature of apoptosis is the cleavage of a set of cellular proteins by effector caspase 3, 6, and 7,[154] leading to morphological and biochemical changes of the cell and ultimately cell death. Clearance of apoptotic cells is nonimmunogenic and induces the production of immune regulatory cytokines such as IL-10 and TGF-β by phagocytes. Apoptotic cells are usually efficiently recognized and engulfed by phagocytes. Deficiency in this process, however, can cause the release of autoAgs and generate autoimmune responses and inflammation. In support of this hypothesis, deficiency in the clearance of cell debris in mice is often associated with lupus-like disease.

Apoptotic cells release "find me" signals which recruit phagocytes and express "eat me" signals which allow the recognition and engulfment by phagocytes. Several "find me" molecules have been identified. Activation of caspase 3 in apoptotic cells causes the release of lysophosphotidylcholine (LPC),[155] which is recognized by its receptor G2A (G2 accumulation protein or G protein-coupled receptor 132) on phagocytes and triggers chemotaxis.[156] Other molecules, such as sphingosine-1-phosphate (S1P),[157] CX3CL1/fractalkine,[158] ATP, and UTP[159] have also been proposed to act as "find me" signals. After reaching the apoptotic site, phagocytes distinguish apoptotic cells from healthy cells by recognizing the "eat me" signals on the surface of apoptotic cells.[151,160] Phosphotidylserine (PS), the best characterized "eat me" signal, is kept exclusively in the inner leaflet of cell membrane of healthy cells.[161] However, PS is exposed on the cell surface of apoptotic cells in a caspase-dependent manner.[162,163] The indispensable role of PS in the clearance of apoptotic cells was demonstrated by the fact that, red blood cells that incorporate exogenously supplied PS in their membrane can be recognized and engulfed by macrophages,[164] whereas masking of PS inhibits the clearance of apoptotic cells and causes autoAb production.[165] PS can be recognized by phagocytes either directly or via bridging molecules which recognize PS on the surface of apoptotic cells and molecules on the surface of phagocytes.

Growth arrested-specific 6 (Gas-6), a molecule abundant in plasma, is a ligand of the TAM family members Tyro3, Axl, and Mer,[166–168] which are receptor tyrosine kinase expressed on the surface of phagocytes.

Defective expression of TAM receptors, especially Mer,[169] has been shown to result in a lupus-like phenotype in mice.[170] Similarly, milk fat globule-EGF factor 8 (MFG-E8), a protein secreted by macrophages, bridges PS on apoptotic cells to integrins on phagocytes,[171] thereby stimulating the engulfment of apoptotic cells.[172] MEG-F8 deficiency in mice leads to the defective engulfment of apoptotic GC B cells by tangible body macrophages. As a result, MFG-E8$^{-/-}$ mice developed splenomegaly, with the formation of numerous GCs, and autoAb-mediated GN.[37] Furthermore, a congenic analysis of the BXSB/Yaa model has identified *Bxs2* as a susceptibility locus associated with ANA production. *Marco*, the candidate gene of *Bxs2*, encodes a type A scavenger receptor expressed on macrophage. The BXSB allele of Marco leads to reduced expression of Marco in macrophages at both mRNA and protein level, leading to a defect in clearing apoptotic cells.[173] NZB/W F1 and MRL/*lpr* mice have fewer macrophages in response to thioglycolate compared to lupus-resistant strains. Their macrophages are also smaller and have lower phagocytic ability.[174] In the pristane-induced lupus model, the hyperproduction of IFN-I prevents the maturation of the peritoneal immature Ly6C^{high} monocytes, leading to impaired cell debris clearance and chronic inflammation.[45]

Apoptotic cells that are not efficiently cleared enter secondary necrosis, which may generate an autoimmune response. How these cells are removed remains elusive. It has been proposed that the complement system is important in clearing late apoptotic cells, thereby serving as a backup system for uncompleted early clearance.[175] C1q can bind to apoptotic cells either directly[175,176] or via complement-fixing IgM. In the latter pathway, intracellular events during apoptosis lead to the exposure of neoantigenic determinants such as lysophosphatidylcholine, phophorylcholine, and malondiadlehyde, which can be recognized by IgM natural autoAbs.[177,178] IgM is able to fix C1q and initiate the activation of the classic pathway.[179–182] C1q and other complement components are then recognized by phagocyte receptors, such as CD91, which facilitates phagocytosis.[175] Notably, instead of inducing pro-inflammatory cytokine production, C1q inhibits NF-κB and promotes the production of immune regulatory cytokines such as IL-10.[183] Mannose-binding lectin (MBL) also binds late apoptotic cells and serves as a ligand for CD91 on macrophages to facilitate clearance of apoptotic cells.[184]

Complement deficiencies are closely associated with lupus in both human and mice. Almost all C1q-deficient individuals develop SLE. Deficiency in C4 and C2 also predisposes SLE in human.[185] In mice, although C1q deficiency does not lead to a lupus phenotype in a B6 background, it causes full expression of lupus disease in a 129 × B6 mixed background and exacerbates disease in the MRL/*lpr* mice.[35,36] Moreover, mice deficient in serum IgM, which is important for the fixation of C1q by apoptotic cells, also develop a lupus-like disease.[186] C4 deficiency also leads to defective apoptotic cell clearance and anti-dsDNA Ab production in a mixed 129 × B6 background.[187,188] Interestingly, C1q deficiency in both mice and humans leads to mice with more severe defects in apoptotic cell clearance and lupus phenotypes than C4 deficiency indicating a hierarchy of importance of complement components in the clearance of apoptotic cells and the etiology of SLE.[188] It is important to note, however, that deficiency in the clearance of apoptotic cells does not necessarily cause lupus symptoms. For example, MBL-deficient mice displayed defective clearance of apoptotic cells without evident autoimmunity, even on a lupus-prone genetic background.[189] This indicates that defective cell clearance needs to interplay with other factors in order to generate lupus symptoms.

IV. Role of Sex Hormones

SLE is a gender-biased disease with 90% of the affected individuals being females.[1] The observation that the high hormone milieu of pregnancy is associated with a high frequency of SLE flares, while the low hormone state of menopause is affiliated with remission of disease led to the hypothesis that female sex hormones played a pathogenic role and were largely responsible for the gender bias in SLE.[190,191]

In order to study the role of sex hormones in lupus, lupus-prone mice were castrated and then treated with either androgen or estrogen. Castrated NZB/W F1 female mice that were given a subcutaneous implant that released a physiologic level of testosterone or dihydrotestosterone over time died at a slower rate than untreated, noncastrated female mice.[192] When the external source of androgen tapered off 8 months postimplant, both the rate of mortality and level of IgG anti-DNA Abs increased in the castrated, androgen-treated female NZB/W F1 mice,[192] showing a direct causative effect between high androgen levels and disease protection. The protective nature of androgen against murine lupus was also demonstrated in both female and male MRL/*lpr* mice.[193] The protective role of androgen was also shown in the cGVHD-induced model of lupus. Ovariectomized cGVHD female mice treated with testosterone displayed a drastically reduced level of urine albumin when compared to untreated mice.[194] On the other hand, estrogen is pathogenic in the context of lupus. Castrated female NZB/W F1 mice treated with estrogen showed an increased mortality compared to untreated sham control.[195] In addition, estrogen also accelerated the production of high titers of anti-DNA IgG autoAbs and the class-switch from IgM to more pathogenic IgG Abs, and increased glomerular lymphocyte infiltration in castrated male NZB/W F1 mice as compared to untreated sham controls.[192,195] Overall, androgen and estrogens have opposite effects on the rate of mortality, autoAb production, and nephritis in various models of murine lupus.

Estrogen exacerbates murine lupus by binding to estrogen receptor-α (ERα). Selective activation of ERα by a specific agonist in ovariectomized NZB/W F1 mice was linked to an elevated titer of anti-DNA IgG Abs, early onset of albuminuria, and expedited mortality.[196] Meanwhile, specific activation of ERβ in ovariectomized NZB/W F1 mice did not alter lupus-like phenotype except for lowering the serum level of IgG2b anti-DNA.[196] Further evidence in support of the idea that estrogen activation of ERα contributes to lupus was shown in ERα KO NZB/W F1 mice, which presented lower titers of anti-histone and anti-DNA IgG, less severe GN, and enhanced survival.[197]

Although murine models of lupus have depicted estrogen as pathogenic, attempts to block estrogen receptor with the estrogen antagonist, tamoxifen, have failed to alleviate disease in SLE patients.[198] Murine studies have clearly shown that prolactin, another female hormone, promotes the development of autoreactive B cells and lupus-like-phenotypes, as reviewed recently.[199] Further studies of how sex hormones regulate lupus and its various pathogenic pathways are necessary to clarify the contribution of sex hormones to gender bias.

In addition to sex hormones, a large number of studies have shown that the X chromosome itself contributes to the female gender bias in autoimmunity, including SLE.[200] Defects in X-inactivation or X-linked gene dosage have been proposed to promote autoimmunity through several mechanisms, including loss of mosaicism or reactivation of immune-related genes on the inactivated X.[200] Most of these studies have been conducted with human subjects. The critical role of X-linked gene dosage has been, however, demonstrated in the mouse, first with the translocation of a group of X-linked genes, including *Tlr7*, to the Y chromosome resulting in spontaneous lupus in Yaa males.[31] An elegant model was created to dissociate the sex chromosomes from sex hormones by deleting the testes-determining gene *Sry* from the Y chromosome, resulting in XX and XY^- ovary-bearing mice. Furthermore, autosomal Tg expression of *Sry* resulted in XX*Sry* and XY^- *Sry* testes-bearing mice.[201] In both Tg and non-Tg mice, the castrated XX-carrying mice were more susceptible to pristane-induced lupus as compared with castrated XY-carrying mice. The same result was obtained for experimental autoimmune encephalomyelitis (EAE), providing the first experimental evidence that the XX sex chromosome complement, as compared with XY, confers a greater susceptibility to autoimmune diseases independently from sex hormones.

V. Role of Cytokines

A. Interferons

A hallmark of human SLE is the IFN signature, which refers to the upregulated expression of type I IFN (IFNα and IFNβ)-induced genes by peripheral blood mononuclear cells (PBMCs) that directly correlates with disease severity.[202] Genetic studies have also strongly associated the type I IFN pathway with lupus susceptibility.[203] The serum from SLE patients contains high level of IFNα, and this type I IFN has pleiotropic effects on the immune system, including the induction of monocyte differentiation into DCs, which in turn present Ags from apoptotic cells to activate autoreactive $CD4^+$ T cells.[204,205] Current clinical trials are being conducted in SLE patients with anti-IFNα treatment.[206]

The type I IFN signature is best recapitulated in pristane-induced lupus. In this model, the elevated level of type I IFN is produced by immature $Ly6C^{hi}$ monocytes found in the peritoneal cavity.[207] This differs from human SLE in which the major producers of type I IFN are pDCs stimulated by activated neutrophils.[208] IFN regulatory factor 9 (IRF9) and signal transducers and activators of transcription 1 (STAT1) are crucial links between type 1 IFN signaling and TLR expression in autoreactive B cells from pristane-treated BALB/c mice.[209] While IRF9 is implicated in the activation of B cells through TLR7, STAT1 is required for the activation of B cells through both TLR7 and TLR9.[209] Type I IFN-mediated signaling in B cells upregulates the expression of TLR7 and TLR9, both of which can recognize autoAgs in the form of nucleic acid-containing ICs, and then facilitates production of autoAbs by the TLR-activated B cells. In the pristane model, production of IFN I may come before production of autoAbs and therefore constitute a primary driver of the autoimmune response.[210] High levels of type I IFN are pathogenic in that these cytokines promote the expansion of autoAb-secreting cells. IFNα treatment of NZB/W F1 mice accelerated disease[211] and induced a sustained increase of the number of GC B cells and plasmablasts, which developed into autoreactive short-lived plasma cells.[212] This effect of IFNα treatment was not seen in BALB/c mice,[211] indicating that high levels of IFNα are not by themselves sufficient to activate autoreactive B cells. Paradoxically, IFNα/β played a protective role in MRL/*lpr* mice,[213] demonstrating that the effect of type I IFN on murine lupus varies depending on the genetic background.

In addition to the presence of an IFN signature in many SLE patients, the IFNα-induced expansion of DNA-specific plasma cells demonstrated in murine studies portrays IFNα as a promising target for SLE therapy. Furthermore, IFNα may be indirectly involved in the activation of B cells. Since human DCs respond to IFNα stimulation by producing B-cell-activating factor (BAFF),[214] and IFNα treatment increases circulating BAFF in NZB/W F1 mice,[211] IFNα may induce DC production of BAFF, which in turn activates B cells. Introduction of adenoviral vector carrying the gene for murine IFNα into NZM2328.Baff$^{-/-}$ mice failed to generate a difference in level of IgG autoAbs compared to that present in the empty vector-treated NZM2328.Baff$^{-/-}$ mice.[215] This study suggests that both IFNα and BAFF are needed for establishment of lupus.

Several studies have also documented elevated levels of serum IFNγ that correlated with disease activity in SLE patients, either directly[216,217] or indirectly through the presence of IFNγ-induced chemokines.[218] IFNγ had been consistently proven pathogenic in both MRL/*lpr* and NZB/W F1 mice,[219,220] and the genetic KO of IFNγ from MRL/*lpr* background delayed onset of GN and death.[220] The role of IFNγ in lupus will be reviewed in more detail in the T cell section.

Regardless of the lupus-prone murine model in which type I or type II IFN is associated with disease, it is typically linked to the production of autoAbs and to the severity of lupus nephritis. For example, type I IFN promotes production of IgG autoAbs in NZB mice. NZB mice in which IFNAR, the gene encoding for the α-chain of the receptor shared by both IFNα and IFNβ, has been knocked out showed a reduced titer of serum IgG anti-dsDNA in both homozygous IFNAR$^{-/-}$ and heterozygous IFNAR$^{+/-}$ mice.[221] A similar association of IFNα/β with autoAbs was also observed in pristane-induced murine lupus.[46] On the other hand, IFNγ is required for production of autoAbs in MRL/*lpr* mice. This was demonstrated in MRL/*lpr*.IFNγ$^{-/-}$ mice which displayed lower titer dsDNA or snRNP-specific IgG2b and IgG2a autoAbs than that observed in wild-type littermates.[222] In addition to promoting autoAbs, IFNγ is also required for development of IC-mediated GN, since NZB/W F1 mice that have been knocked out for the IFNγ receptor showed a delayed onset and reduced severity of IC-mediated GN.[223]

B. B-Cell-Modulating Cytokines

1. BAFF

BAFF is a member of the TNF family that is needed for B-cell development and survival.[224] BAFF is produced by innate immune cells and stromal cells in response to IFN stimulation or TLR ligation.[224] Soluble BAFF binds uniquely to BAFF receptor (BAFFR) and to two receptors it shares with APRIL (a proliferation-inducing ligand): B-cell maturation antigen receptor (BCMA) and transmembrane activator and calcium modulator and cyclophilin ligand interactor (TACI).[224] BCR-Tg mice have shown that positively selected autoreactive B cells are expanded by excess BAFF, resulting either from its overexpression in BAFF-Tg mice or from its reduced competition with non-autoreactive B cells in lymphopenic settings.[225,226] Many SLE patients present with high levels of circulating BAFF that parallel with titers of anti-dsDNA autoAbs.[227] Recently, an anti-BAFF biologic has been approved for the treatment of SLE (see below).

Tg overexpression of BAFF in an otherwise non-autoimmune murine background leads to development of lupus-like phenotypes, such as the expansion of mature B cells, notably MZ B cells, high-titer autoAbs, IC deposition in the kidneys, and proteinuria.[228,229] Such BAFF-induced autoimmunity, especially the production of autoAbs, is largely due to T-cell-independent, but TLR-dependent, proliferation of low-affinity self-reactive B cells.[230] The effects of BAFF overexpression are intensified in the presence of lupus-susceptibility loci. For instance, both B6.*Sle1*.BAFF-Tg and B6.*Nba2*.BAFF-Tg mice showed accelerated GN as compared to B6.BAFF-Tg mice.[231] Renal pathology was, however, uncoupled from the production of antichromatin autoAbs induced by the *Sle1* and *Nba2* lupus-susceptibility loci, implying that other factors induced by these loci amplify the effects of BAFF overexpression. Moreover, BAFF overexpression was not able to overcome the presence of lupus-resistance loci, such as MHC-II deficiency or *Sles1* expression.[232] These complex interactions revealed in BAFF-Tg mice may explain some of the clinical heterogeneity observed in SLE patients in regard with the circulating BAFF levels, as well as their variable response to BAFF-neutralizing agents.

Blocking BAFF signaling with the soluble receptor TACI-Fc alleviated disease in NZB/W F1 mice.[233] This study could not conclude, however, to a specific role of BAFF in the NZB/W F1 pathogenesis because of the dual binding of TACI to BAFF and APRIL. The specific pathogenic role of BAFF in lupus was nonetheless demonstrated by the treatment of NZB/W F1 mice with the BAFFR-Fc soluble receptor.[234] Neutralization of BAFF in NZB/W F1 mice reduced the B-cell level of p52, a breakdown product of transcription factor NF-κB2/p100, which is essential in the BAFFR signaling pathway. The reduced expression of p52 paralleled enhanced survival, lower anti-dsDNA Abs, and reduced kidney damage.[234] A direct comparison of selective BAFF blockade with BAFFR-Ig and combined BAFF and APRIL blockade with TACI-Ig performed in the NZB/W F1 and NZM2410 models showed very similar outcomes, except that the reduction of the number of plasma cells in the spleen, a preeminent future of NZM2410, but not of NZB/W F1 mice, was observed only with TACI-Ig.[235,236] These results suggested that treatments leading to either selective BAFF neutralization or combined BAFF and APRIL neutralizing should be tailored to patients depending on the nature of their B-cell differentiation pathway.

2. IL-6

Pro-inflammatory cytokine interleukin 6 (IL-6) is found at high levels in the serum of SLE patients, in which it may serve multiple functions, including the promotion of Ab production by plasma cells.[237] Several studies have demonstrated a pathogenic role for IL-6 in murine lupus. IL-6 is required for the generation of anti-DNA Abs in pristane-induced lupus.[238] In the NZM2410 mice, IL-6 produced by DCs contributes to B-cell activation, proliferation, and Ab production.[239] In the same NZM2410 model, IL-6 produced by DCs also impaired regulatory $CD4^+$ T cell (Treg) function, contributing to the T-cell hyperactivation associated with lupus.[239] Interestingly, this study found that increased IL-6 production mapped to the *Sle1* susceptibility locus, implying that primary genetic defect contributes to this pro-inflammatory cytokine imbalance.

Because IL-6 contributes to murine lupus in multiple ways, blocking IL-6 signaling has been evaluated in lupus therapy. Treatment of MRL/*lpr* mice with an anti-IL6-R Ab transiently reduced anti-dsDNA Ab levels.[240] IL-6 inhibition in human SLE using Tocilizumab, an Ab specific for the α chain of IL6-R, led to decreased numbers of circulating plasma cells and lowered titers of anti-dsDNA Abs.[241]

C. T-Cell Cytokines

Multiple evidence clearly show that T cells play an indispensable role in the pathogenesis of SLE in mice. Ab-mediated depletion of $CD4^+$ T cells reversed advanced lupus in NZB/W F1 mice.[242] Athymic NZB/W F1 mice do not develop lupus, and thymus transplantation to reconstitute T cells restored lupus symptoms in these mice.[243] T cells contribute to lupus in multiple ways, but largely through cytokines. Autoreactive T cells provide help to autoreactive B cells and facilitate the production of high-affinity, class-switched autoAbs. Different subsets of T-helper cells produce different cytokines, many of which are mediators of autoimmune responses and inflammation. Moreover, T cells infiltrate into end organs, such as the kidney, and function as effectors mediating tissue injury.

1. T Cells Provide Aberrant Help to B Cells in SLE

B cells activated by Ag in the periphery migrate to the T-cell zone in the secondary lymphoid tissues, where they receive help from cognate $CD4^+$ T-helper cells through the pairing of costimulatory membrane-bound molecules (including CD28-B7.1/B7.2, CD40L-CD40, and ICOS-ICOSL) and cytokines such as IFNγ, IL-4, IL-5, and IL-21. T-cell help can induce B cells either to differentiate extrafollicularly and become short-lived plasma cells that produce low-affinity Abs (extrafollicular pathway) or to migrate into the follicle and form GCs (intrafollicular/GC pathway). GC B cells undergo class-switching and somatic hypermutations and become long-lived plasma cells or memory cells that produce high-affinity and class-switched Abs. A distinct T-cell subset expressing Bcl-6, follicular helper T (Tfh) cells, is critical for the differentiation and functions of GC B cells.[244]

ANA production, the hallmark of SLE, has been shown to be both T-cell-independent and T-cell-dependent. T-cell-independent autoAb production occurs in lupus models in which strong signals in B cells override the need for T-cell help. This is the case for the MRL/*lpr* AM14 Tg mice in which the activation of autoreactive B cells is strongly dependent on TLR7 and TLR9 signaling[245,246] and the BAFF-Tg mice.[230] These autoAbs are generally low-affinity IgM produced by short-lived B cells outside the follicles. Most autoAbs in SLE, however, have high affinities and are skewed to T-cell-dependent isotypes, IgG1 in humans and IgG2a in mice, showing that Tfh cells and the GC response play an important role in autoAb production. This was directly demonstrated by the *Sanroque* mutant mice which develop lupus-like autoimmunity including anti-DNA IgG, autoimmune thrombocytopenia, lymphoid hyperplasia, and GN.[247] This mutation inactivates the RING-type ubiquitin ligase, Roquin, leading to the overexpression of ICOS and IL-21, large accumulation of Tfh cells, and spontaneous GC development. The *Sanroque* Tfh cells are responsible for the spontaneous GC formation, as transfer of *Sanroque* Tfh cells to normal mice caused GC formation. The suppression of the GC response abolishes the lupus-like disease in *Sanroque* mice, indicating a role for overreactive and overabundant Tfh cells in systemic autoimmunity,[248] as well as the critical need to keep Tfh cells tightly regulated.

A subset of $CD8^+$ T cells can suppress Tfh function through the interaction between the TCR/CD8 complex on $CD8^+$ cells and the Qa-1 molecules on Tfh cells. Qa-1 knock-in mice, which contain a Qa-1 amino acid exchange mutation that disrupts Qa-1 binding to the TCR/CD8 coreceptor, develop a lupus-like disease.[249] Moreover, the lupus-prone B6-Yaa mice have increased number of Tfh cells, and their $CD8^+$ regulatory T cells are deficient in suppressing $CD4^+$ T cells.[250] These findings clearly show that the dysregulation of Tfh cells is an important pathway in the development of SLE.

2. Many Cytokines Produced by Helper T Cell Mediate Autoimmune Response

T cells other than Tfh cells are also important in SLE, although the mechanisms by which they contribute to pathogenesis is less clear. According to the master transcriptional factors they express and the different cytokines they secrete, T cells are divided into several subsets. Transcription factor *T-bet* controls Th1 cell differentiation, which is induced by IL-12. The key cytokine produced by Th1 cells is IFNγ. Th2 cell differentiation is under the control of the transcription factor *GATA-3*, which is induced by IL-4. The key cytokines produced by Th2 cells are IL-4, IL-5, and IL-13. Th17 cell differentiation is under the control of transcription factor *RORγT*, which is induced by TGF-β and IL-6. The key cytokines produced by Th17 cells are IL-17 and IL-21. Development of each T-helper lineage is self-reinforcing and mutually antagonistic. For example, IFNγ induces the differentiation of Th1 cells and inhibits the differentiation of Th2 cells. Both IFNγ and IL-4 inhibit the differentiation of Th17 cells. We will discuss below how imbalances among these cytokines contribute to SLE in mouse models.

a. IFNγ. IFNγ is produced not only by Th1 cells but also by $CD8^+$ T cells, NK cells, and NKT cells. Multiple studies have shown that MRL/*lpr* mice produce more IFNγ than other mice.[251–253] Interestingly, peritoneal cells from MRL/*lpr* mice produced markedly higher concentrations of IL-12 than those from MRL/+ or BALB/c mice. Injection of IL-12 led to increased serum levels of IFNγ and accelerated GN in young MRL/*lpr* mice.[254] These results imply an association between high IL-12 levels, IFNγ hyperproduction, and disease in MRL/*lpr* mice. In addition, IFNγ treatment accelerated disease in NZW/B F1 mice, whereas inhibition of IFNγ signaling delayed disease and improved survival.[255] Genetic disruption of IFNγ or IFNγ receptor in MRL/*lpr*, NZW/B F1, and pristane-treated BALB/c mice protected them from the development of lupus.[220,222,223,256,257] Overall, evidence from mouse models uniformly identifies IFNγ as a detrimental factor and a potential therapeutic target in SLE.

How does IFNγ contribute to SLE? Macrophages from IFNγ-deficient MRL/*lpr* mice express markedly decreased levels of MHC class I and II molecules as compared with controls, which indicates that IFNγ hyperproduction may increase autoAg presentation to otherwise quiescent autoreactive T cells.[220] IFNγ may also contribute to lupus by facilitating isotype switching to IgG2a, which is efficient in fixing complement and mediating inflammation, as IFNγ-deficient MRL/*lpr* mice have a dramatic decrease in IgG2a immunoglobulins, serum autoAb, and renal pathology.[220,222] In MRL/*lpr* kidneys, IFNγ is required for the production of CSF-1 and TNFα, which induce apoptosis of renal parenchymal cells, a feature of human lupus nephritis.[256] Moreover, IFNγ-deficiency reduced the heightened expression of MHC-II molecules in the proximal tubules of MRL/*lpr* mice, indicating its role in mediating local inflammatory processes.[257] Thus, IFNγ is important in both global immune tolerance and local inflammation.

b. IL-4. The role of IL-4 in SLE is still elusive. Treatment of NZB/W F1 mice with anti-IL-4 Ab reduced the production of anti-dsDNA Ab and prevented lupus nephritis.[258] Soluble IL-4 receptor treatment and anti-IL-4 Ab treatment in MRL/*lpr* mice and in the cGVHD model also reduced mortality and disease.[259] In the cGVHD model, B cells from $CD4^+$ T-cell-deficient mice did not respond to alloreactive $CD4^+$ T cells to produce autoAbs, indicating that autoreactive B cells require T cell "nurturing" during early ontogeny. Treatment of $CD4^+$ T-cell-deficient recipient mice with IL-4, however, restored the B cell's ability to generate a cGVHD response. Collectively, these results show that IL-4 is required in SLE. However, IL-4 had no effect or was even protective in other mouse models. Although IL-4 deficiency reduced IgG1 Ab levels in BXSB/Yaa mice, it did not decrease overall disease and mortality.[260] IL-4 deficiency had also no effect on either autoAb production or kidney disease in pristane-induced lupus.[257] Moreover, Tg expression of IL-4 in B cells completely prevented the development of GN in (NZW × B6-Yaa) F1 males, accompanied by a shift of autoAb isotype from IgG2a and IgG3 to IgG1.[261]

c. IL-17. IL-17 plays an indispensable role in the pathogenesis of mouse models of induced encephalomyelitis and arthritis.[262] Their function in SLE became a field of interest more recently.[263–265] In SLE patients, $CD3^+$ $CD4^-$ $CD8^-$ double negative T (DNT) cells produce significant amounts of IL-17. Expanded numbers of these DNT cells can infiltrate the kidneys and mediate lupus nephritis.[266] Interestingly, the MRL/*lpr* strain is characterized by the accumulation of DNT cells in secondary lymphoid organs. It has been shown that their DNT cells also produce large amounts of IL-17 and mediate lupus nephritis.[267] Renal pathology has also been associated with infiltration of IL-17-producing $CD4^+$ T cells in the B6.*Sle2c1.lpr* strain that combined an NZM2410 susceptibility locus with the *lpr* mutation.[268] Moreover, the interruption of IL-23 signaling, which is essential for Th17 development and stabilization, abrogated the lupus phenotypes in B6/*lpr* mice with a reduction of DNT cells and their concomitant IL-17 production, showing a role of IL-23/IL-17 axis in lupus.[269] In the NZM2328 model, the disruption of TNFα signaling exacerbated lupus, which was associated with an expansion of IL-17-producing effector/memory $CD4^+$ T cells.[270] Overall, these results demonstrate a pathogenic role of IL-17 in SLE, suggesting the intervention of IL-17-producing cells as a potential therapeutic strategy.

VI. Tolerance Models

A breakdown in tolerance to self-Ags by B cells and T cells leads to the production of pathogenic autoAbs that is characteristic of SLE. Various Tg mouse models expressing autoreactive BCRs or TCRs have been developed to study the mechanisms of tolerance and how they are disrupted in SLE.

A. B-Cell Transgenic Models

Mouse models which express Tg as a rearranged immunoglobulin heavy chain (HC) with or without a rearranged immunoglobulin light chain (LC) to produce a BCR specific for an autoAg have been key to dissect out the mechanisms of B-cell tolerance, and to define a series of tolerance checkpoints along B-cell development and differentiation.[271] More specifically for lupus, Tg mice have been created to express BCRs specific to lupus-associated autoAgs. Some of the models that have made the most significant contributions to the field will be reviewed below.

1. Anti-DNA Tolerance: The 3H9 Model

A Tg mouse was created to study the fate of anti-DNA autoreactive B cells in both lupus-prone and non-autoimmune backgrounds. The 3H9 strain expresses as a Tg the HC of a monoclonal anti-DNA Ab derived from the MRL/*lpr* strain. The 3H9 HC in combination with various endogenous LCs produces Abs reactive to either ssDNA or dsDNA.[272] The regulation of these anti-DNA B cells was shown to differ in an autoimmune versus non-autoimmune backgrounds. In non-autoimmune BALB/c mice, 3H9 Tg B cells expressed high levels of anti-ssDNA BCRs but were unable to secrete anti-ssDNA Abs. Furthermore, in the double Tg 3H9/Vκ8 mice, most B cells were anti-ssDNA specific, but showed a decreased proliferation despite being long lived.[273] Meanwhile, anti-dsDNA 3H9 B cells were found to be Ag-experienced and accumulated at the T–B interface of the splenic follicle; where they appeared to be developmentally arrested and had a high turnover rate.[274] In contrast, anti-dsDNA 3H9 B cells were not developmentally arrested and were able to enter the B-cell follicles in MRL/*lpr* mice.[275] These results showed that anti-DNA autoreactive B cells are functionally inactivated in a non-autoimmune mouse, most likely because of their poor ability to compete with non-autoreactive polyclonal B cells for survival niches. In contrast, the autoimmune MRL/*lpr* background licensed the anti-DNA B cells to differentiate and survive into Ab-producing cells, indicating that the some of the checkpoints of peripheral tolerance are defective in that strain.

Site-directed 3H9/Vκ8 mice in the MRL/*lpr* background, in which Abs can undergo class-switching and somatic hypermutation, produced activated anti-ssDNA B cells that were able to acquire specificity to other autoAbs.[276] A mutation from aspartate to arginine at position 56 in the CDR2 region of the *3H9* gene (3H9/56R) resulted in higher affinity anti-DNA BCR that could bind to dsDNA with a larger number of LCs. In the non-autoimmune BALB/c background, 3H9/56R B cells undergo receptor editing to generate a BCR that does not bind to DNA. However, in the MRL/*lpr* background, most Tg B cells express LCs that bind to DNA.[277,278] This showed that the ability of undergoing receptor editing, a key feature of B-cell tolerance,[279] is impaired on the autoimmune MRL genetic background. This was confirmed in *in vitro* experiments using another BCR model expressed in MRL mice.[280] Therefore, the 3H9 model has identified multiple aspects of the dysregulation in B-cell tolerance toward an SLE-relevant self-Ag that occurs in a predisposed lupus background.

2. Anti-Rheumatoid Factor (RF) Tolerance: The AM14 Model

RFs are Abs directed against self-immunoglobulins (IgM or IgG) that are found in multiple autoimmune diseases including SLE. The most characterized RF Tg model expresses the AM14 HC which was derived from an IgM MRL/*lpr* hybridoma. The AM14 HC when combined with endogenous Vκ8 LCs form a BCR with specificity for IgG2a[a] Abs. The specificity of the AM14 RF for Abs of the "a" allotype allowed for the production of Tg mice expressing or lacking the IgG2a[a] Ag in either a non-autoimmune or lupus-prone background.[281]

AM14 HC Tg mice in the BALB/c background do not produce any RF either in the presence or absence of IgG2a[a]. Their AM14 Tg B cells are clonally ignorant rather than anergized or tolerized as they were capable of making primary and secondary immune responses to autoAg immunization.[282] However, in the MRL/*lpr* background, the AM14 B cells are spontaneously activated and differentiate into Ab-secreting cells (ASCs) in the presence of the IgG2a[a] autoAg.[283] In addition, the Tg B cells in the MRL/*lpr* background bypassed the GC reaction and instead proliferated into short-lived plasmablasts and underwent somatic hypermutation in extrafollicular zones.[284,285] A site-directed AM14 HC model confirmed that in the MRL/*lpr*, but not BALB/c background, activated class-switched RF B cells were present in extrafollicular clusters as well as in the spleen and bone marrow.[286]

Activation of the RF B cells in the lupus-prone background occurred due to the presence of ICs containing the IgG2a[a] autoAg.[287] Furthermore, it was shown that activation of the RF B cells required dual ligation of their BCR and TLR7/TLR9.[134,288] Administration of IgG2a[a] antichromatin hybridomas led to the activation of RF B cells in both MRL/*lpr* and non-autoimmune BALB/c mice.[289] In the presence of antichromatin IgG2a[a] Abs, AM14 B cells did not require T-cell help to differentiate into ASCs, class-switch, and undergo somatic hypermutation at extrafollicular sites.[245,246] However, T cells were able to modulate the response as the absence of T cells resulted in fewer RF plasmablasts and the number of somatic hypermutations was decreased. Furthermore, CD40L interactions as well as IL-21 signaling were shown to play a role in the T-cell-mediated expansion of the RF response. Therefore, the AM14 model has identified the mechanisms by which TRL7/TLR9-dependent autoAgs present in SLE could induce the activation of clonally ignorant autoreactive B cells.

3. Anti-Sm: The 2–12 Model

Abs against snRNPs, known as anti-Sm, are often found in SLE patients. The 2–12 HC derived from an anti-Sm MRL/*lpr* hybridoma was utilized to create a Tg mouse with a repertoire enriched for anti-Sm B cells.[290] The combination of the 2–12 HC with endogenous LCs also produced anti-ssDNA B cells as well as non-autoreactive B cells. 2–12 HC Tg mice with a non-autoimmune background did not secrete anti-Sm Abs. Sm-specific B cells were developmentally arrested as transitional B cells, although some B cells were completely differentiated but anergized, as they secreted Abs when immunized with snRNPs. In contrast, 2–12 HC Tg mice in the MRL/*lpr* background secreted large amounts of anti-Sm Abs in comparison to non-Tg MRL/*lpr* mice.[291] Therefore, the autoimmune background accelerates the anti-Sm response in 2–12 HC Tg mice, which is tolerized in a non-autoimmune background.

Characterization of the B-cell subsets in the 2–12HC Tg mice showed that non-autoimmune mice contained a high frequency of peritoneal anti-Sm B-1 B cells while these IgM-producing B cells were absent in the MRL/*lpr* Tg mice.[291,292] The differentiation of anti-Sm B cells to B-1 cells is dependent on a high BCR signaling strength as CD19 deficiency, which lowers BCR signaling, resulted in its differentiation to conventional B-2 cells. Furthermore, in non-autoimmune mice expressing the double 2–12/Vκ8 Tg, which produce low-affinity anti-Sm Abs, the anti-Sm B cells were B-2 cells and tolerized by anergy.[293] Meanwhile, non-autoimmune 2–12 HC Tg mice defective in the clearance of apoptotic cells had decreased levels of peritoneal anti-Sm B-1 cells as these cells were activated, differentiated into ASCs, and migrated out of the peritoneum.[294] Therefore, the 2–12 anti-Sm model has shown how the signaling threshold as well as the availability of the autoAg determine whether anti-Sm B cells differentiate into B-1 or B-2 cells as well as whether tolerance is maintained or broken.

B. T-Cell Tolerance

Autoreactive T cells are usually eliminated either by central tolerance in the thymus and peripheral tolerance through multiple mechanisms. In the thymus, negative selection induces apoptosis of autoreactive T cells. However, not all the self-Ags are expressed in the thymus. Therefore, autoreactive T cells that escape central tolerance must be suppressed in the periphery. Peripheral tolerance is usually maintained by anergy and dominant suppression by Tregs. How do autoreactive T cells remain at large in lupus? Most murine studies have concluded that central T-cell tolerance in SLE is intact. Singer and Abbas[295] bred a Tg TCR specific for a peptide of pigeon cytochrome *C* (PCC) into MRL/*lpr* and control $MRL^{+/+}$ mice. Subsequent administration of PCC to the mice caused the deletion of thymic T cells to a similar extent in both MRL/*lpr* and $MRL^{+/+}$ strains, indicating Fas deficiency does not affect central tolerance in MRL mice. Other studies have showed that central T cell tolerance is also normal in the NZB and MRL strains.[296,297]

Peripheral tolerance, however, is breached in lupus-prone mice. The same Tg system described above showed that PCC-induced deletion of peripheral T cells occurs in the $MRL^{+/+}$ but not in the MRL/*lpr* strain, indicating that peripheral T cell tolerance is Fas dependent.[295] Bouzahzah and colleagues generated PCC-specific TCR Tg mice in the Fas-intact MRL strain and H-2^k-matched non-autoimmune control CBA/CaJ and B10.BR mice.[298] When T cells from these strains were transferred into (MRL × CBA) F1 or (MRL × B10) F1 recipients that expressed membrane-bound PCC (mPCC) as a Tg self-Ag, only CBA and B10 T cells showed the typical phenotype of anergic cells, whereas MRL T cells avoided anergy. These results demonstrate that T-cell anergy induction is defective in the MRL background, which allows the maintenance and activation of autoreactive T cells.

$CD4^+$ $CD25^+$ $FoxP3^+$ T regulatory (Treg) cells play a critical role in maintaining peripheral tolerance. Natural Tregs (nTreg) develop in the thymus, whereas inducible Tregs (iTreg) develop in the periphery in response to T-cell stimulation in the presence of IL-2 and TGFβ. Although the surface markers that can distinguish between nTreg and iTreg are not clearly defined, it has been shown that they have similar distribution and function, but may differ in their stability.[299] Several mechanisms by which Treg suppression of T effector (Teff) cells have been proposed, including the secretion of immunoregulatory cytokines such as IL-10 and TGFβ, and the competition for growth factors, such as IL-2, with Teffs.

A number of studies have shown that lupus-prone mice have a lower percentage of Tregs.[300] While young NZW/B F1 mice have a reduced percentage of Tregs compared to age-matched BALB/c mice, this difference, however, disappears when mice get older, suggesting that the expansion of activated Teff drives the expansion of Tregs.[301] Depletion of Tregs in NZB/W F1 mice 3 days after birth by the administration of anti-CD25 Ab accelerates disease,[302] suggesting that Tregs still exert a functional suppression in this lupus model. Furthermore, systemic reduction of IL-2 levels in NZB/W F1 mice altered the homeostatic balance between Treg and Teff in favor of the latter, resulting in a sharp disease acceleration.[303] Several studies have shown a normal suppression by Tregs in lupus-prone mice, but MRL/*lpr* Teffs are resistant to Treg suppression[304] and intrinsic defects in both Teff and APCs impair Treg suppression in that strain.[305] The NZM2410-derived lupus-susceptibility locus *Sle1a* is associated with both a lower number of Treg and a lower Treg suppressive ability, which correlate with an expanded population of autoreactive and activated memory $CD4^+$ T cells in mice expressing the *Sle1a* locus.[306,307] These latter results demonstrated that Treg impairments associated with SLE are genetically determined.

Given the important role of Tregs in peripheral tolerance, it has been proposed that adoptive transfer of *ex vivo*-expanded Tregs may have a therapeutic effect on SLE. Two studies in NZB/W F1 mice have shown that transfer of exogenously expanded Tregs can significantly slow down disease progression and improve survival.[301,308] This further suggested that a defect in Treg numbers or function may contribute to lupus pathogenesis. Moreover, it has been found that administration of tolerogenic peptides, such as pCons from anti-DNA IgG in NZB/W F1 mice[309] or histone peptide H471 in SNF1 mice,[310] ameliorated disease by expanding the Treg subset. Clearly, modulating Treg number and function might be a promising way to treat human SLE.

VII. Therapies

Murine models of lupus have been essential for the development of new therapies for the treatment of SLE. These models have led to the discovery and validation of therapeutic targets as they allow for a thorough analysis of molecular pathways and their alteration in the disease state. Therapeutic translations have been mentioned throughout this review for the relevant targets. Recently, two biologics, Rituximab and Belimumab, have been tested in extensive clinical trials in SLE patients with different outcomes. Murine models were indispensable in their development as well as in the interpretation of results obtained from these trials.

A. Rituximab

Rituximab is an anti-CD20 monoclonal Ab that targets and depletes B lymphocytes. Binding of Rituximab to target cells induces cell death via complement and FcγR-mediated pathways.[311] Rituximab was viewed as a potential SLE treatment by eliminating the cells that produce the pathogenic autoAbs characteristic of this disease. Rituximab treatment was shown to ameliorate disease activity in two other autoimmune diseases, rheumatoid arthritis and multiple sclerosis,[312,313] although it may not be through the elimination of autoAbs.

Two Rituximab clinical trials have been performed in SLE patients, but have failed to meet primary clinical endpoints. EXPLORER, a randomized, Phase II/III trial of nonrenal lupus patients, showed no difference in clinical disease when comparing treated versus placebo groups as determined using the British Isles Lupus Assessment Group (BILAG) index.[314] However, a clinical response was apparent in the African American/Hispanic group, suggesting the efficiency of Rituximab varies among ethnic groups. The Rituximab-treated group did show decreased levels of anti-dsDNA Abs and increased complement C3 and C4 levels. LUNAR, a clinical trial of lupus nephritis patients, was designed to investigate if Rituximab improved renal pathology. Nonetheless, the LUNAR trial also failed to meet its primary endpoint although a positive trend was seen in African Americans and ANAs were decreased.[315]

A potential drawback of Rituximab treatment for B-cell depletion in SLE patients is the absence of CD20 expression on mature plasma cells that are already secreting pathogenic autoAbs. Murine models have been used to model Rituximab therapy and the reasons for its lack of success in SLE patients. Anti-CD20 treatments in MRL/*lpr* and NZB/W F1 mice expressing humanized CD20 were not as efficient as in non-autoimmune strains.[316] Although depletion of circulating blood B cells was successful in the lupus-prone strains, depletion in their lymphoid tissues was incomplete, implicating that the autoimmune genetic background may impact the efficiency of B-cell depletion. However, anti-CD20 treatments did reduce but did not eliminate autoimmune pathology, with a decrease in GN and autoAbs production. Furthermore, a recent study showed that the timing for the initiation of anti-CD20 treatment played a major role in the efficacy of the treatment.[317] Anti-CD20 administration to NZB/W F1 mice prior to disease onset resulted in decreased survival and accelerated the emergence of proteinuria, most likely due to the depletion of regulatory B cells. Initiation of treatment at the onset of disease symptom proved to be the most beneficial, resulting in enhanced survival and delayed proteinuria. However, treatment initiated late after disease onset had no effect, indicating that B cell depletion is not effective once disease symptoms are well developed. Therefore, results obtained from murine models serve to illustrate the complexities of B-cell depletion as a treatment for SLE.

B. Belimumab

Belimumab is a monoclonal Ab that binds soluble BAFF with high affinity, and prevents it from ligating any of its three receptors (TACI, BCMA, BAFFR). In part, based on the results obtained in the mouse (see above), clinical trials were conducted in SLE patients. An initial Phase I trial of patients with mild to moderate SLE demonstrated that Belimumab led to decreased numbers of $CD20^+$ B cells.[318] The Phase II trial of Belimumab did not meet its primary endpoint, which was a significant change in the Safety of Estrogens in Lupus Erythematosus National Assessment-SLE Disease Activity Index (SELENA-SLEDAI). However, a significant decrease in SELENA-SLEDAI was seen in ANA positive patients.[319] Two Phase III Belimumab clinical trials, BLISS-52 and BLISS-76, met their primary endpoints. Unlike the Phase II trial, these trials only contained SLE patients that were ANA positive at the start of the study. The primary endpoint for the BLISS-52 trial was improvement in the Systemic Lupus Responder Index (SRI) by week 52. In this trial, a dose-dependent effect was seen when comparing Belimumab-treated SLE patients versus placebo. Treated patients showed a reduction in flares, anti-dsDNA autoAbs, and presented decreased hypergammaglobulinemia and elevated complement levels.[320] The BLISS-76 clinical trial had the same primary endpoint at week 52 as BLISS-52, but was extended by an additional 24 weeks to determine if the response changed with time. The secondary endpoint at week 76 also showed a higher response in Belimumab-treated patients; however, it did not reach statistical significance.[315] Nevertheless, the positive outcome of Belimumab in two different clinical trials demonstrated that the addition of Belimumab to standard of care was beneficial in ameliorating disease symptoms. In March 2011, Belimumab became the first drug in more than 50 years to be approved by the US Food and Drug Administration (FDA) for the treatment of SLE.

VIII. Conclusions

SLE is a complex disease for which current therapies only alleviate symptoms, but at the cost of significant deleterious side effects. In addition to these clinical considerations, SLE is the prototypic disease resulting from systemic autoimmunity. Therefore, the elucidation of many aspects of SLE pathogenesis has contributed greatly to the understanding of the mechanisms of systemic B and T cell tolerance. The availability of an array of diverse mouse models of SLE, spontaneous, induced, or genetically engineered, has greatly contributed to the large body of data that has accumulated in the past 20 years, deciphering the alterations of the immune system and target organs that lead to the break of tolerance to nucleic acid–protein complexes and ensuing clinical manifestations. Although none of the murine models reproduce faithfully SLE in its vast clinical heterogeneity, they have collectively delineated cellular and genetic pathogenic pathways that remarkably shared a great level identity with human SLE.[18]

Research on human SLE has advanced in great strides in the past few years, with many genetic determinants identified and a plethora of clinical trials being conducted on various targets. This does not call for a more limited role of murine models, but redefines some of the ways in which they will be used. More specifically, murine models will be increasingly used to specifically dissect the mechanisms of cellular observations that have been made in humans, increasing the interdependence between the two types of research. In addition, the refinement of humanized mouse models will most likely provide in the near future venues in which the immune cells from SLE patients can be directly manipulated and characterized in the mouse.

References

1. Pons-Estel GJ, Alarcón GS, Scofield L, Reinlib L, Cooper GS. Understanding the epidemiology and progression of systemic lupus erythematosus. *Semin Arthritis Rheum* 2010;**39**:257–68.
2. Kotzin BL. Systemic lupus erythematosus. *Cell* 1996;**85**:303–6.
3. Helyer BJ, Howie JB. Renal disease associated with positive lupus erythematosus tests in a cross-bred strain of mice. *Nature* 1963;**197**:197.

4. Rudofsky U, Evans B, Balaban S, Mottironi V, Gabrielsen A. Differences in expression of lupus nephritis in New Zealand mixed H-2z homozygous inbred strains of mice derived from New Zealand black and New Zealand white mice. Origins and initial characterization. *Lab Invest* 1993;**68**:419–26.
5. Morel L, Rudofsky UH, Longmate JA, Schiffenbauer J, Wakeland EK. Polygenic control of susceptibility to murine systemic lupus erythematosus. *Immunity* 1994;**1**:219–29.
6. Waters ST, Fu SM, Gaskin F, Deshmukh US, Sung SS, Kannapell CC, et al. NZM2328: a new mouse model of systemic lupus erythematosus with unique genetic susceptibility loci. *Clin Immunol* 2001;**100**:372–83.
7. Andrews BS, Eisenberg RA, Theofilopoulos AN, Izui S, Wilson CB, McConahey PJ, et al. Spontaneous murine lupus-like syndromes. Clinical and immunopathological manifestations in several strains. *J Exp Med* 1978;**148**:1198–215.
8. Murphy ED, Roths JB. A y chromosome associated factor in strain bxsb producing accelerated autoimmunity and lymphoproliferation. *Arthritis Rheum* 1979;**22**:1188–94.
9. Bygrave AE, Rose KL, Cortes-Hernandez J, Warren J, Rigby RJ, Cook HT, et al. Spontaneous autoimmunity in 129 and C57BL/6 mice-implications for autoimmunity described in gene-targeted mice. *PLoS Biol* 2004;**2**:E243.
10. Satoh M, Reeves W. Induction of lupus-associated autoantibodies in BALB/c mice by intraperitoneal injection of pristane. *J Exp Med* 1994;**180**:2341–6.
11. Via CS. Advances in lupus stemming from the parent-into-F1 model. *Trends Immunol* 2010;**31**:236–45.
12. Morris SC, Cohen PL, Eisenberg RA. Experimental induction of systemic lupus erythematosus by recognition of foreign Ia. *Clin Immunol Immunopathol* 1990;**57**(2):263–73.
13. Obata Y, Tanaka T, Stockert E, Good RA. Autoimmune and lymphoproliferative disease in (B6-GIX + X 129)F1 mice: relation to naturally occurring antibodies against murine leukemia virus-related cell surface antigens. *Proc Natl Acad Sci USA* 1979;**76**:5289–93.
14. Morel L. Mouse models of human autoimmune diseases: essential tools that require the proper controls. *PLoS Biol* 2004;**2**:E241.
15. Theofilopoulos A, Dixon F. Murine models of systemic lupus erythematosus. *Adv Immunol* 1985;**37**:269–390.
16. Kono D, Theofilopoulos A. Genetics of SLE in mice. *Springer Semin Immunopathol* 2006;**28**:83–96.
17. Perry D, Sang A, Yin Y, Zheng YY, Morel L. Murine models of systemic lupus erythematosus. *J Biomed Biotechnol* 2011;**2011**:271694.
18. Morel L. Genetics of SLE: evidence from mouse models. *Nat Rev Rheumatol* 2010;**6**:348–57.
19. Chu J, Drappa J, Parnassa A, Elkon K. The defect in Fas mRNA expression in MRL/lpr mice is associated with insertion of the retrotransposon, ETn. *J Exp Med* 1993;**178**:723–30.
20. Watson M, Rao J, Gilkeson G, Ruiz P, Eicher E, Pisetsky D, et al. Genetic analysis of MRL-lpr mice: relationship of the Fas apoptosis gene to disease manifestations and renal disease-modifying loci. *J Exp Med* 1992;**176**:1645–56.
21. Watanabe-Fukunaga R, Brannan C, Copeland N, Jenkins N, Nagata S. Lymphoproliferation disorder in mice explained by defects in Fas antigen that mediates apoptosis. *Nature* 1992;**356**:314–7.
22. Strasser A, Jost P, Nagata S. The many roles of FAS receptor signaling in the immune system. *Immunity* 2009;**30**:180–92.
23. Takahashi T, Tanaka M, Brannan C, Jenkins N, Copeland N, Suda T, et al. Generalized lymphoproliferative disease in mice, caused by a point mutation in the Fas ligand. *Cell* 1994;**76**:969–76.
24. Lynch D, Watson M, Alderson M, Baum P, Miller R, Tough T, et al. The mouse Fas-ligand gene is mutated in gld mice and is part of a TNF family gene cluster. *Immunity* 1994;**1**:131–6.

25. Kelley VE, Roths JB. Interaction of mutant lpr gene with background strain influences renal disease. *Clin Immunol Immunopathol* 1985;**37**:220–9.
26. Maibaum MA, Haywood MEK, Walport MJ, Morley BJ. Lupus susceptibility loci map within regions of BXSB derived from the SB/Le parental strain. *Immunogenetics* 2000;**51**:370–2.
27. Eisenberg R, Dixon F. Effect of castration on male-determined acceleration of autoimmune disease in BXSB mice. *J Immunol* 1980;**125**:1959–61.
28. Hudgins C, Steinberg R, Klinman D, Reeves M, Steinberg A. Studies of consomic mice bearing the Y chromosome of the BXSB mouse. *J Immunol* 1985;**134**:3849–54.
29. Merino R, Shibata T, De Kossodo S, Izui S. Differential effect of the autoimmune Yaa and lpr genes on the acceleration of lupus-like syndrome in MRL/MpJ mice. *Eur J Immunol* 1989;**19**:2131–7.
30. Pisitkun P, Deane JA, Difilippantonio MJ, Tarasenko T, Satterthwaite AB, Bolland S. Autoreactive B cell responses to RNA-related antigens due to TLR7 gene duplication. *Science* 2006;**312**:1669–72.
31. Subramanian S, Tus K, Li QZ, Wang A, Tian XH, Zhou J, et al. A Tlr7 translocation accelerates systemic autoimmunity in murine lupus. *Proc Natl Acad Sci USA* 2006;**103**:9970–5.
32. Fairhurst AM, Hwang SH, Wang A, Tian XH, Boudreaux C, Zhou XJ, et al. Yaa autoimmune phenotypes are conferred by overexpression of TLR7. *Eur J Immunol* 2008;**38**:1971–8.
33. Deane JA, Pisitkun P, Barrett RS, Feigenbaum L, Town T, Ward JM, et al. Control of toll-like receptor 7 expression is essential to restrict autoimmunity and dendritic cell proliferation. *Immunity* 2007;**27**:801–10.
34. Santiago-Raber ML, Kikuchi S, Borel P, Uematsu S, Akira S, Kotzin BL, et al. Evidence for genes in addition to Tlr7 in the Yaa translocation linked with acceleration of systemic lupus erythematosus. *J Immunol* 2008;**181**:1556–62.
35. Botto M, Dell'Agnola C, Bygrave AE, Thompson EM, Cook HT, Petry F, et al. Homozygous C1q deficiency causes glomerulonephritis associated with multiple apoptotic bodies. *Nat Genet* 1998;**19**:56–9.
36. Mitchell DA, Pickering MC, Warren J, Fossati-Jimack L, Cortes-Hernandez J, Cook HT, et al. C1q deficiency and autoimmunity: the effects of genetic background on disease expression. *J Immunol* 2002;**168**:2538–43.
37. Hanayama R, Tanaka M, Miyasaka K, Aozasa K, Koike M, Uchiyama Y, et al. Autoimmune disease and impaired uptake of apoptotic cells in MFG-E8-deficient mice. *Science* 2004;**304**:1147–50.
38. Peng Y, Elkon KB. Autoimmunity in MFG-E8-deficient mice is associated with altered trafficking and enhanced cross-presentation of apoptotic cell antigens. *J Clin Invest* 2011;**121**:2221–41.
39. Carlucci F, Cortes-Hernandez J, Fossati-Jimack L, Bygrave AE, Walport MJ, Vyse TJ, et al. Genetic dissection of spontaneous autoimmunity driven by 129-derived chromosome 1 Loci when expressed on C57BL/6 mice. *J Immunol* 2007;**178**:2352–60.
40. Heidari Y, Bygrave AE, Rigby RJ, Rose KL, Walport MJ, Cook HT, et al. Identification of chromosome intervals from 129 and C57BL/6 mouse strains linked to the development of systemic lupus erythematosus. *Genes Immun* 2006;**7**:592–9.
41. Carlucci F, Fossati-Jimack L, Dumitriu IE, Heidari Y, Walport MJ, Szajna M, et al. Identification and characterization of a lupus suppressor 129 locus on chromosome 3. *J Immunol* 2010;**184**:6256–65.
42. Heidari Y, Fossati-Jimack L, Carlucci F, Walport MJ, Cook HT, Botto M. A lupus-susceptibility C57BL/6 locus on chromosome 3 (Sle18) contributes to autoantibody production in 129 mice. *Genes Immun* 2009;**10**:47–55.

43. Satoh M, Richards H, Shaheen V, Yoshida H, Shaw M, Naim J, et al. Widespread susceptibility among inbred mouse strains to the induction of lupus autoantibodies by pristane. *Clin Exp Immunol* 2000;**121**:399–405.
44. Smith D, Dong X, Du S, Oh M, Singh R, Voskuhl R. A female preponderance for chemically induced lupus in SJL/J mice. *Clin Immunol* 2007;**122**:101–7.
45. Lee P, Li Y, Kumagai Y, Xu Y, Weinstein J, Kellner E, et al. Type I interferon modulates monocyte recruitment and maturation in chronic inflammation. *Am J Pathol* 2009;**175**:2023–33.
46. Nacionales DC, Kelly-Scumpia KM, Lee PY, Weinstein JS, Lyons R, Sobel E, et al. Deficiency of the type I interferon receptor protects mice from experimental lupus. *Arthritis Rheum* 2007;**56**:3770–83.
47. Van Rappard-Van Der Veen FM, Radaszkiewicz T, Terraneo L, Gleichmann E. Attempts at standardization of lupus-like graft-vs-host disease: inadvertent repopulation by DBA/2 spleen cells of H-2-different nonirradiated F1 mice. *J Immunol* 1983;**130**:2693–701.
48. Rus V, Nguyen V, Puliaev R, Puliaeva I, Zernetkina V, Luzina I, et al. T cell TRAIL promotes murine lupus by sustaining effector CD4 Th cell numbers and by inhibiting CD8 CTL activity. *J Immunol* 2007;**178**:3962–72.
49. Grader-Beck T, Casciola-Rosen L, Lang TJ, Puliaev R, Rosen A, Via CS. Apoptotic splenocytes drive the autoimmune response to poly(ADP-ribose) polymerase 1 in a murine model of lupus. *J Immunol* 2007;**178**:95–102.
50. Lang TJ, Nguyen P, Papadimitriou JC, Via CS. Increased severity of murine lupus in female mice is due to enhanced expansion of pathogenic T cells. *J Immunol* 2003;**171**:5795–801.
51. Jevnikar AM, Grusby MJ, Glimcher LH. Prevention of nephritis in major histocompatibility complex class II-deficient MRL-lpr mice. *J Exp Med* 1994;**179**:1137–43.
52. Christianson GJ, Blankenburg RL, Duffy TM, Panka D, Roths JB, Marshak-Rothstein A, et al. beta2-microglobulin dependence of the lupus-like autoimmune syndrome of MRL-lpr mice. *J Immunol* 1996;**156**:4932–9.
53. Taylor KE, Chung SA, Graham RR, Ortmann WA, Lee AT, Langefeld CD, et al. Risk alleles for systemic lupus erythematosus in a large case–control collection and associations with clinical subphenotypes. *PLoS Genet* 2011;**7**:e1001311.
54. Ramos PS, Langefeld CD, Bera LA, Gaffney PM, Noble JA, Moser KL. Variation in the ATP-binding cassette transporter 2 gene is a separate risk factor for systemic lupus erythematosus within the MHC. *Genes Immun* 2009;**10**:350–5.
55. Babcock SK, Appel VB, Schiff M, Palmer E, Kotzin BL. Genetic analysis of the imperfect association of H-2 haplotype with lupus-like autoimmune disease. *Proc Natl Acad Sci USA* 1989;**86**:7552–5.
56. Kotzin BL, Palmer E. The contribution of NZW genes to lupus-like disease in (NZB x NZW) F1 mice. *J Exp Med* 1987;**165**:1237–51.
57. Kono DH, Burlingame RW, Owens DG, Kuramochi A, Balderas RS, Balomenos D, et al. Lupus susceptibility loci in New Zealand mice. *Proc Natl Acad Sci USA* 1994;**91**:10168–72.
58. Morel L, Tian X, Croker B, Wakeland E. Epistatic modifiers of autoimmunity in a murine model of lupus nephritis. *Immunity* 1999;**11**:131–9.
59. Subramanian S, Yim YS, Liu K, Tus K, Zhou XJ, Wakeland EK. Epistatic suppression of systemic lupus erythematosus: fine mapping of Sles1 to less than 1 mb. *J Immunol* 2005;**175**:1062–72.
60. Vyse TJ, Drake CG, Rozzo SJ, Roper E, Izui S, Kotzin BL. Genetic linkage of IgG autoantibody production in relation to lupus nephritis in New Zealand hybrid mice. *J Clin Invest* 1996;**98**:1762–72.
61. Hogarth MB, Slingsby JH, Allen PJ, Thompson EM, Chandler P, Davies KA, et al. Multiple lupus susceptibility loci map to chromosome 1 in BXSB mice. *J Immunol* 1998;**161**:2753–61.

62. Sekine H, Graham KL, Zhao S, Elliott MK, Ruiz P, Utz PJ, et al. Role of MHC-linked genes in autoantigen selection and renal disease in a murine model of systemic lupus erythematosus. *J Immunol* 2006;**177**:7423–34.
63. Hannestad K, Scott H. The MHC haplotype H2b converts two pure nonlupus mouse strains to producers of antinuclear antibodies. *J Immunol* 2009;**183**:3542–50.
64. Ibnou-Zekri N, Iwamoto M, Fossati L, McConahey PJ, Izui S. Role of the major histocompatibility complex class II Ea gene in lupus susceptibility in mice. *Proc Natl Acad Sci USA* 1997;**94**:14654–9.
65. Merino R, Iwamoto M, Fossati L, Muniesa P, Araki K, Takahashi S, et al. Prevention of systemic lupus erythematosus in autoimmune BXSB mice by a transgene encoding I-E alpha chain. *J Exp Med* 1993;**178**:1189–97.
66. Ibnou-Zekri N, Iwamoto M, Gershwin ME, Izui S. Protection of murine lupus by the Ead transgene is MHC haplotype-dependent. *J Immunol* 2000;**164**:505–11.
67. Martínez-Soria E, Santiago-Raber ML, Ho L, Moll T, Izui S. Protection of murine systemic lupus by the Ea transgene without expression of I-E heterodimers. *J Immunol* 2008;**181**:3651–7.
68. Zhang D, Fujio K, Jiang Y, Zhao J, Tada N, Sudo K, et al. Dissection of the role of MHC class II A and E genes in autoimmune susceptibility in murine lupus models with intragenic recombination. *Proc Natl Acad Sci USA* 2004;**101**:13838–43.
69. Vyse TJ, Rozzo SJ, Drake CG, Appel VB, Lemeur M, Izui S, et al. Contributions of Ea(z) and Eb(z) MHC genes to lupus susceptibility in New Zealand mice. *J Immunol* 1998;**160**:2757–66.
70. Rozzo SJ, Vyse TJ, David CS, Palmer E, Izui S, Kotzin BL. Analysis of MHC class II genes in the susceptibility to lupus in New Zealand mice. *J Immunol* 1999;**162**:2623–30.
71. Paisansinsup T, Vallejo AN, Luthra H, David CS. HLA-DR modulates autoantibody repertoire, but not mortality, in a humanized mouse model of systemic lupus erythematosus. *J Immunol* 2001;**167**:4083–90.
72. Stohl W, Xu D, Metzger TE, Kim KS, Morel L, Kotzin BL. Dichotomous effects of complete versus partial class II major histocompatibility complex deficiency on circulating autoantibody levels in autoimmune-prone mice. *Arthritis Rheum* 2004;**50**:2227–39.
73. Jacob CO, McDevitt HO. Tumour necrosis factor-alpha in murine autoimmune 'lupus' nephritis. *Nature* 1988;**331**:356–8.
74. Jacob CO, Hwang F, Lewis GD, Stall AM. Tumor necrosis factor alpha in murine systemic lupus erythematosus disease models: implications for genetic predisposition and immune regulation. *Cytokine* 1991;**3**:551–61.
75. Jongeneel CV, Acha-Orbea H, Blankenstein T. A polymorphic microsatellite in the tumor necrosis factor alpha promoter identifies an allele unique to the NZW mouse strain. *J Exp Med* 1990;**171**:2141–6.
76. Fujimura T, Hirose S, Jiang Y, Kodera S, Ohmuro H, Zhang D, et al. Dissection of the effects of tumor necrosis factor-alpha and class II gene polymorphisms within the MHC on murine systemic lupus erythematosus (SLE). *Int Immunol* 1998;**10**:1467–72.
77. Jacob CO, Fronek Z, Lewis GD, Koo M, Hansen JA, McDevitt HO. Heritable major histocompatibility complex class II-associated differences in production of tumor necrosis factor alpha: relevance to genetic predisposition to systemic lupus erythematosus. *Proc Natl Acad Sci USA* 1990;**87**:1233–7.
78. Perl A, Fernandez DR, Telarico T, Doherty E, Francis L, Phillips PE. T-cell and B-cell signaling biomarkers and treatment targets in lupus. *Curr Opin Rheumatol* 2009;**21**:454–64.
79. Lipsky PE. Systemic lupus erythematosus: an autoimmune disease of B cell hyperactivity. *Nat Immunol* 2001;**2**:764–6.

80. Liossis SN, Kovacs B, Dennis G, Kammer GM, Tsokos GC. B cells from patients with systemic lupus erythematosus display abnormal antigen receptor-mediated early signal transduction events. *J Clin Invest* 1996;**98**:2549–57.
81. Wu T, Qin X, Kurepa Z, Kumar KR, Liu K, Kanta H, et al. Shared signaling networks active in B cells isolated from genetically distinct mouse models of lupus. *J Clin Invest* 2007;**117**:2186–96.
82. Liossis SN, Solomou EE, Dimopoulos MA, Panayiotidis P, Mavrikakis MM, Sfikakis PP. B-cell kinase lyn deficiency in patients with systemic lupus erythematosus. *J Investig Med* 2001;**49**:157–65.
83. Xu Y, Harder KW, Huntington ND, Hibbs ML, Tarlinton DM. Lyn tyrosine kinase: accentuating the positive and the negative. *Immunity* 2005;**22**:9–18.
84. Hibbs ML, Tarlinton DM, Armes J, Grail D, Hodgson G, Maglitto R, et al. Multiple defects in the immune system of Lyn-deficient mice, culminating in autoimmune disease. *Cell* 1995;**83**:301–11.
85. Nishizumi H, Taniuchi I, Yamanashi Y, Kitamura D, Ilic D, Mori S, et al. Impaired proliferation of peripheral B cells and indication of autoimmune disease in lyn-deficient mice. *Immunity* 1995;**3**:549–60.
86. Takeshita H, Taniuchi I, Kato J, Watanabe T. Abrogation of autoimmune disease in Lyn-deficient mice by the mutation of the Btk gene. *Int Immunol* 1998;**10**:435–44.
87. Gutierrez T, Halcomb KE, Coughran AJ, Li QZ, Satterthwaite AB. Separate checkpoints regulate splenic plasma cell accumulation and IgG autoantibody production in Lyn-deficient mice. *Eur J Immunol* 2010;**40**:1897–905.
88. Tsantikos E, Oracki SA, Quilici C, Anderson GP, Tarlinton DM, Hibbs ML. Autoimmune disease in Lyn-deficient mice is dependent on an inflammatory environment established by IL-6. *J Immunol* 2010;**184**:1348–60.
89. Flores-Borja F, Kabouridis PS, Jury EC, Isenberg DA, Mageed RA. Decreased Lyn expression and translocation to lipid raft signaling domains in B lymphocytes from patients with systemic lupus erythematosus. *Arthritis Rheum* 2005;**52**:3955–65.
90. O'Keefe TL, Williams GT, Batista FD, Neuberger MS. Deficiency in CD22, a B cell-specific inhibitory receptor, is sufficient to predispose to development of high affinity autoantibodies. *J Exp Med* 1999;**189**:1307–13.
91. Sato S, Miller AS, Inaoki M, Bock CB, Jansen PJ, Tang ML, et al. CD22 is both a positive and negative regulator of B lymphocyte antigen receptor signal transduction: altered signaling in CD22-deficient mice. *Immunity* 1996;**5**:551–62.
92. Nadler MJ, McLean PA, Neel BG, Wortis HH. B cell antigen receptor-evoked calcium influx is enhanced in CD22-deficient B cell lines. *J Immunol* 1997;**159**:4233–43.
93. Nitschke L, Lajaunias F, Moll T, Ho L, Martinez-Soria E, Kikuchi S, et al. Expression of aberrant forms of CD22 on B lymphocytes in Cd22a lupus-prone mice affects ligand binding. *Int Immunol* 2006;**18**:59–68.
94. Mary C, Laporte C, Parzy D, Santiago ML, Stefani F, Lajaunias F, et al. Dysregulated expression of the Cd22 gene as a result of a short interspersed nucleotide element insertion in Cd22a lupus-prone mice. *J Immunol* 2000;**165**:2987–96.
95. Pillai S, Cariappa A, Pirnie SP. Esterases and autoimmunity: the sialic acid acetylesterase pathway and the regulation of peripheral B cell tolerance. *Trends Immunol* 2009;**30**:488–93.
96. Cariappa A, Takematsu H, Liu H, Diaz S, Haider K, Boboila C, et al. B cell antigen receptor signal strength and peripheral B cell development are regulated by a 9-O-acetyl sialic acid esterase. *J Exp Med* 2009;**206**:125–38.
97. Huck S, Le Corre R, Youinou P, Zouali M. Expression of B cell receptor-associated signaling molecules in human lupus. *Autoimmunity* 2001;**33**:213–24.

98. Traczewski P, Rudnicka L. Treatment of systemic lupus erythematosus with epratuzumab. *Br J Clin Pharmacol* 2011;**71**:175–82.
99. Daridon C, Blassfeld D, Reiter K, Mei HE, Giesecke C, Goldenberg DM, et al. Epratuzumab targeting of CD22 affects adhesion molecule expression and migration of B-cells in systemic lupus erythematosus. *Arthritis Res Ther* 2010;**12**:R204.
100. Surolia I, Pirnie SP, Chellappa V, Taylor KN, Cariappa A, Moya J, et al. Functionally defective germline variants of sialic acid acetylesterase in autoimmunity. *Nature* 2010;**466**:243–7.
101. Ravetch JV, Bolland S. IgG Fc receptors. *Annu Rev Immunol* 2001;**19**:275–90.
102. Smith KG, Clatworthy MR. FcgammaRIIB in autoimmunity and infection: evolutionary and therapeutic implications. *Nat Rev Immunol* 2010;**10**:328–43.
103. Kalergis AM, Ravetch JV. Inducing tumor immunity through the selective engagement of activating Fcgamma receptors on dendritic cells. *J Exp Med* 2002;**195**:1653–9.
104. Jiang Y, Hirose S, Abe M, Sanokawa-Akakura R, Ohtsuji M, Mi X, et al. Polymorphisms in IgG Fc receptor IIB regulatory regions associated with autoimmune susceptibility. *Immunogenetics* 2000;**51**:429–35.
105. Rahman ZSM, Niu H, Perry D, Wakeland E, Manser T, Morel L. Expression of the autoimmune Fcgr2b NZW allele fails to be upregulated in germinal center B cells and is associated with increased IgG production. *Genes Immun* 2007;**8**:604–12.
106. Bolland S, Ravetch JV. Spontaneous autoimmune disease in Fc(gamma)RIIB-deficient mice results from strain-specific epistasis. *Immunity* 2000;**13**:277–85.
107. Fukuyama H, Nimmerjahn F, Ravetch JV. The inhibitory Fcgamma receptor modulates autoimmunity by limiting the accumulation of immunoglobulin G+ anti-DNA plasma cells. *Nat Immunol* 2005;**6**:99–106.
108. Tiller T, Kofer J, Kreschel C, Busse CE, Riebel S, Wickert S, et al. Development of self-reactive germinal center B cells and plasma cells in autoimmune Fc gammaRIIB-deficient mice. *J Exp Med* 2010;**207**:2767–78.
109. Lee YH, Ji JD, Song GG. Fcgamma receptor IIB and IIIB polymorphisms and susceptibility to systemic lupus erythematosus and lupus nephritis: a meta-analysis. *Lupus* 2009;**18**:727–34.
110. Mackay M, Stanevsky A, Wang T, Aranow C, Li M, Koenig S, et al. Selective dysregulation of the FcgammaIIB receptor on memory B cells in SLE. *J Exp Med* 2006;**203**:2157–64.
111. Su K, Yang H, Li X, Gibson AW, Cafardi JM, Zhou T, et al. Expression profile of Fcgamma-RIIb on leukocytes and its dysregulation in systemic lupus erythematosus. *J Immunol* 2007;**178**:3272–80.
112. Crispín JC, Kyttaris VC, Terhorst C, Tsokos GC. T cells as therapeutic targets in SLE. *Nat Rev Rheumatol* 2010;**6**:317–25.
113. Liossis SN, Ding XZ, Dennis GJ, Tsokos GC. Altered pattern of TCR/CD3-mediated protein-tyrosyl phosphorylation in T cells from patients with systemic lupus erythematosus. Deficient expression of the T cell receptor zeta chain. *J Clin Invest* 1998;**101**:1448–57.
114. Enyedy EJ, Nambiar MP, Liossis SN, Dennis G, Kammer GM, Tsokos GC. Fc epsilon receptor type I gamma chain replaces the deficient T cell receptor zeta chain in T cells of patients with systemic lupus erythematosus. *Arthritis Rheum* 2001;**44**:1114–21.
115. Krishnan S, Juang YT, Chowdhury B, Magilavy A, Fisher CU, Nguyen H, et al. Differential expression and molecular associations of Syk in systemic lupus erythematosus T cells. *J Immunol* 2008;**181**:8145–52.
116. Deng GM, Liu L, Bahjat FR, Pine PR, Tsokos GC. Suppression of skin and kidney disease by inhibition of spleen tyrosine kinase in lupus-prone mice. *Arthritis Rheum* 2010;**62**:2086–92.
117. Bahjat FR, Pine PR, Reitsma A, Cassafer G, Baluom M, Grillo S, et al. An orally bioavailable spleen tyrosine kinase inhibitor delays disease progression and prolongs survival in murine lupus. *Arthritis Rheum* 2008;**58**:1433–44.

118. Deng GM, Tsokos GC. Cholera toxin B accelerates disease progression in lupus-prone mice by promoting lipid raft aggregation. *J Immunol* 2008;**181**:4019–26.
119. Krishnan S, Nambiar MP, Warke VG, Fisher CU, Mitchell J, Delaney N, et al. Alterations in lipid raft composition and dynamics contribute to abnormal T cell responses in systemic lupus erythematosus. *J Immunol* 2004;**172**:7821–31.
120. Scalapino KJ, Daikh DI. CTLA-4: a key regulatory point in the control of autoimmune disease. *Immunol Rev* 2008;**223**:143–55.
121. Tivol EA, Borriello F, Schweitzer AN, Lynch WP, Bluestone JA, Sharpe AH. Loss of CTLA-4 leads to massive lymphoproliferation and fatal multiorgan tissue destruction, revealing a critical negative regulatory role of CTLA-4. *Immunity* 1995;**3**:541–7.
122. Liu MF, Liu HS, Wang CR, Lei HY. Expression of CTLA-4 molecule in peripheral blood T lymphocytes from patients with systemic lupus erythematosus. *J Clin Immunol* 1998;**18**:392–8.
123. Jury EC, Flores-Borja F, Kalsi HS, Lazarus M, Isenberg DA, Mauri C, et al. Abnormal CTLA-4 function in T cells from patients with systemic lupus erythematosus. *Eur J Immunol* 2010;**40**:569–78.
124. Lee YH, Harley JB, Nath SK. CTLA-4 polymorphisms and systemic lupus erythematosus (SLE): a meta-analysis. *Hum Genet* 2005;**116**:361–7.
125. Tivol EA, Boyd SD, McKeon S, Borriello F, Nickerson P, Strom TB, et al. CTLA4Ig prevents lymphoproliferation and fatal multiorgan tissue destruction in CTLA-4-deficient mice. *J Immunol* 1997;**158**:5091–4.
126. Davidson A, Diamond B, Wofsy D, Daikh D. Block and tackle: CTLA4Ig takes on lupus. *Lupus* 2005;**14**:197–203.
127. Merrill JT, Burgos-Vargas R, Westhovens R, Chalmers A, D'Cruz D, Wallace DJ, et al. The efficacy and safety of abatacept in patients with non-life-threatening manifestations of systemic lupus erythematosus: results of a twelve-month, multicenter, exploratory, phase IIb, randomized, double-blind, placebo-controlled trial. *Arthritis Rheum* 2010;**62**:3077–87.
128. Liu Z, Bethunaickan R, Huang W, Ramanujam M, Madaio MP, Davidson A. IFN-{alpha} confers resistance of systemic lupus erythematosus nephritis to therapy in NZB/W F1 mice. *J Immunol* 2011;**187**:1506–13.
129. Takeda K, Kaisho T, Akira S. Toll-like receptors. *Annu Rev Immunol* 2003;**21**:335–76.
130. Rahman AH, Eisenberg RA. The role of toll-like receptors in systemic lupus erythematosus. *Springer Semin Immunopathol* 2006;**28**:131–43.
131. Shen N, Fu Q, Deng Y, Qian X, Zhao J, Kaufman KM, et al. Sex-specific association of X-linked Toll-like receptor 7 (TLR7) with male systemic lupus erythematosus. *Proc Natl Acad Sci USA* 2010;**107**:15838–43.
132. Huang CM, Huang PH, Chen CL, Lin YJ, Tsai CH, Huang WL, et al. Association of toll-like receptor 9 gene polymorphism in Chinese patients with systemic lupus erythematosus in Taiwan. *Rheumatol Int.* 2011; PMID: 21499878 [Epub ahead of print].
133. De Jager PL, Richardson A, Vyse TJ, Rioux JD. Genetic variation in toll-like receptor 9 and susceptibility to systemic lupus erythematosus. *Arthritis Rheum* 2006;**54**:1279–82.
134. Leadbetter EA, Rifkin IR, Hohlbaum AM, Beaudette BC, Shlomchik MJ, Marshak-Rothstein A. Chromatin-IgG complexes activate B cells by dual engagement of IgM and Toll-like receptors. *Nature* 2002;**416**:603–7.
135. Viglianti GA, Lau CM, Hanley TM, Miko BA, Shlomchik MJ, Marshak-Rothstein A. Activation of autoreactive B cells by CpG dsDNA. *Immunity* 2003;**19**:837–47.
136. Boulé MW, Broughton C, Mackay F, Akira S, Marshak-Rothstein A, Rifkin IR. Toll-like receptor 9-dependent and -independent dendritic cell activation by chromatin-immunoglobulin G complexes. *J Exp Med* 2004;**199**:1631–40.

137. Means TK, Latz E, Hayashi F, Murali MR, Golenbock DT, Luster AD. Human lupus autoantibody-DNA complexes activate DCs through cooperation of CD32 and TLR9. *J Clin Invest* 2005;**115**:407–17.
138. Kelly KM, Zhuang H, Nacionales DC, Scumpia PO, Lyons R, Akaogi J, et al. "Endogenous adjuvant" activity of the RNA components of lupus autoantigens Sm/RNP and Ro 60. *Arthritis Rheum* 2006;**54**:1557–67.
139. Green NM, Laws A, Kiefer K, Busconi L, Kim YM, Brinkmann MM, et al. Murine B cell response to TLR7 ligands depends on an IFN-beta feedback loop. *J Immunol* 2009;**183**:1569–76.
140. Christensen SR, Shupe J, Nickerson K, Kashgarian M, Flavell RA, Shlomchik MJ. Toll-like receptor 7 and TLR9 dictate autoantibody specificity and have opposing inflammatory and regulatory roles in a murine model of lupus. *Immunity* 2006;**25**:417–28.
141. Wu X, Peng SL. Toll-like receptor 9 signaling protects against murine lupus. *Arthritis Rheum* 2006;**54**:336–42.
142. Nickerson KM, Christensen SR, Shupe J, Kashgarian M, Kim D, Elkon K, et al. TLR9 regulates TLR7- and MyD88-dependent autoantibody production and disease in a murine model of lupus. *J Immunol* 2010;**184**:1840–8.
143. Guiducci C, Gong M, Xu Z, Gill M, Chaussabel D, Meeker T, et al. TLR recognition of self nucleic acids hampers glucocorticoid activity in lupus. *Nature* 2010;**465**:937–41.
144. Guiducci C, Tripodo C, Gong M, Sangaletti S, Colombo MP, Coffman RL, et al. Autoimmune skin inflammation is dependent on plasmacytoid dendritic cell activation by nucleic acids via TLR7 and TLR9. *J Exp Med* 2010;**207**:2931–42.
145. Hennessy EJ, Parker AE, O'Neill LA. Targeting Toll-like receptors: emerging therapeutics? *Nat Rev Drug Discov* 2010;**9**:293–307.
146. Jabs D, Kuppers R, Saboori A, Burek C, Enger C, Lee B, et al. Effects of early and late treatment with anti-CD4 monoclonal antibody on autoimmune disease in MRL/MP-lpr/lpr mice. *Cell Immunol* 1994;**154**:66–76.
147. Shlomchik M, Madaio M, Ni D, Trounstein M, Huszar D. The role of B cells in lpr/lpr-induced autoimmunity. *J Exp Med* 1994;**180**:1295–306.
148. Hao Z, Hampel B, Yagita H, Rajewsky K. T cell-specific ablation of Fas leads to Fas ligand-mediated lymphocyte depletion and inflammatory pulmonary fibrosis. *J Exp Med* 2004;**199**:1355–65.
149. Stranges P, Watson J, Cooper C, Choisy-Rossi C, Stonebraker A, Beighton R, et al. Elimination of antigen-presenting cells and autoreactive T cells by Fas contributes to prevention of autoimmunity. *Immunity* 2007;**26**:629–41.
150. Teachey D, Seif A, Grupp S. Advances in the management and understanding of autoimmune lymphoproliferative syndrome (ALPS). *Br J Haematol* 2010;**148**:205–16.
151. Nagata S, Hanayama R, Kawane K. Autoimmunity and the clearance of dead cells. *Cell* 2010;**140**:619–30.
152. Casciola-Rosen LA, Anhalt G, Rosen A. Autoantigens targeted in systemic lupus erythematosus are clustered in two populations of surface structures on apoptotic keratinocytes. *J Exp Med* 1994;**179**:1317–30.
153. Mevorach D, Zhou JL, Song X, Elkon KB. Systemic exposure to irradiated apoptotic cells induces autoantibody production. *J Exp Med* 1998;**188**:387–92.
154. Timmer JC, Salvesen GS. Caspase substrates. *Cell Death Differ* 2007;**14**:66–72.
155. Lauber K, Bohn E, Kröber SM, Xiao YJ, Blumenthal SG, Lindemann RK, et al. Apoptotic cells induce migration of phagocytes via caspase-3-mediated release of a lipid attraction signal. *Cell* 2003;**113**:717–30.

156. Peter C, Waibel M, Radu CG, Yang LV, Witte ON, Schulze-Osthoff K, et al. Migration to apoptotic "find-me" signals is mediated via the phagocyte receptor G2A. *J Biol Chem* 2008;**283**:5296–305.
157. Gude DR, Alvarez SE, Paugh SW, Mitra P, Yu J, Griffiths R, et al. Apoptosis induces expression of sphingosine kinase 1 to release sphingosine-1-phosphate as a "come-and-get-me" signal. *FASEB J* 2008;**22**:2629–38.
158. Truman LA, Ford CA, Pasikowska M, Pound JD, Wilkinson SJ, Dumitriu IE, et al. CX3CL1/fractalkine is released from apoptotic lymphocytes to stimulate macrophage chemotaxis. *Blood* 2008;**112**:5026–36.
159. Elliott MR, Chekeni FB, Trampont PC, Lazarowski ER, Kadl A, Walk SF, et al. Nucleotides released by apoptotic cells act as a find-me signal to promote phagocytic clearance. *Nature* 2009;**461**:282–6.
160. Muñoz LE, Lauber K, Schiller M, Manfredi AA, Herrmann M. The role of defective clearance of apoptotic cells in systemic autoimmunity. *Nat Rev Rheumatol* 2010;**6**:280–9.
161. Balasubramanian K, Schroit AJ. Aminophospholipid asymmetry: a matter of life and death. *Annu Rev Physiol* 2003;**65**:701–34.
162. Martin SJ, Finucane DM, Amarante-Mendes GP, O'Brien GA, Green DR. Phosphatidylserine externalization during CD95-induced apoptosis of cells and cytoplasts requires ICE/CED-3 protease activity. *J Biol Chem* 1996;**271**:28753–6.
163. Fadok VA, Voelker DR, Campbell PA, Cohen JJ, Bratton DL, Henson PM. Exposure of phosphatidylserine on the surface of apoptotic lymphocytes triggers specific recognition and removal by macrophages. *J Immunol* 1992;**148**:2207–16.
164. Tanaka Y, Schroit AJ. Insertion of fluorescent phosphatidylserine into the plasma membrane of red blood cells. Recognition by autologous macrophages. *J Biol Chem* 1983;**258**:11335–43.
165. Asano K, Miwa M, Miwa K, Hanayama R, Nagase H, Nagata S, et al. Masking of phosphatidylserine inhibits apoptotic cell engulfment and induces autoantibody production in mice. *J Exp Med* 2004;**200**:459–67.
166. Godowski PJ, Mark MR, Chen J, Sadick MD, Raab H, Hammonds RG. Reevaluation of the roles of protein S and Gas6 as ligands for the receptor tyrosine kinase Rse/Tyro 3. *Cell* 1995;**82**:355–8.
167. Varnum BC, Young C, Elliott G, Garcia A, Bartley TD, Fridell YW, et al. Axl receptor tyrosine kinase stimulated by the vitamin K-dependent protein encoded by growth-arrest-specific gene 6. *Nature* 1995;**373**:623–6.
168. Nagata K, Ohashi K, Nakano T, Arita H, Zong C, Hanafusa H, et al. Identification of the product of growth arrest-specific gene 6 as a common ligand for Axl, Sky, and Mer receptor tyrosine kinases. *J Biol Chem* 1996;**271**:30022–7.
169. Cohen PL, Caricchio R, Abraham V, Camenisch TD, Jennette JC, Roubey RA, et al. Delayed apoptotic cell clearance and lupus-like autoimmunity in mice lacking the c-mer membrane tyrosine kinase. *J Exp Med* 2002;**196**:135–40.
170. Rothlin CV, Lemke G. TAM receptor signaling and autoimmune disease. *Curr Opin Immunol* 2010;**22**:740–6.
171. Yamaguchi H, Takagi J, Miyamae T, Yokota S, Fujimoto T, Nakamura S, et al. Milk fat globule EGF factor 8 in the serum of human patients of systemic lupus erythematosus. *J Leukoc Biol* 2008;**83**:1300–7.
172. Hanayama R, Tanaka M, Miwa K, Shinohara A, Iwamatsu A, Nagata S. Identification of a factor that links apoptotic cells to phagocytes. *Nature* 2002;**417**:182–7.
173. Rogers NJ, Lees MJ, Gabriel L, Maniati E, Rose SJ, Potter PK, et al. A defect in Marco expression contributes to systemic lupus erythematosus development via failure to clear apoptotic cells. *J Immunol* 2009;**182**:1982–90.

174. Potter PK, Cortes-Hernandez J, Quartier P, Botto M, Walport MJ. Lupus-prone mice have an abnormal response to thioglycolate and an impaired clearance of apoptotic cells. *J Immunol* 2003;**170**:3223–32.

175. Trouw LA, Blom AM, Gasque P. Role of complement and complement regulators in the removal of apoptotic cells. *Mol Immunol* 2008;**45**:1199–207.

176. Korb LC, Ahearn JM. C1q binds directly and specifically to surface blebs of apoptotic human keratinocytes: complement deficiency and systemic lupus erythematosus revisited. *J Immunol* 1997;**158**:4525–8.

177. Kim SJ, Gershov D, Ma X, Brot N, Elkon KB. I-PLA(2) activation during apoptosis promotes the exposure of membrane lysophosphatidylcholine leading to binding by natural immunoglobulin M antibodies and complement activation. *J Exp Med* 2002;**196**:655–65.

178. Chen Y, Park YB, Patel E, Silverman GJ. IgM antibodies to apoptosis-associated determinants recruit C1q and enhance dendritic cell phagocytosis of apoptotic cells. *J Immunol* 2009;**182**:6031–43.

179. Zwart B, Ciurana C, Rensink I, Manoe R, Hack CE, Aarden LA. Complement activation by apoptotic cells occurs predominantly via IgM and is limited to late apoptotic (secondary necrotic) cells. *Autoimmunity* 2004;**37**:95–102.

180. Ogden CA, Kowalewski R, Peng Y, Montenegro V, Elkon KB. IGM is required for efficient complement mediated phagocytosis of apoptotic cells in vivo. *Autoimmunity* 2005;**38**:259–64.

181. Ciurana CL, Zwart B, van Mierlo G, Hack CE. Complement activation by necrotic cells in normal plasma environment compares to that by late apoptotic cells and involves predominantly IgM. *Eur J Immunol* 2004;**34**:2609–19.

182. Quartier P, Potter PK, Ehrenstein MR, Walport MJ, Botto M. Predominant role of IgM-dependent activation of the classical pathway in the clearance of dying cells by murine bone marrow-derived macrophages in vitro. *Eur J Immunol* 2005;**35**:252–60.

183. Fraser DA, Arora M, Bohlson SS, Lozano E, Tenner AJ. Generation of inhibitory NFkappaB complexes and phosphorylated cAMP response element-binding protein correlates with the anti-inflammatory activity of complement protein C1q in human monocytes. *J Biol Chem* 2007;**282**:7360–7.

184. Ogden CA, deCathelineau A, Hoffmann PR, Bratton D, Ghebrehiwet B, Fadok VA, et al. C1q and mannose binding lectin engagement of cell surface calreticulin and CD91 initiates macropinocytosis and uptake of apoptotic cells. *J Exp Med* 2001;**194**:781–95.

185. Morgan BP, Walport MJ. Complement deficiency and disease. *Immunol Today* 1991;**12**:301–6.

186. Ehrenstein MR, Cook HT, Neuberger MS. Deficiency in serum immunoglobulin (Ig)M predisposes to development of IgG autoantibodies. *J Exp Med* 2000;**191**:1253–8.

187. Paul E, Pozdnyakova OO, Mitchell E, Carroll MC. Anti-DNA autoreactivity in C4-deficient mice. *Eur J Immunol* 2002;**32**:2672–9.

188. Taylor PR, Carugati A, Fadok VA, Cook HT, Andrews M, Carroll MC, et al. A hierarchical role for classical pathway complement proteins in the clearance of apoptotic cells in vivo. *J Exp Med* 2000;**192**:359–66.

189. Stuart LM, Takahashi K, Shi L, Savill J, Ezekowitz RA. Mannose-binding lectin-deficient mice display defective apoptotic cell clearance but no autoimmune phenotype. *J Immunol* 2005;**174**:3220–6.

190. Cutolo M, Sulli A, Capellino S, Villaggio B, Montagna P, Seriolo B, et al. Sex hormones influence on the immune system: basic and clinical aspects in autoimmunity. *Lupus* 2004;**13**:635–8.

191. Mok CC, Lau CS, Ho CT, Wong RW. Do flares of systemic lupus erythematosus decline after menopause? *Scand J Rheumatol* 1999;**28**:357–62.

192. Roubinian J, Talal N, Siiteri PK, Sadakian JA. Sex hormone modulation of autoimmunity in NZB/NZW mice. *Arthritis Rheum* 1979;**22**:1162–9.
193. Steinberg AD, Roths JB, Murphy ED, Steinberg RT, Raveche ES. Effects of thymectomy or androgen administration upon the autoimmune disease of MRL/Mp-lpr/lpr mice. *J Immunol* 1980;**125**:871–3.
194. Van Griensven M, Bergijk EC, Baelde JJ, De Heer E, Bruijn JA. Differential effects of sex hormones on autoantibody production and proteinuria in chronic graft-versus-host disease-induced experimental lupus nephritis. *Clin Exp Immunol* 1997;**107**:254–60.
195. Roubinian JR, Talal N, Greenspan JS, Goodman JR, Siiteri PK. Effect of castration and sex hormone treatment on survival, anti-nucleic acid antibodies, and glomerulonephritis in NZB/NZW F1 mice. *J Exp Med* 1978;**147**:1568–83.
196. Li J, McMurray RW. Effects of estrogen receptor subtype-selective agonists on autoimmune disease in lupus-prone NZB/NZW F1 mouse model. *Clin Immunol* 2007;**123**:219–26.
197. Bynoté KK, Hackenberg JM, Korach KS, Lubahn DB, Lane PH, Gould KA. Estrogen receptor-alpha deficiency attenuates autoimmune disease in (NZB x NZW)F1 mice. *Genes Immun* 2008;**9**:137–52.
198. Sturgess AD, Evans DT, Mackay IR, Riglar A. Effects of the oestrogen antagonist tamoxifen on disease indices in systemic lupus erythematosus. *J Clin Lab Immunol* 1984;**13**:11–4.
199. Saha S, Tieng A, Pepeljugoski KP, Zandamn-Goddard G, Peeva E. Prolactin, systemic lupus erythematosus, and autoreactive B cells: lessons learnt from murine models. *Clin Rev Allergy Immunol* 2011;**40**:8–15.
200. Libert C, Dejager L, Pinheiro I. The X chromosome in immune functions: when a chromosome makes the difference. *Nat Rev Immunol* 2010;**10**:594–604.
201. Smith-Bouvier DL, Divekar AA, Sasidhar M, Du S, Tiwari-Woodruff SK, King JK, et al. A role for sex chromosome complement in the female bias in autoimmune disease. *J Exp Med* 2008;**205**:1099–108.
202. Banchereau J, Pascual V. Type I interferon in systemic lupus erythematosus and other autoimmune diseases. *Immunity* 2006;**25**:383–92.
203. Deng Y, Tsao BP. Genetic susceptibility to systemic lupus erythematosus in the genomic era. *Nat Rev Rheumatol* 2010;**6**:683–92.
204. Preble OT, Black RJ, Friedman RM, Klippel JH, Vilcek J. Systemic lupus erythematosus: presence in human serum of an unusual acid-labile leukocyte interferon. *Science* 1982;**216**:429–31.
205. Blanco P, Palucka AK, Gill M, Pascual V, Banchereau J. Induction of dendritic cell differentiation by IFN-alpha in systemic lupus erythematosus. *Science* 2001;**294**:1540–3.
206. Yao Y, Richman L, Higgs BW, Morehouse CA, de los Reyes M, Brohawn P, et al. Neutralization of interferon-alpha/beta-inducible genes and downstream effect in a phase I trial of an anti-interferon-alpha monoclonal antibody in systemic lupus erythematosus. *Arthritis Rheum* 2009;**60**:1785–96.
207. Lee PY, Weinstein JS, Nacionales DC, Scumpia PO, Li Y, Butfiloski E, et al. A novel type I IFN-producing cell subset in murine lupus. *J Immunol* 2008;**180**:5101–8.
208. Garcia-Romo GS, Caielli S, Vega B, Connolly J, Allantaz F, Xu Z, et al. Netting neutrophils are major inducers of type I IFN production in pediatric systemic lupus erythematosus. *Sci Transl Med* 2011;**3**:73ra20.
209. Thibault DL, Chu AD, Graham KL, Balboni I, Lee LY, Kohlmoos C, et al. IRF9 and STAT1 are required for IgG autoantibody production and B cell expression of TLR7 in mice. *J Clin Invest* 2008;**118**:1417–26.
210. Reeves WH, Lee PY, Weinstein JS, Satoh M, Lu L. Induction of autoimmunity by pristane and other naturally occurring hydrocarbons. *Trends Immunol* 2009;**30**:455–64.

211. Mathian A, Weinberg A, Gallegos M, Banchereau J, Koutouzov S. IFN-alpha induces early lethal lupus in preautoimmune (New Zealand Black x New Zealand White) F1 but not in BALB/c mice. *J Immunol* 2005;**174**:2499–506.
212. Mathian A, Gallegos M, Pascual V, Banchereau J, Koutouzov S. Interferon-α induces unabated production of short-lived plasma cells in pre-autoimmune lupus-prone (NZB×NZW)F1 mice but not in BALB/c mice. *Eur J Immunol* 2011;**41**:863–72.
213. Hron JD, Peng SL. Type I IFN protects against murine lupus. *J Immunol* 2004;**173**:2134–42.
214. Litinskiy MB, Nardelli B, Hilbert DM, He B, Schaffer A, Casali P, et al. DCs induce CD40-independent immunoglobulin class switching through BLyS and APRIL. *Nat Immunol* 2002;**3**:822–9.
215. Jacob N, Guo S, Mathian A, Koss MN, Gindea S, Putterman C, et al. B Cell and BAFF dependence of IFN-α-exaggerated disease in systemic lupus erythematosus-prone NZM 2328 mice. *J Immunol* 2011;**186**:4984–93.
216. Hervier B, Beziat V, Haroche J, Mathian A, Lebon P, Ghillani-Dalbin P, et al. Phenotype and function of natural killer cells in systemic lupus erythematosus: excess interferon-γ production in patients with active disease. *Arthritis Rheum* 2011;**63**:1698–706.
217. Karonitsch T, Feierl E, Steiner CW, Dalwigk K, Korb A, Binder N, et al. Activation of the interferon-gamma signaling pathway in systemic lupus erythematosus peripheral blood mononuclear cells. *Arthritis Rheum* 2009;**60**:1463–71.
218. Bauer JW, Petri M, Batliwalla FM, Koeuth T, Wilson J, Slattery C, et al. Interferon-regulated chemokines as biomarkers of systemic lupus erythematosus disease activity: a validation study. *Arthritis Rheum* 2009;**60**:3098–107.
219. Jacob CO, van der Meide PH, McDevitt HO. In vivo treatment of (NZB X NZW)F1 lupus-like nephritis with monoclonal antibody to gamma interferon. *J Exp Med* 1987;**166**:798–803.
220. Balomenos D, Rumold R, Theofilopoulos AN. Interferon-gamma is required for lupus-like disease and lymphoaccumulation in MRL-lpr mice. *J Clin Invest* 1998;**101**:364–71.
221. Santiago-Raber ML, Baccala R, Haraldsson KM, Choubey D, Stewart TA, Kono DH, et al. Type-I interferon receptor deficiency reduces lupus-like disease in NZB mice. *J Exp Med* 2003;**197**:777–88.
222. Peng SL, Moslehi J, Craft J. Roles of interferon-gamma and interleukin-4 in murine lupus. *J Clin Invest* 1997;**99**:1936–46.
223. Haas C, Ryffel B, Le Hir M. IFN-gamma receptor deletion prevents autoantibody production and glomerulonephritis in lupus-prone (NZB x NZW)F1 mice. *J Immunol* 1998;**160**:3713–8.
224. Mackay F, Schneider P. Cracking the BAFF code. *Nat Rev Immunol* 2009;**9**:491–502.
225. Meyer-Bahlburg A, Andrews SF, Yu KO, Porcelli SA, Rawlings DJ. Characterization of a late transitional B cell population highly sensitive to BAFF-mediated homeostatic proliferation. *J Exp Med* 2008;**205**:155–68.
226. Thien M, Phan TG, Gardam S, Amesbury M, Basten A, Mackay F, et al. Excess BAFF rescues self-reactive B cells from peripheral deletion and allows them to enter forbidden follicular and marginal zone niches. *Immunity* 2004;**20**:785–98.
227. Cheema GS, Roschke V, Hilbert DM, Stohl W. Elevated serum B lymphocyte stimulator levels in patients with systemic immune-based rheumatic diseases. *Arthritis Rheum* 2001;**44**:1313–9.
228. Mackay F, Woodcock SA, Lawton P, Ambrose C, Baetscher M, Schneider P, et al. Mice transgenic for BAFF develop lymphocytic disorders along with autoimmune manifestations. *J Exp Med* 1999;**190**:1697–710.
229. Khare SD, Sarosi I, Xia XZ, McCabe S, Miner K, Solovyev I, et al. Severe B cell hyperplasia and autoimmune disease in TALL-1 transgenic mice. *Proc Natl Acad Sci USA* 2000;**97**:3370–5.
230. Groom JR, Fletcher CA, Walters SN, Grey ST, Watt SV, Sweet MJ, et al. BAFF and MyD88 signals promote a lupuslike disease independent of T cells. *J Exp Med* 2007;**204**:1959–71.

231. Stohl W, Xu D, Kim KS, Koss MN, Jorgensen TN, Deocharan B, et al. BAFF overexpression and accelerated glomerular disease in mice with an incomplete genetic predisposition to systemic lupus erythematosus. *Arthritis Rheum* 2005;**52**:2080–91.
232. Stohl W, Jacob N, Guo S, Morel L. Constitutive overexpression of BAFF in autoimmune-resistant mice drives only some aspects of systemic lupus erythematosus-like autoimmunity. *Arthritis Rheum* 2010;**62**:2432–42.
233. Gross JA, Johnston J, Mudri S, Enselman R, Dillon SR, Madden K, et al. TACI and BCMA are receptors for a TNF homologue implicated in B-cell autoimmune disease. *Nature* 2000;**404**:995–9.
234. Kayagaki N, Yan M, Seshasayee D, Wang H, Lee W, French DM, et al. BAFF/BLyS receptor 3 binds the B cell survival factor BAFF ligand through a discrete surface loop and promotes processing of NF-kappaB2. *Immunity* 2002;**17**:515–24.
235. Ramanujam M, Bethunaickan R, Huang W, Tao H, Madaio MP, Davidson A. Selective blockade of BAFF for the prevention and treatment of systemic lupus erythematosus nephritis in NZM2410 mice. *Arthritis Rheum* 2010;**62**:1457–68.
236. Ramanujam M, Wang X, Huang W, Liu Z, Schiffer L, Tao H, et al. Similarities and differences between selective and nonselective BAFF blockade in murine SLE. *J Clin Invest* 2006;**116**:724–34.
237. Rönnblom L, Elkon KB. Cytokines as therapeutic targets in SLE. *Nat Rev Rheumatol* 2010;**6**:339–47.
238. Richards HB, Satoh M, Shaw M, Libert C, Poli V, Reeves WH. Interleukin 6 dependence of anti-DNA antibody production: evidence for two pathways of autoantibody formation in pristane-induced lupus. *J Exp Med* 1998;**188**:985–90.
239. Wan S, Xia C, Morel L. IL-6 produced by dendritic cells from lupus-prone mice inhibits CD4 +CD25+ T cell regulatory functions. *J Immunol* 2007;**178**:271–9.
240. Kiberd BA. Interleukin-6 receptor blockage ameliorates murine lupus nephritis. *J Am Soc Nephrol* 1993;**4**:58–61.
241. Illei GG, Shirota Y, Yarboro CH, Daruwalla J, Tackey E, Takada K, et al. Tocilizumab in systemic lupus erythematosus: data on safety, preliminary efficacy, and impact on circulating plasma cells from an open-label phase I dosage-escalation study. *Arthritis Rheum* 2010;**62**:542–52.
242. Wofsy D, Seaman WE. Reversal of advanced murine lupus in NZB/NZW F1 mice by treatment with monoclonal antibody to L3T4. *J Immunol* 1987;**138**:3247–53.
243. Mihara M, Ohsugi Y, Saito K, Miyai T, Togashi M, Ono S, et al. Immunologic abnormality in NZB/NZW F1 mice. Thymus-independent occurrence of B cell abnormality and requirement for T cells in the development of autoimmune disease, as evidenced by an analysis of the athymic nude individuals. *J Immunol* 1988;**141**:85–90.
244. Nutt SL, Tarlinton DM. Germinal center B and follicular helper T cells: siblings, cousins or just good friends? *Nat Immunol* 2011;**12**:472–7.
245. Herlands R, Christensen S, Sweet R, Hershberg U, Shlomchik M. T cell-independent and toll-like receptor-dependent antigen-driven activation of autoreactive B cells. *Immunity* 2008;**29**:249–60.
246. Sweet RA, Ols ML, Cullen JL, Milam AV, Yagita H, Shlomchik MJ. Facultative role for T cells in extrafollicular Toll-like receptor-dependent autoreactive B-cell responses in vivo. *Proc Natl Acad Sci USA* 2011;**108**:7932–7.
247. Vinuesa CG, Cook MC, Angelucci C, Athanasopoulos V, Rui L, Hill KM, et al. A RING-type ubiquitin ligase family member required to repress follicular helper T cells and autoimmunity. *Nature* 2005;**435**:452–8.
248. Linterman MA, Rigby RJ, Wong RK, Yu D, Brink R, Cannons JL, et al. Follicular helper T cells are required for systemic autoimmunity. *J Exp Med* 2009;**206**:561–76.

249. Kim HJ, Verbinnen B, Tang X, Lu L, Cantor H. Inhibition of follicular T-helper cells by CD8 (+) regulatory T cells is essential for self tolerance. *Nature* 2010;**467**:328–32.
250. Kim HJ, Wang X, Radfar S, Sproule TJ, Roopenian DC, Cantor H. CD8 + T regulatory cells express the Ly49 Class I MHC receptor and are defective in autoimmune prone B6-Yaa mice. *Proc Natl Acad Sci USA* 2011;**108**:2010–5.
251. Fan X, Wüthrich RP. Upregulation of lymphoid and renal interferon-gamma mRNA in autoimmune MRL-Fas(lpr) mice with lupus nephritis. *Inflammation* 1997;**21**:105–12.
252. Shirai A, Conover J, Klinman DM. Increased activation and altered ratio of interferon-gamma: interleukin-4 secreting cells in MRL-lpr/lpr mice. *Autoimmunity* 1995;**21**:107–16.
253. Prud'homme GJ, Kono DH, Theofilopoulos AN. Quantitative polymerase chain reaction analysis reveals marked overexpression of interleukin-1 beta, interleukin-1 and interferon-gamma mRNA in the lymph nodes of lupus-prone mice. *Mol Immunol* 1995;**32**:495–503.
254. Huang FP, Feng GJ, Lindop G, Stott DI, Liew FY. The role of interleukin 12 and nitric oxide in the development of spontaneous autoimmune disease in MRL/MP-lpr/lpr mice. *J Exp Med* 1996;**183**:1447–59.
255. Ozmen L, Roman D, Fountoulakis M, Schmid G, Ryffel B, Garotta G. Experimental therapy of systemic lupus erythematosus: the treatment of NZB/W mice with mouse soluble interferon-gamma receptor inhibits the onset of glomerulonephritis. *Eur J Immunol* 1995;**25**:6–12.
256. Schwarting A, Wada T, Kinoshita K, Tesch G, Kelley VR. IFN-gamma receptor signaling is essential for the initiation, acceleration, and destruction of autoimmune kidney disease in MRL-Fas(lpr) mice. *J Immunol* 1998;**161**:494–503.
257. Richards HB, Satoh M, Jennette JC, Croker BP, Yoshida H, Reeves WH. Interferon-gamma is required for lupus nephritis in mice treated with the hydrocarbon oil pristane. *Kidney Int* 2001;**60**:2173–80.
258. Nakajima A, Hirose S, Yagita H, Okumura K. Roles of IL-4 and IL-12 in the development of lupus in NZB/W F1 mice. *J Immunol* 1997;**158**:1466–72.
259. Schorlemmer HU, Dickneite G, Kanzy EJ, Enssle KH. Modulation of the immunoglobulin dysregulation in GvH- and SLE-like diseases by the murine IL-4 receptor (IL-4-R). *Inflamm Res* 1995;**44**(Suppl. 2):S194–6.
260. Kono DH, Balomenos D, Park MS, Theofilopoulos AN. Development of lupus in BXSB mice is independent of IL-4. *J Immunol* 2000;**164**:38–42.
261. Santiago ML, Fossati L, Jacquet C, Müller W, Izui S, Reininger L. Interleukin-4 protects against a genetically linked lupus-like autoimmune syndrome. *J Exp Med* 1997;**185**:65–70.
262. Korn T, Bettelli E, Oukka M, Kuchroo VK. IL-17 and Th17 Cells. *Annu Rev Immunol* 2009;**27**:485–517.
263. Crispín JC, Tsokos GC. IL-17 in systemic lupus erythematosus. *J Biomed Biotechnol* 2010;**2010**:943254.
264. Crispín JC, Tsokos GC. Interleukin-17-producing T cells in lupus. *Curr Opin Rheumatol* 2010;**22**:499–503.
265. Garrett-Sinha LA, John S, Gaffen SL. IL-17 and the Th17 lineage in systemic lupus erythematosus. *Curr Opin Rheumatol* 2008;**20**:519–25.
266. Crispín JC, Oukka M, Bayliss G, Cohen RA, Van Beek CA, Stillman IE, et al. Expanded double negative T cells in patients with systemic lupus erythematosus produce IL-17 and infiltrate the kidneys. *J Immunol* 2008;**181**:8761–6.
267. Zhang Z, Kyttaris VC, Tsokos GC. The role of IL-23/IL-17 axis in lupus nephritis. *J Immunol* 2009;**183**:3160–9.
268. Xu Z, Cuda CM, Croker BP, Morel L. The NZM2410-derived lupus susceptibility locus Sle2c1 increases Th17 polarization and induces nephritis in fas-deficient mice. *Arthritis Rheum* 2011;**63**:764–74.

269. Kyttaris VC, Zhang Z, Kuchroo VK, Oukka M, Tsokos GC. Cutting edge: IL-23 receptor deficiency prevents the development of lupus nephritis in C57BL/6-lpr/lpr mice. *J Immunol* 2010;**184**:4605–9.
270. Jacob N, Yang H, Pricop L, Liu Y, Gao X, Zheng SG, et al. Accelerated pathological and clinical nephritis in systemic lupus erythematosus-prone New Zealand Mixed 2328 mice doubly deficient in TNF receptor 1 and TNF receptor 2 via a Th17-associated pathway. *J Immunol* 2009;**182**:2532–41.
271. Fields ML, Erikson J. The regulation of lupus-associated autoantibodies: immunoglobulin transgenic models. *Curr Opin Immunol* 2003;**15**:709–17.
272. Erikson J, Radic MZ, Camper SA, Hardy RR, Carmack C, Weigert M. Expression of anti-DNA immunoglobulin transgenes in non-autoimmune mice. *Nature* 1991;**349**:331–4.
273. Nguyen KA, Mandik L, Bui A, Kavaler J, Norvell A, Monroe JG, et al. Characterization of anti-single-stranded DNA B cells in a non-autoimmune background. *J Immunol* 1997;**159**:2633–44.
274. Mandik-Nayak L, Bui A, Noorchashm H, Eaton A, Erikson J. Regulation of anti-double-stranded DNA B cells in nonautoimmune mice: localization to the T-B interface of the splenic follicle. *J Exp Med* 1997;**186**:1257–67.
275. Mandik-Nayak L, Seo SJ, Sokol C, Potts KM, Bui A, Erikson J. MRL-lpr/lpr mice exhibit a defect in maintaining developmental arrest and follicular exclusion of anti-double-stranded DNA B cells. *J Exp Med* 1999;**189**:1799–814.
276. Brard F, Shannon M, Prak EL, Litwin S, Weigert M. Somatic mutation and light chain rearrangement generate autoimmunity in anti-single-stranded DNA transgenic MRL/lpr mice. *J Exp Med* 1999;**190**:691–704.
277. Li H, Jiang Y, Prak EL, Radic M, Weigert M. Editors and editing of anti-DNA receptors. *Immunity* 2001;**15**:947–57.
278. Li Y, Li H, Ni D, Weigert M. Anti-DNA B cells in MRL/lpr mice show altered differentiation and editing pattern. *J Exp Med* 2002;**196**:1543–52.
279. Nemazee D. Receptor editing in lymphocyte development and central tolerance. *Nat Rev Immunol* 2006;**6**:728–40.
280. Watson LC, Moffatt-Blue CS, McDonald RZ, Kompfner E, Ait-Azzouzene D, Nemazee D, et al. Paucity of V-D-D-J rearrangements and VH replacement events in lupus prone and nonautoimmune TdT−/− and TdT+/+ mice. *J Immunol* 2006;**177**:1120–8.
281. Shlomchik M, Zharhary D, Saunders T, Camper S, Weigert M. A rheumatoid factor transgenic mouse model of autoantibody regulation. *Int Immunol* 1993;**5**:1329–41.
282. Hannum LG, Ni D, Haberman AM, Weigert MG, Shlomchik MJ. A disease-related rheumatoid factor autoantibody is not tolerized in a normal mouse: implications for the origins of autoantibodies in autoimmune disease. *J Exp Med* 1996;**184**:1269–78.
283. Wang H, Shlomchik M. Autoantigen-specific B cell activation in Fas-deficient rheumatoid factor immunoglobulin transgenic mice. *J Exp Med* 1999;**190**:639–49.
284. William J, Euler C, Christensen S, Shlomchik M. Evolution of autoantibody responses via somatic hypermutation outside of germinal centers. *Science* 2002;**297**:2066–70.
285. William J, Euler C, Shlomchik M. Short-lived plasmablasts dominate the early spontaneous rheumatoid factor response: differentiation pathways, hypermutating cell types, and affinity maturation outside the germinal center. *J Immunol* 2005;**174**:6879–87.
286. Sweet RA, Christensen SR, Harris ML, Shupe J, Sutherland JL, Shlomchik MJ. A new site-directed transgenic rheumatoid factor mouse model demonstrates extrafollicular class switch and plasmablast formation. *Autoimmunity* 2010;**43**:607–18.
287. Rifkin I, Leadbetter E, Beaudette B, Kiani C, Monestier M, Shlomchik M, et al. Immune complexes present in the sera of autoimmune mice activate rheumatoid factor B cells. *J Immunol* 2000;**165**:1626–33.

288. Lau C, Broughton C, Tabor A, Akira S, Flavell R, Mamula M, et al. RNA-associated autoantigens activate B cells by combined B cell antigen receptor/Toll-like receptor 7 engagement. *J Exp Med* 2005;**202**:1171–7.
289. Herlands R, William J, Hershberg U, Shlomchik M. Anti-chromatin antibodies drive in vivo antigen-specific activation and somatic hypermutation of rheumatoid factor B cells at extrafollicular sites. *Eur J Immunol* 2007;**37**:3339–51.
290. Santulli-Marotto S, Retter MW, Gee R, Mamula MJ, Clarke SH. Autoreactive B cell regulation: peripheral induction of developmental arrest by lupus-associated autoantigens. *Immunity* 1998;**8**:209–19.
291. Santulli-Marotto S, Qian Y, Ferguson S, Clarke SH. Anti-Sm B cell differentiation in Ig transgenic MRL/Mp-lpr/lpr mice: altered differentiation and an accelerated response. *J Immunol* 2001;**166**:5292–9.
292. Qian Y, Santiago C, Borrero M, Tedder TF, Clarke SH. Lupus-specific antiribonucleoprotein B cell tolerance in nonautoimmune mice is maintained by differentiation to B-1 and governed by B cell receptor signaling thresholds. *J Immunol* 2001;**166**:2412–9.
293. Borrero M, Clarke SH. Low-affinity anti-Smith antigen B cells are regulated by anergy as opposed to developmental arrest or differentiation to B-1. *J Immunol* 2002;**168**:13–21.
294. Qian Y, Wang H, Clarke SH. Impaired clearance of apoptotic cells induces the activation of autoreactive anti-Sm marginal zone and B-1 B cells. *J Immunol* 2004;**172**:625–35.
295. Singer GG, Abbas AK. The fas antigen is involved in peripheral but not thymic deletion of T lymphocytes in T cell receptor transgenic mice. *Immunity* 1994;**1**:365–71.
296. Wither J, Vukusic B. Autoimmunity develops in lupus-prone NZB mice despite normal T cell tolerance. *J Immunol* 1998;**161**:4555–62.
297. Fatenejad S, Peng SL, Disorbo O, Craft J. Central T cell tolerance in lupus-prone mice: influence of autoimmune background and the lpr mutation. *J Immunol* 1998;**161**:6427–32.
298. Bouzahzah F, Jung S, Craft J. CD4+ T cells from lupus-prone mice avoid antigen-specific tolerance induction in vivo. *J Immunol* 2003;**170**:741–8.
299. Bailey-Bucktrout SL, Bluestone JA. Regulatory T cells: stability revisited. *Trends Immunol* 2011;**32**:301–6.
300. La Cava A. T-regulatory cells in systemic lupus erythematosus. *Lupus* 2008;**17**:421–5.
301. Scalapino K, Tang Q, Bluestone J, Bonyhadi M, Daikh D. Suppression of disease in New Zealand Black/New Zealand White lupus-prone mice by adoptive transfer of ex vivo expanded regulatory T cells. *J Immunol* 2006;**177**:1451–9.
302. Hayashi T, Hasegawa K, Adachi C. Elimination of CD4(+)CD25(+) T cell accelerates the development of glomerulonephritis during the preactive phase in autoimmune-prone female NZB x NZW F mice. *Int J Exp Pathol* 2005;**86**:289–96.
303. Humrich JY, Morbach H, Undeutsch R, Enghard P, Rosenberger S, Weigert O, et al. Homeostatic imbalance of regulatory and effector T cells due to IL-2 deprivation amplifies murine lupus. *Proc Natl Acad Sci USA* 2010;**107**:204–9.
304. Monk CR, Spachidou M, Rovis F, Leung E, Botto M, Lechler RI, et al. MRL/Mp CD4+, CD25- T cells show reduced sensitivity to suppression by CD4+, CD25+ regulatory T cells in vitro: a novel defect of T cell regulation in systemic lupus erythematosus. *Arthritis Rheum* 2005;**52**:1180–4.
305. Parietti V, Monneaux F, Décossas M, Muller S. Function of CD4+, CD25+ Treg cells in MRL/lpr mice is compromised by intrinsic defects in antigen-presenting cells and effector T cells. *Arthritis Rheum* 2008;**58**:1751–61.
306. Chen Y, Cuda C, Morel L. Genetic determination of T cell help in loss of tolerance to nuclear antigens. *J Immunol* 2005;**174**:7692–702.

307. Cuda CM, Wan S, Sobel ES, Croker BP, Morel L. Murine lupus susceptibility locus Sle1a controls regulatory T cell number and function through multiple mechanisms. *J Immunol* 2007;**179**:7439–47.
308. Scalapino K, Daikh D. Suppression of glomerulonephritis in NZB/NZW lupus prone mice by adoptive transfer of ex vivo expanded regulatory T cells. *PLoS One* 2009;**4**:e6031.
309. La Cava A, Ebling FM, Hahn BH. Ig-reactive CD4+CD25+ T cells from tolerized (New Zealand Black x New Zealand White)F1 mice suppress in vitro production of antibodies to DNA. *J Immunol* 2004;**173**:3542–8.
310. Michaels MA, Kang HK, Kaliyaperumal A, Satyaraj E, Shi Y, Datta SK. A defect in deletion of nucleosome-specific autoimmune T cells in lupus-prone thymus: role of thymic dendritic cells. *J Immunol* 2005;**175**:5857–65.
311. Townsend MJ, Monroe JG, Chan AC. B-cell targeted therapies in human autoimmune diseases: an updated perspective. *Immunol Rev* 2010;**237**:264–83.
312. Hauser S, Waubant E, Arnold D, Vollmer T, Antel J, Fox R, et al. B-cell depletion with rituximab in relapsing-remitting multiple sclerosis. *N Engl J Med* 2008;**358**:676–88.
313. Cohen SB, Emery P, Greenwald MW, Dougados M, Furie RA, Genovese MC, et al. Rituximab for rheumatoid arthritis refractory to anti-tumor necrosis factor therapy: results of a multicenter, randomized, double-blind, placebo-controlled, phase III trial evaluating primary efficacy and safety at twenty-four weeks. *Arthritis Rheum* 2006;**54**:2793–806.
314. Merrill JT, Neuwelt CM, Wallace DJ, Shanahan JC, Latinis KM, Oates JC, et al. Efficacy and safety of rituximab in moderately-to-severely active systemic lupus erythematosus: the randomized, double-blind, phase II/III systemic lupus erythematosus evaluation of rituximab trial. *Arthritis Rheum* 2010;**62**:222–33.
315. Looney RJ. B cell-targeted therapies for systemic lupus erythematosus: an update on clinical trial data. *Drugs* 2010;**70**:529–40.
316. Ahuja A, Shupe J, Dunn R, Kashgarian M, Kehry MR, Shlomchik MJ. Depletion of B cells in murine lupus: efficacy and resistance. *J Immunol* 2007;**179**:3351–61.
317. Haas K, Watanabe R, Matsushita T, Nakashima H, Ishiura N, Okochi H, et al. Protective and pathogenic roles for B cells during systemic autoimmunity in NZB/W F1 mice. *J Immunol* 2010;**184**:4789–800.
318. Furie R, Stohl W, Ginzler EM, Becker M, Mishra N, Chatham W, et al. Biologic activity and safety of belimumab, a neutralizing anti-B-lymphocyte stimulator (BLyS) monoclonal antibody: a phase I trial in patients with systemic lupus erythematosus. *Arthritis Res Ther* 2008;**10**:R109.
319. Wallace DJ, Stohl W, Furie RA, Lisse JR, McKay JD, Merrill JT, et al. A phase II, randomized, double-blind, placebo-controlled, dose-ranging study of belimumab in patients with active systemic lupus erythematosus. *Arthritis Rheum* 2009;**61**:1168–78.
320. Navarra SV, Guzmán RM, Gallacher AE, Hall S, Levy RA, Jimenez RE, et al. Efficacy and safety of belimumab in patients with active systemic lupus erythematosus: a randomised, placebo-controlled, phase 3 trial. *Lancet* 2011;**377**:721–31.

Animal Models of Cutaneous and Hepatic Fibrosis

GIDEON P. SMITH

Department of Dermatology, Massachusetts General Hospital, Harvard University, Boston, Massachusetts, USA

Fibrosis occurs as a part of normal wound healing. However, excessive or dysregulated fibrosis can lead to severe organ dysfunction and is a feature of a variety of diseases. Due to its insidious onset, fibrosis tends to go undetected in its early stages. This is in part why these diseases remain so poorly understood. Animal models have provided a means to examine these early stages and to isolate and understand the effect of perturbations in signaling pathways, chemokines, and cytokines. Here, we summarize recent progress in the understanding of the molecular pathogenesis of fibrosis, both its initiation and its maintenance phases, from animal models of fibrosis in the skin and liver. Due to these organs' properties, modeling fibrosis in them poses unique challenges. Elegant solutions have therefore been developed for modeling fibrosis in each, and now, great potential for animal models to contribute to our understanding appears scientifically imminent.

Progress in Molecular Biology
and Translational Science, Vol. 105
DOI: 10.1016/B978-0-12-394596-9.00011-1

1877-1173/12 $35.00

I. Liver Fibrosis

A. Introduction

Fibrosis is the end-point outcome of a variety of chronic liver diseases, usually resulting from chronic or repetitive activation of the wound healing response.[1] Although this end stage disease state is superficially similar, the triggers and pathways by which it is reached vary between disease states. In addition, subtle differences such as the architectural location of the fibrosis within the liver are important differences between these seemingly similar entities. Animal models are useful both for understanding these pathomechanisms and in the development and testing of antifibrotic therapies. In the former, they are important as they allow the systematic analysis of set time-points within the disease process. Possibly, most significantly, animal models allow examination of the very earliest stages of disease, a stage almost universally missed in human disease due to the inevitable time lag between disease onset and presentation and diagnosis. However, weaknesses exist in all animal models. First, in order to elicit a fibrotic response, a trigger or target for dysregulation must be chosen. This can at times produce the phenotype of a disease, without actually reflecting the disease process at all. More commonly, however, only parts of the disease phenotype are reproduced raising questions about the validity of the model. Even when the etiology is known, it sometimes cannot be reproduced in the chosen animal, for example, mice are not susceptible to hepatitis C infection. Indeed, for some fibrotic disease states, no truly adequate model exists. In addition, disease states that take years to manifest in humans are often modeled by animal models over a time course of weeks to be amenable with time and money considerations in experimental design. Finally, and perhaps most obviously, interspecies differences can sometimes be profound and are reflected not only in the physical size and dose responses, but also in the very presence and absence of genes, their expression and regulation, and the resultant presence or absence of critical cell types, cell interactions, chemokines, and immune responses. Despite these caveats, great progress has been made in understanding liver fibrosis through the application of animal models. Here, we review the different models of liver fibrosis that have been developed over the years and relate what we have learnt from them about the pathomechanism of liver fibrosis, as well as highlight the strengths and weaknesses of these models in recapitulating the features of different liver disease states.

B. Hepatotoxin- and Diet-Induced Liver Fibrosis Models

The most obvious example of hepatotoxin-mediated liver disease in humans is that of alcohol-induced liver fibrosis. Alcohol-induced fibrosis is caused by acetaldehyde, an oxidized metabolite of alcohol. Although the exact histological features of the hepatic injury induced can vary ranging from steatosis (fatty change), through lobular inflammation and periportal fibrosis to end stage cirrhosis, the fibrosis is characterized by deposition of extracellular matrix material in the space of Disse around sinusoids or hepatocytes.[2] Reproduction of this in animals is problematic. The most obvious approach is introduction of alcohol to the animal's diet. This is the approach of the Lieber–DeCarli model. In this model, a liquid diet in which alcohol isocalorically substitutes for carbohydrates up to 50% is given *ad libitum* as the only nutritional source. However, due to a natural aversion for alcohol intake in animals, the level of alcohol that can be administered in this manner remains limited. In rodents, the maximum level of alcohol is 36% of the caloric intake and must be increased to this level gradually. As a result, it does not result in advanced liver fibrosis.[3] Only in baboons has it been possible to raise this to 50%, a level where the features of human disease including pericellular and perivenular fibrosis, progressively forming septa is observed. However, this model still only reproduces these effects in approximately one-third of all test animals and requires an investment of 1–3 years before results are seen.[4] To circumvent animals' natural aversion for alcohol intake, the Tsukamoto–French rat model uses a continuous intragastric infusion of alcohol with a high-fat liquid diet. This model successfully increases the levels of alcohol consumption obtainable and produces sustained elevations in alcohol blood levels.[5] Liver fibrosis is reproducible over a short period of time; however, it is only moderate and does not result in end stage cirrhosis. Despite its advantages for the study of low-level liver fibrosis, this method requires users to have extensive surgical skill and knowledge, and facility for sustained animal anesthesia, for the initial implantation of an intragastric catheter to the test animal.

Despite the limitations of these models, toxic insult remains one of the most common and reproducible methods of inducing a liver injury. The most commonly used agents include carbon tetrachloride (CCl_4), thioacetamide (TAA), and dimethyl or diethyl nitrosamine (DMN or DEN). The resulting fibrosis develops first in pericentral areas, forming bridges between central areas and then secondarily begins between central and portal areas eventuating in cirrhosis.

Such models are both relatively reproducible and easy to perform. While these toxins can be administered *ad libitum* in the diet, reproducibility can be improved via direct administration, for example, intraperitoneal injection.

CYP2E1, an enzyme expressed in perivenular hepatocytes, transforms CCl_4 into a $CCl^-{}_3$ radical which leads to lipid peroxidation and membrane damage. It thus results in an acute centrolobular necrosis, with a progressive fibrosis in the liver, but induces minimal extrahepatic damage, as CYP2E1 is poorly expressed outside the liver. Rate of progression depends on such factors as animal species, strain, dose, route, and frequency of administration of the toxin but can be upregulated by coadministration of a CYP2E1 inducer such as phenobarbital. In mice, even without such attempts to accelerate the process, fibrosis is generally significant after just 2–4 weeks, severe bridging fibrosis occurs after 5–7 weeks, cirrhosis appears within 8–9 weeks of administration, and micronodular cirrhosis, portal hypertension, and ascities develop after 10–20 weeks.[6,7]

Targeted gene knockouts in mice have been combined with a variety of fibrosis models to help clarify the profibrotic pathways induced in each. This has allowed the targeting of specific genes of interest, which can be constitutively disrupted or in which disruption can be triggered. If liver- or cell type-specific gene promoters such as albumin to target hepatocytes are used, such disruption occurs exclusively in the organ or cell type of interest. By fusing *Cre* with a mutant estrogen receptor (*Cre-ERT2*) in which Cre recombinase is induced by the injection of the estrogen analog tamoxifen, it is also possible to control the temporal expression or silencing of genes.[8]

One of the cytokines of common interest in fibrosis is transforming growth factor (TGF)-β. CCl_4-treated mice have been used in the analysis of the TGF-β profibrotic pathway. TGF-β is important in the regulation of a variety of processes active in fibrosis, including hepatocyte apoptosis, inflammatory cell recruitment and activation, and transdifferentiation of quiescent hepatic stem cells (HSCs), into collagen-producing myofibroblasts.[9,10] Although three TGF-β isoforms (β1, β2, and β3) have been identified, only TGF-β1 is pathogenic in liver fibrogenesis.[11] TGF-β binds to cell surface TGF-β receptors which, in turn, phosphorylate the transcription factor Smads, which upon translocation to the nucleus regulates gene expression.[12–14] Knockout of Smad7, an inhibitory Smad that downregulates TGF-β signaling, produces increased liver damage and fibrosis after injury by CCl_4 compared with wild types.[15] Conversely, overexpression of Smad7 attenuates TGF-β-mediated Smad signaling, thus reducing TGF-β-dependent transdifferentiation of HSCs and fibrosis in response to CCl_4.[16]

Liver injury specifically upregulates levels of a molecule called amphiregulin. *In vitro* it has been shown that activation of the epidermal growth factor receptor by amphiregulin stimulates hepatocyte proliferation and inhibits apoptosis of hepatic myofibroblasts and HSCs. Similarly, amphiregulin-null mice treated with CCL_4 have significantly lower levels of tissue inhibitor of matrix metalloproteinase (TIMP)-1, connective tissue growth factor (CTGF) and develop less fibrosis compared with wild-type controls.[17]

The renin–angiotensin system, more commonly associated with kidney function and the regulation of both blood pressure and salt:water balance, has also been implicated in liver fibrosis. Angiotensin II is both produced by, and acts on, activated HSCs[18,19] where it stimulates secretion of proinflammatory cytokines and cellular proliferation, migration, and collagen synthesis.[20–22] There are two forms of the angiotensin II receptor with disparate effects.

Angiotensin II type 1A receptor signaling is profibrotic, with disruption of the receptor reducing the levels of hepatic inflammation and fibrosis produced by CCl_4 treatments.[23,24] In contrast, knockout of the angiotensin II type 2 receptor increases ROS production and the liver damage and fibrosis induced by CCl_4 treatments.[25]

Tumor necrosis factor (TNF)-α can activate HSCs *in vitro* through stress-activated protein kinase pathways and is important in the regulation of cell death.[26,27] Consistent with this, TNF is elevated in the liver after acute injury induced by CCl_4 treatment and disruption of the TNF type 1 receptor reduces post-CCl_4 liver fibrosis.[28,29]

Comparison of models of fibrosis can often also lead to interesting insights as to the differences between cells. For example, interleukin (IL)-6 knockout mice experience downregulation of phosphorylation of the signal transducer and activator of transcription 3 (STAT3) with subsequent increased biliary fibrosis compared to controls after bile duct ligation.[30] However, in CCl_4-treated mice, while gp130 knockout in all cell types also leads to greater liver fibrosis, knockout of gp130 from hepatocytes alone does not affect liver fibrogenesis. These findings indicate that the IL-6/gp130-dependent pathway is highly cell specific, and that in nonparenchymal liver cells, it may play a protective role in the fibrotic injury of CCl_4.[31,32]

Equally as important in fibrosis as cytokines, such as IL-6 or IL-13, are chemokines. These factors are upregulated during liver injury and have unique roles in the profibrotic cascade, for example, CC Chemokine Receptor (CCR) 2 recruits macrophages.[33] During HSC activation, expression levels of monocyte chemoattractant protein (MCP)-1, macrophage inflammatory protein-2, and IL-8/CINC and their receptors, including CXC chemokine receptor 3, CCR 5, and CCR7, are upregulated.[34–36] CCl_4 treatment of CCR1 or CCR5 knockout mice produces less liver fibrosis compared with controls.[37,38] Predictably, CCR2 knockouts activate fewer HSCs, recruit less macrophages, and show less liver fibrogenesis when treated with CCl_4 or bile duct ligation.[39]

In specific types of liver fibrosis, the complement system may play a more important role, for example, in chronic hepatitis C virus infection.[40] In response to injury, hepatic complement 5 (C5) levels and C5 receptor 1 expression increase in HSCs.[40–43] The *Hc* gene, which encodes complement factor C5, is a quantitative trait gene that modifies liver fibrogenesis by experimental intercrosses between fibrosis-susceptible BALB/cJ and fibrosis-resistant A/L inbred mice.[42] CCl_4-treated Hc−/− strains show less liver fibrosis than Hc+/+ strains. Administration of C5 receptor 1 antagonist to *Hc*+/+ BALB/cJ mice reduces liver fibrogenesis but does not benefit *Hc*-null A/L mice.[42]

Complications of intraperitoneal, subcutaneous, or intragastric administration include an injection site reaction ranging from mild inflammation to necrosis which can adversely affect animal survival in extreme cases. This can be avoided by inhalational administration.[7,44] Animals are sacrificed at 2–3 days. Earlier sacrifice results in samples where necrosis and inflammation are prominent, but beyond 3 days remodeling dominates.

CYP2E1 metabolism of DMN and DEN also results in toxic metabolites. Active methylating moieties form adducts with nucleic acid resulting in perivenular and periportal necrosis and resultant centroportal fibrotic septa within a few weeks of IP or PO administration. One important difference of this model is that the fibrosis can be progressive for months after discontinuation of the intoxication and can result in hepatocellular carcinoma complicating interpretation in terms of fibrotic processes where this does not occur.[45,46]

TAA damages zones 1 and 3 hepatocytes resulting in more prominent periportal injury than is obtained with other toxins. It can be administered intraperitoneally or via drinking water, but requires extended administration prior to fibrosis and can result in test animal loss due to induction of cholangiocarcinoma and hepatocellular carcinoma.[45,47]

Choline is essential for cell membrane integrity and transmembrane signaling. A choline-deficient L-amino acid (CDAA)-defined diet results in homogeneous severe macrovesicular steatosis with inflammation in rats. After 6–8 weeks, fibrosis develops rapidly with bridging between vascular areas at week 12.[48] Unfortunately, the fibrosis develops in a highly nonlinear time frame making experimental analysis difficult. Hepatocellular carcinoma again complicates long-term treatment. In contrast, in mice, CDAA diet induces only a moderate pericellular fibrosis more consistent with nonalcoholic steatohepatitis (NASH).[49]

C. Nonalcoholic Fatty Liver Disease

Nonalcoholic fatty liver disease (NAFLD) represents a spectrum of fibrotic liver disease ranging from simple steatosis (fatty change), through NASH to nonalcoholic cirrhosis. NASH is part of the metabolic syndrome. Hepatic fibrosis in this condition ranges from pericellular to bridging fibrosis, and in some patients, fully developed cirrhosis. Administration of the estrogen antagonist Tamoxifen or the CPT1 inhibitor etomoxir has been shown to produce this histologic pattern via impairment of b-oxidation.[50] A similar mechanism is also taken advantage of with modifications to the levels of essential nutrients such as choline and methionine in the diet. As already mentioned, mice fed a CDAA diet CDAA exhibit hepatic fibrosis somewhat consistent with NASH. Unlike NASH in humans, however, these mice are otherwise healthy with no weight gain or insulin insensitivity.

Another diet-defined model is that of rats or mice fed a diet deficient in methionine. This amino acid is an essential amino acid, required for methyl metabolism and glutathione synthesis. It has been suggested that liver injury is promoted via subsequent adenosine triphosphate deficiency and inhibition of mRNA synthesis through replacement of *S*-adenosylmethionine with *S*-adenosylethionine and adenine trapping.[51] After 6–8 weeks in mice, or 12 weeks in rats, animals develop steatohepatitis with progressive pericellular fibrosis due to the induced oxidative stress. In rats, extended administration of this diet can result in cirrhosis.[52] However, in contrast to human NASH, they exhibit weight loss up to 40% of the starting body weight and a hypersensitivity to insulin.[53,54]

129S6/SvEvTac mice develop a microsteatotic phenotype when fed a high-fat diet (HFD).[55] This effect is even more pronounced in C57BL/6 mice which even with a normal diet tend with age to develop obesity, hyperinsulinemia, and glucose intolerance, possibly due to a spontaneous mutation in the nicotinamide nucleotide transhydrogenase gene which induces glucose intolerance.[56] In C57BL/6 mice, a mixed micro- and macrovesicular steatosis may be provoked by feeding a 55% HFD for 6 months. In both mice, expression of SREBP-1c, SREBP-2, and stearoyl-CoA desaturase 1 was increased by this diet.[55] Increased hepatic expression of the endotoxin-induced macrophage receptor with a collagenous structure (MARCO) and portal endotoxin levels have also been noted in some mouse strains fed a HFD. This suggests that this diet may not only make the liver more prone to lipopolysaccharide (LPS) through its affects on the immune system but also increase portal delivery of LPS.[57]

Interestingly, a similar phenotype including obesity, insulin resistance, and macrovesicular steatosis can be evoked in C57BL/6 mice by a 65% sucrose diet administered for 8 weeks.[58]

1. *ob/ob* Mouse

The autosomal recessive *ob/ob* mouse has a mutation in the *leptin* gene on chromosome 6.[59–61] Leptin is produced by white adipose tissue and has anorexic effects via action on the hypothalamic ventral median nucleus. ob/ob mice therefore continuously have a sense of starvation[61] and are thus hyperphagic, inactive, obese, and diabetic with marked hyperglycemia.[62–64] Histologically, there is hepatic fat deposition with enlarged islets of Langerhans.[65] Increased leptin levels lead to SREBP-1c activation which accumulates in hepatocyte nuclei promoting fatty acid synthase activity and *de novo* synthesis.[66] In addition, increased levels of TNF-α secondary to adipose volume lead to lipolysis releasing long-chain fatty acids which are also delivered to the liver via the circulation. Demonstration of stable PPARa activity levels[50] suggests that compensatory increases in hepatocyte b-oxidation[67,68] are insufficient to dispose of the increased fatty acid load leading to activation of the alternate microsomal o-oxidation pathways.[69] Combined, these effects increase storage of lipid in the liver. While ob/ob mice do not spontaneously progress from steatosis to steatohepatitis, this can be induced by ethanol administration, hepatic ischemia–reperfusion, or exposure to subpathologic dosages of LPS endotoxin.[50,70–72]

However, models of leptin deficiency are limited in the degree of fibrosis and are best suited to studies of steatohepatitis with minimal fibrosis. This is in part because leptin deficiency suppresses monocytes, T-cell activation, and Natural Killer cells, thus affecting both the innate and adaptive immune systems. While this favors a proinflammatory milieu, it is shifted toward the antifibrotic Th-1. In addition, leptin has antifibrotic effects preventing expression of TGFβ-dependent genes including procollagen type 1.[73–75] These effects can be overcome by administration of leptin, or of norepinephrine which directly activates hepatic stellate cells and shifts toward a profibrotic Th-2 response with increased IL-4, IL-10, and TGFβ levels.[76,77]

2. *db/db* Mouse

The *db/db* mice (and the *fa/fa* rat) have normal to elevated levels of leptin but are resistant to its effects, thus producing a phenotype very similar to the *ob/ob* mouse.[78] The autosomal recessive mutation mapped to mouse chromosome 4 results in errors in the leptin receptor (OB-R).[79] The OB-Rb isoform of the receptor has a long intracytoplasmic region that contains signal transduction motifs that activate the JAK/STAT protein kinase signal transduction cascade.[80] C57BL/Ks *db/db* mice have a sequence insertion at the 30 end of the mRNA transcript that contains a stop codon resulting in truncation of the OB-Rb long intracellular signaling domain.[81]

Hepatic lipogenesis can be evoked via targeted overexpression of the insulin-controlled transcription factor SREBP-1. Two nonobese transgenic mouse models with severe hepatic steatosis, PEPCK-nSREBP-1a and aP2-nSREBP-1c, have been developed.[82,83] PEPCK-nSREBP-1a mice exhibit histological steatosis but not steatohepatitis or dyslipidemia. The mechanism is thought to be through overexpression of a truncated SREBP-1a in the liver under the control of a phosphoenolpyruvate carboxykinase promoter.[82] In contrast, aP2-nSREBP-1c mice overexpress SREBP-1c in adipose tissue. These animals have steatosis with lipoatrophy.[83]

Hepatic lipogenesis can also be evoked via targeted disruption of the *ChREBP* gene. This gene is responsible for transcriptional regulation of carbohydrate-mediated lipogenesis,[84] and ChREBP-deficient mice have a 50% reduction in enzymes that promote lipogenesis.[85]

The importance of adequate functioning of peroxisomal b-oxidation in prevention of steatohepatitis is suggested by several mouse models. The juvenile visceral steatosis mouse a mutation in the carnitine transporter gene *Octn2* results in systemic carnitine deficiency resulting in loss of fatty acid transport into mitochondria for b-oxidation.[86] Extreme steatosis results several days after birth. Acyl-CoA oxidase (*AOX−/−*)-deficient mice carry a deletion in a key enzyme in peroxisomal b-oxidation. Although initially phenotypically normal over 8-weeks the mice develop severe steatosis, which spontaneously resolves as hepatocytes are replaced. Homozygote PPARa knockout mice do fail to upregulate fatty acid oxidation and thus develop steatosis when fatty acid delivery to the liver is increased by diet modification or fasting.[50,87] In addition, while homozygous mutation of MTPa, a key enzyme mediating mitochondrial b-oxidation, causes early death, heterozygotes survive to develop hepatic steatosis by 9 months of age, as well as insulin resistance with an associated hyperinsulinemia.[43,88] They also have evidence of increased hepatic oxidative stress with increased CYP2E1 expression, elevated antioxidant activity of total superoxide dismutase, and measurably reduced glutathione levels.

D. Biliary Fibrosis

Biliary fibrosis has most commonly been modeled in animals by ligation of the common bile duct. Animals undergo mid ventral laparotomy with isolation of the common bile duct above the duodenum and ligature above the pancreatic duct and below the junction of the hepatic ducts. The intervening section of the bile duct is resected to avoid reanastomization.[89] Control animals undergo laparotomy but without ligation. Mice postligation are prone to dilatation of the gall bladder. This increases mortality primarily secondary to increased rates of perforation and bilioperitoneum. However, even in those mice that survive, there exits an interanimal variability in gall bladder dilation resulting in a variable parenchymal response. This effect may be ameliorated by cholecystectomy or surgical clipping of the cystic duct.[90] In contrast, rats lack a gall bladder and so do not suffer such complications and are thus better suited to this type of study.

The use of the model requires surgical training, and early mortality is often secondary to poor surgical technique including bile leakage, or rupture of the gall bladder, or a biliary cyst. However, even with the best surgical skills, mortality in test animals is high secondary to the resultant bile acid toxicity with mortality exceeding 20% in rats at 6 and 4 weeks in mice. However, this same bile acid toxicity is responsible for stimulation of bile duct epithelial cell proliferation which in concert with periportal myofibroblasts leads to portal inflammation and fibrosis within weeks.[91] Although histologic cirrhosis is not apparent,[91] portal hypertension and ascities are found in rats after 6–8 weeks.[92,93]

The *ABCB4* gene encodes the P-glycoprotein ABCB4 (Mdr2), responsible for phospholipid transport across the canalicular membrane. *Abcb4*−/−mice lack this transporter resulting in the absence of micelle forming phospholipids in bile, leading to chronic cholangitis and subsequent liver injury.[94] The response is highly reproducible, and *Abcb4*−/− mice demonstrate both macroscopic and microscopic features typical of sclerosing cholangitis in humans including bile duct strictures, dilatation of large bile ducts, "onion skin" pericholangitis, and periductular fibrosis.[95] This is perhaps unsurprising given that mutations of the human *ABCB4* gene result in progressive familial intrahepatic cholestasis (PFIC type 3) and biliary liver cirrhosis, and this is an excellent model for studying these human diseases.[96]

Another model of chronic biliary diseases in humans is via repeated i.p. administration over 7–13 weeks of D-galactosamine, a hepatotoxin. This leads to expansion of the portal tracts, bile duct proliferation, and portoportal fibrous septa. With extended treatment regimes, cirrhosis occurs. The D-galactosamine works by reducing uridine nucleotides and thereby RNA and protein synthesis.[97]

The renin–angiotensin system is also dysregulated by bile duct ligation. As with the CCl_4 treatment model, angiotensin II type 1A receptor signaling is found to be profibrotic, with disruption of the receptor reducing the levels of hepatic inflammation and fibrosis produced by bile duct ligation.[23,24] These effects are mediated by ROS generated by a nonphagocytic form of nicotinamide adenine dinucleotidephosphate (NADPH) oxidase.[98] Bile duct ligation of NADPH subunit p47phox knockouts shows reduced angiotensin II-mediated liver fibrosis and reduced expression of the TGF-β.[22] TGF-β has also been implicated in other knockout mouse models as important in biliary fibrosis. In response to liver injury, latent TGF-β complexes are converted into active TGF-β. In the liver, this conversion is in part controlled by local expression of integrin alphavβ6. Bile duct ligation has been shown to increase integrin alphavβ6 on cholangioepithelial cells. In addition, periductal fibrogenesis is attenuated and TGF-β signaling decreased in integrin β6-null mice.[99]

TGF-β signaling and hepatic fibrosis are enhanced by Toll-like receptor (TLR)4.[100] TLR is a component of the innate immune system. These receptors recognize pathogen-associated molecular patterns. In particular, TLR4 recognizes LPS, a cell wall component of Gram-negative bacteria which have been implicated as a mediator of hepatic fibrosis.[101] The levels of plasma LPS increase in response to liver injury induced by bile duct ligation (or CCl_4 or TAA). Mature monocytes, macrophages, and neutrophils possess CD14 surface receptors, a 55-kDa glycosylphosphatidylinositol-anchored glycoprotein, which recognizes LPS. Knockout of the CD14 receptor leads to decreased macrophage and monocyte infiltration of the liver and lower levels of liver fibrosis after bile duct ligation.

E. Immunologically Induced Fibrosis

Hepatic fibrosis may result both from immune and autoimmune injuries. Just as with many other autoimmune diseases, it is believed that autoimmune hepatitis (AIH) develops in individuals with a genetic predisposition once triggered by one or probably several environmental factors. The disease is poorly understood in part because AIH is often not diagnosed during its early stages, and it is therefore difficult to obtain information about the pathological triggers and early changes that lead to the disease. A second problem is that it is in fact difficult to evoke an autoimmune response against hepatocytes, at least when part of the whole liver *in vivo*.[102] This is evidenced by the low prevalence of autoimmune liver diseases such as AIH when compared with autoimmune diseases against other organ systems cells such as the thyroid, melanocytes, or bone and tendon. In addition, liver transplants show a lower rejection rate than other organ transplants,[103] and even with allograft transplant of liver to rodents, some strains show complete tolerance.[104,105]

This is perhaps at first surprising as the liver is exposed to many antigens via portal drainage of breakdown products of materials passing through the gut and gut-flora. In disease states of the gut where there is compromise of the barrier, this exposure would likely be even greater. In addition, the liver acts to detoxify the blood and is therefore also prone to cellular damage or even destruction, with accompanying inflammation. However, it is also probably these very facts that from an evolutionary standpoint have required the liver to evolve as a rapidly healing and minimally reactive organ.

A variety of factors appear to contribute to this general lack of reactivity in hepatocytes. As in many anergic conditions, there is an upregulation of regulatory T cells (Tregs), in this case by liver sinusoidal endothelial cells (LSECs).[106] Indeed, when antigen myelin basic protein (MBP) was ectopically expressed, MBP-specific Tregs were shown to be generated that blocked the proliferation of MBP-specific effector T cells. Other effects contributing to immune tolerance include T-cell inactivation by antigenic priming in the liver,[107] tolerance induction via cross-presentation by LSECs, and hepatic stellate cell-induced T-cell apoptosis.[108]

Early models for AIH used fractions of liver homogenates, most commonly of heterologous origin, in combination with an activating adjuvant (e. g., complete Freund's adjuvant (CFA)). Most of these early models produced only limited or short-lived hepatitis, insufficient to evoke fibrosis. Those models that did produce a prolonged hepatitis, although still not a chronic fibrotic phenotype, required strong and repetitive stimulations with allogeneic or xenogeneic liver.[109]

One example of this is the experimental autoimmune hepatitis model.[110] In this model, mice with syngeneic liver homogenate of liver-specific lipoproteins were injected with smooth muscle antibody and an adjuvant composed of the polysaccharide of Klebsiella pneumoniae 03:K1. These mice developed portal infiltration of predominantly mononuclear cells.[110] In addition, transfer of splenocytes from these mice to unaffected recipient mice resulted in similar histologic findings with necrosis of parenchymal cells in the recipients.[110] C57BL/6 mice administered an S-100, crude supernatant of syngeneic liver homogenate, emulsified in CFA, developed identical findings with the addition of S-100 protein-specific T cells.[111]

Inbred A/J mice with thymectomy injected subcutaneously with syngeneic S-100 extracts showed frank hepatitis, production of autoantibodies to liver-specific protein (LSP), and delayed-type hypersensitivity to LSP[112] Given that it has been shown that neonatal thymectomy prevents the normal development of the CD4+ CD25+ population of Tregs. These data suggest the likely importance of Treg dysfunction in development of AIH.[113]

Using similar principles, various transgenic mouse models were developed. These models express target antigens in the liver, and the mice have to be transfused with target antigen-specific lymphocytes or infected with a pathogen carrying a similar antigen that will then evoke an immune response. Again in these models, the hepatitis was usually short lived with only mild cellular infiltrations of the periportal region.

An early example of this was a Tg mice which expressed the hepatitis B virus surface antigen (HBsAg) specifically in hepatocytes, under the control of the mouse albumin promoter. An adoptive transfer of activated T cells from an HBsAg-primed donor mouse led to cytotoxic T lymphocyte-driven apoptosis of hepatocytes expressing HBsAg and release of interferon (IFN)-γ on antigen encounter.[114] This, however, led to only 3 weeks of active inflammation which resembled a delayed-type hypersensitivity reaction.

In the CRP-Kb mouse, alloantigen H-2Kb is expressed under the inducible complement-reactive protein (CRP) promoter. In these mice, even low, uninduced levels of Kb expression caused tolerance to allogeneic skin. Similarly, tolerance was present in CRP-Kb mice crossed with mice transgenically expressing the Kb-reactive Des-T-cell receptor (TCR) on most of their CD8+T cells.[115] This tolerance was mediated at least in part by downregulation of the TCR. Indeed, application of LPS to induce expression of antigen resulted in even more downregulation of the TCR and subsequent deletion of the antigen-specific cells.[116] The presence of Tg T cells with high levels of TCR expression suggests that some degree of anergy was involved, but that this anergy and the downregulation of the TCR were lost if contact with the antigen was not maintained.[117] Tolerance was also broken if additional cells expressing the H-2Kb target antigen were transferred with IL-2 or when mice were infected with *Listeria monocytogenes*.[118] This work suggests that autoimmune disease can only occur in the context of a preexisting proinflammatory milieu. Another way to break through the livers resistance to autoimmunity was shown by administration of immunostimulatory cytosine phosphate guanine-rich oligodeoxynucleotides (CpG-ODNs) in H-2Kb/TCR-double-Tg mice. The H-2Kb-specific T cells were found in the liver within 2 h of adoptive transfer.[119] Despite this rapid onset, the inflammatory response was only transient and chronic hepatic damage could only be obtained via repetitive administration of the CpG-ODN.

In Met-Kb mice, H-2Kb expression occurs under the control of the metallothionein promoter. This showed tolerance secondary to negative selection by thymic expression,[120] but thymectomized Met-Kb mice engrafted with a non-Tg thymus and reconstituted with Des-TCR-Tg bone marrow provided hepatic infiltrates 3–5 weeks after bone marrow reconstitution.[121] In these mice, Des-TCR-Tg T cells in the periphery were diminished, even controlling for their number in the thymus itself, suggesting preferential deletion of these cells. As in the CRP-Kb mouse, adoptive transfer of postthymic Des-TCR-Tg T cells into Met-Kb mice produced an initial expansion of TCR-Tg T cells accompanied by hepatic infiltrates and resolution of the inflammation which paralleled the subsequent TCR-Tg T-cell deletion.[122]

Tg TFOVA expresses the model antigen ovalbumin (OVA) under control of the transferrin (TF) promoter specifically in hepatocytes.[123] Similarly, ASBT-OVA mice express the model antigen OVA under control of the apical sodium-dependent bile transporter (ASBT) promoter specifically in cholangiocytes. Transfer of TCR-Tg CD8 T cells (OT-I) or CD4 T cells (OT-II) in these models showed that only OT-I T cells migrated to the liver, where they proliferated causing inflammation and acute liver injury.[123] Again the inflammation was time limited to the acute stages and serum markers elevated only transiently.

Several models take advantage of liver-specific infections, whether viral, bacterial, or parasitic to induce inflammation. 6–8 weeks after subcutaneous injection of schistosomis mansoni cercariae, mice develop granulomas in the liver. They develop 20 times more collagen than a control liver and show a correlated reduction in albumin synthesis capacity.[124]

Others combine such approaches with use of an adoptive transfer of target antigen-specific TCR-Tg T cells. In these models, the pathogen infection primes and targets the liver leading to enhanced inflammation.[125] One example of this is a Tg mice expressing exclusively in the liver the immunodominant CTL epitope GP33 of the glycoprotein (GP) of the lymphocytic choriomeningitis virus (LCMV) under the control of the Alb promoter. Only when infected with LCMV after adoptive transfer of TCR-Tg cells, does a transient form of hepatitis develops.[126] Such models highlight the importance of the immunologic component in liver fibrogenesis.

A different approach involves either systemic or local elicitation of inflammation using induction of cytokines or injection of agents designed to evoke an inflammatory response. Repeated i.p. administration of heterologous pig or horse serum, egg albumin, or human gamma globulin over 8–12 weeks produces hepatic fibrosis with minimal hepatocellular injury, through an allergen-induced late phase reaction. Predictably given the allergen nature of the reaction eosinophilic inflammation is prominent and mast cells infiltrate the liver. Histologically, thin fibrotic bridging bundles are seen with severity dependent on both strain and species.

TNF-α and IFN-γ positive cells within biopsy specimens of AIH patients correlate with disease activity.[127] Indeed, regardless of the trigger, high levels of IFN-γ and TNF-α in animal models can lead to inflammatory hepatocyte apoptosis.[128] Serum amyloid P promoter controlled Tg expression of IFN-γ in the liver results in elevated serum transaminases and eventual chronic hepatitis.[129] A variety of other cytokine-mediated liver injury mouse models exist. Nonspecific T-cell stimulation and severe liver injury, with leukocytes adhering to the endothelial cells, are observed with systemic application of the leptin concanavalin A (ConA).[130] Although this does not directly raise TNF-α or IFN-γ levels, neutralizing antibodies to either cytokine can inhibit the disease indicating they are key parts of the pathway.[131,132] In addition, in mice with severe combined immunodeficiency syndrome, liver injury does not occur as interaction between CD4+ cells and macrophages is required.[133]

LSECs that bind ConA preferentially are both early mediators and later targets for CD4+ cells.[134] CD1d-deficient mice lack natural killer T (NKT) cells and do not produce liver injury in response to ConA[135] unless adoptive transfer of hepatic NKT cells from wild-type mice is performed. Of note, adoptive transfer of hepatic NKT cells from FasL-deficient gld mice to CD1d-deficient mice does not produce hepatitis, suggesting a role of FasL on NK cells in ConA-induced liver inflammation.[135] TNF-α exposure in mice sensitized with D-galactosamine leads to hepatocyte apoptosis and liver inflammation, as does evocation of a TNF-α response via stimulation with anti-CD3 antibodies or the superantigen staphylococcal enterotoxin B.[136,137]

Due to its key role in the development of other autoimmune diseases, such as type 1 diabetes and autoimmune thyroiditis, in experimental models IL-12 has attracted attention for study in AIH.[138–140] Direct intraperitoneal injection of IL-12 or DNA vaccination in combination with the expression of a target antigen in the liver can also evoke hepatitis.[141–143]

In transthyretin (TTR)-LCMV mice, which express the nucleoprotein (NP) of LCMV as a target antigen under the control of the TTR promoter,[142] elevated serum aminotransferase levels and hepatitis were induced by DNA vaccination with plasmids encoding the NP and IL-12 genes. They found that coexpression of IL-12 was required for the response. Although an NP-specific CTL response was detectable early, serum aminotransferase levels were not increased prior to month 5. Liver inflammation was also mild.[142] More significant cellular infiltrates with minimal elevations in serum alanine aminotransferase were noted with coexpression of IL-12 with CYP2D6 but not with CYP2D6 plasmid alone.[143]

Wild-type C57BL/6 mice vaccinated with dendritic cell (DC) loaded with well-differentiated hepatocellular carcinoma (Hepa)1–6 cells (DC/Hepa1–6) with intraperitoneal injection of recombinant human IL-12 showed moderate hepatitis. Adoptive transfer of DC/Hepa1–6-activated splenocytes into nonvaccinated mice resulted in a similar picture of liver damage when recipients were pretreated with IL-12.[141]

Another interesting model for the examination of AIH is the CYP2D6 mouse. In this model, infection of wild-type FVB mice with an adenovirus expressing human CYP2D6 (Ad-CYP2D6) resulted in the breakdown of natural tolerance to the mouse CYP homologues.[144] This mouse was developed on the theory that the transient nature of liver inflammation in the models discussed to this point may be due to an inappropriate selection of the target antigen for the liver, and that molecular mimicry, which uses similar antigens as target and trigger, may be more successful in inducing autoimmunity in the liver than molecular identity using exactly matching antigens. For this reason, they selected CYP2D6, which is the immunodominant antigen in human AIH type 2.[145,146] It confers molecular mimicry to the mouse CYP P450 isoforms. Although elevated serum aminotransferase levels were transient, as predicted, chronic hepatic damage with extensive subcapsular liver fibrosis was observed. B cells, CD4+ and CD8+ T cells, macrophages, and DCs were demonstrated infiltrating the subcapsular and perivascular areas of the liver.[147] Perhaps most significantly, the Ad-2D6-infected mice developed high titers of anti-CYP2D6 antibodies and recognized the same immunodominant epitope recognized by human AIH type 2 patients.

Chronic hepatitis has also been evoked via infection with an adenovirus expressing the human antigen FTCD. It is notable that while hepatitis could be elicited via this mechanism in nonobese diabetic (NOD) mouse strain, it did not occur in identically treated FVB/N, C57/BL6, or BALB/c.[144] This perhaps reflects the well-accepted view that for autoimmunity to occur, there must not only be the appropriate trigger but also be the correct genetic predisposition to it.

F. Viral Infection and Hepatic Fibrosis

In HCV, about 80% of individuals who become infected will develop a chronic infection rather than clearing it.[148,149] These patients are at increased risk for developing liver fibrosis, cirrhosis, and hepatocellular carcinoma. In man, viral infection is the second most common cause of liver fibrosis and thus an important area for study. Despite this, study of virally induced fibrosis is complicated by several factors. First, even in humans, differences between individual hosts can lead to chronic HCV/HBV infection or clearing of the disease. Perhaps even more importantly human hepatitis viruses are not infectious in mice or rats, the most common lab animal models. For this reason, while cell-based *en vitro* studies advanced at a rapid rate, results from animal models were slower relying on the chimpanzee as its test subject, thus necessitating longer time frames, larger facilities, and research budgets for results. Indeed, inoculation of chimpanzees provided many important discoveries, including identification of the hepatitis-causing agent by cloning of the virus[150] which allowed diagnostic detection of HCV which has since been used to screen and improve safety of blood products. Despite such insights into the early antiviral immune responses and pathogenesis, the chimpanzee remains a poor model for chronic infection with fibrosis. This is because the clinical course of infection in chimpanzees is milder compared to humans. Although following acute infection necroinflammatory lesions in the liver and increased serum aminotransferase levels are seen, liver cirrhosis or fibrosis does not develop, although one case with a non-A/non-B hepatitis-associated hepatocellular carcinoma was reported.[151]

As noted earlier, HCV does not infect rodent species. As such most rodent models have relied on xenografting of human cells into rodents with immune dysfunction such that the engrafted cells are not rejected. One of the few models where this is not the case is the immunotolerized rat model. In this model, fetal rats *in utero* are exposed to human hepatoma cells during the period of immune maturation. Hepatoma cells transplanted in the spleen postnatally result in engraftment making up to 6% of the liver at 14 weeks age. Intrasplenic inoculation resulted in a 62% infectivity rate; however, although elevated serum aminotransferase levels and mononuclear cell infiltrations of the liver are seen on histopathology during the acute phase, viremia remains low and fibrosis did not develop.[152]

Several mouse models exist based on heterotopic human liver grafts. In these models, human liver fragments are transplanted in immunodeficient mice in the areas of good vascular supply such as under the kidney capsule or behind the ear pinna.[153,154] While interesting from the standpoint of studying the evolution of the viral infection, the liver fragments fibrose and necrose secondary to the lack of a portal circulation, not the evolution of the infection and so are not relevant to study fibrosis of HCV.

One of the more successful models in terms of establishing prolonged HCV infection and studying possible targets for treatment is based on a mouse model with overexpression of a urokinase plasminogen activator (uPA) transgene.[155] These mice develop liver failure, but human liver cells can be used to repopulate the liver in uPA transgenic mice backcrossed with an appropriate immunodeficient background.[156] These mice can support prolonged HCV infections with high viral titers and evidence for active replication of the virus.[157] Despite these successes, fibrosis is not a common feature in these mice.

To date, therefore, no truly adequate animal model has emerged to study the fibrosis of long-term HCV infection. Despite that fact, this area remains one of active research and new models continue to emerge. Perhaps most promisingly recently, a genetically humanized mouse was developed which was not only susceptible to infection by a variety of HCV genotypes but was also fully immunocompetent.[158] This model was based on prior work that showed that CD81 and OCLN were the minimal human factors required for HCV uptake by rodent cells. Recombinant adenoviruses encoding human CD81, SCARB1, CLDN1, and OCLN were constructed, and intravenous delivery of these vectors resulted in

18–25% of murine hepatocytes expressing human CD81 and OCLN together, and 5% of cells expressing all four heterologous genes. It was found via inoculation of cell culture-derived HCV-CRE into Rosa26-Fluc animals expressing these human entry factors that all mice expressing at least human CD81 and OCLN could be successfully infected.[158] This is an important development as being immunocompetent it provides the first small animal platform suitable for combined immunization and challenge studies. In addition, in terms of study of the fibrosis that accompanies chronic hepatitis C infection, it is a step toward more closely recapitulating the life cycle of the virus. Although not reproducing all of the features of human HCV infection, this new model does show proof of concept for this approach and perhaps with inclusion of further human factors, or transgenic expression of the HCV factors rather than adenoviral infection, to increase the expression level, we can get close enough to permit a model meaningful for study of the fibrotic pathway in this disease.

G. Conclusions

No experimental model exactly reproduces human liver fibrosis by etiology. Nonetheless, each of the models reviewed here has served to enhance our understanding of pathogenetic mechanisms of liver fibrosis. In addition, we have highlighted models that have sought to circumvent some of the unique problems encountered when working with the liver, such as the liver's resistance to sustained autoimmunity or the adaptation of mouse models to HCV infection. Indeed, while models of alcohol-induced cirrhosis, NASH and biliary fibrosis are already well established and closely resemble the human disease, the area of AIH modeling is currently rapidly advancing, and that of HCV induced fibrosis, is still in its relative infancy compared to the other disease models. Nevertheless, there have also been important common findings derived from several different models. One example is the role of TGF-β in liver fibrogenesis, and its involvement via a plethora of different pathways. This is a common theme in models of fibrosis and allows investigators in one field of fibrosis to profit from the findings of fibrosis results in another. As indicated by the work discussed here, the future no doubt holds animal models, ever closer to recapitulating the fibrosis found in human disease.

II. Skin Fibrosis and Scleroderma

A. Introduction

Scleroderma or systemic sclerosis (SSc) is a systemic autoimmune connective tissue disease. Although the lungs and skin are most commonly affected, there can also be prominent involvement of the heart, kidney, and gastrointestinal tract. The triggering event and early pathologic changes of this disease remain poorly understood in part due to the diseases slow onset resulting in patients being at a fairly advanced stage of disease at the time of diagnosis. Despite this, early events are believed to consist of vascular injury and immune dysfunction, although there is debate as to which occurs first. The fibrotic changes with which patients present are believed to reflect a much later stage. Although scleroderma has a wide variety of presentations, it is usually broadly categorized as either systemic or local. Systemic sclerosis or scleroderma is systemic (multiple organs) and diffuse (a variety of skin–body areas). This contrasts with generalized morphea which although affecting a variety of skin–body areas do not have visceral involvement. In contrast to both of these, localized sclerosis or morphea is usually limited to a contiguous region but may affect underlying joints if it extends over these. However, there are also other, more tightly defined subtypes such as CREST syndrome, a variety of systemic sclerosis, consisting of calcinosis cutis, Raynaud's phenomenon, esophageal dysmotility, sclerodactyly, and telangiectasias. Finally, there exist subsets which have been named for their unique clinical presentations. These include atrophoderma of Pasini and Pierini, pansclerotic morphea where the sclerosis extends down to the bone, morphea profunda which extends to the subcutaneous fat and fascia and linear morphea which is more common in children. In addition, a number of scleroderma-like diseases exist which tend to present similarly and are thus often misdiagnosed initially as scleroderma. These entities include eosinophilic fasciitis, eosinophilia myalgia syndrome, and other sclerodermatous conditions which are often related to a toxic exposure, including nephrogenic systemic fibrosis from gadolinium contrast media in patients with severe renal disease. Forming an accurate animal model of skin fibrosis is thus complicated by this diversity of clinical entities and even in the presentations of scleroderma itself, the prototypical skin fibrosis disorder. However, even when examining only one subtype of sclerodermatous disease, animal models often fail to reproduce all of the clinical and histopathological features of the subtype and so one-for-one mapping of pathogenetic events from animal model to human disease is often not possible. Animal models do, however, provide the opportunity to examine the early stages of disease, which are often missed *in vivo* and allow isolated study of molecules of interest, analysis of pathways, and the effect of their dysregulation. This can be accomplished through therapeutic intervention or genetic models such as knockout experiments. Here, we discuss the insights provided by studies into the early molecular and cellular changes that may promote the initial fibrotic response, and those that maintain it.

B. Models with Vascular Dysfunction

As noted earlier, it is believed that one of the earliest changes in scleroderma is a vascular dysfunction. Unfortunately, very few models recapitulate this part of the pathologic process. Although University of California at Davis

(UCD) lines 200 and 206 chickens are not currently widely used, this model produced some of the earliest publications examining the fibrosis-associated vasculopathy of SSc. One to two weeks after hatching, UCD-200/206 chickens develop microvascular occlusions, followed by intense perivascular lymphocytic infiltration with mainly TCR $\gamma/\delta+$/CD3+/MHC (major histocompatibility complex—*B* complex in chickens) class II^{-}(B-L^{-}) T lymphocytes.[159] Both the skin and internal organs are involved. In a process analogous to Raynaud's phenomenon in SSc, they develop erythema, edema and necrosis of the comb. With time there is also pronounced collagen deposition with skin fibrosis and thickening. This model therefore represents the hypothesized sequence of vasculopathy, inflammation, and subsequent fibrosis believed to occur in human scleroderma. Frozen skin sections analyzed for apoptosis by terminal deoxynucleotidyl transferase-mediated FITC-dUTP nick end labeling and indirect immunofluorescence staining of cell markers with tetramethylrhodamine isothiocyanate conjugates showed that endothelial cells were the first cells to undergo apoptosis in the skin of this model.[160] Despite this, the process occurs over a much shorter time period and can involve frank necrosis of the comb even in its earliest stages.

One of the most commonly used mouse models for study of skin fibrosis is the tight skin (*Tsk*) mouse model. While the homozygous *Tsk* mouse *Tsk/Tsk* genotypes die *in utero* within 1 week of conception, *Tsk/+* mice survive, developing interscapular *Tsk*, firmly bound to underlying structures. Due to this phenotype, the *Tsk/+* genotype has therefore been used as an animal model to represent the skin changes of SSc, even though it lacks an early inflammatory infiltrate which is one of the key pathologic features usually seen in early SSc.

While the *Tsk* mouse lacks a vasculopathy, it may give us some insight into why patients with scleroderma do not recover from the vasculopathy via neoangiogenesis. In a recent study, human umbilical vein endothelial cells were seeded on microfibrils isolated from skin of control mice (C57BL/6J) and *Tsk*(−/+) mice with and without chronic pretreatment with the apolipoprotein Apo A-I mimetic D-4F. In this study, Western blot analysis showed that *Tsk*(−/+) mice demonstrated abnormal enlarged fibrillin expression compared with control. In addition, endothelial cells cultured on microfibrils prepared from *Tsk*(−/+) mice demonstrated reduced proliferation, a pro-oxidant state, increased apoptosis, and collagen-related protein expression associated with mesenchymal transition. The reduced proliferation of endothelial cells is particularly interesting as it suggests at least one mechanism for the lack of an angiogenic response in scleroderma patients.[161]

Caveolin-1 is a member of the caveolin group of proteins which are responsible for development of endothelial invaginations, receptor-independent endocytosis, and signal transduction. Upregulation of caveolin-1 leads to internalization of TGF-β receptors and subsequent reduction in TGF-β signaling. In contrast, caveolin-1 knockouts develop a fibrotic phenotype secondary to increased TGF-β.[162] The fibrosis affects both skin and lungs as well as producing a vasculopathy and alterations in lipid profiles.

The fos-related antigen (fra-2) is part of the activator protein 1 (AP-1) signaling pathway. This pathway is involved in regulation of wound healing and as such contributes to cell proliferation, apoptosis, and inflammation. Mice with ectopic expression of fra-2 developed generalized fibrosis, inflammation, and vasculopathy. These effects were greatest in the lungs but were also present in other organs including the skin.[163]

The MPR/lpr mouse model produces autoimmune antibodies and develops arthritis, vasculitis, cutaneous lesions, and lethal glomerulonephritis. Modified versions with deficiency of IFN-γ or its receptor do not show the glomerulonephritis but instead have been shown to have a vasculopathy characterized by intimal thickening and accumulation of collagen.[164]

The newest model to recapitulate the vasculopathy is based on Fli-1, a member of the Ets family of transcription factors, which is expressed in both fibroblasts and endothelial cells, although more heavily in the latter. Fli1 knockout mice (Fli1 ECKO mice) generated by crossing *Fli1$^{flox/flox}$* mice with Tie2-*Cre* transgenic mice[165] develop poorly organized vascular networks with arteriole stenosis, microaneurysm formation, and capillary dilation. This vascular phenotype is due to a decreased expression of vascular endothelial (VE)-cadherin, platelet/endothelial cell adhesion molecule (PECAM)-1, platelet-derived growth factor (PDGF)-B, and sphingosine-1-phosphate $(S1P)_1$ receptors. This absence of VE-cadherin and PECAM-1 results in a weak endothelial cell–cell interaction, and thus capillary dilation, vascular fragility, and increased vascular permeability. This explanation of the vascular fragility in SSc is interesting as in humans; Fli1 is greatly reduced in endothelial and periendothelial cells in SSc skin compared to healthy controls.[166,167]

C. Inflammatory Dysfunction and Autoantibodies

Autoantibodies are a normal feature of SSc, with indirect immunofluorescence antinuclear antibodies (IIF-ANA) being detected in approximately 90% of scleroderma patients.[168] In addition to a vasculopathy, the UCD-200/206 chicken develops autoantibodies, including ANA, antiendothelial cell antibodies (AECA), anticardiolipin antibodies, and rheumatoid factors. The development of AECA is interesting as some authors have suggested that endothelial cells are the primary target in scleroderma pathogenesis and are targeted by AECA-dependent cellular cytotoxicity via Fas/Fas ligand interaction.[160] This hypothesis suggests that AECA induces microvascular alterations via upregulation of endothelial adhesion molecules such as E-selectin, P-selectin, the Ig-supergene family members intercellular adhesion molecule 1, and vascular cell adhesion molecule 1.[169] However, although AECA has been noted in human SSc, it only appears to occur in localized morphea and in the very early stages of SSc thus posing the question if this is indeed the initiating factor, why the disease does not always become systemic and what maintains the ongoing fibrosis.

In the UCD-200 chicken, the inflammatory phase is composed of TCR γ/δ +/CD3+/MHC class II-T cells in the stratum papillare and TCR α/β+/CD3+/CD4+/MHC class II + T cells in the deeper dermis.[170] Skin thickening is due to excessive accumulation of collagen types I, III, and VI, although no mutation of the collagen genes is noted on RFLP studies.[171]

It remains unclear if autoimmunity is a part of the pathogenesis of fibrosis initiation or maintenance in Tsk mice. In support of a role for immune activation in Tsk mice, transplantation of Tsk bone marrow into normal mice resulted in skin tissue fibrosis and increased collagen gene expression in the recipients. In addition, Tsk/+ mice with no functioning IL-4 or Stat6 genes,[172] as well as mice with a targeted mutation in the IL-4 receptor α chain,[173] did not develop dermal fibrosis, suggesting that IL-4 is involved in regulating fibrosis in *Tsk* mice. CD19 knockout mice also showed less dermal fibrosis.[174] However, conflicting evidence for immune regulation exists. For example, it has been shown that mouse strains lacking α/β receptor T cells, B cells, or even mice lacking all T and B cells, such as the scid or Rag knockout mouse, still developed Tsk when combined with the Tsk mutation.

A variety of transgenic mice with mutant fibrillin genes have been produced with a range of resulting phenotypes. Transgenic mice that demonstrated persistent cutaneous thickening and positive antitopoisomerase antibodies were produced by the use of a plasmid vector, pdFbn-1, carrying the *Tsk Fbn1* sequence.[175] However, injection of the same plasmid to newborn mice produced only a transient skin thickening.

A related model is the *Tsk2* mouse. The tight skin 2 (*Tsk2*) mutation is a chemically (ethylnitrosourea) induced autosomal dominant mutation in mouse chromosome 1. Although this produces similar phenotypic skin changes to *Tsk1*, the advantage of this model is that the histological changes are closer to that of human SSc with dermal fibrosis and loss of adipose tissue. Perhaps more importantly, there are positive antinuclear antibodies in 88% of *Tsk2/+* mice compared with 10% of controls and, as in SSc, anticentromere, antitopoisomerase-I, and anti-RNA polymerase II autoantibodies.[176] Although vessels appear thickened in *Tsk2*, they still lack the frank vasculopathy of SSc.

As a result of the observation that cancer patients treated with bleomycin develop pulmonary fibrosis, this compound has been used to produce fibrosis in mouse models. Indeed, this has become one of the most popular models of SSc in the mouse. In models of lung fibrosis, fibrosis can be produced via a one-time intratracheal injection. In contrast, skin fibrosis is produced by repeated subcutaneous injections for at least 3 weeks.[177] This technique produces a small peri-injection area of induration with histologic changes showing epidermal hypertrophy, dermal fibrosis, consisting of accumulation of myofibroblasts, collagen and dense extracellular matrix material, and adipose atrophy. This effect varies by mouse strain with the C3H/He and B10.A giving the most pronounced histologic changes.[178]

Fibrosis is usually associated with a T-helper (Th)-2 immune response, in part due to downregulation of the antifibrotic Th1 cytokine IFN-γ. Mice deficient for T-bet, a regulator essential for initiating Th1, showed increased and more rapid fibrosis with an exaggerated immune response to bleomycin in comparison to wild-type littermate mice.[179] Other factors shown to be important features in the increased fibrosis of T-bet null mice were extensive eosinophil degranulation and constitutively elevated IL-13.[180] As in scleroderma, autoimmune antibodies have been detected in the mouse sera. In addition, transfer of CD4(+) T cells from bleomycin-treated BALB/c mice induced the same pathological changes and antibody production in untreated BALB/c nude mice supporting the immunologic basis of disease in this model.[181]

D. Fibrosis and Dysregulation of Signaling

The final stage of scleroderma is believed to be the fibrosis resulting from the vasculopathy, immune dysfunction, and resultant dysregulated signaling. This end stage of fibrosis in the animal models is attained via a variety of pathways. The *Tsk* mouse model is a spontaneous autosomal dominant mutation. It possesses a large in frame intragenic duplication of the fibrillin-1 gene (*Fbn1*) from exons 17–40.[182] RNA extracts prepared from *Tsk*/+ mice demonstrate transcripts for both a normal 11 kb Fbn1 and a larger 14 kb version. In addition, both a normal 350-kDa Fbn1 and a 450-kDa variant have been shown to be secreted by *Tsk*/+ dermal fibroblasts. Although the definitive mechanism for fibrogenesis by this altered fibrillin molecule has not been found, two possible mechanisms have been postulated. First, it has been established that fibrillin-1 is a component of connective tissue microfibrils and has an important role in correct elastic fiber assembly. It is likely therefore that alteration in fibrillin-1 would also secondarily alter the production of elastic fibers. Second, fibrillin can in part control the bioavailability of the profibrotic molecule TGF-β via interaction with the latent TGF-β binding proteins 1 and 4. The observation that the altered fibrillin binds more TGF-β than normal fibrillin also appears to be consistent with this explanation. This may support the hypothesis that similarities in the duplicated regions of the fibrillin molecule to TGF-β-binding protein domains are at least in part responsible for altering the TGF-β profibrotic cascade.

In addition, to upregulation of the foregoing microfibrillar proteins gene expression, analysis of 6-week-old *Tsk/+* mice using GeneChip arrays has also shown upregulation in collagen, bone morphogenic protein (BMP), secreted frizzle-related protein (SRFP4), CTGF, and Wnt signaling proteins, including Wnt3a, which have been implicated in the regulation of cell fate and integration of signals from other pathways such as TGF-β, FGF, and BMP.[183] This is interesting as elevated levels of SRFP4 were noted in tissue from SSc patients and application of Wnt3a to cultured fibroblasts produced increased levels of Fbn-1 and collagen.

Cell culture of fibroblasts from *Tsk/+* mice also show evidence of persistent activation. Dermal fibroblast cultures from *Tsk/+* mice show increased collagen deposition associated with elevated steady-state levels of collagen types I and III messenger RNA (mRNA).[184] Consistent with this, probes to types I and III collagen show upregulation in *Tsk* fibroblasts when compared to normal controls.

While the *Tsk/+* mouse may phenotypically resemble clinical findings of SSc, as a model for SSc, it clearly therefore has some weaknesses. For example, SSc typically begins with mild inflammation and fibrosis in the deep dermis. However, histological analysis of *Tsk* mice shows that there is neither inflammation nor deep fibrosis, but instead skin tethering to the subjacent muscle occurs. In addition, the pulmonary fibrosis seen in SSc is not seen in *Tsk/+* mice. Although the *Tsk/+* mouse does show pulmonary changes, these are of an emphysema-like pattern with increased alveolar spaces. In contrast to SSc, *Tsk/+* mice do not exhibit a vasculopathy. However, despite these weaknesses, the *Tsk/+* mouse model has provided insight into the possible role of Wnt dysregulation, and TGF-β in skin fibrosis, as well as clues regarding the interaction with the immune system, and is proof that extensive skin fibrosis may develop from a single mutation. Finally, the *Tsk* mouse may be important in part for what it is not; it provides a model of noninflammatory fibrosis and is therefore useful as a comparison to inflammatory models. This was taken advantage of in a recent study that showed that Stat4 knockout could ameliorate the fibrosis of the bleomycin mouse, but not the *Tsk* mouse.[185] Similarly, being some of the best studied models, new studies continue to compare the effect of different potential treatments on both bleomycin and *Tsk* mice.[186–189]

Cultures of *Tsk2* fibroblasts show elevated steady-state levels of Col1a1 mRNA with transcriptional activation of *Col1a1* and *Col3a1* mediated by Sp1 and NF-1.[190] Increased collagen gene expression was also shown by use of a collagen transgene expressing green fluorescent protein. This showed an age-dependent increase in GFP-expressing fibroblasts when compared to wild-type littermates.[191]

Other mutant fibrillin transgene experiments have resulted in the production of mice that show anomalies in TGF-β signaling but no skin thickening.[175] Perhaps the most intriguing model of this kind was a conditional mutant fibrillin transgene model. In embryonic fibroblast cultures, fibrillin fibers had a distorted structure with increased collagen deposition despite a lack of change in Col1a1 mRNA or procollagen production. Instead, elevated levels of microfibril-associated glycoprotein 2 (MAGP-2) which stimulates elastic fiber assembly were detected, which is interesting as this has also been shown to be present in the tissue of SSc patients.

In the conditional constitutively active TGF-β receptor 1 transgenic mouse model, a knockin mouse strain with a constitutively active mutant of the ALK5 TGF-β type I receptor (*TβR1ca*) inserted in a transcription stop cassette, flanked by two loxP sites, was targeted to the *ROSA26* locus promoter. The receptor was thus fused to a ubiquitously expressed promoter that could only be activated by a *Cre*-mediated rearrangement of the LoxP sites. This mouse strain was then bred with a transgenic mouse with a tamoxifen inducible Cre (*CreER*) controlled by the *Col1a2* promoter. Injection of tamoxifen for 5 consecutive days to the offspring produced TβR1 activation in fibroblasts[192] and the offspring recapitulated many of the clinical skin features of SSc. These included progressive dermal fibrosis, epidermal thinning, loss of adnexal, and subcutaneous adipose structures, as well as extracutaneous findings such as intramural collagen accumulation in small pulmonary vessels. Subsequent analysis of these fibroblasts showed ligand-independent Smad2/3 phosphorylation and nuclear accumulation. In addition, there was elevated gene expression of downstream targets of TGF-β, including CTGF, TIMP-1, type 1 collagen, fibronectin, plasminogen activator inhibitor (PAI)-1, and secreted protein, acidic and rich in cysteines (SPARC). P38 and mitogen-activated protein kinase (MAPK) pathway activity was also demonstrated. This model therefore showed successful production of an SSc phenotype, even without initial inflammation, purely by TGF-β signal dysregulation.

The type 2 TGF-β receptor mutant transgenic mouse model, which also resulted in fibrosis, used fibroblast-restricted expression of a kinase-deficient type 2TGF-β receptor (TβRIIδk).[193] This result was surprising given that TβRIIδk in cultured fibroblasts had been reported to be a dominant negative inhibitor of TGF-β signaling. More detailed analysis showed upregulation of PAI-1, CTGF, Smad3, Smad4, and Smad7, and that Smad2/3 phosphorylation was increased in transgenic fibroblasts. Although seemingly contradictory to the results of prior experiments, this model provided dramatic evidence of the ability of even small perturbations in TGF-β signaling to result in the formation of a fibrotic phenotype. Interestingly, these mice also develop a vasculopathy consistent with that seen in systemic scleroderma. In addition, 25% of these mice develop spontaneous lung fibrosis similar to that found in human scleroderma. Even in those mice that do not spontaneously develop lung fibrosis, it can be provoked with minor lung injury using intratracheal saline, resembling the changes seen with bleomycin-treated wild-type mice. Intratracheal bleomycin results in even more pronounced lung fibrosis. However, in a subset of these mice, a fibrotic cardiomyopathy was also seen, that does not represent typical SSc findings.[163,194,195]

It has been shown that the hormone peptide relaxin is effective in blocking TGF-β activity in either normal or SSc skin fibroblasts. It thus reduces collagen production and fibrosis. Although relaxin-null mice develop spontaneous fibrosis of the skin and lungs, which could be attenuated by exogenous supplementation of relaxin,[196] and *in vitro* studies showed higher collagen production in relaxin-null fibroblasts than wild type, none of the vascular or immunological features of SSc were evident in these mice.

Biopsy of lesional skin in the bleomycin mouse model shows upregulated TGF-β production, phosphorylated Smad2/3,[197] coactivators of the TGF-β/Smad pathway, such as histone acetyltransferase p300,[198] and CTGF.[199] In contrast, Smad3-null mice treated with bleomycin showed significantly less fibrosis[200] emphasizing the importance of this pathway. However, it is clear that in this model, TGF-β-stimulated collagen production is also upregulated via Smad-independent signal transduction. This is demonstrated by the Smad-independent chronic upregulation of the early growth response factor (Egr)-1 in bleomycin-treated mice, a factor that is normally rapidly but only transiently elevated in injury or hypoxia.[201] Both pathways, Smad-dependent and -independent stimulation of TGF-β, including downregulation of Egr-1 and hepatocyte growth factor (HGF) are inhibited by ligation by PPAR-γ, a ubiquitous nuclear hormone receptor.[202] In addition, treatment of C3H mice with 15-deoxyprostaglandin J2 (PGJ2), an endogenous ligand of PPAR-γ, or rosiglitazone, a pharmacologic ligand,[203] showed attenuation of bleomycin-induced fibrosis. Part of the antifibrotic effect may also be attributable to the downregulation of HGF, a factor secreted by mesenchymal cells but acting on epithelium via the c-Met receptor coupled by Gab1 and Grb2 to downstream tyrosine kinases. This is thought to play an antifibrotic role in acute injury repair and has been shown to downregulate collagen synthesis in normal and SSc fibroblasts *in vitro*.[204] In support of this intramuscular injection of cDNA, coding for HGF gave partial protection against skin fibrosis in bleomycin-treated mice.[205]

Similar to Egr-1, adenosine levels rise rapidly in normal tissue but transiently in injury or hypoxia, reflecting both its anti-inflammatory role[206] and its role in wound healing where it stimulates fibroblast collagen production via the A_{2A} cell surface receptor[207] and downstream mitogen-activated protein kinase/extracellular-regulated kinase (MEK1/ERK) signal transduction pathway.[208] Attenuation of bleomycin-induced fibrosis is seen in A_{2A} receptor knockout mice.[208] Low tissue adenosine can be achieved by either double knockout CD39/CD73 mice or knockouts of the individual enzymes CD39 or CD73 responsible for adenosine formation. Such knockout mice have also been shown to be resistant to the fibrosing effects of bleomycin (Gideon Smith, in preparation, 2011). Mice deficient for the adenosine-catabolizing enzyme adenosine deaminase (ADA) with resulting adenosine accumulation in blood and tissues exhibit increased dermal fibrosis with enhanced expression of IL-13 and CTGF, as also occurs in bleomycin-treated mice.[209]

Chemokines have also been shown to be important. MCP-1 also known as chemokine ligand (CCL)-2 and its receptor (CCR)-2 have been shown to be upregulated in SSc patients. In addition, stimulation with MCP-1 was shown to increase collagen production in cultured fibroblasts. For this reason, the effect of bleomycin on MCP-1 knockout mice was examined and showed a dramatically reduced early inflammatory infiltrate, less fibrosis, and more regular collagen.[210] In comparison, wild-type mice had smaller-sized irregular collagen, with unregulated expression of MCP-1 and CCR-2 on early infiltrating monocytes and late stage fibroblasts.

Activated adhesive signaling is seen in fibroblasts from SSc patients and is believed to be a key step in formation of fibrosis. This is supported by observations that show that the deletion of β1 integrin results in resistance to bleomycin-induced fibrosis in mouse models.[211]

In addition to its vasculopathic properties, Fli-1 represses transcription of collagen genes via an Sp1-dependent pathway.[212] Indeed, embryonic fibroblasts from *Fli1−/−* mice showed significantly increased collagen type I mRNA and protein levels. This is consistent with the results from examination of cultured SSc fibroblasts which show downregulation of Fli1 expression levels.[166] This downregulation of Fli1 is associated with increased CTGF and decreased MMP1[213] and may be due to Fli1 degradation when TGF-β initiates a PCAF-dependent acetylation of Fli1.[214] Interestingly, this same upregulation of Fli-1 is demonstrated in bleomycin-treated C57/BL6 mice but is abrogated in bleomycin-treated CD39/CD73 knockouts, suggesting that the fibrosis seen in the bleomycin models discussed earlier are also controlled by Fli-1 regulation (Gideon Smith, in preparation, 2011).

The observation of vasculopathy, arthritis, and fibrosis in the MPR/lpr mouse model is curious given that IFN-γ is classically associated with a TH1 antifibrotic phenotype. The exact role of IFN-γ in fibrosis in these models is therefore not clear. Although accumulation of thick collagen bundles in the kidneys, liver, lungs, and salivary glands in IFN-γ receptor-deficient mice was in contrast to the normal collagen bundles seen in the organs of MPR/lpr/γR+/+ mice, both lines showed fibrosis in skin lesions. One potential explanation is that the IFN-γ mutation, by protecting against glomerulonephritis in this animal model, allows longer life expectancy. This, in turn, provided sufficient time for the fibrosis to develop in different organs, and that skin is fibrotic in both as it is the first organ affected. Thus it seems that an alternative, as yet unexplained, mechanism of fibrosis in this model is likely also at work.

The sclerodermatous graft-versus-host mouse model is produced by transplantation of immunologically incompatible spleen and bone marrow cells. Female Balb/c mice are irradiated and receive donor cells from male B10.D2 mice, allowing tracking of donor cells via the Y-specific sequences. These mice strains differ at minor histocompatibility loci which set up a sequence of immune activation and subsequent fibrosis. In the initial stage, elevated levels of mRNA for Th1 and Th2 cytokines, TGF-β1, macrophage chemoattractant protein (MCP)-1, macrophage inflammatory protein (MIP)-1α, and RANTES are produced, and tissue is infiltrated by donor T cells and monocytes/macrophages.[215] This first stage is followed by skin and lung fibrosis, which occurs at 2–3 weeks posttransplantation, accompanied by elevated levels of type I collagen RNA synthesis. Of note, this cascade could be blocked by administration of anti-TGF-β antibodies indicating a key role for TGF-β in the fibrotic process. In addition, one study identified the key role of downregulation of TNF-α in producing the sclerodermatous phenotype.[216]

A similar model, that uses the recombination activating gene (*Rag*) 2 mouse has been produced that thus avoids the need for irradiation.[217] This mouse is immune deficient secondary to the loss of the ability to rearrange immunoglobulin and T-cell receptor genes, and the donor cells can therefore be injected directly. This produced a mouse with similar skin phenotype between weeks 3 and 5, as well as producing fibrosis and VE changes in the kidney. It did not, however, reproduce the lung pathology

E. Conclusions

The experiments to date with animal models of skin fibrosis point to a number of potentially important factors in the pathogenesis of this disease state and continue to identify new pathways, and thus new potential therapeutic targets for the modulation of fibrosis. Indeed, what is most clear from these experiments is that fibrosis is an extremely complex process which depends on the interactions of many molecules and different pathways. While on the one hand that makes our job of understanding this process more complex, on the other, it informs us that there are many potential therapeutic targets to develop that might potentially improve the prognosis of patients with these terrible disease states. Although none of the currently available models mimic exactly the development of human fibrosing diseases such as SSc or NSF, they do allow us to attempt to see the early stages of fibrosis, a stage frequently missed in the evaluation of human disease due to the slow, almost asymptomatic initial onset. They are also allowing the testing of both hypotheses and potential therapies.

References

1. Iredale JP. Models of liver fibrosis: exploring the dynamic nature of inflammation and repair in a solid organ. *J Clin Invest* 2007;**117**:539–48.
2. Lefkowitch JH. Morphology of alcoholic liver disease. *Clin Liver Dis* 2005;**9**:37–53.
3. Lieber CS, Jones DP, Decarli LM. Effects of prolonged ethanol intake: production of fatty liver despite adequate diets. *J Clin Invest* 1965;**44**:1009–21.
4. Rubin E, Lieber CS. Fatty liver, alcoholic hepatitis and cirrhosis produced by alcohol in primates. *N Engl J Med* 1974;**290**:128–35.
5. Tsukamoto H, Towner SJ, Ciofalo LM, French SW. Ethanol-induced liver fibrosis in rats fed high fat diet. *Hepatology* 1986;**6**:814–22.
6. Constandinou C, Henderson N, Iredale JP. Modeling liver fibrosis in rodents. *Methods Mol Med* 2005;**117**:237–50.
7. Domenicali M, Caraceni P, Giannone F, Baldassarre M, Lucchetti G, Quarta C, et al. A novel model of CCl4-induced cirrhosis with ascites in the mouse. *J Hepatol* 2009;**51**:991–9.
8. Indra AK, Warot X, Brocard J, Bornert JM, Xiao JH, Chambon P, et al. Temporally-controlled site-specific mutagenesis in the basal layer of the epidermis: comparison of the recombinase activity of the tamoxifen-inducible Cre-ER(T) and Cre-ER(T2) recombinases. *Nucleic Acids Res* 1999;**27**:4324–7.
9. Schuster N, Krieglstein K. Mechanisms of TGF-beta-mediated apoptosis. *Cell Tissue Res* 2002;**307**:1–14.
10. Bataller R, Brenner DA. Liver fibrosis. *J Clin Invest* 2005;**115**:209–18.
11. Hellerbrand C, Stefanovic B, Giordano F, Burchardt ER, Brenner DA. The role of TGFbeta1 in initiating hepatic stellate cell activation *in vivo*. *J Hepatol* 1999;**30**:77–87.
12. Massague J, Seoane J, Wotton D. Smad transcription factors. *Genes Dev* 2005;**19**:2783–810.
13. Moustakas A, Heldin CH. The regulation of TGFbeta signal transduction. *Development* 2009;**136**:3699–714.
14. Moustakas A, Souchelnytskyi S, Heldin CH. Smad regulation in TGF-beta signal transduction. *J Cell Sci* 2001;**114**:4359–69.
15. Hamzavi J, Ehnert S, Godoy P, Ciuclan L, Weng H, Mertens PR, et al. Disruption of the Smad7 gene enhances CCI4-dependent liver damage and fibrogenesis in mice. *J Cell Mol Med* 2008;**12**:2130–44.
16. Dooley S, Hamzavi J, Ciuclan L, Godoy P, Ilkavets I, Ehnert S, et al. Hepatocyte-specific Smad7 expression attenuates TGF-beta-mediated fibrogenesis and protects against liver damage. *Gastroenterology* 2008;**135**:642–59.
17. Perugorria MJ, Latasa MU, Nicou A, Cartagena-Lirola H, Castillo J, Goni S, et al. The epidermal growth factor receptor ligand amphiregulin participates in the development of mouse liver fibrosis. *Hepatology* 2008;**48**:1251–61.
18. Paizis G, Cooper ME, Schembri JM, Tikellis C, Burrell LM, Angus PW. Up-regulation of components of the renin-angiotensin system in the bile duct-ligated rat liver. *Gastroenterology* 2002;**123**:1667–76.
19. Bataller R, Sancho-Bru P, Gines P, Lora JM, Al-Garawi A, Sole M, et al. Activated human hepatic stellate cells express the renin-angiotensin system and synthesize angiotensin II. *Gastroenterology* 2003;**125**:117–25.
20. Bataller R, Gabele E, Schoonhoven R, Morris T, Lehnert M, Yang L, et al. Prolonged infusion of angiotensin II into normal rats induces stellate cell activation and proinflammatory events in liver. *Am J Physiol Gastrointest Liver Physiol* 2003;**285**:G642–51.
21. Bataller R, Gines P, Nicolas JM, Gorbig MN, Garcia-Ramallo E, Gasull X, et al. Angiotensin II induces contraction and proliferation of human hepatic stellate cells. *Gastroenterology* 2000;**118**:1149–56.

22. Bataller R, Schwabe RF, Choi YH, Yang L, Paik YH, Lindquist J, et al. NADPH oxidase signal transduces angiotensin II in hepatic stellate cells and is critical in hepatic fibrosis. *J Clin Invest* 2003;**112**:1383–94.
23. Kanno K, Tazuma S, Chayama K. AT1A-deficient mice show less severe progression of liver fibrosis induced by CCl(4). *Biochem Biophys Res Commun* 2003;**308**:177–83.
24. Yang L, Bataller R, Dulyx J, Coffman TM, Gines P, Rippe RA, et al. Attenuated hepatic inflammation and fibrosis in angiotensin type 1a receptor deficient mice. *J Hepatol* 2005;**43**:317–23.
25. Nabeshima Y, Tazuma S, Kanno K, Hyogo H, Iwai M, Horiuchi M, et al. Anti-fibrogenic function of angiotensin II type 2 receptor in CCl4-induced liver fibrosis. *Biochem Biophys Res Commun* 2006;**346**:658–64.
26. Baker SJ, Reddy EP. Transducers of life and death: TNF receptor superfamily and associated proteins. *Oncogene* 1996;**12**:1–9.
27. Reeves HL, Dack CL, Peak M, Burt AD, Day CP. Stress-activated protein kinases in the activation of rat hepatic stellate cells in culture. *J Hepatol* 2000;**32**:465–72.
28. Sudo K, Yamada Y, Moriwaki H, Saito K, Seishima M. Lack of tumor necrosis factor receptor type 1 inhibits liver fibrosis induced by carbon tetrachloride in mice. *Cytokine* 2005;**29**:236–44.
29. Yamada Y, Fausto N. Deficient liver regeneration after carbon tetrachloride injury in mice lacking type 1 but not type 2 tumor necrosis factor receptor. *Am J Pathol* 1998;**152**:1577–89.
30. Ezure T, Sakamoto T, Tsuji H, Lunz 3rd JG, Murase N, Fung JJ, et al. The development and compensation of biliary cirrhosis in interleukin-6-deficient mice. *Am J Pathol* 2000;**156**:1627–39.
31. Streetz KL, Tacke F, Leifeld L, Wustefeld T, Graw A, Klein C, et al. Interleukin 6/gp130-dependent pathways are protective during chronic liver diseases. *Hepatology* 2003;**38**:218–29.
32. Hayashi H, Sakai T. Animal models for the study of liver fibrosis: new insights from knockout mouse models. *Am J Physiol Gastrointest Liver Physiol* 2011;**300**:G729–38.
33. Afford SC, Fisher NC, Neil DA, Fear J, Brun P, Hubscher SG, et al. Distinct patterns of chemokine expression are associated with leukocyte recruitment in alcoholic hepatitis and alcoholic cirrhosis. *J Pathol* 1998;**186**:82–9.
34. Bonacchi A, Petrai I, Defranco RM, Lazzeri E, Annunziato F, Efsen E, et al. The chemokine CCL21 modulates lymphocyte recruitment and fibrosis in chronic hepatitis C. *Gastroenterology* 2003;**125**:1060–76.
35. Bonacchi A, Romagnani P, Romanelli RG, Efsen E, Annunziato F, Lasagni L, et al. Signal transduction by the chemokine receptor CXCR3: activation of Ras/ERK, Src, and phosphatidylinositol 3-kinase/Akt controls cell migration and proliferation in human vascular pericytes. *J Biol Chem* 2001;**276**:9945–54.
36. Marra F, DeFranco R, Grappone C, Milani S, Pastacaldi S, Pinzani M, et al. Increased expression of monocyte chemotactic protein-1 during active hepatic fibrogenesis: correlation with monocyte infiltration. *Am J Pathol* 1998;**152**:423–30.
37. Schwabe RF, Bataller R, Brenner DA. Human hepatic stellate cells express CCR5 and RANTES to induce proliferation and migration. *Am J Physiol Gastrointest Liver Physiol* 2003;**285**:G949–58.
38. Seki E, De Minicis S, Gwak GY, Kluwe J, Inokuchi S, Bursill CA, et al. CCR1 and CCR5 promote hepatic fibrosis in mice. *J Clin Invest* 2009;**119**:1858–70.
39. Seki E, de Minicis S, Inokuchi S, Taura K, Miyai K, van Rooijen N, et al. CCR2 promotes hepatic fibrosis in mice. *Hepatology* 2009;**50**:185–97.
40. Walport MJ. Complement. Second of two parts. *N Engl J Med* 2001;**344**:1140–4.
41. Addis-Lieser E, Kohl J, Chiaramonte MG. Opposing regulatory roles of complement factor 5 in the development of bleomycin-induced pulmonary fibrosis. *J Immunol* 2005;**175**:1894–902.

42. Hillebrandt S, Wasmuth HE, Weiskirchen R, Hellerbrand C, Keppeler H, Werth A, et al. Complement factor 5 is a quantitative trait gene that modifies liver fibrogenesis in mice and humans. *Nat Genet* 2005;**37**:835–43.
43. Ibdah JA, Paul H, Zhao Y, Binford S, Salleng K, Cline M, et al. Lack of mitochondrial trifunctional protein in mice causes neonatal hypoglycemia and sudden death. *J Clin Invest* 2001;**107**:1403–9.
44. Jimenez W, Claria J, Arroyo V, Rodes J. Carbon tetrachloride induced cirrhosis in rats: a useful tool for investigating the pathogenesis of ascites in chronic liver disease. *J Gastroenterol Hepatol* 1992;**7**:90–7.
45. Newell P, Villanueva A, Friedman SL, Koike K, Llovet JM. Experimental models of hepatocellular carcinoma. *J Hepatol* 2008;**48**:858–79.
46. Poirier LA. Hepatocarcinogenesis by diethylnitrosamine in rats fed high dietary levels of lipotropes. *J Natl Cancer Inst* 1975;**54**:137–40.
47. Salguero Palacios R, Roderfeld M, Hemmann S, Rath T, Atanasova S, Tschuschner A, et al. Activation of hepatic stellate cells is associated with cytokine expression in thioacetamide-induced hepatic fibrosis in mice. *Lab Invest* 2008;**88**:1192–203.
48. Sakaida I, Matsumura Y, Kubota M, Kayano K, Takenaka K, Okita K. The prolyl 4-hydroxylase inhibitor HOE 077 prevents activation of Ito cells, reducing procollagen gene expression in rat liver fibrosis induced by choline-deficient L-amino acid-defined diet. *Hepatology* 1996;**23**:755–63.
49. Kodama Y, Kisseleva T, Iwaisako K, Miura K, Taura K, De Minicis S, et al. c-Jun N-terminal kinase-1 from hematopoietic cells mediates progression from hepatic steatosis to steatohepatitis and fibrosis in mice. *Gastroenterology* 2009;**137**:1467–1477.e5.
50. Koteish A, Diehl AM. Animal models of steatosis. *Semin Liver Dis* 2001;**21**:89–104.
51. Farber E, Shull KH, Villa-Trevino S, Lombardi B, Thomas M. Biochemical pathology of acute hepatic adenosinetriphosphate deficiency. *Nature* 1964;**203**:34–40.
52. George J, Pera N, Phung N, Leclercq I, Yun Hou J, Farrell G. Lipid peroxidation, stellate cell activation and hepatic fibrogenesis in a rat model of chronic steatohepatitis. *J Hepatol* 2003;**39**:756–64.
53. Rinella ME, Green RM. The methionine-choline deficient dietary model of steatohepatitis does not exhibit insulin resistance. *J Hepatol* 2004;**40**:47–51.
54. Leclercq IA, Lebrun VA, Starkel P, Horsmans YJ. Intrahepatic insulin resistance in a murine model of steatohepatitis: effect of PPARgamma agonist pioglitazone. *Lab Invest* 2007;**87**:56–65.
55. Biddinger SB, Almind K, Miyazaki M, Kokkotou E, Ntambi JM, Kahn CR. Effects of diet and genetic background on sterol regulatory element-binding protein-1c, stearoyl-CoA desaturase 1, and the development of the metabolic syndrome. *Diabetes* 2005;**54**:1314–23.
56. Toye AA, Lippiat JD, Proks P, Shimomura K, Bentley L, Hugill A, et al. A genetic and physiological study of impaired glucose homeostasis control in C57BL/6J mice. *Diabetologia* 2005;**48**:675–86.
57. Yoshimatsu M, Terasaki Y, Sakashita N, Kiyota E, Sato H, van der Laan LJ, et al. Induction of macrophage scavenger receptor MARCO in nonalcoholic steatohepatitis indicates possible involvement of endotoxin in its pathogenic process. *Int J Exp Pathol* 2004;**85**:335–43.
58. Feldstein AE, Canbay A, Guicciardi ME, Higuchi H, Bronk SF, Gores GJ. Diet associated hepatic steatosis sensitizes to Fas mediated liver injury in mice. *J Hepatol* 2003;**39**:978–83.
59. Friedman JM, Leibel RL, Siegel DS, Walsh J, Bahary N. Molecular mapping of the mouse ob mutation. *Genomics* 1991;**11**:1054–62.
60. Zhang Y, Proenca R, Maffei M, Barone M, Leopold L, Friedman JM. Positional cloning of the mouse obese gene and its human homologue. *Nature* 1994;**372**:425–32.

61. Fantuzzi G, Faggioni R. Leptin in the regulation of immunity, inflammation, and hematopoiesis. *J Leukoc Biol* 2000;**68**:437–46.
62. Ingalls AM, Dickie MM, Snell GD. Obese, a new mutation in the house mouse. *J Hered* 1950;**41**:317–8.
63. Mayer J, Bates MW, Dickie MM. Hereditary diabetes in genetically obese mice. *Science* 1951;**113**:746–7.
64. Mayer J, Dickie MM, Bates MW, Vitale JJ. Free selection of nutrients by hereditarily obese mice. *Science* 1951;**113**:745–6.
65. Bleisch VR, Mayer J, Dickie MM. Familial diabetes mellitus in mice, associated with insulin resistance, obesity, and hyperplasia of the islands of langerhans. *Am J Pathol* 1952;**28**:369–85.
66. Shimomura I, Bashmakov Y, Horton JD. Increased levels of nuclear SREBP-1c associated with fatty livers in two mouse models of diabetes mellitus. *J Biol Chem* 1999;**274**:30028–32.
67. Brix AE, Elgavish A, Nagy TR, Gower BA, Rhead WJ, Wood PA. Evaluation of liver fatty acid oxidation in the leptin-deficient obese mouse. *Mol Genet Metab* 2002;**75**:219–26.
68. Edvardsson U, von Lowenhielm HB, Panfilov O, Nystrom AC, Nilsson F, Dahllof B. Hepatic protein expression of lean mice and obese diabetic mice treated with peroxisome proliferator-activated receptor activators. *Proteomics* 2003;**3**:468–78.
69. Enriquez A, Leclercq I, Farrell GC, Robertson G. Altered expression of hepatic CYP2E1 and CYP4A in obese, diabetic ob/ob mice, and fa/fa Zucker rats. *Biochem Biophys Res Commun* 1999;**255**:300–6.
70. Yang SQ, Lin HZ, Lane MD, Clemens M, Diehl AM. Obesity increases sensitivity to endotoxin liver injury: implications for the pathogenesis of steatohepatitis. *Proc Natl Acad Sci USA* 1997;**94**:2557–62.
71. Faggioni R, Fantuzzi G, Gabay C, Moser A, Dinarello CA, Feingold KR, et al. Leptin deficiency enhances sensitivity to endotoxin-induced lethality. *Am J Physiol* 1999;**276**: R136–42.
72. Chavin KD, Yang S, Lin HZ, Chatham J, Chacko VP, Hoek JB, et al. Obesity induces expression of uncoupling protein-2 in hepatocytes and promotes liver ATP depletion. *J Biol Chem* 1999;**274**:5692–700.
73. Honda H, Ikejima K, Hirose M, Yoshikawa M, Lang T, Enomoto N, et al. Leptin is required for fibrogenic responses induced by thioacetamide in the murine liver. *Hepatology* 2002;**36**:12–21.
74. Leclercq IA, Farrell GC, Schriemer R, Robertson GR. Leptin is essential for the hepatic fibrogenic response to chronic liver injury. *J Hepatol* 2002;**37**:206–13.
75. Leclercq IA, Field J, Farrell GC. Leptin-specific mechanisms for impaired liver regeneration in ob/ob mice after toxic injury. *Gastroenterology* 2003;**124**:1451–64.
76. Oben JA, Roskams T, Yang S, Lin H, Sinelli N, Li Z, et al. Norepinephrine induces hepatic fibrogenesis in leptin deficient ob/ob mice. *Biochem Biophys Res Commun* 2003;**308**:284–92.
77. Li Z, Oben JA, Yang S, Lin H, Stafford EA, Soloski MJ, et al. Norepinephrine regulates hepatic innate immune system in leptin-deficient mice with nonalcoholic steatohepatitis. *Hepatology* 2004;**40**:434–41.
78. Hummel KP, Dickie MM, Coleman DL. Diabetes, a new mutation in the mouse. *Science* 1966;**153**:1127–8.
79. Tartaglia LA, Dembski M, Weng X, Deng N, Culpepper J, Devos R, et al. Identification and expression cloning of a leptin receptor, OB-R. *Cell* 1995;**83**:1263–71.
80. Ghilardi N, Ziegler S, Wiestner A, Stoffel R, Heim MH, Skoda RC. Defective STAT signaling by the leptin receptor in diabetic mice. *Proc Natl Acad Sci USA* 1996;**93**:6231–5.
81. Chen H, Charlat O, Tartaglia LA, Woolf EA, Weng X, Ellis SJ, et al. Evidence that the diabetes gene encodes the leptin receptor: identification of a mutation in the leptin receptor gene in db/db mice. *Cell* 1996;**84**:491–5.

82. Shimano H, Horton JD, Hammer RE, Shimomura I, Brown MS, Goldstein JL. Overproduction of cholesterol and fatty acids causes massive liver enlargement in transgenic mice expressing truncated SREBP-1a. *J Clin Invest* 1996;**98**:1575–84.
83. Shimomura I, Shimano H, Korn BS, Bashmakov Y, Horton JD. Nuclear sterol regulatory element-binding proteins activate genes responsible for the entire program of unsaturated fatty acid biosynthesis in transgenic mouse liver. *J Biol Chem* 1998;**273**:35299–306.
84. Yamashita H, Takenoshita M, Sakurai M, Bruick RK, Henzel WJ, Shillinglaw W, et al. A glucose-responsive transcription factor that regulates carbohydrate metabolism in the liver. *Proc Natl Acad Sci USA* 2001;**98**:9116–21.
85. Iizuka K, Bruick RK, Liang G, Horton JD, Uyeda K. Deficiency of carbohydrate response element-binding protein (ChREBP) reduces lipogenesis as well as glycolysis. *Proc Natl Acad Sci USA* 2004;**101**:7281–6.
86. Kuwajima M, Kono N, Horiuchi M, Imamura Y, Ono A, Inui Y, et al. Animal model of systemic carnitine deficiency: analysis in C3H-H-2 degrees strain of mouse associated with juvenile visceral steatosis. *Biochem Biophys Res Commun* 1991;**174**:1090–4.
87. Kersten S, Seydoux J, Peters JM, Gonzalez FJ, Desvergne B, Wahli W. Peroxisome proliferator-activated receptor alpha mediates the adaptive response to fasting. *J Clin Invest* 1999;**103**:1489–98.
88. Ibdah JA, Perlegas P, Zhao Y, Angdisen J, Borgerink H, Shadoan MK, et al. Mice heterozygous for a defect in mitochondrial trifunctional protein develop hepatic steatosis and insulin resistance. *Gastroenterology* 2005;**128**:1381–90.
89. Starkel P, Leclercq IA. Animal models for the study of hepatic fibrosis. *Best Pract Res Clin Gastroenterol* 2011;**25**:319–33.
90. Fickert P, Zollner G, Fuchsbichler A, Stumptner C, Weiglein AH, Lammert F, et al. Ursodeoxycholic acid aggravates bile infarcts in bile duct-ligated and Mdr2 knockout mice via disruption of cholangioles. *Gastroenterology* 2002;**123**:1238–51.
91. Georgiev P, Jochum W, Heinrich S, Jang JH, Nocito A, Dahm F, et al. Characterization of time-related changes after experimental bile duct ligation. *Br J Surg* 2008;**95**:646–56.
92. Popov Y, Sverdlov DY, Bhaskar KR, Sharma AK, Millonig G, Patsenker E, et al. Macrophage-mediated phagocytosis of apoptotic cholangiocytes contributes to reversal of experimental biliary fibrosis. *Am J Physiol Gastrointest Liver Physiol* 2010;**298**:G323–34.
93. Geerts AM, Vanheule E, Praet M, Van Vlierberghe H, De Vos M, Colle I. Comparison of three research models of portal hypertension in mice: macroscopic, histological and portal pressure evaluation. *Int J Exp Pathol* 2008;**89**:251–63.
94. Fickert P, Fuchsbichler A, Wagner M, Zollner G, Kaser A, Tilg H, et al. Regurgitation of bile acids from leaky bile ducts causes sclerosing cholangitis in Mdr2 (Abcb4) knockout mice. *Gastroenterology* 2004;**127**:261–74.
95. Roderfeld M, Rath T, Voswinckel R, Dierkes C, Dietrich H, Zahner D, et al. Bone marrow transplantation demonstrates medullar origin of CD34+ fibrocytes and ameliorates hepatic fibrosis in Abcb4−/− mice. *Hepatology* 2010;**51**:267–76.
96. Wasmuth HE, Glantz A, Keppeler H, Simon E, Bartz C, Rath W, et al. Intrahepatic cholestasis of pregnancy: the severe form is associated with common variants of the hepatobiliary phospholipid transporter ABCB4 gene. *Gut* 2007;**56**:265–70.
97. Jonker AM, Dijkhuis FW, Hardonk MJ, Moerkerk P, Ten Kate J, Grond J. Immunohistochemical study of hepatic fibrosis induced in rats by multiple galactosamine injections. *Hepatology* 1994;**19**:775–81.
98. De Minicis S, Brenner DA. NOX in liver fibrosis. *Arch Biochem Biophys* 2007;**462**:266–72.
99. Wang B, Dolinski BM, Kikuchi N, Leone DR, Peters MG, Weinreb PH, et al. Role of alphavbeta6 integrin in acute biliary fibrosis. *Hepatology* 2007;**46**:1404–12.

100. Seki E, De Minicis S, Osterreicher CH, Kluwe J, Osawa Y, Brenner DA, et al. TLR4 enhances TGF-beta signaling and hepatic fibrosis. *Nat Med* 2007;**13**:1324–32.
101. Isayama F, Hines IN, Kremer M, Milton RJ, Byrd CL, Perry AW, et al. LPS signaling enhances hepatic fibrogenesis caused by experimental cholestasis in mice. *Am J Physiol Gastrointest Liver Physiol* 2006;**290**:G1318–28.
102. Bumgardner GL, Li J, Heininger M, Ferguson RM, Orosz CG. *In vivo* immunogenicity of purified allogeneic hepatocytes in a murine hepatocyte transplant model. *Transplantation* 1998;**65**:47–52.
103. Calne RY, Sells RA, Pena JR, Davis DR, Millard PR, Herbertson BM, et al. Induction of immunological tolerance by porcine liver allografts. *Nature* 1969;**223**:472–6.
104. Qian S, Demetris AJ, Murase N, Rao AS, Fung JJ, Starzl TE. Murine liver allograft transplantation: tolerance and donor cell chimerism. *Hepatology* 1994;**19**:916–24.
105. Kamada N. The immunology of experimental liver transplantation in the rat. *Immunology* 1985;**55**:369–89.
106. Crispe IN. Hepatic T cells and liver tolerance. *Nat Rev Immunol* 2003;**3**:51–62.
107. Bowen DG, Zen M, Holz L, Davis T, McCaughan GW, Bertolino P. The site of primary T cell activation is a determinant of the balance between intrahepatic tolerance and immunity. *J Clin Invest* 2004;**114**:701–12.
108. Chen CH, Kuo LM, Chang Y, Wu W, Goldbach C, Ross MA, et al. *In vivo* immune modulatory activity of hepatic stellate cells in mice. *Hepatology* 2006;**44**:1171–81.
109. Buschenfelde KH, Kossling FK, Miescher PA. Experimental chronic active hepatitis in rabbits following immunization with human liver proteins. *Clin Exp Immunol* 1972;**11**:99–108.
110. Kuriki J, Murakami H, Kakumu S, Sakamoto N, Yokochi T, Nakashima I, et al. Experimental autoimmune hepatitis in mice after immunization with syngeneic liver proteins together with the polysaccharide of Klebsiella pneumoniae. *Gastroenterology* 1983;**84**:596–603.
111. Lohse AW, Manns M, Dienes HP, Meyer zum Buschenfelde KH, Cohen IR. Experimental autoimmune hepatitis: disease induction, time course and T-cell reactivity. *Hepatology* 1990;**11**:24–30.
112. Watanabe Y, Kawakami H, Kawamoto H, Ikemoto Y, Masuda K, Takezaki E, et al. Effect of neonatal thymectomy on experimental autoimmune hepatitis in mice. *Clin Exp Immunol* 1987;**67**:105–13.
113. Asano M, Toda M, Sakaguchi N, Sakaguchi S. Autoimmune disease as a consequence of developmental abnormality of a T cell subpopulation. *J Exp Med* 1996;**184**:387–96.
114. Ando K, Guidotti LG, Wirth S, Ishikawa T, Missale G, Moriyama T, et al. Class I-restricted cytotoxic T lymphocytes are directly cytopathic for their target cells *in vivo*. *J Immunol* 1994;**152**:3245–53.
115. Ferber I, Schonrich G, Schenkel J, Mellor AL, Hammerling GJ, Arnold B. Levels of peripheral T cell tolerance induced by different doses of tolerogen. *Science* 1994;**263**:674–6.
116. Schonrich G, Alferink J, Klevenz A, Kublbeck G, Auphan N, Schmitt-Verhulst AM, et al. Tolerance induction as a multi-step process. *Eur J Immunol* 1994;**24**:285–93.
117. Alferink J, Schittek B, Schonrich G, Hammerling GJ, Arnold B. Long life span of tolerant T cells and the role of antigen in maintenance of peripheral tolerance. *Int Immunol* 1995;**7**:331–6.
118. Limmer A, Sacher T, Alferink J, Kretschmar M, Schonrich G, Nichterlein T, et al. Failure to induce organ-specific autoimmunity by breaking of tolerance: importance of the microenvironment. *Eur J Immunol* 1998;**28**:2395–406.
119. Bertolino P, Bowen DG, McCaughan GW, Fazekas de St Groth B. Antigen-specific primary activation of CD8 + T cells within the liver. *J Immunol* 2001;**166**:5430–8.

120. Morahan G, Brennan FE, Bhathal PS, Allison J, Cox KO, Miller JF. Expression in transgenic mice of class I histocompatibility antigens controlled by the metallothionein promoter. *Proc Natl Acad Sci USA* 1989;**86**:3782–6.
121. Bertolino P, Heath WR, Hardy CL, Morahan G, Miller JF. Peripheral deletion of autoreactive CD8+ T cells in transgenic mice expressing H-2Kb in the liver. *Eur J Immunol* 1995;**25**:1932–42.
122. Rocha B, von Boehmer H. Peripheral selection of the T cell repertoire. *Science* 1991;**251**:1225–8.
123. Derkow K, Loddenkemper C, Mintern J, Kruse N, Klugewitz K, Berg T, et al. Differential priming of CD8 and CD4 T-cells in animal models of autoimmune hepatitis and cholangitis. *Hepatology* 2007;**46**:1155–65.
124. Tsukamoto H, Matsuoka M, French SW. Experimental models of hepatic fibrosis: a review. *Semin Liver Dis* 1990;**10**:56–65.
125. von Herrath MG, Fujinami RS, Whitton JL. Microorganisms and autoimmunity: making the barren field fertile? *Nat Rev Microbiol* 2003;**1**:151–7.
126. Voehringer D, Blaser C, Grawitz AB, Chisari FV, Buerki K, Pircher H. Break of T cell ignorance to a viral antigen in the liver induces hepatitis. *J Immunol* 2000;**165**:2415–22.
127. Hussain MJ, Mustafa A, Gallati H, Mowat AP, Mieli-Vergani G, Vergani D. Cellular expression of tumour necrosis factor-alpha and interferon-gamma in the liver biopsies of children with chronic liver disease. *J Hepatol* 1994;**21**:816–21.
128. Tiegs G. Experimental hepatitis and role of cytokines. *Acta Gastroenterol Belg* 1997;**60**:176–9.
129. Toyonaga T, Hino O, Sugai S, Wakasugi S, Abe K, Shichiri M, et al. Chronic active hepatitis in transgenic mice expressing interferon-gamma in the liver. *Proc Natl Acad Sci USA* 1994;**91**:614–8.
130. Tiegs G, Hentschel J, Wendel A. A T cell-dependent experimental liver injury in mice inducible by concanavalin A. *J Clin Invest* 1992;**90**:196–203.
131. Kusters S, Gantner F, Kunstle G, Tiegs G. Interferon gamma plays a critical role in T cell-dependent liver injury in mice initiated by concanavalin A. *Gastroenterology* 1996;**111**:462–71.
132. Gantner F, Leist M, Lohse AW, Germann PG, Tiegs G. Concanavalin A-induced T-cell-mediated hepatic injury in mice: the role of tumor necrosis factor. *Hepatology* 1995;**21**:190–8.
133. Gantner F, Leist M, Kusters S, Vogt K, Volk HD, Tiegs G. T cell stimulus-induced crosstalk between lymphocytes and liver macrophages results in augmented cytokine release. *Exp Cell Res* 1996;**229**:137–46.
134. Knolle PA, Gerken G, Loser E, Dienes HP, Gantner F, Tiegs G, et al. Role of sinusoidal endothelial cells of the liver in concanavalin A-induced hepatic injury in mice. *Hepatology* 1996;**24**:824–9.
135. Takeda K, Hayakawa Y, Van Kaer L, Matsuda H, Yagita H, Okumura K. Critical contribution of liver natural killer T cells to a murine model of hepatitis. *Proc Natl Acad Sci USA* 2000;**97**:5498–503.
136. Gantner F, Leist M, Jilg S, Germann PG, Freudenberg MA, Tiegs G. Tumor necrosis factor-induced hepatic DNA fragmentation as an early marker of T cell-dependent liver injury in mice. *Gastroenterology* 1995;**109**:166–76.
137. Leist M, Gantner F, Bohlinger I, Germann PG, Tiegs G, Wendel A. Murine hepatocyte apoptosis induced *in vitro* and *in vivo* by TNF-alpha requires transcriptional arrest. *J Immunol* 1994;**153**:1778–88.
138. Trembleau S, Penna G, Gregori S, Chapman HD, Serreze DV, Magram J, et al. Pancreas-infiltrating Th1 cells and diabetes develop in IL-12-deficient nonobese diabetic mice. *J Immunol* 1999;**163**:2960–8.

139. Trembleau S, Penna G, Gregori S, Giarratana N, Adorini L. IL-12 administration accelerates autoimmune diabetes in both wild-type and IFN-gamma-deficient nonobese diabetic mice, revealing pathogenic and protective effects of IL-12-induced IFN-gamma. *J Immunol* 2003;**170**:5491–501.
140. Zaccone P, Hutchings P, Nicoletti F, Penna G, Adorini L, Cooke A. The involvement of IL-12 in murine experimentally induced autoimmune thyroid disease. *Eur J Immunol* 1999;**29**:1933–42.
141. Tamaki S, Homma S, Enomoto Y, Komita H, Zeniya M, Ohno T, et al. Autoimmune hepatic inflammation by vaccination of mice with dendritic cells loaded with well-differentiated hepatocellular carcinoma cells and administration of interleukin-12. *Clin Immunol* 2005;**117**:280–93.
142. Djilali-Saiah I, Lapierre P, Vittozi S, Alvarez F. DNA vaccination breaks tolerance for a neo-self antigen in liver: a transgenic murine model of autoimmune hepatitis. *J Immunol* 2002;**169**:4889–96.
143. Lapierre P, Djilali-Saiah I, Vitozzi S, Alvarez F. A murine model of type 2 autoimmune hepatitis: xenoimmunization with human antigens. *Hepatology* 2004;**39**:1066–74.
144. Christen U, Hintermann E, Jaeckel E. New animal models for autoimmune hepatitis. *Semin Liver Dis* 2009;**29**:262–72.
145. Manns MP, Johnson EF, Griffin KJ, Tan EM, Sullivan KF. Major antigen of liver kidney microsomal autoantibodies in idiopathic autoimmune hepatitis is cytochrome P450db1. *J Clin Invest* 1989;**83**:1066–72.
146. Zanger UM, Hauri HP, Loeper J, Homberg JC, Meyer UA. Antibodies against human cytochrome P-450db1 in autoimmune hepatitis type II. *Proc Natl Acad Sci USA* 1988;**85**:8256–60.
147. Holdener M, Hintermann E, Bayer M, Rhode A, Rodrigo E, Hintereder G, et al. Breaking tolerance to the natural human liver autoantigen cytochrome P450 2D6 by virus infection. *J Exp Med* 2008;**205**:1409–22.
148. Conry-Cantilena C, VanRaden M, Gibble J, Melpolder J, Shakil AO, Viladomiu L, et al. Routes of infection, viremia, and liver disease in blood donors found to have hepatitis C virus infection. *N Engl J Med* 1996;**334**:1691–6.
149. Armstrong GL, Wasley A, Simard EP, McQuillan GM, Kuhnert WL, Alter MJ. The prevalence of hepatitis C virus infection in the United States, 1999 through 2002. *Ann Intern Med* 2006;**144**:705–14.
150. Choo QL, Kuo G, Weiner AJ, Overby LR, Bradley DW, Houghton M. Isolation of a cDNA clone derived from a blood-borne non-A, non-B viral hepatitis genome. *Science* 1989;**244**:359–62.
151. Muchmore E, Popper H, Peterson DA, Miller MF, Lieberman HM. Non-A, non-B hepatitis-related hepatocellular carcinoma in a chimpanzee. *J Med Primatol* 1988;**17**:235–46.
152. Wu GY, Konishi M, Walton CM, Olive D, Hayashi K, Wu CH. A novel immunocompetent rat model of HCV infection and hepatitis. *Gastroenterology* 2005;**128**:1416–23.
153. Galun E, Burakova T, Ketzinel M, Lubin I, Shezen E, Kahana Y, et al. Hepatitis C virus viremia in SCID->BNX mouse chimera. *J Infect Dis* 1995;**172**:25–30.
154. Maeda N, Watanabe M, Okamoto S, Kanai T, Yamada T, Hata J, et al. Hepatitis C virus infection in human liver tissue engrafted in mice with an infectious molecular clone. *Liver Int* 2004;**24**:259–67.
155. Heckel JL, Sandgren EP, Degen JL, Palmiter RD, Brinster RL. Neonatal bleeding in transgenic mice expressing urokinase-type plasminogen activator. *Cell* 1990;**62**:447–56.
156. Meuleman P, Libbrecht L, De Vos R, de Hemptinne B, Gevaert K, Vandekerckhove J, et al. Morphological and biochemical characterization of a human liver in a uPA-SCID mouse chimera. *Hepatology* 2005;**41**:847–56.

157. Mercer DF, Schiller DE, Elliott JF, Douglas DN, Hao C, Rinfret A, et al. Hepatitis C virus replication in mice with chimeric human livers. *Nat Med* 2001;**7**:927–33.
158. Dorner M, Horwitz JA, Robbins JB, Barry WT, Feng Q, Mu K, et al. A genetically humanized mouse model for hepatitis C virus infection. *Nature* 2011;**474**:208–11.
159. Sgonc R, Dietrich H, Gershwin ME, Colombatti A, Wick G. Genomic analysis of collagen and endogenous virus loci in the UCD-200 and 206 lines of chickens, animal models for scleroderma. *J Autoimmun* 1995;**8**:763–70.
160. Sgonc R, Gruschwitz MS, Dietrich H, Recheis H, Gershwin ME, Wick G. Endothelial cell apoptosis is a primary pathogenetic event underlying skin lesions in avian and human scleroderma. *J Clin Invest* 1996;**98**:785–92.
161. Xu H, Zaidi M, Struve J, Jones DW, Krolikowski JG, Nandedkar S, et al. Abnormal fibrillin-1 expression and chronic oxidative stress mediate endothelial mesenchymal transition in a murine model of systemic sclerosis. *Am J Physiol Cell Physiol* 2011;**300**:C550–6.
162. Del Galdo F, Sotgia F, de Almeida CJ, Jasmin JF, Musick M, Lisanti MP, et al. Decreased expression of caveolin 1 in patients with systemic sclerosis: crucial role in the pathogenesis of tissue fibrosis. *Arthritis Rheum* 2008;**58**:2854–65.
163. Derrett-Smith EC, Denton CP, Sonnylal S. Animal models of scleroderma: lessons from transgenic and knockout mice. *Curr Opin Rheumatol* 2009;**21**:630–5.
164. Le Hir M, Martin M, Haas C. A syndrome resembling human systemic sclerosis (scleroderma) in MRL/lpr mice lacking interferon-gamma (IFN-gamma) receptor (MRL/lprgammaR−/−). *Clin Exp Immunol* 1999;**115**:281–7.
165. Asano Y, Stawski L, Hant F, Highland K, Silver R, Szalai G, et al. Endothelial Fli1 deficiency impairs vascular homeostasis: a role in scleroderma vasculopathy. *Am J Pathol* 2010;**176**:1983–98.
166. Kubo M, Czuwara-Ladykowska J, Moussa O, Markiewicz M, Smith E, Silver RM, et al. Persistent down-regulation of Fli1, a suppressor of collagen transcription, in fibrotic scleroderma skin. *Am J Pathol* 2003;**163**:571–81.
167. Hollenhorst PC, Jones DA, Graves BJ. Expression profiles frame the promoter specificity dilemma of the ETS family of transcription factors. *Nucleic Acids Res* 2004;**32**:5693–702.
168. Shanmugam VK, Swistowski DR, Saddic N, Wang H, Steen VD. Comparison of indirect immunofluorescence and multiplex antinuclear antibody screening in systemic sclerosis. *Clin Rheumatol* 2011;**30**:1363–8.
169. Carvalho D, Savage CO, Black CM, Pearson JD. IgG antiendothelial cell autoantibodies from scleroderma patients induce leukocyte adhesion to human vascular endothelial cells *in vitro*. Induction of adhesion molecule expression and involvement of endothelium-derived cytokines. *J Clin Invest* 1996;**97**:111–9.
170. Gruschwitz MS, Moormann S, Kromer G, Sgonc R, Dietrich H, Boeck G, et al. Phenotypic analysis of skin infiltrates in comparison with peripheral blood lymphocytes, spleen cells and thymocytes in early avian scleroderma. *J Autoimmun* 1991;**4**:577–93.
171. Sgonc R. The vascular perspective of systemic sclerosis: of chickens, mice and men. *Int Arch Allergy Immunol* 1999;**120**:169–76.
172. Ong CJ, Ip S, Teh SJ, Wong C, Jirik FR, Grusby MJ, et al. A role for T helper 2 cells in mediating skin fibrosis in tight-skin mice. *Cell Immunol* 1999;**196**:60–8.
173. McGaha T, Saito S, Phelps RG, Gordon R, Noben-Trauth N, Paul WE, et al. Lack of skin fibrosis in tight skin (TSK) mice with targeted mutation in the interleukin-4R alpha and transforming growth factor-beta genes. *J Invest Dermatol* 2001;**116**:136–43.
174. Saito E, Fujimoto M, Hasegawa M, Komura K, Hamaguchi Y, Kaburagi Y, et al. CD19-dependent B lymphocyte signaling thresholds influence skin fibrosis and autoimmunity in the tight-skin mouse. *J Clin Invest* 2002;**109**:1453–62.

175. Saito S, Nishimura H, Phelps RG, Wolf I, Suzuki M, Honjo T, et al. Induction of skin fibrosis in mice expressing a mutated fibrillin-1 gene. *Mol Med* 2000;**6**:825–36.
176. Gentiletti J, McCloskey LJ, Artlett CM, Peters J, Jimenez SA, Christner PJ. Demonstration of autoimmunity in the tight skin-2 mouse: a model for scleroderma. *J Immunol* 2005;**175**:2418–26.
177. Yamamoto T, Takagawa S, Katayama I, Nishioka K. Anti-sclerotic effect of transforming growth factor-beta antibody in a mouse model of bleomycin-induced scleroderma. *Clin Immunol* 1999;**92**:6–13.
178. Yamamoto T, Kuroda M, Nishioka K. Animal model of sclerotic skin. III: histopathological comparison of bleomycin-induced scleroderma in various mice strains. *Arch Dermatol Res* 2000;**292**:535–41.
179. Lakos G, Melichian D, Wu M, Varga J. Increased bleomycin-induced skin fibrosis in mice lacking the Th1-specific transcription factor T-bet. *Pathobiology* 2006;**73**:224–37.
180. Aliprantis AO, Wang J, Fathman JW, Lemaire R, Dorfman DM, Lafyatis R, et al. Transcription factor T-bet regulates skin sclerosis through its function in innate immunity and via IL-13. *Proc Natl Acad Sci USA* 2007;**104**:2827–30.
181. Ishikawa H, Takeda K, Okamoto A, Matsuo S, Isobe K. Induction of autoimmunity in a bleomycin-induced murine model of experimental systemic sclerosis: an important role for CD4+ T cells. *J Invest Dermatol* 2009;**129**:1688–95.
182. Siracusa LD, McGrath R, Ma Q, Moskow JJ, Manne J, Christner PJ, et al. A tandem duplication within the fibrillin 1 gene is associated with the mouse tight skin mutation. *Genome Res* 1996;**6**:300–13.
183. Bayle J, Fitch J, Jacobsen K, Kumar R, Lafyatis R, Lemaire R. Increased expression of Wnt2 and SFRP4 in Tsk mouse skin: role of Wnt signaling in altered dermal fibrillin deposition and systemic sclerosis. *J Invest Dermatol* 2008;**128**:871–81.
184. Jimenez SA, Williams CJ, Myers JC, Bashey RI. Increased collagen biosynthesis and increased expression of type I and type III procollagen genes in tight skin (TSK) mouse fibroblasts. *J Biol Chem* 1986;**261**:657–62.
185. Avouac J, Furnrohr BG, Tomcik M, Palumbo K, Zerr P, Horn A, et al. Inactivation of the transcription factor STAT-4 prevents inflammation-driven fibrosis in animal models of systemic sclerosis. *Arthritis Rheum* 2011;**63**:800–9.
186. Yoshizaki A, Yanaba K, Ogawa A, Iwata Y, Ogawa F, Takenaka M, et al. The specific free radical scavenger edaravone suppresses fibrosis in tight-skin and bleomycin-induced mouse models of systemic sclerosis. *Arthritis Rheum* 2011;**63**:3086–97.
187. Asai T, Ohno Y, Minatoguchi S, Funaguchi N, Yuhgetsu H, Sawada M, et al. The specific free radical scavenger edaravone suppresses bleomycin-induced acute pulmonary injury in rabbits. *Clin Exp Pharmacol Physiol* 2007;**34**:22–6.
188. Yoshizaki A, Yanaba K, Iwata Y, Komura K, Ogawa F, Takenaka M, et al. Treatment with rapamycin prevents fibrosis in tight-skin and bleomycin-induced mouse models of systemic sclerosis. *Arthritis Rheum* 2010;**62**:2476–87.
189. Akhmetshina A, Venalis P, Dees C, Busch N, Zwerina J, Schett G, et al. Treatment with imatinib prevents fibrosis in different preclinical models of systemic sclerosis and induces regression of established fibrosis. *Arthritis Rheum* 2009;**60**:219–24.
190. Christner PJ, Hitraya EG, Peters J, McGrath R, Jimenez SA. Transcriptional activation of the alpha1(I) procollagen gene and up-regulation of alpha1(I) and alpha1(III) procollagen messenger RNA in dermal fibroblasts from tight skin 2 mice. *Arthritis Rheum* 1998;**41**:2132–42.
191. Barisic-Dujmovic T, Boban I, Clark SH. Regulation of collagen gene expression in the Tsk2 mouse. *J Cell Physiol* 2008;**215**:464–71.

192. Sonnylal S, Denton CP, Zheng B, Keene DR, He R, Adams HP, et al. Postnatal induction of transforming growth factor beta signaling in fibroblasts of mice recapitulates clinical, histologic, and biochemical features of scleroderma. *Arthritis Rheum* 2007;**56**:334–44.
193. Denton CP, Zheng B, Evans LA, Shi-wen X, Ong VH, Fisher I, et al. Fibroblast-specific expression of a kinase-deficient type II transforming growth factor beta (TGFbeta) receptor leads to paradoxical activation of TGFbeta signaling pathways with fibrosis in transgenic mice. *J Biol Chem* 2003;**278**:25109–19.
194. Hoyles RK, Khan K, Shiwen X, Howat SL, Lindahl GE, Leoni P, et al. Fibroblast-specific perturbation of transforming growth factor beta signaling provides insight into potential pathogenic mechanisms of scleroderma-associated lung fibrosis: exaggerated response to alveolar epithelial injury in a novel mouse model. *Arthritis Rheum* 2008;**58**:1175–88.
195. Denton CP, Lindahl GE, Khan K, Shiwen X, Ong VH, Gaspar NJ, et al. Activation of key profibrotic mechanisms in transgenic fibroblasts expressing kinase-deficient type II transforming growth factor-{beta} receptor (T{beta}RII{delta}k). *J Biol Chem* 2005;**280**:16053–65.
196. Samuel CS, Zhao C, Yang Q, Wang H, Tian H, Tregear GW, et al. The relaxin gene knockout mouse: a model of progressive scleroderma. *J Invest Dermatol* 2005;**125**:692–9.
197. Takagawa S, Lakos G, Mori Y, Yamamoto T, Nishioka K, Varga J. Sustained activation of fibroblast transforming growth factor-beta/Smad signaling in a murine model of scleroderma. *J Invest Dermatol* 2003;**121**:41–50.
198. Bhattacharyya S, Ghosh AK, Pannu J, Mori Y, Takagawa S, Chen G, et al. Fibroblast expression of the coactivator p300 governs the intensity of profibrotic response to transforming growth factor beta. *Arthritis Rheum* 2005;**52**:1248–58.
199. Mori Y, Hinchcliff M, Wu M, Warner-Blankenship M, Lyons KM, Varga J. Connective tissue growth factor/CCN2-null mouse embryonic fibroblasts retain intact transforming growth factor-beta responsiveness. *Exp Cell Res* 2008;**314**:1094–104.
200. Lakos G, Takagawa S, Chen SJ, Ferreira AM, Han G, Masuda K, et al. Targeted disruption of TGF-beta/Smad3 signaling modulates skin fibrosis in a mouse model of scleroderma. *Am J Pathol* 2004;**165**:203–17.
201. Chen SJ, Ning H, Ishida W, Sodin-Semrl S, Takagawa S, Mori Y, et al. The early-immediate gene EGR-1 is induced by transforming growth factor-beta and mediates stimulation of collagen gene expression. *J Biol Chem* 2006;**281**:21183–97.
202. Ghosh AK, Bhattacharyya S, Lakos G, Chen SJ, Mori Y, Varga J. Disruption of transforming growth factor beta signaling and profibrotic responses in normal skin fibroblasts by peroxisome proliferator-activated receptor gamma. *Arthritis Rheum* 2004;**50**:1305–18.
203. Wu M, Melichian DS, Chang E, Warner-Blankenship M, Ghosh AK, Varga J. Rosiglitazone abrogates bleomycin-induced scleroderma and blocks profibrotic responses through peroxisome proliferator-activated receptor-gamma. *Am J Pathol* 2009;**174**:519–33.
204. Jinnin M, Ihn H, Mimura Y, Asano Y, Yamane K, Tamaki K. Effects of hepatocyte growth factor on the expression of type I collagen and matrix metalloproteinase-1 in normal and scleroderma dermal fibroblasts. *J Invest Dermatol* 2005;**124**:324–30.
205. Wu MH, Yokozeki H, Takagawa S, Yamamoto T, Satoh T, Kaneda Y, et al. Hepatocyte growth factor both prevents and ameliorates the symptoms of dermal sclerosis in a mouse model of scleroderma. *Gene Ther* 2004;**11**:170–80.
206. Cronstein BN. Adenosine, an endogenous anti-inflammatory agent. *J Appl Physiol* 1994;**76**:5–13.
207. Montesinos MC, Desai A, Chen JF, Yee H, Schwarzschild MA, Fink JS, et al. Adenosine promotes wound healing and mediates angiogenesis in response to tissue injury via occupancy of A(2A) receptors. *Am J Pathol* 2002;**160**:2009–18.

208. Chan ES, Fernandez P, Merchant AA, Montesinos MC, Trzaska S, Desai A, et al. Adenosine A2A receptors in diffuse dermal fibrosis: pathogenic role in human dermal fibroblasts and in a murine model of scleroderma. *Arthritis Rheum* 2006;**54**:2632–42.
209. Fernandez P, Trzaska S, Wilder T, Chiriboga L, Blackburn MR, Cronstein BN, et al. Pharmacological blockade of A2A receptors prevents dermal fibrosis in a model of elevated tissue adenosine. *Am J Pathol* 2008;**172**:1675–82.
210. Ferreira AM, Takagawa S, Fresco R, Zhu X, Varga J, DiPietro LA. Diminished induction of skin fibrosis in mice with MCP-1 deficiency. *J Invest Dermatol* 2006;**126**:1900–8.
211. Liu S, Kapoor M, Denton CP, Abraham DJ, Leask A. Loss of beta1 integrin in mouse fibroblasts results in resistance to skin scleroderma in a mouse model. *Arthritis Rheum* 2009;**60**:2817–21.
212. Czuwara-Ladykowska J, Shirasaki F, Jackers P, Watson DK, Trojanowska M. Fli-1 inhibits collagen type I production in dermal fibroblasts via an Sp1-dependent pathway. *J Biol Chem* 2001;**276**:20839–48.
213. Nakerakanti SS, Kapanadze B, Yamasaki M, Markiewicz M, Trojanowska M. Fli1 and Ets1 have distinct roles in connective tissue growth factor/CCN2 gene regulation and induction of the profibrotic gene program. *J Biol Chem* 2006;**281**:25259–69.
214. Asano Y, Czuwara J, Trojanowska M. Transforming growth factor-beta regulates DNA binding activity of transcription factor Fli1 by p300/CREB-binding protein-associated factor-dependent acetylation. *J Biol Chem* 2007;**282**:34672–83.
215. Zhang Y, McCormick LL, Desai SR, Wu C, Gilliam AC. Murine sclerodermatous graft-versus-host disease, a model for human scleroderma: cutaneous cytokines, chemokines, and immune cell activation. *J Immunol* 2002;**168**:3088–98.
216. Askew D, Zhou L, Wu C, Chen G, Gilliam AC. Absence of cutaneous TNFalpha-producing CD4+ T cells and TNFalpha may allow for fibrosis rather than epithelial cytotoxicity in murine sclerodermatous graft-versus-host disease, a model for human scleroderma. *J Invest Dermatol* 2007;**127**:1905–14.
217. Ruzek MC, Jha S, Ledbetter S, Richards SM, Garman RD. A modified model of graft-versus-host-induced systemic sclerosis (scleroderma) exhibits all major aspects of the human disease. *Arthritis Rheum* 2004;**50**:1319–31.

Animal Models of Schizophrenia: Emphasizing Construct Validity

MASAYA YANAGI, SARAH SOUTHCOTT, JACOB LISTER, AND CAROL A. TAMMINGA

Department of Psychiatry, UT Southwestern Medical School, Dallas, Texas, USA

Achieving animal models of schizophrenia which are representative of clear aspects of the illness is critical to understanding pathophysiology and developing novel treatments for the complex syndrome. This chapter reviews the various approaches that have been used in the past to create animal models of schizophrenia, including pharmacological approaches, environmental risk conditions and schizophrenia risk genes. In addition, we present a new animal model which derives directly from human tissue and brain imaging data used to develop a human schizophrenia model. This chapter emphasizes the crucial need for construct validity and of modeling discrete elements of schizophrenia's illness presentation as the way to successful advances.

Progress in Molecular Biology
and Translational Science, Vol. 105
DOI: 10.1016/B978-0-12-394596-9.00012-3

1877-1173/12 $35.00

I. Introduction

Animal models of human brain diseases—especially those which are stringently based on disease features and have strong construct validity—are critical to fully testing disease pathophysiology and, after molecular discovery, to stimulate novel treatment development.[1,2] Attention to discovering these models is most important for complex, syndromic illnesses like schizophrenia. However, since our knowledge of the mechanistic basis of schizophrenia is seriously limited, our ability to develop verified animal models has been challenged.[3] Current animal models of schizophrenia are primitive and often mechanistically speculative, without a forward vision of how to verify models; there is often insufficient indication that these models are related to disease pathophysiology in human constructs. However, there have been developments in the conceptualization of complex diseases like schizophrenia that will allow broader types of animal models to be relevant and their verification more direct.[4]

Schizophrenia is a chronic psychotic illness that includes complex manifestations of cognitive dysfunction, depression, and negative symptoms in addition to the phenomenon of psychosis.[5,6] A recent initiative to understand the specific and independent targets within this complex disease—the MATRIC initiative—conceptualized the illness as a syndrome with distinguishable symptom dimensions, for example, psychosis, cognitive dysfunction, and negative symptoms.[7–9] MATRICS formulated these symptom dimensions as independent from each other with respect to etiology, course, and biology, based on the distinct symptom course, specific content, and pharmacology of each of these symptom dimensions. With this reappraisal, the MATRICS group created a new opportunity to identify animal models, not for schizophrenia overall, but, preferentially, for each of the dimensional manifestations of illness.[2,10] This formulation raises the possibility of creating a single set of animal models for psychosis (as proposed here), another set for cognitive dysfunction,[11] and still another for negative symptoms.[12] This formulation also implies that treatments will be distinct. Having distinct treatments for co-occurring but independent disease manifestations means that drug cotreatments will be indicated in the future, with each treatment directed to an independent dimensions of illness.

These ideas alter the conceptualization, presentation, and the development of animal models for schizophrenia. Therefore, in this chapter, while we review animal models for all aspects of schizophrenia, it will be obvious where those models predominantly or exclusively are directed toward a distinct dimension of the illness. In newer animal models of the illness, we expect dimensional models to predominate. Further, we will propose an animal model ourselves, based on reverse translation, which we are testing as an animal model within the domain of psychosis.

We emphasize from the beginning that our understanding of schizophrenia is based only on observations of disease phenomenology; the phenomenologic characteristics are extensive and have been honed for many decades, even millennium.[13] Nonetheless, confirmed molecular markers of diagnosis or dimensions of illness, much less pathophysiology, are simply not available.[3] Psychiatric disease definitions as currently available were created to organize phenomenology, not with an eye to discovering neural mechanisms, but primarily to generate clinical consistency.[14] Now is the time, while we pursue disease and dimension definitions, to keep a sight on how to organize disease manifestations based on ideas of neural organization.[4,15,16] To do this, one of the greatest needs in the field of functional brain disease is to identify the pathophysiology of our diseases and dimensions; this will provide a molecular basis for disease understanding, diagnosis, and treatment development. Therefore, our goal here is to review animal models which have been (and are being) generated to identify and characterize molecular mechanisms of schizophrenic psychosis, cognitive dysfunction, and negative symptoms, hence an emphasis on animal models with strong construct validity.

In order to provide animal models with strong construct validity, we need to consider to what extent the model can reflect the putative essential biology of the human disease,[3] using all available elements of forward and reverse translation to test this, once a construct has been formulated. Construct validity has been substantiated in the past by using drugs in animals that have been shown to generate symptoms of psychosis in humans.[1] It is also done by creating genetically altered animals based on known schizophrenia risk genes[17,18] and by including in the model early adversity conditions which are known to be risk factors for the illness.[19] In addition to all these approaches, and based on the growth in phenotypic characterization of schizophrenia, there have been advanced models based on human endophenotypes of schizophrenia,[20] brain imaging,[21] and human postmortem characteristics[22,23] to validate disease models and molecules.

It is also necessary to consider the specific application and goals of any specific animal model, especially those with strong construct validity. Of course, animal models are necessary for drug development; this goes without saying. But the emphasis of this review will be on the ability of an animal model to expand and extend the description of a biological process which is discovered to be relevant to schizophrenia, in a dynamic and tractable animal system. It is the process of partial discovery in humans, with fuller definition in animals, and final confirmation in the human disease model, which will ultimately be able to generate a valid understanding of disease biology, and which will generate animal models which are accurate enough for drug discovery.

II. The Schizophrenia Phenotype

A. Clinical Manifestations

Schizophrenia is a chronic psychotic illness.[24,25] Characteristically, psychosis is persistent and serious. The psychosis has its onset typically in late adolescence/early adulthood, with only 5% occurring in childhood or in later midlife. Once a first florid psychotic episode has occurred, the psychosis becomes entrenched, although there are very rare examples of full psychosis reversal.[5,6] Antipsychotic treatment is chronic, and if withdrawn, the annual risk of relapse is 70%. Psychosis is the most prominent symptom dimension in schizophrenia, with hallucinations, delusions, paranoia, and thought disorder including loose and novel associations. Accompanying the psychosis, and sometimes preceding its onset, is cognitive dysfunction.[26] Cognitive dysfunction in schizophrenia is neither pervasive across all cognition domains nor progressive. The cognitive aspects most affected in the illness are attention, working memory, declarative memory, executive function, and social cognition.[8] It is not unusual to see aspects of these cognitive dimensions altered in unaffected nonpsychotic family members. There is considerable psychosocial deterioration with schizophrenia, as the illness progresses, not necessarily accompanied by intellectual deterioration. The prospect of modeling aspects of the clinical manifestations of schizophrenia in animals is daunting because of its complexity, variability, and changing manifestations over the life course. Therefore, the task of modeling symptom dimensions—for example, psychosis or cognitive dysfunction—is much more manageable.

B. Phenotypic Characteristics

Direct and spontaneous measures of brain function in schizophrenia include specific types of cognition, electrophysiological features, eye movements, and human brain imaging.[27] It is hypothesized that these intermediate or endophenotypes are more directly impacted by brain activity than is behavior, therefore, should more directly reflect brain characteristics. The current importance of intermediate phenotypes is the potential for better genetic association with phenotypes than with diagnosis. Also, for these same reasons, the phenotypes could be highly useful constructs around which to develop animal models.

Cognition is one of the easiest phenotypes to translate. It is the domains of working memory, declarative memory, attention, executive function, and social cognition that are primarily affected in the illness. Keeler and Robbins[11] have already strategized to develop the tasks for testing these cognitive domains in animals. In addition, assessments of EEG responses to sensory stimuli test aspects of the brain's sensory reception, integration, and management of sensory input. In schizophrenia, prepulse inhibition (PPI) fails as well as P50 and P300 are altered in probands and in unaffected relatives.[27] Eye movements have been evaluated in schizophrenia for many years, with smooth pursuit movements consistently identified as abnormal in probands and relatives with the illness.[28] Finally, there are clear and consistent structural alterations, albeit subtle, of schizophrenia brains which include reduced overall gray matter loss and reduction in hippocampal size. Animal models built around these known deficits in schizophrenia would be useful to enhance construct validity.[20] There is no question of the existence of these "brain" phenotypes in schizophrenia. Nonetheless, their presence does not extend molecular or systems knowledge of disease pathophysiology as yet, even though this might be anticipated in the future.

C. Pharmacology

Even though schizophrenia as a syndrome has been present throughout recorded history, a successful pharmacological treatment has only been available since the mid-1950s. In the late 1940s, two French psychiatrists observed strong sedative properties of chlorpromazine in normal people, used as a preanesthetic agent at that time; they decided to test this drug for behavioral sedation in floridly psychotic patients.[29] Delay and Deniker observed an unexpected outcome, which was a specific antipsychotic action with less sedation than in normal individuals. They directly followed up their informal observation with controlled clinical trials using chlorpromazine and documented a selective antipsychotic action.[29] They documented the antipsychotic action of chlorpromazine.

It took another 10 years for Carlsson and Lindquist to discover that chlorpromazine blocks dopamine and other catecholamine receptors.[30] Dopamine receptor blockade was put forth as the mechanism of action of chlorpromazine, providing pharmaceutical companies a molecular target for drug development. Over the next 50 years, many dopamine receptor antagonists have been developed and successfully tested in the illness.[31] Among antipsychotics, the clinical potency in reducing psychotic symptoms was shown to be directly proportional to dopamine receptor affinity, supporting dopamine receptor blockade as the mechanism of antipsychotic action. Currently, there is a strong push to indentify novel antipsychotics to improve the manifestations of psychosis.[32]

Several candidates have presented themselves but with only preliminary efficacy data. LY404039 (a prodrug for LY2140023) is an mGluR2/3 agonist which has already shown some preliminary evidence of antipsychotic action.[33] Postmortem tissue evidence in schizophrenia of a reduction in mGluR3 receptors in PFC provides a molecular basis for drug efficacy.[34] Also, xanomeline is a muscarinic cholinergic agonist which is being tested for cognitive enhancement in schizophrenia.[35] However, there is one small published study testing xanomeline in Alzheimer's dementia, showing positive outcomes on cognition despite limiting peripheral gastrointestinal side effects; the single study in schizophrenia is interesting in that it shows positive outcomes on both psychosis and cognition.[36] Other novel candidates are modifications of dopamine antagonism and not altogether novel.

The dopamine hypothesis of schizophrenia proposed that the molecular pathophysiology of the illness is based on increased dopamine transmission at some critical CNS neuronal populations and was a popular explanation of schizophrenia for several decades. However, the evidence of alterations in dopamine transmission never materialized from studies done in the past 60 years, with the exception of some isolated dopamine release data,[37] leading scientists to propose that the dopamine hypothesis of schizophrenia has been falsified.[38] This negation of the dopamine-based disease concept does not falsify the firm knowledge that dopamine blockade is involved in the antipsychotic action of these psychotropic drugs.

D. Brain Circuits Involved in Psychosis and Cognitive Dysfunction

There is substantial evidence that areas of both prefrontal cortex (PFC) and hippocampus are dysfunctional in schizophrenia, although the exact anatomic and molecular basis of that dysfunction is not entirely clear in either region. In PFC, the parvalbumin-containing GABAergic interneurons that modulate pyramidal neurons in neocortex show reduced levels of the GABA-synthesizing enzyme glutamic acid decarboxylase (GAD) and[39] both of which may account for the reduction in gamma synchrony that is observed on EEG in individuals with schizophrenia.[40,41] It is clear that this kind of a molecular alteration and its putative functional sequelae could underlie the cognitive dysfunction apparent in persons with the illness, especially the deficits in working memory and attention.[8,42–44] Based on the careful description of these PFC alterations in schizophrenia tissue, animal models which demonstrate these same kind of pathological changes are sought after in the study of cognitive dysfunction in animal models of the illness.

In hippocampus, several observations have been published noting increased neuronal activity that could be subfield specific[21,45–51] and diminished glutamatergic transmission possibly specific to dentate gyrus (DG), although the specific localization and diversity of these defects require additional study. Our own research (presented below as the basis for a strong construct validity animal model) suggests that a glutamate defect in DG functions to generate an altered homeostatic plasticity state in its target subfield, CA3; and that "run away" overactivity in CA3 could function to generate psychotic symptoms. Other regions that have been solidly implicated in schizophrenia include the superior temporal gyrus,[52] the thalamus,[53] and cerebellum.[54]

III. Existing Animal Models for Schizophrenia

The *Schizophrenia Research Forum* (www.schizophreniaforum.org) has listed 87 putative rodent models for aspects of schizophrenia. These models can be broken down into three main categories and are models not only for psychosis in schizophrenia but also increasingly for cognitive dysfunction and for negative symptoms, all dimensions of illness in schizophrenia. These animal models are based on (A) psychotomimetic drug stimulation, (B) early adversity and lesion models, and (C) schizophrenia risk gene models. We evaluate these existing models by predominance for modeling psychosis, cognition, or negative symptoms. Moreover, each is evaluated for strength of construct validity, the emphasis in this review.

A. Psychotomimetic Drug Stimulation

Some pharmacological agents are described as psychotomimetic. These are drugs which cause psychosis in humans, often including schizophrenia-like symptoms and tend to exacerbate the psychotic manifestations in people with schizophrenia. An understanding of the mechanism(s) of these psychosis-inducing actions might contribute to our fundamental understanding of the illness. These drugs have been administered to animals both for disease pathophysiology studies, but even more importantly, in order to model pharmacological antagonist approaches as treatments.

1. Amphetamine Stimulation

Amphetamine is a drug which releases dopamine and other catecholamines from presynaptic stores and blocks reuptake. The drug causes paranoid manifestations when administered to normal humans. The usual course of drug-induced psychotic symptoms is the appearance of frank and persistent paranoid ideation after subchronic amphetamine administration.[55,56] Moreover, rhesus monkey behaviors generated by subchronic amphetamine suggest hallucinatory phenomena.[10] The putative mechanism of this psychotomimetic action, postulated as a dopamine-mediated phenomenon, fits scientifically into the early hypothesis of schizophrenia as a disease of dopamine hyperactivity,[56] an idea based primarily on the antidopaminergic action of antipsychotic medications. Therefore, amphetamine and other direct and indirect dopamine agonists, like apomorphine, were used in animal models to screen early antipsychotic drugs for antipsychotic action. Animal behaviors believed to resemble psychosis, like locomotor behavior, stereotyped movements, passive-avoidance responding, and even PPI, were tested with and without amphetamine and the outcomes helped to generate refined antidopaminergic drugs like haloperidol from the earlier more complex antipsychotics like chlorpromazine. This animal application was useful in generating antipsychotic compounds with dopamine antagonist actions (since all centrally active antidopaminergic drugs have been antipsychotic) but not suggesting strategies with novel molecular actions. The early confidence in the construct validity of amphetamine as an animal model of schizophrenia weakened once the dopamine hypothesis of schizophrenia faded as a disease model.[38] Nonetheless, developing drugs which block dopamine agonist-induced actions in animals was a strategy that stocked our shelves with antipsychotic drugs for decades, albeit not ones with novel mechanisms of action. And, because antipsychotic drugs are very effective for ameliorating psychosis but not for treating other schizophrenia symptoms like cognition dysfunction and negative symptoms, the use of amphetamine has become a limiting model. Currently, this model remains useful in testing new antipsychotic drugs for their motor side effect potential, which is clearly an undesirable set of side effects, and for clarifying if any action of a new drug is related to dopamine (i.e., blocks amphetamine-induced behaviors). Amphetamine is a model which can be useful in refining side effect profiles of a drug, but lacks substantive construct validity and is not likely to be useful in identifying pathophysiology.

2. Phencyclidine Stimulation

Phencyclidine (PCP) is a noncompetitive antagonist of the voltage-gated NMDA ionophore at its own PCP receptor, acting to decrease glutamate signaling. This NMDA antagonist was originally developed as a safe anesthetic agent; however, its use was discontinued because of serious psychotomimetic side effects. PCP causes mental status changes in normal humans which are schizophrenia-like and resemble positive and negative symptoms.[57] Based on the observations that PCP generates a schizophrenia-like psychosis in humans, the PCP-exposed animal has been one of the most widely used schizophrenia models and has been repeatedly examined. PCP-treated animals show an increase in locomotor activity and stereotyped movements, a decrease in social interaction, alterations in memory and spatial function, and impairments in PPI, mimicking this same change seen in schizophrenia.[58] Another psychotomimetic NMDA antagonist and a weaker congener of PCP is ketamine, a drug which is still in use as an anesthetic agent in limited situations.[59] In controlled human studies, ketamine has been observed to generate mild psychosis in normal individuals[60,61] and worsen psychotic symptoms in schizophrenia.[62,63] In addition, ketamine causes alterations in normal and schizophrenia cognition, namely, prediction errors, decreased working memory capacity, altered response inhibition, and reduced executive processing.[64] Thus, ketamine has been proposed as an agent to generate a human model of schizophrenia, as well as to use in animals in order to create schizophrenia models. Because the symptoms evoked with PCP and ketamine have such a striking resemblance to phenomenology in schizophrenia, the drug has become very broadly used, sometimes alone and at other times, with other risk elements (environmental or genetic features of illness). The psychotomimetic behavioral manifestations of PCP were known before PCP was identified as an NMDA receptor antagonist. Therefore, when PCP was identified as an NMDA receptor antagonist, this observation implicated glutamate system dysfunction in schizophrenia; then, the glutamate hypofunction hypothesis of schizophrenia was articulated and has been pursued, thereafter, generating evidence of glutamate dysfunction in brain tissue from schizophrenia individuals.[65,66] This sequence of discovery has suggested both NMDA insufficiency in schizophrenia and a therapeutic action of NMDA enhancement for psychosis and cognition. Therefore, PCP has been pursued as a model of schizophrenia both for hypothesis-driven reasons and phenomenological ones. The idea that glutamate dysfunction might be regional and not necessarily present at all glutamatergic synapses has been more recently understood.[21] However, the clear observations of glutamate hypofunction in schizophrenia itself have made this modeling approach a direction with strong construct validity. These lines of evidence suggest the NMDA antagonists model—administration of PCP, ketamine, or MK801 (all direct antagonists of the NMDAR and congeners of PCP)—as a solidly supported construct to translate to small animals with behavioral assessment.

B. Early Environmental Adversity and Brain Lesions

The approach of introducing environmental adversity to animals early in development to model schizophrenia grew out of observations and epidemiological data showing that early childhood adversity, including prenatal, perinatal, and developmental adverse events, predisposes individual to schizophrenia, including childhood trauma, abuse, or lesions as well as perinatal illness, infection, or famine.[19,67] How these childhood adversities predispose individuals to develop illness in adulthood is not known but could be associated with epigenetic alterations which develop in childhood and pose risks during adulthood. The idea that early changes in brain structure, function, or epigenetic signaling contribute to an adult-onset illness has been a generative concept in the field of schizophrenia pathophysiology. Some specific examples of these animal models which introduce early adversity and test for behaviors of psychosis during adult years are described below. The resulting early adversity abnormalities are said to be nonreversible; however, given the potency of plasticity-driven treatments on brain reorganization, we could guess that there could be treatments to mitigate adult outcomes, especially those which target plasticity approaches.

1. Methyl-azoxy-methanol Model

Methyl-azoxy-methanol (MAM) is an antimitotic/antiproliferative agent that methylates DNA specifically in neuroblasts while sparing glia.[68,69] It generates epigenetic signaling changes and can modify expression of critical proteins in brain development, altering the developmental trajectories of neurons. Therefore, MAM actions have distinct sequelae depending on the timing of its administration during development. For example, when administered on gestational day (GD) 15 or earlier, MAM decreases whole brain cerebellar and hippocampal volume. However, MAM exposure at GD17 has a more selective effect on reducing the volume of frontal and temporal cortices regions.[70] This gestational timing to GD17 is important as this is when the majority of subcortical regions have completed neuronal proliferation (development) and neonatal mice have their peak of neuronal propagation in cortical regions. This MAM GD17 administration causes some schizophrenia-like anatomical and behavioral phenotypes in the rodent offspring when administered to the pregnant dam. MAM-treated offspring (GD17) shows several schizophrenia-like pathological aspects such as the reduced hippocampal volume, hyperdopaminergic neuronal activity, and decreased numbers of parvalbumin-positive interneurons in the PFC and hippocampus. Behaviorally, the offspring shows enhanced locomotor response to amphetamine and PCP and schizophrenia-like impairments in PPI, latent inhibition, working memory, and social interaction.[18] Several investigators describe the usefulness of this animal model as a developmental disruption, which generates schizophrenia-like anatomical, behavioral, and neurophysiological deficits to resemble pathology observed in schizophrenia, although gestational MAM administration does not reflect the specific molecular correlates of schizophrenia. Because this model is empiric and is not generated to reflect a known molecular or anatomic lesion in the illness, its construct validity is not as strong as more direct models and it is fully dependent on the pathological constructs that the MAM generates. Although it could be the case that, as epigenetic observations are generated and tested in schizophrenia, this model will become more useful.[67]

2. Prenatal Infection Model

The prenatal infection model is based on observations of an association between maternal infection during pregnancy, especially during the second trimester, with an increased risk for schizophrenia in the offspring.[19,71] Although this only represents a risk condition and is not a predictive risk, the idea that it might combine with other etiologies to create the adult-onset illness has been articulated. In prospective study, epidemiological approaches have followed cohorts of mothers and children from birth through age of risk for schizophrenia and have found a higher incidence of illness in infants who had viral titers in their cord blood. No molecular mechanism has yet been discovered to explain this although the strength of the clinical observations is strong enough to support the co-use of this technique to develop schizophrenia models in animals. To develop this model in animals, pregnant rat dams are injected within 9–15 days of birth with polyribo-inosinice–polyribo-cytidylic acid (poly I:C), a cytokine, which mimics aspects of the immune response to infection; then the rat pups are grown to adulthood and tested for a schizophrenia phenotype as adults. This double-stranded RNA mimics the immune cytokine response to viral infection. Pointkewitz and his collaborators describe alterations in frontal and hippocampal pyramidal neuronal morphology, loss of parvalbumin-containing GABAergic interneurons in the hippocampus and PFC, increased dopaminergic activity, and disrupted neurogenesis in the adult rodents born to the infected mother.[72] Prenatal poly I:C results in the adult behaviors of impaired PPI and increased susceptibility to amphetamine and MK-801 hyperlocomotion.[73] This model is now used to potentiate the model-generating effects of other etiological agents, and the strength of the clinical evidence that it generates a risk environment is strong enough to support its construct validity as a model.

Brown points out that it is not only intrauterine infection but also numbers of additional early environmental events which set up a risk environment for schizophrenia. These risks can include obstetrical complications, early malnutrition, winter birth, immigration and social disruption, poverty, and parental age, although the risk estimate for any single one of these may be very low.[67] While it is the maternal administration of poly I:C administration during gestation that is concrete enough for a model, we can nonetheless recognize a cluster of environmental factors that appear to be important as risk factors. It would suggest that additional environmental characteristics, like early maternal deprivation, early stress, repeatedly new rearing environments, etc., might all be characteristics of small effect that would, nonetheless, predispose to schizophrenia. It would be optimal if the neural mechanism(s) underlying these environmental adversities could be discovered. Already scientists have speculated that gene–environment interactions, gene antecedents of environmental events, and/or epigenetics could be involved.

3. Hippocampal Lesion Model

A popular model of schizophrenia has developed around the use of a ventral hippocampal lesion in a small animal,[74] a lesion which can be induced chemically with an excitotoxin or sterically with a knife. Such a lesion is based on similar lesions, although more restricted ones, done in nonhuman primates which aimed to create models of autism and changes in primate social behavior.

The lesions are generally inclusive of large regions of hippocampus or at least discrete subregions of the cerebral structure. In a small animal, such a lesion applied during early life is said to generate an animal with several characteristics of schizophrenia in adulthood. These include the ventricular enlargement and hippocampal alterations often reported in schizophrenia, with behavior manifestations which include deficits in social interaction, increased aggression, and impairment in spatial learning and working memory, all appearing during early animal adulthood and each characteristically repaired by antipsychotic drug treatment, even though this model does not include alterations in dopamine transmission. As currently applied, this model does not have strong construct validity because hippocampal lesions are not described in schizophrenia, even though hippocampal pathology is clearly present. Indeed, one could argue that the presence of an intact hyperactive hippocampus is critical to the manifestations of schizophrenia. Yet, the neonatal hippocampally lesioned animals, when they are adult rats, show some molecular abnormalities in neocortex which are reminiscent of schizophrenia and could result from aberrant projections to cortex from hippocampus: changes like reduced PFC *N*-acetyl-L-aspartic acid (NAA) levels, reduced GAD67 mRNA, lower spine density and dendritic length of the cortical pyramidal neurons, and potassium-stimulated glutamate release, along with increased $GABA_A$ receptors.[18] Acute injections of picrotoxin, or NMDA stimulation, in this adult lesion model lead to short bouts of hyperactivity, attenuation of PPI, and enhanced DA release. This model does show that serious brain lesions that are inflicted during childhood can have adult phenotypes and generate regional brain changes that appear during adulthood and are potentially reminiscent of brain changes in schizophrenia.

4. Early Adolescent Cannabis Use

A large body of literature shows an association between adolescent cannabis use and adult onset of psychosis.[75–79] A recent systematic review of longitudinal studies of cannabis use and subsequent psychotic outcomes reported a 40% increased risk of psychotic outcome in individuals who had ever used cannabis (pooled adjusted OR = 1.41, 95% CI 1.20 ± 1.65).[80] These human studies demonstrate associations and do not demonstrate causality. There is, however, evidence from rodent studies suggesting that Δ^9-tetrahydrocannabinol or cannabinoid receptor 1 agonists administered peripubertally lead to behavioral deficits in adult rats.[81–84] We have conducted a series of experiments in mice that demonstrate an age dependent, long-term effect of early cannabinoid administration on hippocampal-dependent contextual learning. "Adolescent" (5-week-old) mice show significant deficits with an associated upregulation of endocannabinoid signaling while mice treated as adults (9 weeks old) did not show any long-term deficits. These data indicate that there is a window of vulnerability when endocannabinoid system disruptions can lead to long-term schizophrenia-like behavioral effects.

C. Risk Genes for Schizophrenia

Schizophrenia is known to be influenced by genetic architecture. Relatives of people with schizophrenia show a higher risk of the illness than unrelated persons, a phenomenon which is especially clear when comparing identical twins reared together and apart.[60] The closer a person is related to a proband with schizophrenia, the stronger is the risk of illness; the risk is approximately 1% in the general population, 5% with a schizophrenia sibling, 12% with a schizophrenic parent, and 40–50% with an identical twin with the illness. While this later figure shows 40–50% concordance in identical twins, it also shows that having a "schizophrenia genome" does not inevitably result in the illness. Early genetic studies of populations[61] and families[62,64] have strongly established the heritability of the illness. Yet, recent genetic analyses of high numbers of families with the illness have had an unexpectedly difficult time finding specific and general genetic associations within populations.[85] It is not only synonymous and nonsynonymous single-nucleotide polymorphisms (SNPs) which have been associated with the illness, but spontaneous mutations manifest as CNVs, even increasing the genetic diversity of these risk genes. Despite these difficulties, there have been described mutations within many different genes, often at multiple different sites within the gene generating complex allele patterns, which are taken to be risk genes for schizophrenia. Neither are these genes altered in every person with the illness, nor does the illness occur whenever a characteristic risk mutation is present. But, a mutation in one of the risk genes for schizophrenia is thought to generate molecular changes in brain regions which will predispose to the development of psychosis.[86] Curiously, the risk genes for schizophrenia overlap to a large degree with the risk genes for other complex psychiatric disorders, especially with risk genes for bipolar disorder[27] and autism.[87] The development of genetically modified animals with mutations in schizophrenia risk genes has often been used to generate animals which will inform schizophrenia studies. Animals which bear genetic alterations which have been associated with schizophrenia, however weakly, have been examined behaviorally and molecularly to understand the outcomes associated with specific potentially schizophrenia-relevant genetic lesions. While this is a popular and unquestionably contributory approach, the caution stems from the diversity of the genetic background for the illness and the probability of many different spontaneous mutations causing the phenotypic profile of schizophrenia.[86] Indeed, recent DNA sequencing of nonfamilial schizophrenia suggested that up to 50% of people with the illness may have spontaneous causative mutations associated with their illness.[88] Multiple genes (perhaps thousands) of small effect are thought to exist for the illness, given its status as a complex genetic illness.[85] Nonetheless, we assume that tissue effects in animals with risk mutations are associated in some way with schizophrenia pathophysiology; and where risk genes generalize across categorical disease groups, we assume that the genetically based tissue changes underlie common disease dimensions. Therefore, the biological characterization of animals with risk genes is highly relevant to ultimately discovering disease pathophysiology, even though every aspect of the biology may not be critical.

1. NRG1

Neuregulin was first identified as a schizophrenia risk gene from a large multigenerational Icelandic pedigree within the 8p12–8p21 region.[62,64] The original core Icelandic "at risk haplotype" included five SNPs and two microsatellietes in a region spanning the 5′ end of NRG1 and extending into the second intron. It was not long before other pedigrees were found to associate with NRG1 alles,[89–92] suggesting NRG1 as a major risk gene in schizophrenia, albeit at multiple distinct regions in the gene with many SNPs. Neuregulin 1 is a trophic factor that contains an epidermal growth factor-like domain and signals by stimulating ErbB receptor protein tyrosine kinases. NRG1 generates six types of proteins and has at least 31 isoforms, each with a distinct amino-terminal region. NRG1 and ErbB4 signaling functions importantly in neural development, through an involvement with radial neuron migration, axon guidance, myelination, oligodendrocyte developments, and synapse formation.[93,94] They are both importantly involved in synaptic plasticity, sometimes enhancing and often inhibiting long-term potentiation (LTP) in a complex fashion with distinct regional actions in brain.[95,96] NRG1 can regulate activity-dependent synaptic plasticity by recruiting tyrosine kinases that regulate NMDAR function. Neuregulin effects glutamatergic, nicotinic, and GABAergic transmission.[97–100] The question of whether or not NRG1 is altered in human schizophrenia brain has been variously answered.[18,101] Several studies suggest that both NRG1 and ErbB proteins are altered in schizophrenia tissue, especially hippocampus,[102–105] although there is disagreement in the literature. A particularly well-done postmortem study reports no change at all in NRG1 or ErbB protein levels in prefrontal tissue; however, it showed an increase in NRG1-induced ErbB activation.[106]

NRG1 mutant mice have "schizophrenia-like" behaviors as well as alterations in glutamate function. Null mutant mice have low NR1 expression and altered tyrosine phosphorylation of the NMDA channel. NRG1 heterozygotes have impaired novel object recognition and contextual fear conditioning. Other relevant behavior is social withdrawal and increased sensitivity to environmental manipulation. These data allow us to infer that partial depletion of neuregulin could cause some deficit in recognition memory. Spatial memory does not seem to be affected.[107] Clozapine ameliorates the hyperactivity but not the PPI attenuation. Interestingly, mice that overexpress NRG1 have similar hyperactive locomotion, impaired context fear conditioning, and increased aggressiveness. The use of NRG1-specific genetically modified mice as animal models of schizophrenia has proven useful to identify a range of potential tissue substrates for the illness; but the extensive breadth of affected systems and proteins fails to pinpoint a psychosis mechanism. Rather, the use of this risk mouse within a given schizophrenia paradigm might be a productive approach to examine molecular systems related to pathophysiology.

2. DISC1

The Disrupted-in-Schizophrenia-1 gene was discovered in a large Scottish family, by its segregation with the psychosis diagnoses of schizophrenia, depression, and mania. This gene is disrupted in a t(1;11)(q42.1;q14.3) translocation. Alternate transcriptional splice variants, encoding different isoforms, have also been characterized. The DISC1 protein is predicted to contain multiple blocks of coiled-coil motifs and two leucine zippers at the C-terminal domains, domains which may mediate critical protein–protein interactions.[108] The N-terminal protein contains a globular domain and two putative nuclear localization signals. Mitochondria are the primary site of DISC1 expression, but the gene has also been found to locate to the nucleus, centrosome, cytoplasm, axons, and synapses. DISC1's role is highly diverse and its functions seem to depend on its cellular location. DISC1 in mitochondria have a role in oxidative phosphorylation, calcium homeostasis, and apoptosis all of which may be relevant for the neuronal mechanisms believed to be involved in schizophrenia. In addition, DISC1 is associated with cytoskeleton protein and microtubule function, as well as with the membrane trafficking of receptors.[109] The DISC1 protein regulates morphogenesis, maturation, migration, and synaptic integration of neurons. It also regulates the proliferation of progenitor cells, specifically the rate of neurogenesis at the point where progenitor cells mature into neurons.

Seven different transgenic mouse strains of the DISC1 gene have been characterized.[18] These mice exhibit enlarged lateral ventricles and reduced cortical thickness and brain volume, mimicking some characteristics of schizophrenia. Some mutant mice have reduced parvalbumin in the PFC and hippocampus. Other mutants display reductions in hippocampal dendritic complexity, structure, and density. Some DISC1 mice have PPI deficits that are attenuated with antipsychotics (haloperidol and clozapine). Other behavioral abnormalities seen in some mutants are hyperactivity, reduced sociability, working memory, and executive function impairments, all which can be seen in the human illness. Talbot *et al.*[110] has suggested that an abnormality in the DISC1 gene could negatively impact cognition and working memory processes, possibly through altering the PFC and hippocampus molecular environment during development. DISC1 polymorphisms are associated with reduced hippocampal gray matter volume and the NAA signal and abnormal engagement of the hippocampus during several cognitive tasks. The mechanisms may therefore involve hippocampal development and ensuing function.[111] While the use of DISC1 mice has proven useful as an animal model of schizophrenia to identify a range of potential cellular mechanisms related to the illness, the use of DISC1 mutants has not contributed to an identification of affected systems and proteins which define a psychosis mechanism. This DISC1 KO mouse coupled with risk environments for schizophrenia could generate true disease models with high construct validity. Meanwhile, a study of DISC1 biology may well inform future disease models and general brain biology.

3. DTNP1

Multipoint linkage analysis of Irish high-density pedigrees initially associated dystrobrevin-binding protein 1 (dysbindin) with schizophrenia in region 6p24-21 and identified it as a risk gene for SZ. Further analysis of the family-based simple sequence-length-polymorphism markers and SNP markers showed that the 6p22.3 SNPs of dysbindin are strongly associated with schizophrenia.[112] Dysbindin protein is well distributed in the brain and binds to alpha- and beta-dystrobrevin, a component of the dystrophin-associate complex (DPX). DPX is concentrated within the postsynaptic signaling area, at the postsynaptic density (PSD). Thus, it has been proposed that dysbindin may regulate trafficking or tethering of receptors and signal transduction proteins at the PSD in the glutamate synapse.[99] Dysbindin is also concentrated in the presynaptic glutamatergic terminals, where it interacts with SNAP and synapsin 1 and modulates vesicular release of glutamate.[113,114] Overexpression of dysbindin results in increased glutamate release by pyramidal neurons.[115] Dysbindin protein levels are inversely associated with general cognitive ability and poor premorbid function.[116] Patients with schizophrenia show reduced dysbindin mRNA levels in multiple layers of the dorsolateral PFC. Talbot *et al.*[110] report that both mRNA and protein are decreased in SZ, especially in the hippocampus and expressly presynaptically. Reduced dysbindin, as it is in schizophrenia, increases D2-receptor abundance and regulates glutamatergic signaling and synaptic stability.[110] Dysbindin is also thought to modulate sensitivity to D2 agents. Human genetic studies have not always supported the association of dysbindin and schizophrenia. According to the database of a meta-analysis,[103] two SNPs of this gene are reported to have significant association with schizophrenia (http://www.szgene.org/meta.asp?geneID=1), but the significance is not strong.

Sandy mice have a spontaneous genetic deletion of Dtnbp1 leading to the complete loss of dysbindin. This deletion affects pre- and postsynaptic aspects of dopaminergic, glutamatergic, and GABAergic transmission. These mice display social interaction deficits, long-term memory deficits, novel object-recognition deficits, and working memory deficits (through the DA/D2 mechanism). These deficits are functionally enhanced when the mice are under stress.[117] Sandy mice provide a means of studying the potential contribution of dysbindin to clinical features of schizophrenia. This model has affirmed that this protein has a role in SZ behaviors (PPI and social interaction) and altered aspects of spatial memory, but is unsuccessful in illuminating a mechanism or cause of psychosis.

4. BDNF

Brain-derived neurotrophic factor (BDNF) is a neurotrophin that promotes growth and differentiation of developing neurons during development and neuronal survival in mature tissue in response to stress. Neurotrophins play a significant role in adult brain in activity-dependent neuronal plasticity and synaptic transmission.[118] The gene for BDNF maps to human chromosome 11p13 and is organized in 13 exons. The last exon contains the entire open reading frame of BDNF and encodes two BDNF protein variants. The long form contains 247 amino acids and consists of a 5′ pro-BDNF sequence that is cleaved to form a mature protein, while the short form lacks this 5′ pro-BDNF region.[119] BDNF selectively interacts with TrkB (a tropomyosin-related kinase) on the postsynaptic membrane and with low affinity to p75NTR. Binding of BDNF to TrkB triggers its kinase activity and therefore activates multiple signal pathways (PI3K, MAPK, PLC-) which leads to cell survival. However, pro-BDNF binds with high affinity to p75NTR which initiates apoptotic signaling.[120,121] A common BDNF SNP results in a valine to methionine switch at nucleotide 196. This change occurs at the 5′ end of the pro-BDNF sequence and has been investigated frequently for its role in schizophrenia. A meta-analysis by Gratacos *et al.*[122] found that the met–met genotype is associated with a 19% risk of developing a psychotic disorder. Although the SNP does not affect mature BDNF directly, it alters intracellular tracking and packaging of pro-BDNF and thus affects the regulation of the mature form of BDNF,[123–125] a function reported to be impaired extinction in fear memory paradigms for both humans and mice displaying the Val66Met SNP. Another polymorphism (C281A) results in the reduction of promoter activity in hippocampal neurons and is reported to be associated with late age of the first episode of paranoid schizophrenia in men.[126]

BDNF has been recognized as a candidate gene for schizophrenia, as well as other psychiatric disorders on the basis of several observations. First, it modulates dopaminergic, GABAergic, and serotonergic receptors,[127] systems which can be dysfucntional in schizophrenia; second, it is highly concentrated in the hippocampus and PFC, as well as other regions involved in mood and behavior; and finally, animal studies reveal that BDNF is involved in pharmacologic response to antipsychotics, antidepressents, and mood stabilizers (lithium). Several studies have reported altered BDNF mRNA and protein in PFC postmortem brain tissue of schizophrenics.[128–132] Decreased mRNA expression of TrkB is also lowered in PFC of SZ patients.[133] It is not clear that the full spectrum of BDNF function has been identified yet in normal human brain, nor its critical function in the pathophysiology of schizophrenic illness.

When BDNF is deleted in mouse forebrain, the animals fail to learn the Morris water maze and they display enhanced freezing during fear conditioning. BDNF may have alternative roles when deleted or depleted at different stages in the life cycle. When BDNF is deleted in the dorsal hippocampus only when mice have reached adulthood, it impairs spatial water maze and object-recognition learning. Partial genetic ablation of BDNF in forebrain, again when mice have matured, results in context-only fear conditioning deficits. When disrupted in development, mice are hyperactive and show profound dysfunction in both cue and contextual fear conditioning. Conditional and forebrain-specific TrkB knockout mice are unable to learn the Morris water maze test and impaired in the eight-arm radial maze, further suggesting BDNF's role in learning and memory.[110] Regionally specific alterations in BDNF protein have distinct outcomes, emphasizing the importance of localized circuits and regions as pertinent to BDNF physiology.

5. CHRNA7

The alpha7 nicotinic receptor (CHRNA7) was initially linked to a neurophysiological phenotype commonly found in schizophrenia, marked by a P50 auditory-evoked potential deficit. The genetic marker is located at chromosome 15q13-14, a linkage based on nine multiply affected, European–American high-density psychosis families.[134] The initial investigation led to further study and replication of results and to the finding of significant genotype-wise disequilibrium at D15S165.[134] This is a polymorphic simple sequence marker within 120 kb of the CHRNA7 coding region. The human gene is part of a duplicated cassette (located nearly adjacent to each other) of expressed sequences. The expressed sequence has four novel exons followed by six exons that duplicate exons 5–10 of the full-length CHRNA7.[134]

Decreased CHRNA7 expression in schizophrenia linkage studies was first reported by Freedman *et al.*[134] Because normally CHRNA7 is expressed at its highest levels in normal hippocampus,[135] Freedman sought and found protein reductions of CHRNA7 in schizophrenia hippocampal tissue. Evidence supports a therapeutic effect of nicotinic agonists on cognitive function in schizophrenia.[136] This suggests that CHRNA7 receptors are a potential target for cognitive enhancement. Nicotine is known to acutely reverse diminished auditory sensory gating in people with schizophrenia, but this is not a long-lasting effect, due to the rapid desensitization of the receptor. Heavy smoking, often formulated as a self-medicating behavior, is prominent in schizophrenia patients, consistent with higher concentrations of nicotine needed to reach normal levels of receptor activation in schizophrenia. The alpha7 nicotinic agonist, 3-(2,4 dimethoxy)benzylidene-anabaseine (DMXBA), can also enhance auditory sensory gating in animal models and enhances P50 auditory gating and cognition in humans. Deficits in CHRNA7 can lead to disinhibition of pyramidal cells.[101] Inadequacies in auditory gating are associated with attentional impairment and may contribute to cognitive symptoms and perceptual disturbances. Freedman *et al.*[134] believe that such learning problems could govern delusional thinking and hallucinations. An early interaction between receptor density and choline availability may contribute to this.

6. COMT

Catechol-*O*-methyl transferase (COMT) is an enzyme in abundant concentrations in brain that catabolizes catechols, catalyzing their degradation. Genetic studies of COMT in schizophrenia do not support a significant association of COMT SNPs with schizophrenia. Yet, extensive molecular analysis of the function of the COMT gene in schizophrenia has been done based on the plausible involvement of this gene in schizophrenia susceptibility. COMT is expressed abundantly in PFC and hippocampus[137,138] and inactivates catechols, especially dopamine within extracellular spaces.[139] The COMT knockout mouse shows an increase in dopamine itself and in dopamine turnover in the PFC.[104,105] The human Valine158Methionine allele of the COMT gene is functional in that this mutation is less stable and impacts the activity of the enzyme; specifically, the Val-containing alleles are associated with faster enzyme activity, therefore with low levels of dopamine, and the Met allele is associated with slower enzyme activity, therefore with high dopamine levels.[140,141] The Val allele is associated with diminished dopamine activity in PFC and diminished PFC activation function as seen with fMRI BOLD experiments; thus, this allele mimics PFC dysfunction typical of the illness.[106] Most of the genetic studies of the COMT gene have focused on the Val158Met polymorphism and its effects on tissue dopamine levels. Meanwhile, the analyses of the genetic association studies have failed to find a significant association of this Val/Met COMT SNP with schizophrenia (http://www.szgene.org/meta.asp?geneID=420). Therefore, it appears that the value of the COMT knockout mouse is not based on the genetic link between COMT and schizophrenia; rather, the value of the mouse model is based on its low PFC dopamine levels and the alterations in PFC function that this defines. The model can be used to investigate the effect of dopamine signaling on schizophrenia pathophysiology.

7. NRAS3

The neuronal PAS domain protein 3 (NPAS3) gene is one of the risk factor genes for schizophrenia. NPAS3 is a transcription factor whose expression is observed in the maturing neuronal precursor cells of the hippocampal subgranular zone[142] and is involved in the control of adult neurogenesis.[143] The first report of this risk gene in schizophrenia is the disruption of the gene in a mother and daughter affected with schizophrenia.[144] They carried a t(9;14)(q34;q13) chromosome that had a translocation breakpoint junction within the neuronal PAS3 gene.[3] Further studies have observed the association of the SNPs of the NPAS3 gene with schizophrenia.[145–148] Further, a genome-wide association study pointed out a SNP of the NPAS3 gene as one of the loci that is associated with the response to an antipsychotic treatment.[149] These associations of the NPAS3 gene with schizophrenia are also supported by the animal model. NPAS3 gene knockout mice show abnormal behaviors mimicking the psychosis of schizophrenia; they show hyperlocomotor activity and impaired PPI.[8] Interestingly, NPAS3 gene knockout mice have decreased adult neurogenesis and a small volume of the DG in the hippocampus. A postmortem brain study using schizophrenia tissue shows that adult neurogenesis in the DG is decreased in schizophrenia,[9] and another study shows that mossy fiber synapses, the output from DG, are decreased in schizophrenia postmortem brains.[10,11] These lines of evidence suggest that "small DG," which could be associated with decreased adult neurogenesis and decreased mossy fiber synapses, may be involved in the pathology of schizophrenia.

While these several genes have been most frequently associated with schizophrenia, some more broadly with psychosis, this list is by no means complete and will assuredly grow with future examination. It is through the combination of risk gene animals with additional model features that we may achieve informative animal models for schizophrenia which have strong construct validity.

IV. An Example in Progress: A Psychosis Animal Model for Schizophrenia with High Construct Validity

We have emphasized throughout this review the vital need within this field for a molecular understanding of psychosis in schizophrenia.[3] To realize this goal, fundamental molecular, genetic, anatomic, and/or functional elements of schizophrenia pathophysiology need to be (A) identified in effected human brain systems and (B) replicated in the animal. This orientation has motivated the development of a current animal model directly from a human psychosis construct.[21] Because the animal model is focused on the psychosis dimension, the approach has been tractable, without having to broadly include all dimensions of the illness. Based on both human brain imaging changes in psychosis[45,50,51,150] and postmortem tissue changes[151–156] each imbedded in modern trajectories of study, this human model of psychosis in schizophrenia, described below, emerged as a speculative and testable construct, using available human and animal experimental approaches. Moreover, the translation of the human construct to an animal model is direct and testable, with the possibility of high construct validity.

Data from *in vivo* imaging studies of schizophrenia have reliably suggested increased regional perfusion in hippocampus[21,45–51]; whereas, analyses of human postmortem brain tissue from the abnormal regions (hippocampal subfields) potentially demonstrate molecular alterations involving plasticity markers within the targeted area.[21,157,158] We have used human imaging and postmortem tissue analyses as methodological approaches to generate a novel molecular model of human psychosis. We suggest that an animal model generated to concretely parallel these human data in a reverse-translation direction will have high construct validity, albeit its validity will dependent on the reliability of the human disease construct. This kind of an animal model can be expected to be dynamic and addressable and thereafter, available for forward translation. This putative animal psychosis model will serve as an example of a preparation with high construct validity, in that it is anatomically and molecularly matched to the putative human psychosis pathology.

A. The Human Blueprint for the Animal Model

Observations from human brain imaging and postmortem tissue analysis in schizophrenia implicate the hippocampal formation in psychosis along with alterations in declarative memory functions, as previously presented. Whereas, the PFC (commonly targeted in schizophrenia) is thought to be critically involved in cognitive dysfunction in the illness, the hippocampus is thought to be involved in psychotic manifestations and in declarative memory dysfunction. Observations from *in vivo* imaging studies in schizophrenia show *increased* perfusion in hippocampus[45,159,160] and observations from postmortem schizophrenia tissue studies show *decreased* NR1 mRNA and protein in hippocampal subfields, particularly in DG.[151,161] These are seemingly paradoxical observations, which, however, converge to suggest homeostatic plasticity alterations in schizophrenia hippocampal CA3 subfield as a mechanism for psychosis. The preliminary increases in homeostatic plasticity markers that we have already identified within CA3 underscore the feasibility of this formulation and its testing.

It is the functional effect that this reduced glutamate signaling in DG has on CA3 that could be of interest for psychosis. CA3 is the projection target of DG in the trisynaptic pathway through the mossy fiber projection. Currently, the human construct suggests that the decrease in activity-dependent signaling within DG projected through the mossy fiber pathway sensitizes its target tissue (CA3) to incoming stimuli and generates a lower LTP threshold and increased cellular sensitivity,[157,158] thus generating higher levels of neuronal activity in CA3.[21,45,50,51,150] We suggest that this results in increased synaptic strength in CA3, along with increased associational function, especially within the system of CA3 recurrent collateral synapses, a unique feature of CA3 innervation.[162–164] The literature supports this model, with findings by Kolomeets *et al.*[165] showing a reduction in mossy fiber synapses in CA3 and a Reif study[166] showing reduced neurogenesis in DG, each in schizophrenia tissue. Based on this glutamate-mediated metaplasticity model,[21] we predict an increase in markers of plasticity in CA3, such as an increase in NR2B-containing NMDA receptors (augmenting the sensitivity of NMDA receptors) and increases in BDNF activity in CA3 in SZ. It is a speculation, albeit with increasing support,[167] that increases in CA3 associational function could generate false associations some with psychotic content, which would then be laid down in memory as psychotic thoughts and memories. This idea could explain both the observed reduction in glutamate signaling in DG (represented by NR1 protein reduction) and the increase in hippocampal perfusion that we detect overall in hippocampus with rCBF measures[21] and our finding that BDNF is increased regionally in CA3 in SZ tissue.[168] The evidence that DISC1 alleles and the NRG Icelandic haplotype are risk genes for SZ additionally strengthens this hypothesis, given the intimate involvement of these genes in synaptic plasticity especially in hippocampus and the demonstration of alterations in synaptic plasticity with mutations in these genes.[95,169–171]

B. Reducing NR1 in DG Reduces Glutamate Signaling to CA3

The NMDA receptor is a heteromeric ionotropic receptor containing an essential NR1 subunit and one of the four NR2 subunits (NR2A–NR2D).[172,173] The NR2 subunits each confer distinctive functional and pharmacological characteristics to the receptor. While NR1-null mice do not survive,[174,175] genetically engineered mice with reduced but not absent (5–10% of normal) NR1 protein can survive to adulthood.[1] These NR1 knockdown mice can show PCP-like behavioral changes mimicking schizophrenia, including increased locomotor activity, stereotypy, and impaired social interaction, each reversible with antipsychotic medications.[176]

A selective knockout of NR1 in DG already exists in a mouse model; the model is built on a POMC-Cre recombinase driver mouse with a floxed NR1 mutant to create a mutant mouse with DG-specific NR1 protein reduction and with behavioral and anatomic features reminiscent of schizophrenia.[177] In the POMC-Cre recombinase line, Cre expression was strong and dense within the granule cells of the DG, while expression was less dense in the hypothalamus (arcuate nucleus) and the habenula.[177] Crossing this POMC-Cre line with the

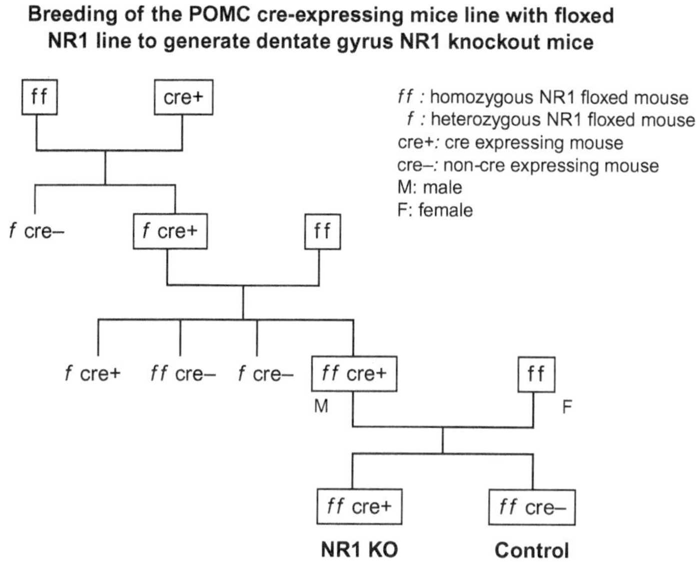

FIG. 1. Breeding of the POMC-cre-expressing mice line with floxed NR1 line to generate dentate gyrus NR1 knockout mice.

floxed NR1 line (Fig. 1) generates a DG-specific NR1 knockout mouse, with the decrease in the DG-NR1 mRNA expression in DG granule cells starting at postnatal days 10–21 and progressing to completion by 16 weeks, with normal hippocampal architecture in DG throughout.[177] McHugh *et al.* showed that the DG NR1 knockout mouse has impaired pattern separation which they demonstrated using a fear conditioning paradigm with two different contexts and an alteration in CA3 with rate reprogramming, but with normal CA1 LTP.[177]

C. Establishing Reduced NR1 Protein in DG as a Relevant Animal Model

To date, we have generated these DG-specific NR1 knockout animals and have begun to examine the effected and the wild-type animals for psychosis-related behaviors, with the DG-NR1 KO offspring contrasted with wild-type littermate controls in behavioral analyses. They have been bred as indicated in Fig. 1. There are no changes in locomotor activity in these animals but reduced PPI. The DG-NR1 KO animals show alterations in pattern separation with fear conditioning, although not in fear conditioning itself, indicating a selective reduction in DG glutamate signaling, as predicted (Fig. 2). Now, in order to establish this model as relevant for schizophrenia, the task is to show consequential increases in CA3 function (neuronal activity) in the mutant animal,

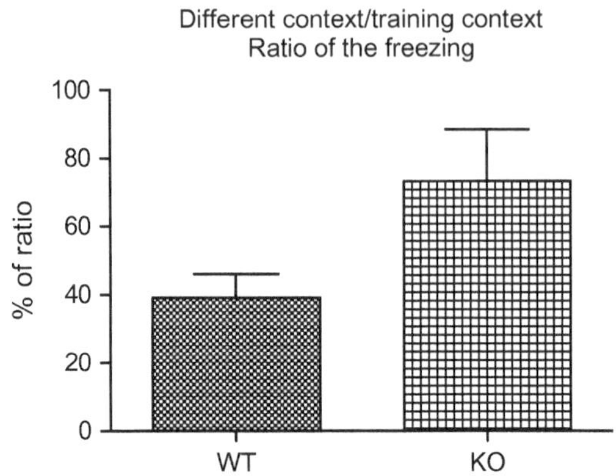

FIG. 2. Poor pattern-separation behavior in DG-NR1 mice using a fear-conditioning paradigm.

possibly using animals with additional risk factors of genetic (DISC1 KO or NRG1 KO) or environmental (e.g., intrauterine cytokines) vulnerabilities. In addition, it is important for us to be able to match tissue changes in the animals with tissue pathology in the schizophrenia postmortem tissue, particularly in DG and CA3. Therefore, we will have to demonstrate increases in markers of homeostatic plasticity in CA3 in the DG-NR1 KO mouse, including increased NR2B/NR1 or P-GluR1/GluR1. Once the mouse model is validated to the human hippocampal pathology, and given the animal status, we can explore the dynamics of LTP in DG-NR1 KO versus wild-type mice, as well. Related studies that we are carrying out in animals with reduced neurogenesis in hippocampus show both reduced NR1 protein in DG and increased NR2B protein in CA3, providing promise that we will be able to match tissue and behavioral pathology in this putative animal model of psychosis with the pathological changes that we have already found in schizophrenia hippocampus. It is the establishment of this close relationship between a human construct and an animal model that will increase confidence in the molecular and anatomic integrity of the animal model.

V. Future Directions for Schizophrenia Models

It will be the goal in the future to develop and use animal models to study schizophrenia pathophysiology, using genetically and environmentally manipulated animals with pharmacological and early adverse stimuli, all matched to schizophrenia illness. The animal models that are verified with human tissue and have clear parallel behaviors will have the strongest construct validity.

REFERENCES

1. Young JW, Zhou X, Geyer MA. Animal models of schizophrenia. *Curr Top Behav Neurosci* 2010;**4**:391–433.
2. Young JW, Powell SB, Risbrough V, Marston HM, Geyer MA. Using the MATRICS to guide development of a preclinical cognitive test battery for research in schizophrenia. *Pharmacol Ther* 2009;**122**:150–202.
3. Nestler EJ, Hyman SE. Animal models of neuropsychiatric disorders. *Nat Neurosci* 2010;**13**:1161–9.
4. Hyman SE, Fenton WS. Medicine: what are the right targets for psychopharmacology? *Science* 2003;**299**:350–1.
5. Carpenter Jr. WT, Buchanan RW. Schizophrenia. *N Engl J Med* 1994;**330**:681–90.
6. Tamminga CA, Holcomb HH. Phenotype of schizophrenia: a review and formulation. *Mol Psychiatry* 2004;**10**:27–39.
7. Green MF, Nuechterlein KH, Gold JM, Barch DM, Cohen J, Essock S, et al. Approaching a consensus cognitive battery for clinical trials in schizophrenia: the NIMH-MATRICS conference to select cognitive domains and test criteria. *Biol Psychiatry* 2004;**56**:301–7.
8. Nuechterlein KH, Robbins TW, Einat H. Distinguishing separable domains of cognition in human and animal studies: what separations are optimal for targeting interventions? A summary of recommendations from breakout group 2 at the measurement and treatment research to improve cognition in schizophrenia new approaches conference. *Schizophr Bull* 2005;**31**:870–4.
9. Stover EL, Brady L, Marder SR. New paradigms for treatment development. *Schizophr Bull* 2007;**33**:1093–9.
10. Fernando ABP, Robbins TW. Animal models of neuropsychiatric disorders. *Annu Rev Clin Psychol* 2011;**7**:39–61.
11. Keeler JF, Robbins TW. Translating cognition from animals to humans. *Biochem Pharmacol* 2011;**81**:1356–66.
12. McGraw LA, Young LJ. The prairie vole: an emerging model organism for understanding the social brain. *Trends Neurosci* 2010;**33**:103–9.
13. Tamminga CA. Development in somatic treatments: introduction. *Am J Psychiatry* 1994;**151**:216–9.
14. Keller WR, Fischer BA, Carpenter Jr. WT. Revisiting the diagnosis of schizophrenia: where have we been and where are we going? *CNS Neurosci Ther* 2011;**17**:83–8.
15. O'Tuathaigh CM, Waddington JL. Mutant mouse models: phenotypic relationships to domains of psychopathology and pathobiology in schizophrenia. *Schizophr Bull* 2010;**36**:243–5.
16. Cuthbert BN, Insel TR. Toward new approaches to psychotic disorders: the NIMH research domain criteria project. *Schizophr Bull* 2010;**36**:1061–2.
17. Harrison PJ, Pritchett D, Stumpenhorst K, Betts JF, Nissen W, Schweimer J, et al. Genetic mouse models relevant to schizophrenia: taking stock and looking forward. *Neuropharmacology* 2011.
18. Jones CA, Watson DJ, Fone K. Animal models of schizophrenia. *Br J Pharmacol* 2011;**164**:1162–94.
19. Brown AS. The environment and susceptibility to schizophrenia. *Prog Neurobiol* 2011;**93**:23–58.
20. Amann LC, Gandal MJ, Halene TB, Ehrlichman RS, White SL, McCarren HS, et al. Mouse behavioral endophenotypes for schizophrenia. *Brain Res Bull* 2010;**83**:147–61.
21. Tamminga CA, Stan AD, Wagner AD. The Hippocampal formation in schizophrenia. *Am J Psychiatry* 2010;**167**:1178–93.

22. Wilkinson MB, Dias C, Magida J, Mazei-Robison M, Lobo M, Kennedy P, et al. A novel role of the WNT-dishevelled-GSK3beta signaling cascade in the mouse nucleus accumbens in a social defeat model of depression. *J Neurosci* 2011;**31**:9084–92.
23. Krishnan V, Han MH, Graham DL, Berton O, Renthal W, Russo SJ, et al. Molecular adaptations underlying susceptibility and resistance to social defeat in brain reward regions. *Cell* 2007;**131**:391–404.
24. Saks ER. *The Center Cannot Hold: My Journey Through Madness*. New York: Hyperion; 2007.
25. Nasar S. *A Beautiful Mind: A Biography of John Forbes Nash, Jr., Winner of the Nobel Prize in economics, 1994*. New York: Simon & Schuster; 1998.
26. Mesholam-Gately RI, Giuliano AJ, Goff KP, Faraone SV, Seidman LJ. Neurocognition in first-episode schizophrenia: a meta-analytic review. *Neuropsychology* 2009;**23**:315–36.
27. Ivleva EI, Morris DW, Moates AF, Suppes T, Thaker GK, Tamminga CA. Genetics and intermediate phenotypes of the schizophrenia-bipolar disorder boundary. *Neurosci Biobehav Rev* 2010;**34**:897–921.
28. Levy DL, Sereno AB, Gooding DC, O'Driscoll GA. Eye tracking dysfunction in schizophrenia: characterization and pathophysiology. *Curr Top Behav Neurosci* 2010;**4**:311–47.
29. Lopez-Munoz F, Alamo C, Cuenca E, Shen WW, Clervoy P, Rubio G. History of the discovery and clinical introduction of chlorpromazine. *Ann Clin Psychiatry* 2005;**17**:113–35.
30. Carlsson A, Lindquist L. Effect of chlorpromazine or haloperidol on formation of 3-methoxytyramine and normetanephrine in mouse brain. *Acta Pharmacol Toxicol* 1963;**20**:140–5.
31. Lepping P, Sambhi RS, Whittington R, Lane S, Poole R. Clinical relevance of findings in trials of antipsychotics: systematic review. *Br J Psychiatry* 2011;**198**:341–5.
32. Ibrahim HM, Tamminga CA. Schizophrenia: treatment targets beyond monoamine systems. *Annu Rev Pharmacol Toxicol* 2011;**51**:189–209.
33. Patil ST, Zhang L, Martenyi F, Lowe SL, Jackson KA, Andreev BV, et al. Activation of mGlu2/3 receptors as a new approach to treat schizophrenia: a randomized phase 2 clinical trial. *Nat Med* 2007;**13**:1102–7.
34. Ghose S, Gleason KA, Potts BW, Lewis-Amezcua K, Tamminga CA. Differential expression of metabotropic glutamate receptors 2 and 3 in schizophrenia: a mechanism for antipsychotic drug action? *Am J Psychiatry* 2009;**166**:812–20.
35. Sellin AK, Shad M, Tamminga C. Muscarinic agonists for the treatment of cognition in schizophrenia. *CNS Spectr* 2008;**13**:985–96.
36. Shekhar A, Potter WZ, Lightfoot J, Lienemann J, Dube S, Mallinckrodt C, et al. Selective muscarinic receptor agonist xanomeline as a novel treatment approach for schizophrenia. *Am J Psychiatry* 2008;**165**:1033–9.
37. Guillin O, Abi-Dargham A, Laruelle M. Neurobiology of dopamine in schizophrenia. *Int Rev Neurobiol* 2007;**78**:1–39.
38. Kendler KS, Schaffner KF. The dopamine hypothesis of schizophrenia: an historical and philosophical analysis. *Philos Psychiatr Psychol* 2011;**18**:41–63.
39. Curley AA, Arion D, Volk DW, Asafu-Adjei JK, Sampson AR, Fish KN, et al. Cortical deficits of glutamic acid decarboxylase 67 expression in schizophrenia: clinical, protein, and cell type-specific features. *Am J Psychiatry* 2011;**168**:921–9.
40. Woo TU, Spencer K, McCarley RW. Gamma oscillation deficits and the onset and early progression of schizophrenia. *Harv Rev Psychiatry* 2010;**18**:173–89.
41. Ford JM, Roach BJ, Hoffman RS, Mathalon DH. The dependence of P300 amplitude on gamma synchrony breaks down in schizophrenia. *Brain Res* 2008;**1235**:133–42.

42. Gur RC, Ragland JD, Moberg PJ, Bilker WB, Kohler C, Siegel SJ, et al. Computerized neurocognitive scanning: II. The profile of schizophrenia. *Neuropsychopharmacology* 2001;**25**:777–88.
43. Dickinson D, Iannone VN, Wilk CM, Gold JM. General and specific cognitive deficits in schizophrenia. *Biol Psychiatry* 2004;**55**:826–33.
44. Flashman LA, Green MF. Review of cognition and brain structure in schizophrenia: profiles, longitudinal course, and effects of treatment. *Psychiatr Clin North Am* 2004;**27**:1–18. vii.
45. Medoff DR, Holcomb HH, Lahti AC, Tamminga CA. Probing the human hippocampus using rCBF: contrasts in schizophrenia. *Hippocampus* 2001;**11**:543–50.
46. Tamminga CA, Thaker GK, Buchanan R, Kirkpatrick B, Alphs LD, Chase TN, et al. Limbic system abnormalities identified in schizophrenia using positron emission tomography with fluorodeoxyglucose and neocortical alterations with deficit syndrome. *Arch Gen Psychiatry* 1992;**49**:522–30.
47. Nordahl TE, Kusubov N, Carter C, Salamat S, Cummings AM, O'Shora-Celaya L, et al. Temporal lobe metabolic differences in medication-free outpatients with schizophrenia via the PET-600. *Neuropsychopharmacology* 1996;**15**:541–54.
48. Liddle PF, Friston KJ, Frith CD, Hirsch SR, Jones T, Frackowiak RS. Patterns of cerebral blood flow in schizophrenia. *Br J Psychiatry* 1992;**160**:179–86.
49. Fletcher P. The missing link: a failure of fronto-hippocampal integration in schizophrenia. *News Views* 1998;**1**:266–7.
50. Gur RE. Left hemisphere dysfunction and left hemisphere overactivation in schizophrenia. *J Abnorm Psychol* 1978;**87**:226–38.
51. Malaspina D, Harkavy-Friedman J, Corcoran C, Mujica-Parodi L, Printz D, Gorman JM, et al. Resting neural activity distinguishes subgroups of schizophrenia patients. *Biol Psychiatry* 2004;**56**:931–7.
52. Sun J, Maller JJ, Guo L, Fitzgerald PB. Superior temporal gyrus volume change in schizophrenia: a review on region of interest volumetric studies. *Brain Res Rev* 2009;**61**:14–32.
53. Cronenwett WJ, Csernansky J. Thalamic pathology in schizophrenia. *Curr Top Behav Neurosci* 2010;**4**:509–28.
54. Andreasen NC, Pierson R. The role of the cerebellum in schizophrenia. *Biol Psychiatry* 2008;**64**:81–8.
55. Snyder SH. Amphetamine psychosis: a "model" schizophrenia mediated by catecholamines. *Am J Psychiatry* 1973;**130**:61–7.
56. Faurbye A. The role of amines in the etiology of schizophrenia. *Compr Psychiatry* 1968;**9**:155 77.
57. Javitt DC, Zukin SR. Recent advances in the phencyclidine model of schizophrenia. *Am J Psychiatry* 1991;**148**:1301–8. [Review].
58. Perrin MA, Butler PD, DiCostanzo J, Forchelli G, Silipo G, Javitt DC. Spatial localization deficits and auditory cortical dysfunction in schizophrenia. *Schizophr Res* 2010;**124**:161–8.
59. van Os J, Kenis G, Rutten BP. The environment and schizophrenia. *Nature* 2010;**468**:203–12.
60. Ingraham LJ, Kety SS. Adoption studies of schizophrenia. *Am J Med Genet* 2000;**97**:18–22.
61. Riley B, Kendler KS. Molecular genetic studies of schizophrenia. *Eur J Hum Genet* 2006;**14**:669–80.
62. Harrison PJ, Owen M. Genes for schizophrenia? Recent findings and their pathophysiological implications. *Lancet* 2003;**361**:417–9.
63. Malhotra AK, Pinals DA, Adler CM, Elman I, Clifton A, Pickar D, et al. Ketamine-induced exacerbation of psychotic symptoms and cognitive impairment in neuroleptic-free schizophrenics. *Neuropsychopharmacology* 1997;**17**:141–50.

64. Stefansson H, Steinthorsdottir V, Thorgeirsson TE, Gulcher JR, Stefansson K. Neuregulin 1 and schizophrenia. *Ann Med* 2004;**36**:62–71.
65. Tamminga CA. Schizophrenia and glutamatergic transmission. *Crit Rev Neurobiol* 1998;**12**:21–36.
66. Coyle JT. Glutamate and schizophrenia: beyond the dopamine hypothesis. *Cell Mol Neurobiol* 2006;**26**:365–84.
67. Akbarian S. Epigenetics of schizophrenia. *Curr Top Behav Neurosci* 2010;**4**:611–28.
68. Matsumoto H, Higa HH. Studies on methylazoxymethanol, the aglycone of cycasin: methylation of nucleic acids in vitro. *Biochem J* 1966;**98**:20C–NaN.
69. Lodge DJ, Grace AA. Hippocampal dysfunction and disruption of dopamine system regulation in an animal model of schizophrenia. *Neurotox Res* 2008;**14**:97–104.
70. Flagstad P, Mork A, Glenthoj BY, van Beek J, Michael-Titus AT, Didriksen M. Disruption of neurogenesis on gestational day 17 in the rat causes behavioral changes relevant to positive and negative schizophrenia symptoms and alters amphetamine-induced dopamine release in nucleus accumbens. *Neuropsychopharmacology* 2004;**29**:2052–64.
71. Bale TL, Baram TZ, Brown AS, Goldstein JM, Insel TR, McCarthy MM, et al. Early life programming and neurodevelopmental disorders. *Biol Psychiatry* 2010;**68**:314–9.
72. Piontkewitz Y, Arad M, Weiner I. Tracing the development of psychosis and its prevention: what can be learned from animal models. *Neuropharmacology* 2011.
73. Meyer U, Nyffeler M, Yee BK, Knuesel I, Feldon J. Adult brain and behavioral pathological markers of prenatal immune challenge during early/middle and late fetal development in mice. *Brain Behav Immun* 2008;**22**:469–86.
74. Lipska BK, Jaskiw GE, Weinberger DR. Postpubertal emergence of hyperresponsiveness to stress and to amphetamine after neonatal excitotoxic hippocampal damage: a potential animal model of schizophrenia. *Neuropsychopharmacology* 1993;**9**:67–75.
75. Andreasson S, Allebeck P, Engstrom A, Rydberg U. Cannabis and schizophrenia. A longitudinal study of Swedish conscripts. *Lancet* 1987;**2**:1483–6.
76. Arseneault L, Cannon M, Poulton R, Murray R, Caspi A, Moffitt TE. Cannabis use in adolescence and risk for adult psychosis: longitudinal prospective study. *BMJ* 2002;**325**:1212–3.
77. van Os J, Bak M, Hanssen M, Bijl RV, de Graaf R, Verdoux H. Cannabis use and psychosis: a longitudinal population-based study. *Am J Epidemiol* 2002;**156**:319–27.
78. Zammit S, Allebeck P, Andreasson S, Lundberg I, Lewis G. Self reported cannabis use as a risk factor for schizophrenia in Swedish conscripts of 1969: historical cohort study. *BMJ* 2002;**325**:1199.
79. Henquet C, Krabbendam L, Spauwen J, Kaplan C, Lieb R, Wittchen HU, et al. Prospective cohort study of cannabis use, predisposition for psychosis, and psychotic symptoms in young people. *BMJ* 2005;**330**:11.
80. Moore TH, Zammit S, Lingford-Hughes A, Barnes TR, Jones PB, Burke M, et al. Cannabis use and risk of psychotic or affective mental health outcomes: a systematic review. *Lancet* 2007;**370**:319–28.
81. Rubino T, Realini N, Braida D, Alberio T, Capurro V, Vigano D, et al. The depressive phenotype induced in adult female rats by adolescent exposure to THC is associated with cognitive impairment and altered neuroplasticity in the prefrontal cortex. *Neurotox Res* 2009;**15**:291–302.
82. Schneider M, Koch M. Chronic pubertal, but not adult chronic cannabinoid treatment impairs sensorimotor gating, recognition memory, and the performance in a progressive ratio task in adult rats. *Neuropsychopharmacology* 2003;**28**:1760–9.
83. Schneider M, Koch M. The effect of chronic peripubertal cannabinoid treatment on deficient object recognition memory in rats after neonatal mPFC lesion. *Eur Neuropsychopharmacol* 2007;**17**:180–6.

84. Wegener N, Koch M. Behavioural disturbances and altered Fos protein expression in adult rats after chronic pubertal cannabinoid treatment. *Brain Res* 2009;**1253**:81–91.
85. Gejman PV, Sanders AR, Kendler KS. Genetics of schizophrenia: new findings and challenges. *Annu Rev Genomics Hum Genet* 2011;**12**:121–44.
86. Jaaro-Peled H, Ayhan Y, Pletnikov MV, Sawa A. Review of pathological hallmarks of schizophrenia: comparison of genetic models with patients and nongenetic models. *Schizophr Bull* 2010;**36**:301–13.
87. Meyer U, Feldon J, Dammann O. Schizophrenia and autism: both shared and disorder-specific pathogenesis via perinatal inflammation? *Pediatr Res* 2011;**69**:26R–3.
88. Xu B, Roos JL, Dexheimer P, Boone B, Plummer B, Levy S, et al. Exome sequencing supports a de novo mutational paradigm for schizophrenia. *Nat Genet* 2011;**43**:864–8.
89. Stefansson H, Sigurdsson E, Steinthorsdottir V, Bjornsdottir S, Sigmundsson T, Ghosh S, et al. Neuregulin 1 and susceptibility to schizophrenia. *Am J Hum Genet* 2002,**71**:877–92.
90. Stefansson H, Sarginson J, Kong A, Yates P, Steinthorsdottir V, Gudfinnsson E, et al. Association of neuregulin 1 with schizophrenia confirmed in a Scottish population. *Am J Hum Genet* 2003;**72**:83–7.
91. Alaerts M, Ceulemans S, Forero D, Moens LN, De Zutter S, Heyrman L, et al. Support for NRG1 as a susceptibility factor for schizophrenia in a northern Swedish isolated population. *Arch Gen Psychiatry* 2009;**66**:828–37.
92. Kao WT, Wang Y, Kleinman JE, Lipska BK, Hyde TM, Weinberger DR, et al. Common genetic variation in Neuregulin 3 (NRG3) influences risk for schizophrenia and impacts NRG3 expression in human brain. *Proc Natl Acad Sci USA* 2010;**107**:15619–24.
93. Mei L, Xiong WC. Neuregulin 1 in neural development, synaptic plasticity and schizophrenia. *Nat Rev Neurosci* 2008;**9**:437–52.
94. Jaaro-Peled H, Hayashi-Takagi A, Seshadri S, Kamiya A, Brandon NJ, Sawa A. Neurodevelopmental mechanisms of schizophrenia: understanding disturbed postnatal brain maturation through neuregulin-1-ErbB4 and DISC1. *Trends Neurosci* 2009;**32**:485–95.
95. Li B, Woo RS, Mei L, Malinow R. The neuregulin-1 receptor erbB4 controls glutamatergic synapse maturation and plasticity. *Neuron* 2007;**54**:583–97.
96. Chen YJ, Zhang M, Yin DM, Wen L, Ting A, Wang P, et al. ErbB4 in parvalbumin-positive interneurons is critical for neuregulin 1 regulation of long-term potentiation. *Proc Natl Acad Sci USA* 2010;**107**:21818–23.
97. Barros CS, Calabrese B, Chamero P, Roberts AJ, Korzus E, Lloyd K, et al. Impaired maturation of dendritic spines without disorganization of cortical cell layers in mice lacking NRG1/ErbB signaling in the central nervous system. *Proc Natl Acad Sci USA* 2009;**106**: 4507–12.
98. Ting AK, Chen Y, Wen L, Yin DM, Shen C, Tao Y, et al. Neuregulin 1 promotes excitatory synapse development and function in GABAergic interneurons. *J Neurosci* 2011;**31**:15–25.
99. Wen L, Lu YS, Zhu XH, Li XM, Woo RS, Chen YJ, et al. Neuregulin 1 regulates pyramidal neuron activity via ErbB4 in parvalbumin-positive interneurons. *Proc Natl Acad Sci USA* 2010;**107**:1211–6.
100. Harrison PJ, Weinberger DR. Schizophrenia genes, gene expression, and neuropathology: on the matter of their convergence. *Mol Psychiatry* 2005;**10**:40–68.
101. Lisman JE, Coyle JT, Green RW, Javitt DC, Benes FM, Heckers S, et al. Circuit-based framework for understanding neurotransmitter and risk gene interactions in schizophrenia. *Trends Neurosci* 2008;**31**(5):234–42.
102. Law AJ, Kleinman JE, Weinberger DR, Weickert CS. Disease-associated intronic variants in the ErbB4 gene are related to altered ErbB4 splice-variant expression in the brain in schizophrenia. *Hum Mol Genet* 2007;**16**:129–41.

103. Schmitt A, Parlapani E, Gruber O, Wobrock T, Falkai P. Impact of neuregulin-1 on the pathophysiology of schizophrenia in human post-mortem studies. *Eur Arch Psychiatry Clin Neurosci* 2008;**258**(Suppl. 5):35–9.
104. Tan W, Wang Y, Gold B, Chen J, Dean M, Harrison PJ, et al. Molecular cloning of a brain-specific, developmentally regulated neuregulin 1 (NRG1) isoform and identification of a functional promoter variant associated with schizophrenia. *J Biol Chem* 2007;**282**:24343–51.
105. Law AJ, Lipska BK, Weickert CS, Hyde TM, Straub RE, Hashimoto R, et al. Neuregulin 1 transcripts are differentially expressed in schizophrenia and regulated by 5′ SNPs associated with the disease. *Proc Natl Acad Sci USA* 2006;**103**:6747–52.
106. Hahn CG, Wang HY, Cho DS, Talbot K, Gur RE, Berrettini WH, et al. Altered neuregulin 1-erbB4 signaling contributes to NMDA receptor hypofunction in schizophrenia. *Nat Med* 2006;**12**:824–8.
107. Duffy L, Cappas E, Lai D, Boucher AA, Karl T. Cognition in transmembrane domain neuregulin 1 mutant mice. *Neuroscience* 2010;**170**:800–7.
108. Taylor MS, Devon RS, Millar JK, Porteous DJ. Evolutionary constraints on the disrupted in Schizophrenia locus. *Genomics* 2003;**81**:67–77.
109. Millar JK, Christie S, Porteous DJ. Yeast two-hybrid screens implicate DISC1 in brain development and function. *Biochem Biophys Res Commun* 2003;**311**:1019–25.
110. Talbot K, Eidem WL, Tinsley CL, Benson MA, Thompson EW, Smith RJ, et al. Dysbindin-1 is reduced in intrinsic, glutamatergic terminals of the hippocampal formation in schizophrenia. *J Clin Invest* 2004;**113**:1353–63.
111. Harrison PJ. The hippocampus in schizophrenia: a review of the neuropathological evidence and its pathophysiological implications. *Psychopharmacology (Berl)* 2004;**174**:151–62.
112. Straub RE, MacLean CJ, Ma Y, Webb BT, Myakishev MV, Harris-Kerr C, et al. Genome-wide scans of three independent sets of 90 Irish multiplex schizophrenia families and follow-up of selected regions in all families provides evidence for multiple susceptibility genes. *Mol Psychiatry* 2002;**7**:542–59.
113. Ghiani CA, Starcevic M, Rodriguez-Fernandez IA, Nazarian R, Cheli VT, Chan LN, et al. The dysbindin-containing complex (BLOC-1) in brain: developmental regulation, interaction with SNARE proteins and role in neurite outgrowth. *Mol Psychiatry* 2010;**15**:204–15 115.
114. Fei E, Ma X, Zhu C, Xue T, Yan J, Xu Y, et al. Nucleocytoplasmic shuttling of dysbindin-1, a schizophrenia-related protein, regulates synapsin I expression. *J Biol Chem* 2010;**285**:38630–40.
115. Numakawa T, Yagasaki Y, Ishimoto T, Okada T, Suzuki T, Iwata N, et al. Evidence of novel neuronal functions of dysbindin, a susceptibility gene for schizophrenia. *Hum Mol Genet* 2004;**13**:2699–708.
116. Burdick KE, Funke B, Goldberg JF, Bates JA, Jaeger J, Kucherlapati R, et al. COMT genotype increases risk for bipolar I disorder and influences neurocognitive performance. *Bipolar Disord* 2007;**9**:370–6.
117. Papaleo F, Lipska BK, Weinberger DR. Mouse models of genetic effects on cognition: relevance to schizophrenia. *Neuropharmacology* 2011.
118. Poo MM. Neurotrophins as synaptic modulators. *Nat Rev Neurosci* 2001;**2**:24–32.
119. Seidah NG, Benjannet S, Pareek S, Chretien M, Murphy RA. Cellular processing of the neurotrophin precursors of NT3 and BDNF by the mammalian proprotein convertases. *FEBS Lett* 1996;**379**:247–50.
120. Huang EJ, Reichardt LF. Trk receptors: roles in neuronal signal transduction. *Annu Rev Biochem* 2003;**72**:609–42.
121. Arevalo JC, Wu SH. Neurotrophin signaling: many exciting surprises!. *Cell Mol Life Sci* 2006;**63**:1523–37.

122. Gratacos M, Gonzalez JR, Mercader JM, de Cid R, Urretavizcaya M, Estivill X. Brain-derived neurotrophic factor Val66Met and psychiatric disorders: meta-analysis of case-control studies confirm association to substance-related disorders, eating disorders, and schizophrenia. *Biol Psychiatry* 2007;**61**:911–22.
123. Chen ZY, Patel PD, Sant G, Meng CX, Teng KK, Hempstead BL, et al. Variant brain-derived neurotrophic factor (BDNF) (Met66) alters the intracellular trafficking and activity-dependent secretion of wild-type BDNF in neurosecretory cells and cortical neurons. *J Neurosci* 2004;**24**:4401–11.
124. Egan MF, Kojima M, Callicott JH, Goldberg TE, Kolachana BS, Bertolino A, et al. The BDNF val66met polymorphism affects activity-dependent secretion of BDNF and human memory and hippocampal function. *Cell* 2003;**112**:257–69.
125. Soliman F, Glatt CE, Bath KG, Levita L, Jones RM, Pattwell SS, et al. A genetic variant BDNF polymorphism alters extinction learning in both mouse and human. *Science* 2010;**327**:863–6.
126. Suchanek R, Owczarek A, Kowalski J. Association study between BDNF C-281A polymorphism and paranoid schizophrenia in polish population. *J Mol Neurosci* 2011;.
127. Bramham CR, Messaoudi E. BDNF function in adult synaptic plasticity: the synaptic consolidation hypothesis. *Prog Neurobiol* 2005;**76**:99–125.
128. Durany N, Michel T, Zochling R, Boissl KW, Cruz-Sanchez FF, Riederer P, et al. Brain-derived neurotrophic factor and neurotrophin 3 in schizophrenic psychoses. *Schizophr Res* 2001;**52**:79–86.
129. Hashimoto T, Bergen SE, Nguyen QL, Xu B, Monteggia LM, Pierri JN, et al. Relationship of brain-derived neurotrophic factor and its receptor TrkB to altered inhibitory prefrontal circuitry in schizophrenia. *J Neurosci* 2005;**25**:372–83.
130. Mellios N, Huang HS, Baker SP, Galdzicka M, Ginns E, Akbarian S. Molecular determinants of dysregulated GABAergic gene expression in the prefrontal cortex of subjects with schizophrenia. *Biol Psychiatry* 2009;**65**:1006–14.
131. Takahashi M, Shirakawa O, Toyooka K, Kitamura N, Hashimoto T, Maeda K, et al. Abnormal expression of brain-derived neurotrophic factor and its receptor in the corticolimbic system of schizophrenic patients. *Mol Psychiatry* 2000;**5**:293–300.
132. Wong J, Hyde TM, Cassano HL, Deep-Soboslay A, Kleinman JE, Weickert CS. Promoter specific alterations of brain-derived neurotrophic factor mRNA in schizophrenia. *Neuroscience* 2010;**169**:1071–84.
133. Weickert CS, Hyde TM, Lipska BK, Herman MM, Weinberger DR, Kleinman JE. Reduced brain-derived neurotrophic factor in prefrontal cortex of patients with schizophrenia. *Mol Psychiatry* 2003;**8**:592–610.
134. Freedman R, Leonard S, Gault JM, Hopkins J, Cloninger CR, Kaufmann CA, et al. Linkage disequilibrium for schizophrenia at the chromosome 15q13-14 locus of the alpha7-nicotinic acetylcholine receptor subunit gene (CHRNA7). *Am J Med Genet* 2001;**105**:20–2.
135. Breese CR, Adams C, Logel J, Drebing C, Rollins Y, Barnhart M, et al. Comparison of the regional expression of nicotinic acetylcholine receptor alpha7 mRNA and [125I]-alpha-bungarotoxin binding in human postmortem brain. *J Comp Neurol* 1997;**387**:385–98.
136. Barr RS, Culhane MA, Jubelt LE, Mufti RS, Dyer MA, Weiss AP, et al. The effects of transdermal nicotine on cognition in nonsmokers with schizophrenia and nonpsychiatric controls. *Neuropsychopharmacology* 2008;**33**:480–90.
137. Matsumoto M, Weickert CS, Akil M, Lipska BK, Hyde TM, Herman MM, et al. Catechol O-methyltransferase mRNA expression in human and rat brain: evidence for a role in cortical neuronal function. *Neuroscience* 2003;**116**:127–37.

138. Kastner A, Anglade P, Bounaix C, Damier P, Javoy-Agid F, Bromet N, et al. Immunohistochemical study of catechol-O-methyltransferase in the human mesostriatal system. *Neuroscience* 1994;**62**:449–57.
139. Karoum F, Chrapusta SJ, Egan MF. 3-Methoxytyramine is the major metabolite of released dopamine in the rat frontal cortex: reassessment of the effects of antipsychotics on the dynamics of dopamine release and metabolism in the frontal cortex, nucleus accumbens, and striatum by a simple two pool model. *J Neurochem* 1994;**63**:972–9.
140. Lotta T, Vidgren J, Tilgmann C, Ulmanen I, Melen K, Julkunen I, et al. Kinetics of human soluble and membrane-bound catechol O-methyltransferase: a revised mechanism and description of the thermolabile variant of the enzyme. *Biochemistry* 1995;**34**:4202–10.
141. Lachman HM, Papolos DF, Saito T, Yu YM, Szumlanski CL, Weinshilboum RM. Human catechol-O-methyltransferase pharmacogenetics: description of a functional polymorphism and its potential application to neuropsychiatric disorders. *Pharmacogenetics* 1996;**6**:243–50.
142. Sha L, Macintyre L, Machell JA, Kelly MP, Porteous DJ, Brandon NJ, et al. Transcriptional regulation of neurodevelopmental and metabolic pathways by NPAS3. *Mol Psychiatry* 2011.
143. Pieper AA, Wu X, Han TW, Estill SJ, Dang Q, Wu LC, et al. The neuronal PAS domain protein 3 transcription factor controls FGF-mediated adult hippocampal neurogenesis in mice. *Proc Natl Acad Sci USA* 2005;**102**:14052–7.
144. Kamnasaran D, Muir WJ, Ferguson-Smith MA, Cox DW. Disruption of the neuronal PAS3 gene in a family affected with schizophrenia. *J Med Genet* 2003;**40**:325–32.
145. Pickard BS, Malloy MP, Porteous DJ, Blackwood DH, Muir WJ. Disruption of a brain transcription factor, NPAS3, is associated with schizophrenia and learning disability. *Am J Med Genet B Neuropsychiatr Genet* 2005;**136B**:26–32.
146. Pickard BS, Christoforou A, Thomson PA, Fawkes A, Evans KL, Morris SW, et al. Interacting haplotypes at the NPAS3 locus alter risk of schizophrenia and bipolar disorder. *Mol Psychiatry* 2009;**14**:874–84.
147. Macintyre G, Alford T, Xiong L, Rouleau GA, Tibbo PG, Cox DW. Association of NPAS3 exonic variation with schizophrenia. *Schizophr Res* 2010;**120**:143–9.
148. Huang J, Perlis RH, Lee PH, Rush AJ, Fava M, Sachs GS, et al. Cross-disorder genomewide analysis of schizophrenia, bipolar disorder, and depression. *Am J Psychiatry* 2010;**167**:1254–63.
149. Lavedan C, Licamele L, Volpi S, Hamilton J, Heaton C, Mack K, et al. Association of the NPAS3 gene and five other loci with response to the antipsychotic iloperidone identified in a whole genome association study. *Mol Psychiatry* 2009;**14**:804–19.
150. Honea R, Crow TJ, Passingham D, Mackay CE. Regional deficits in brain volume in schizophrenia: a meta-analysis of voxel-based morphometry studies. *Am J Psychiatry* 2005;**162**:2233–45.
151. Gao XM, Sakai K, Roberts RC, Conley RR, Dean B, Tamminga CA. Ionotropic glutamate receptors and expression of N-methyl-D-aspartate receptor subunits in subregions of human hippocampus: effects of schizophrenia. *Am J Psychiatry* 2000;**157**:1141–9.
152. Harrison PJ. The hippocampus in schizophrenia: a review of the neuropathological evidence and its pathophysiological implications. *Psychopharmacology* 2004;**174**:151–62.
153. Scarr E, Sundram S, Keriakous D, Dean B. Altered hippocampal muscarinic M4, but not M1, receptor expression from subjects with schizophrenia. *Biol Psychiatry* 2007;**61**:1161–70.
154. Pilowsky LS, Bressan RA, Stone JM, Erlandsson K, Mulligan RS, Krystal JH, et al. First in vivo evidence of an NMDA receptor deficit in medication-free schizophrenic patients. *Mol Psychiatry* 2006;**11**:118–9.
155. Altar CA, Jurata LW, Charles V, Lemire A, Liu P, Bukhman Y, et al. Deficient hippocampal neuron expression of proteasome, ubiquitin, and mitochondrial genes in multiple schizophrenia cohorts. *Biol Psychiatry* 2005;**58**:85–96.

156. Knable MB, Barci BM, Webster MJ, Meador-Woodruff J, Torrey EF. Molecular abnormalities of the hippocampus in severe psychiatric illness: postmortem findings from the Stanley Neuropathology Consortium. *Mol Psychiatry* 2004;**9**:609–20; 544.
157. Abraham WC, Bear MF. Metaplasticity: the plasticity of synaptic plasticity. *Trends Neurosci* 1996;**19**:126–30.
158. Malenka RC, Bear MF. LTP and LTD: an embarrassment of riches. *Neuron* 2004;**44**:5–21.
159. Heckers S, Rauch SL, Goff D, Savage CR, Schacter DLFAJ, Alpert NM. Impaired recruitment of the hippocampus during conscious recollection in schizophrenia. *Nature* 1998;**1**:318–23.
160. Schobel SA, Lewandowski NM, Corcoran CM, Moore H, Brown T, Malaspina D, et al. Differential targeting of the CA1 subfield of the hippocampal formation by schizophrenia and related psychotic disorders. *Arch Gen Psychiatry* 2009;**66**:938–46.
161. Deicken RF, Pegues M, Amend D. Reduced hippocampal N-acetylaspartate without volume loss in schizophrenia. *Schizophr Res* 1999;**37**:217–23.
162. Amaral DG, Insausti R. Hippocampal formation. In: Paxinos G, editor. *The Human Nervous System*. San Diego: Academic Press; 1990. pp. 711–55.
163. Suzuki W, Amaral DG. Perirhinal and parahippocampal cortices of the macaque monkey: cytoarchitectonic and chemoarchitectonic organization. *J Comp Neurol* 2003;**463**:67–91.
164. Lavenex P, Banta LP, Amaral DG. Postnatal development of the primate hippocampal formation. *Dev Neurosci* 2007;**29**:179–92.
165. Kolomeets NS, Orlovskaya DD, Rachmanova VI, Uranova NA. Ultrastructural alterations in hippocampal mossy fiber synapses in schizophrenia: a postmortem morphometric study. *Synapse* 2005;**57**:47–55.
166. Reif A, Fritzen S, Finger M, Strobel A, Lauer M, Schmitt A, et al. Neural stem cell proliferation is decreased in schizophrenia, but not in depression. *Mol Psychiatry* 2006;**11**:514–22.
167. Hoffman RE, Grasemann U, Gueorguieva R, Quinlan D, Lane D, Miikkulainen R. Using computational patients to evaluate illness mechanisms in schizophrenia. *Biol Psychiatry* 2011;**69**:997–1005.
168. Ghose S, Chin R, Lewis-Amezcua K, Frost DO, Roberts R, et al. Distinct gene expression correlations in schizophrenia between NR1 and GAD67 in the anterior hippocampus, In: *Abstract, 36th Annual Meeting of the Society for Neuroscience, Atlanta, GA*; 2006.
169. Duan X, Chang JH, Ge S, Faulkner RL, Kim JY, Kitabatake Y, et al. Disrupted-In-Schizophrenia 1 regulates integration of newly generated neurons in the adult brain. *Cell* 2007;**130**:1146–58.
170. Lichtenstein P, Yip BH, Bjork C, Pawitan Y, Cannon TD, Sullivan PF, et al. Common genetic determinants of schizophrenia and bipolar disorder in Swedish families: a population-based study. *Lancet* 2009;**373**:234–9.
171. Clelland CD, Choi M, Romberg C, Clemenson Jr. GD, Fragniere A, Tyers P, et al. A functional role for adult hippocampal neurogenesis in spatial pattern separation. *Science* 2009;**325**:210–3.
172. Kutsuwada T, Kashiwabuchi N, Mori H, Sakimura K, Kushiya E, Araki K, et al. Molecular diversity of the NMDA receptor channel. *Nature* 1992;**358**:36–41.
173. Monyer H, Sprengel R, Schoepfer R, Herb A, Higuchi M, Lomeli H, et al. Heteromeric NMDA receptors: molecular and functional distinction of subtypes. *Science* 1992;**256**:1217–21.
174. Forrest D, Yuzaki M, Soares HD, Ng L, Luk DC, Sheng M, et al. Targeted disruption of NMDA receptor 1 gene abolishes NMDA response and results in neonatal death. *Neuron* 1994;**13**:325–38.
175. Li Y, Erzurumlu RS, Chen C, Jhaveri S, Tonegawa S. Whisker-related neuronal patterns fail to develop in the trigeminal brainstem nuclei of NMDAR1 knockout mice. *Cell* 1994;**76**:427–37.

176. Gao XM, Elmer GI, Adams-Huet B, Tamminga CA. Social memory in mice: disruption with an NMDA antagonist and attenuation with antipsychotic drugs. *Pharmacol Biochem Behav* 2009;**92**:236–42.
177. McHugh TJ, Jones MW, Quinn JJ, Balthasar N, Coppari R, Elmquist JK, et al. Dentate gyrus NMDA receptors mediate rapid pattern separation in the hippocampal network. *Science* 2007;**317**:94–9.

Index

D

E

F

G

H

I

M

N

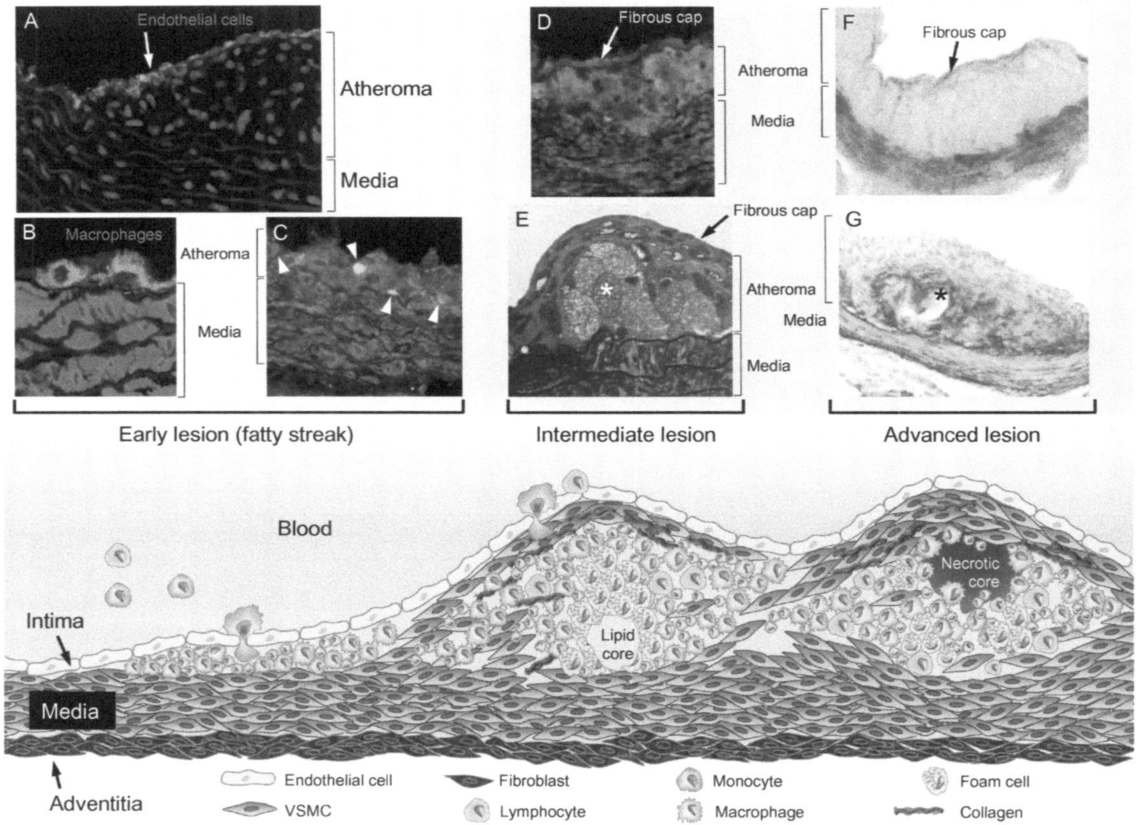

FUSTER *ET AL.*, FIG. 1. Atherosclerotic plaque development. The scheme shows the main events that occur in the vessel wall during atherosclerosis development, starting with a "healthy" artery wall (left) and progression to an advanced atherosclerotic lesion (right). Not shown in the scheme is plaque rupture and thrombus formation at advanced disease stages, which can provoke life-threatening acute ischemic events (e.g., myocardial infarction and stroke). The photomicrographs in A–G show representative immunohistological staining in the aortic sinus and the ascending aorta of apoE-deficient mice,

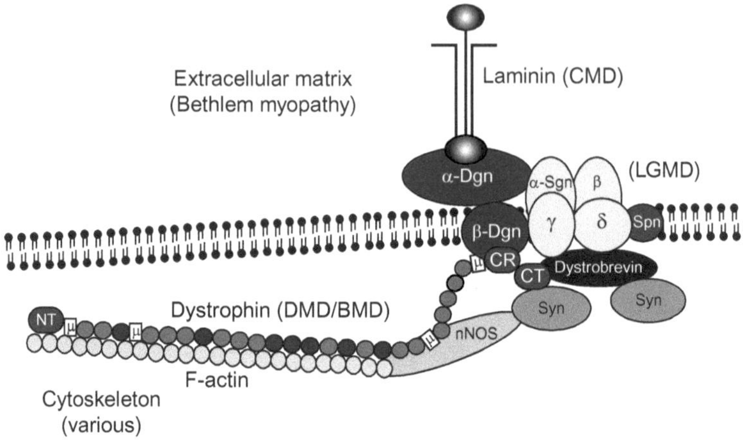

NG *ET AL.*, FIG. 1. The dystrophin–glycoprotein complex (DGC) contains many proteins that, when not expressed, lead to various forms of muscular dystrophy. Shown is the organization of the DGC in skeletal muscle. Dystrophin is defective in DMD and BMD; α-, β-, γ-, and δ- sarcoglycans (Sgn) are defective in four different types of LGMD; laminin α2 is defective in one form of congenital muscular dystrophy; collagen VI in the extracellular matrix is defective in Bethlem myopathy and Ullrich congenital muscular dystrophy; while numerous other cytosolic- and membrane-associated proteins lead to other MDs. Other abbreviations include Syn, syntrophin; SPN, sarcospan; Dgn, dystroglycan; nNOS, neuronal nitric oxide synthase. Dystrophin domains listed include the N-terminal actin-binding domain (NT), various hinges (H), a WW domain (W), a cysteine-rich domain (CR), and a C-terminal domain (CT). Orange and red balls in dystrophin represent spectrin-like repeats, orange ones are acidic, while blue ones are basic and contribute to actin binding.

---

the most widely used mouse model of atherosclerosis. (A) Cell accumulation within the subendothelial space in an early atheroma. Red: nuclei; green: endothelial cells detected by von Willebrand immunofluorescent staining. Elastic laminae in the tunica media exhibit green autofluorescence. (B) Macrophage infiltration into the subendothelial space of the arterial wall. Blue: nuclei; green: neointimal macrophages detected by F4/80 immunofluorescent staining; red: medial VSMCs detected by smooth-muscle-α-actin immunofluorescent staining. (C) Cell proliferation in a fatty streak. Blue: nuclei; red: medial VSMCs detected by smooth-muscle-α-actin immunofluorescent staining; green: neointimal macrophages detected by F4/80 immunofluorescent staining; white: nuclei in proliferating cells detected by Ki67 immunofluorescent staining. Arrows mark proliferating macrophages in the atheroma. (D) Intermediate vascular lesion with a thin fibrous cap. Blue: nuclei; red: VSMCs detected by smooth-muscle-α-actin immunofluorescent staining; green: macrophages detected by F4/80 immunofluorescent staining. (E) Electron microscopy image of an intermediate atherosclerotic lesion with a fibrous cap and lipid core (asterisk). (F) Advanced atherosclerotic plaque with fibrous cap rich in VSMCs (detected by smooth-muscle-α-actin immunohistochemical red staining). (G) Advanced atherosclerotic plaque with necrotic core (asterisk). Macrophages detected by Mac3 immunohistochemical staining are shown in brown. Nuclei are counterstained with hematoxylin.

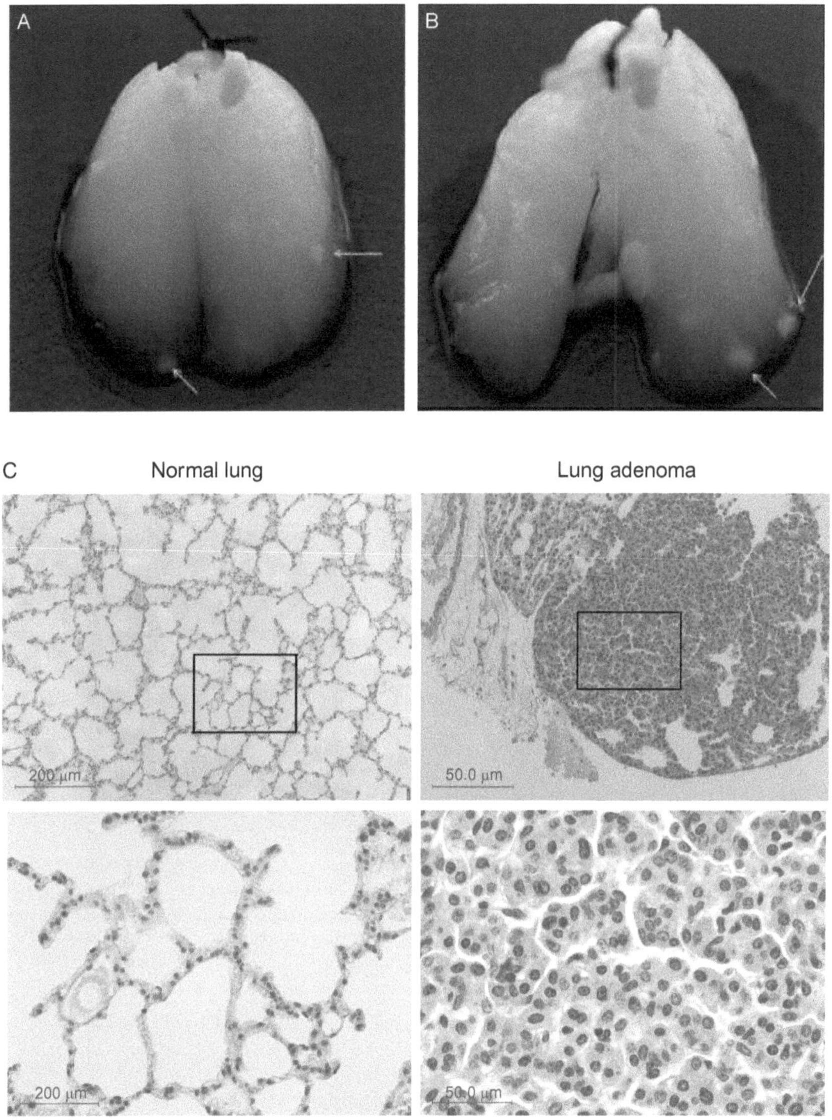

WANG *ET AL.*, FIG. 1. Lung adenoma induced by one dose of B[a]P, 100 mg/kg BW i.p., 20 weeks, A/J female mouse with normal lung as control: (A) gross view of lung with adenoma; (B) gross view of lung with adenocarcinoma induced by vinyl carbamate, 0.32 mg per mouse, i.p. 20 weeks; (C) histopathology of lung adenomas.

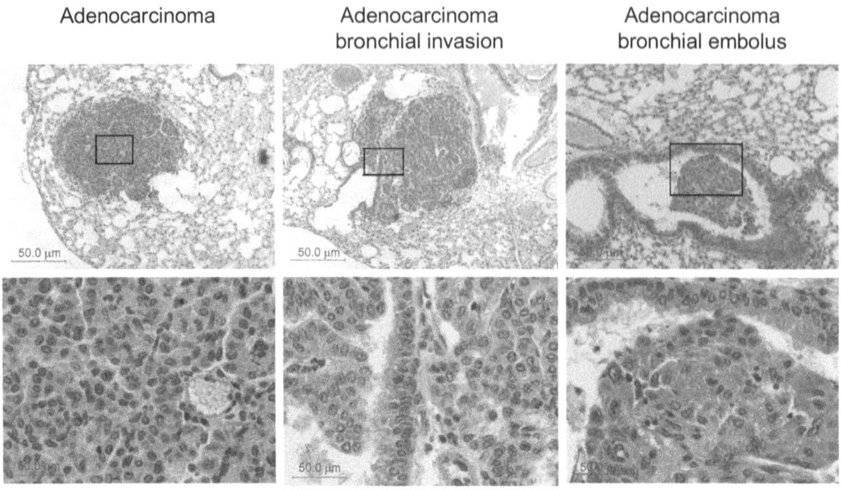

Wang *et al.*, Fig. 2. Histopathology of adenocarcinomas. Histopathology of lung adenocarcinomas induced by vinyl chloride, female A/J mouse.

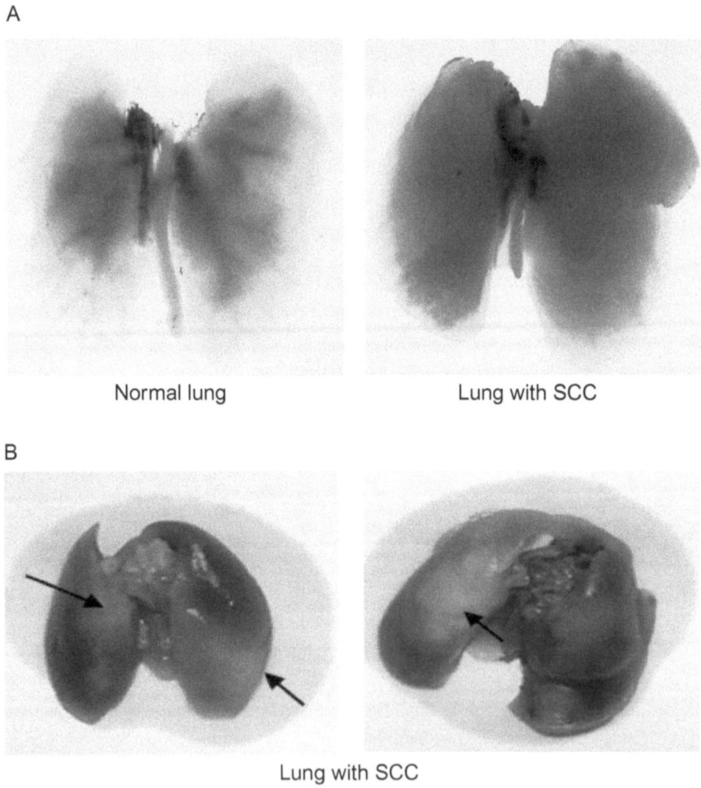

Wang *et al.*, Fig. 3. Gross pathology of squamous cell carcinomas induced by NTCU in A/J mice. Adapted from Ref. 61.

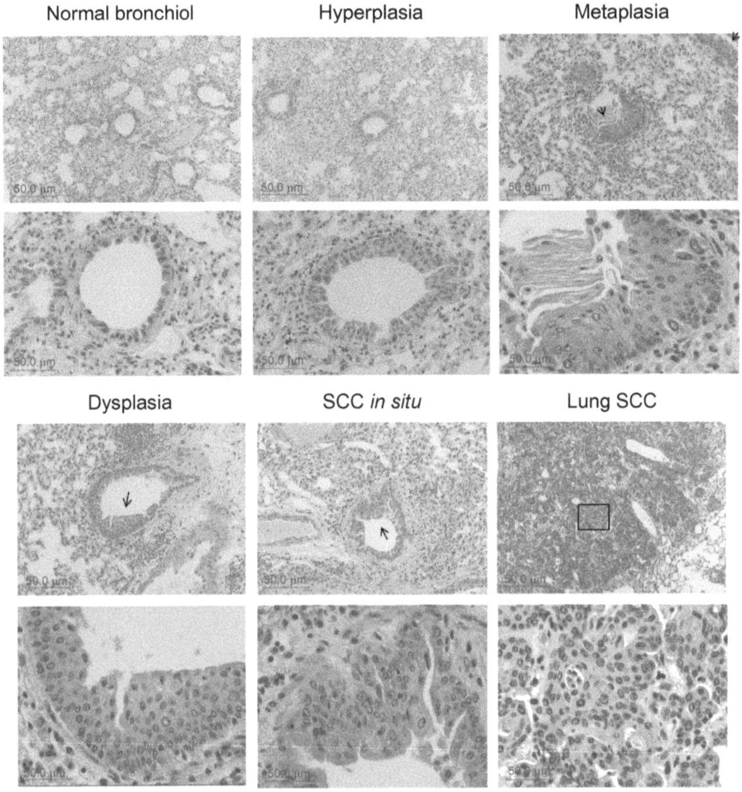

Wang et al., Fig. 4. Histopathology of squamous cell lesions induced by NTCU.

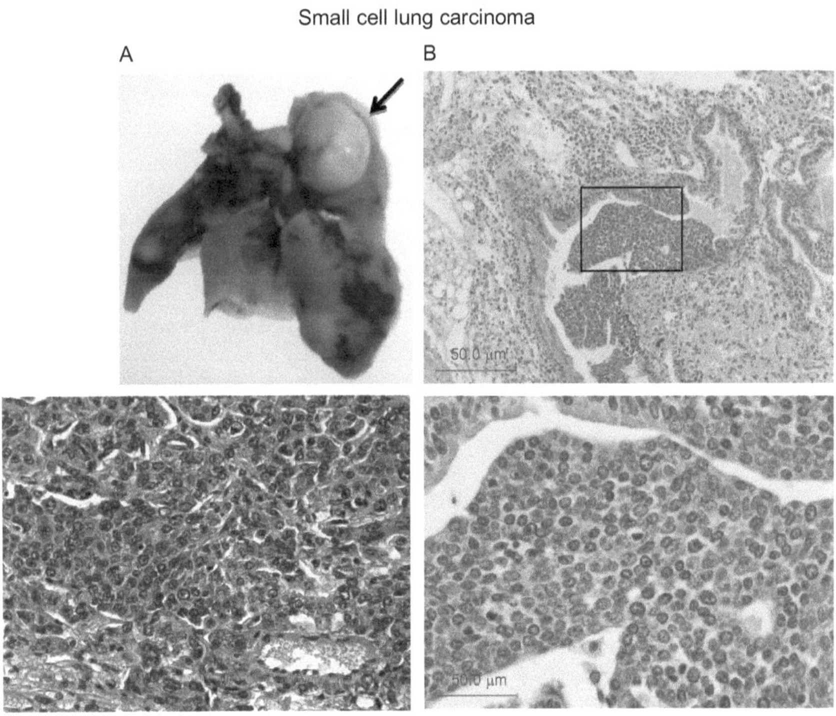

Wang et al., Fig. 5. Small cell carcinoma induced in bitransgenic mice with floxed Rb and p53 tumor suppressor genes.

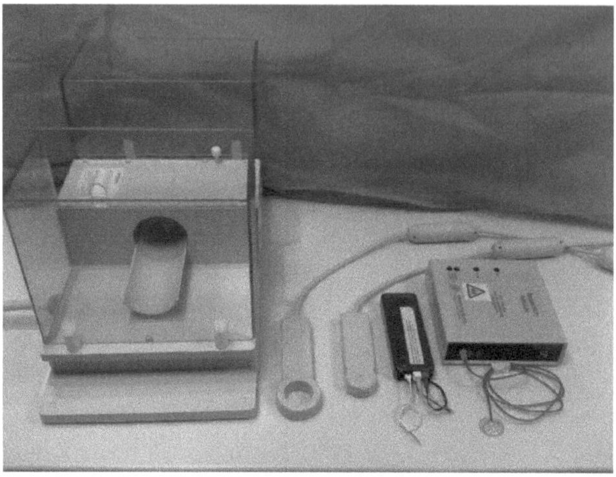

Botnar and Makowski, Fig. 7. Figure shows the different kinds of small animal receiver coils and animal monitoring modules that are available. On the left, a solenoid coil and two microscopy coils, are shown. The type of coils that are displayed are available from Phillips Medical Systems. On the right, an ECG/temperature monitoring module (black box) and a respiratory monitoring module (gray box), are shown. These modules are available from SA Instruments.

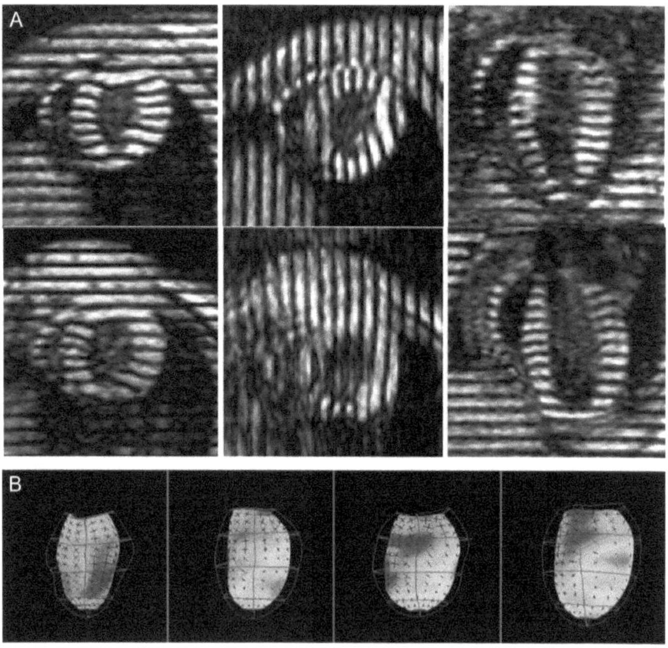

Botnar and Makowski, Fig. 8. (A) High resolution spatial modulation of magnetization (SPAMM) tagged magnetic resonance *in vivo* imaging at baseline and 24 h after myocardial infarction. The SPAMM stripes are 700 μm apart from each other. (B) 3D element model at end-diastole at baseline, day 1, 7, 28 after myocardial infarction. Lines represent model element boundaries. Crosses represent 3D principal strains and directions. Adapted from Ref. 50.

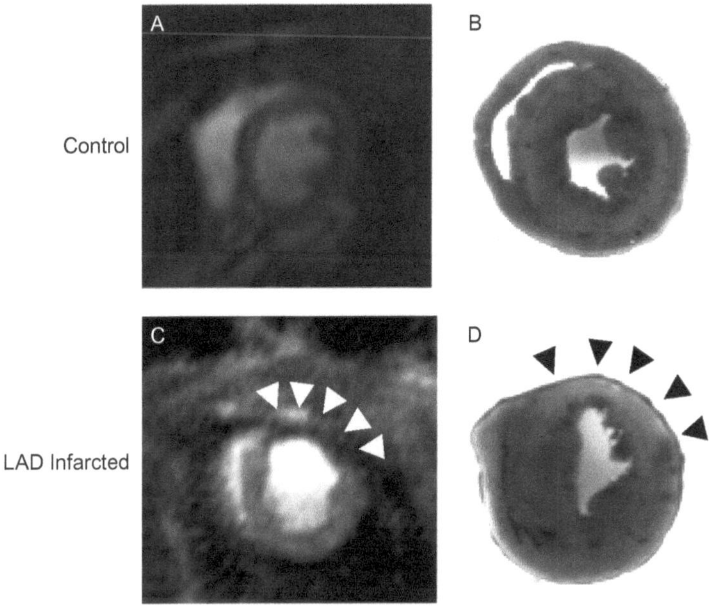

Botnar and Makowski, Fig. 10. First-pass myocardial perfusion images in mice. The images represent matched Evans blue sections (2% w/v, injection into the tail vein) that indicate the area at risk. Note the colocalization of large anterior perfusion defect in first-pass perfusion MRI with the decreased staining in the slice matched Evans blue section (C, D). Adapted from Ref. 75.

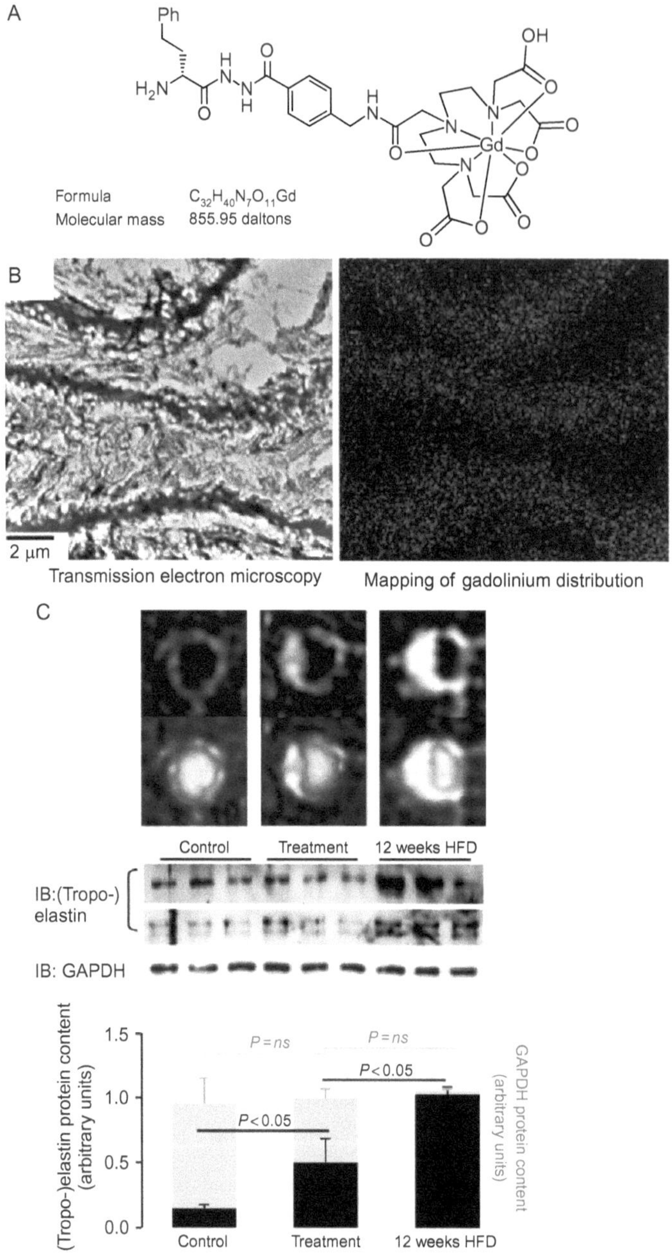

Botnar and Makowski, Fig. 11. Example of a novel MRI elastin-targeting contrast agent. The usefulness of this contrast agent was investigated in mouse model of atherosclerosis. (A) Chemical structure of the elastin-targeting contrast agent. (B) The electron microscopy images show the colocalization of the gadolinium from the contrast agent with dense elastic fibers. (C) Example of delayed enhancement images (upper row) overlaid on TOF images (bottom row) of cross-sectional views of the brachiocephalic artery. On the left a control mouse is shown, in the middle in treated animal (with statin) and on the right animal that has been on the high-fat diet for 3 months. A significant increase in percentage atheroma/media volume could be measured after 3 months of HFD. The treatment with a statin resulted in a significant decrease in the percentage atheroma/media volume. Adapted from Ref. 90.

Kolibri 360 GmbH
Ferdinand-Jühlke-Straße 7
99095 Erfurt, Deutschland
produktsicherheit@kolibri360.de